T0213981

UNITEXT for Physics

UNITEXT for Physics series publishes textbooks in physics and astronomy, characterized by a didactic style and comprehensiveness. The books are addressed to upper-undergraduate and graduate students, but also to scientists and researchers as important resources for their education, knowledge, and teaching.

More information about this series at https://link.springer.com/bookseries/13351

Olaf Stenzel

Light–Matter Interaction

A Crash Course for Students of Optics, Photonics and Materials Science

 Springer

Olaf Stenzel
Fraunhofer IOF
Jena, Thüringen, Germany

ISSN 2198-7882   ISSN 2198-7890   (electronic)
UNITEXT for Physics
ISBN 978-3-030-87146-8   ISBN 978-3-030-87144-4   (eBook)
https://doi.org/10.1007/978-3-030-87144-4

This Springer imprint is published by the registered company Springer Nature Switzerland AG
The registered company address is: Gewerbestrasse 11, 6330 Cham, Switzerland

To the memory of my parents
Gertrud and Günter Stenzel

Preface

Denn die einen sind im Dunkeln. Und die andern sind im Licht.

Und man siehet die im Lichte. Die im Dunkeln sieht man nicht.

Bertolt Brecht, Eine Dreigroschenoper (The Threepenny Opera)

Engl.: *While some are in the darkness, others are in the light.*

What you see are those in the light. Those in darkness, you do not see.

"Licht" ("Light"). Painting and Photograph by Astrid Leiterer, Jena, Germany (www.astrid-art.de). Photograph reproduced with permission

Dear potential reader of this book,

It is my pleasure to recognize that you start reading this preface and maybe continue reading the book afterward. Therefore, I would like to use this preface to provide some background information about this book project, but also to acknowledge the assistance of many people I experienced during writing the manuscript.

Let me start with the background information. From 2017, I am regularly reading lectures on "Structure of Matter" for master students of photonics at the Abbe School of Photonics, Friedrich Schiller University Jena, Germany. This is an international master degree program, and the students come from all over the world, holding a bachelor's degree in engineering or natural sciences usually earned in their home countries. The mentioned course is given in the first semester of the master program and is intended to provide (or repeat) basic knowledge on atomic, molecular, and solid-state physics with emphasis on light–matter interactions within one semester.

Clearly, these topics would fill dozens of books, such that the lecturer has to make a selection of topics and to define a logical path how to arrange the material. After having understood the heterogeneous composition of the audience, and having consulted relevant sources, I decided to design an individual course that is constructed according to the following rules:

- A strong bottom-up approach with regard to the material objects: We start from microscopic objects and finally arrive at continuous media.
- Focus on basics, not on modern developments. The latter are topic of subsequent courses.
- The primary target group are photonics students and not students of physics.
- Where possible, classical models are presented to illustrate the nature of the phenomena. In parallel, quantum mechanical models are developed as a tool for reliable quantitative predictions.
- No thematic demarcation from other disciplines of natural sciences.
- Parallelism to daily life experience is provided by allusions to surrounding nature, architecture, as well as the use of artwork.

In this context, I have to express my thanks to the students that listened to the course and gave me critical feedback in terms of the lecture evaluation procedure. I used a lot of their remarks to improve the course. Finally, I came to the conclusion that this course may be of interest to a broader audience. Therefore, I extended the lecture script to the book you are now holding in hand or have on the computer screen.

Studying physics is like climbing an infinitely high rock in misty weather conditions. When you start, close to the bottom, the fog is densest, and it is challenging to propagate into the correct direction. When you lose orientation, the fog will never clear up, and finally you give up. But if you succeed climbing higher, you will observe that the view becomes better. Now you may see a part of the way in front of you, and you may also see other ways that go into the correct direction.

And even more, you observe connection paths between these ways. But the aim is still in the fog. The frustrating thing is the following: The higher you climb, the more you can see. In particular, you observe that the way in front of you seems to become longer and longer, because although you get an always wider view, the summit still remains in the fog, because of the infinite height of the rock. Love it or leave it—I decided to love it.

With respect to this book, that means: I cannot define objective borders of the physical field I am going to describe in this book. Often enough, I will have to make a subjective decision to stop going into further detail and indicate that a deeper treatment of the topic is outside the scope of this book. In these cases, please remember that the topical seed of the book is a one-semester lecture course. You will find enough advanced literature recommendations at the relevant places.

Naturally, because of the interdisciplinarity of the topic, there appears some overlap to other books I have written before on similar topics. This particularly concerns the book: The Physics of Thin Film Optical Spectra: An Introduction. Some basic stuff from this book is reused here.

So, this is the background information on the origin of this book project. Let me now come to the acknowledgments.

- First of all, I would like to express my thanks to Prof. Andreas Tünnermann, who gave me the possibility to read that lecture course at Abbe School of Photonics.
- My next thanks are to Zachary Evenson and all his co-workers from the Springer book company for their interest to this topic as well as supervision and assistance during the project.
- As in a previous book project, I again collaborated with Astrid Leiterer (www. astrid-art.de). She supplied me with photographs of her subtle paintings and sculptures. I use these illustrations for providing an atmospheric background for the physical phenomena to be presented in the corresponding chapter. I am very grateful to Astrid for having given this possibility to me again.
- Also, I again collaborated with Dr. Alexander Stendal (www.winternet.de/ste ndal/gallery.htm). He prepared tailored cartoons for illustrating important physical concepts in a manner that everybody can understand their essence. I my view, these illustrations are invaluably useful in the book, while in fact their function is twofold: They provide explicit physical information and increase the reading pleasure. Many thanks to Alexander for that.
- Our surrounding nature is rich in examples of illustrative optical phenomena— just remember the rainbow. But even ordinary clouds appearing on the sky may be full of fascinating secrets. During a mountain hike, I recognized strange double formations of clouds as shown in Fig. 7.3 on top. I decided to use one of them for illustration purposes in this book and contacted experts for an assignment of the cloud type. And I had to recognize that similar to spectroscopy, a restricted amount of information makes an unambiguous assignment practically impossible. In the mentioned case, I consulted experts from the Bergische

Universität Wuppertal and Bonn University. I got several assignment proposals that I am listing here: *Cumulus velum, Cumulus pileus*, or a combination of *Cumulus (humilis)* and *Altocumulus lenticularis*. In this context, I am grateful to Marc Krebsbach, Christian Ohlwein, Andreas Hense, Clemens Simmer, Nils Risse, and Victor Venema, who watched over the photograph and provided the mentioned assignment proposals.

- Also, in architecture you may find esthetic examples of the tricky use of symmetry elements. I decided to include several of them into the solid-state physics, Chap. 16. I am grateful to Martin Heider (Doberan Minster, Germany) and Ulf Koischwitz (St. Aegidii Church Quedlinburg, Germany) for giving the permission to use these photographs in this book. In particular, Martin Heider gave me extensive background information on the historical origin of the tessellation shown in Figs. 16.5–16.7, as explained in the figure caption.
- Several of my (former) students provided graphs to the lecture script, and I am reusing some of them here. I do not remember all family names, so that my thanks are simply to my (former) students Fanhui, Rahil, Abrar, and JianYing from the Abbe School of Photonics for their assistance.

Jena, Germany   Olaf Stenzel
August 2021

Light–Matter Interaction

"Narr und Eulenspiegel" (Fool and Eulenspiegel). Detail of a Sculpture assembly by Astrid Leiterer, Jena, Germany (www.astrid-art.de). Photograph by Astrid Leiterer and reproduced with permission

Specular and diffuse reflections of light at metallic surfaces lead to fascinating optical phenomena.

Contents

Part VI Optical Properties of Solids

Main Abbreviations and Symbols

(Note: We have much more physical quantities and parameters than letters. Therefore, this list is neither complete nor unambiguous. Whenever you have doubt on the meaning of a symbol in a concrete formula, please check the accompanying text passages.

Vectors are indicated by bold letters ($\mathbf{Q}$), scalars in italic (Q), operators like $\hat{Q}$ or $\hat{\mathbf{Q}}$).

1D	One-dimensional
2D	Two-dimensional
3D	Three-dimensional
A	Area
A	Rotational constant in Chap. 11
A_{21}	Einstein's coefficient for spontaneous
a	Acceleration
a	Lattice constant
a_0	Bohr's radius
$\hat{a}^+$	Photon creation operator
$\hat{a}$	Photon annihilation operator
α	Eigenvalue of the photon annihilation operator (Chap. 5)
α	Absorption coefficient
α	Fine-structure constant (Chap. 8)
$\mathbf{B}$	Magnetic induction
B	Rotational constant in Chaps. 11 and 14
B	Energy band width parameter in Chaps. 16 and 17
B_{21}	Einstein's coefficient for stimulated emission
B_{12}	Einstein's coefficient for absorption
β	Linear microscopic polarizability
C	Constant
c	Velocity of light in vacuum
$\mathbf{D}$	Electric displacement
D	Density/joint density of quantum states
D	Exchange integral

D	Dissociation energy
D_e	Dissociation energy/(hc)
$\mathbf{d}$	Electric dipole moment
$\mathbf{d}_{\mathbf{perm}}$	Permanent electric dipole moment
$\mathbf{d}_{\mathrm{magn}}$	Magnetic dipole moment
$\mathbf{d}_{\mathrm{ml}}$	Matrix element of the electric dipole moment operator
δ	Phase, phase shift
$\mathbf{E}, E$	Electric field strength
E_0	Field amplitude
E	Energy in quantum mechanics
E_g	Energy band gap in solids
E_n	Energy level in quantum mechanics
$\mathbf{e}$	Unit vector
e	Basis of natural logarithm
e	Elementary charge
ε_0	Permittivity of free space
ε	dielectric function
$\varepsilon_{\mathrm{stat}}$	Static value of the dielectric function
F	Force
f	(Ordinary) frequency
f_j	Relative strength of the absorption lines in classics
f_{ij}	Oscillator strength in quantum mechanics
φ	Angle
G	Spectral term
γ/Γ	Damping constant in angular frequency/wavenumber units
γ_l, γ_s	Gyromagnetic factors
$\mathbf{H}, H$	Magnetic field strength
$\hat{H}$	Hamilton operator, Hamiltonian
h	Planck's constant
$\hbar$	$h/(2\pi)$
I	Intensity
I	Electric current
I	Mass moment of inertia
i	Counting index
$\mathbf{j}$	Probability current density
$\mathbf{j}$	Electric current density
$\mathbf{J}, \mathbf{j}$	Total angular momentum
J	Rotational quantum number
j	Counting index
K	Extinction coefficient
$\mathbf{k}$	Wavevector (general)
$\mathbf{k}_p$	Wavevector of a photon
κ	Response function
κ	Spring constant
k_B	Boltzmann's constant

k	Counting index
$\mathbf{L}$	Angular momentum
l, L	Sometimes used for geometrical dimensions
l	Counting index
λ	Wavelength in vacuum
$\mathbf{M}$	Magnetization
m	Mass
m	Counting index
m_p	Mass at rest of a proton, close to that of a neutron
m_e	Mass at rest of an electron
μ	Reduced mass
μ_B	Bohr's magneton
μ_0	Permeability of free space
MIR	Middle infrared spectral region
N	Concentration
N	Number of objects (where specified)
n	Counting index (in sums, in quantum mechanics); integer number
n	Quantum number
n	Refractive index
$\hat{n}$	Complex index of refraction
NIR	Near-infrared spectral region
NLO	Nonlinear optics
ν	Wavenumber; $\nu = 1/\lambda$
$\mathbf{P}, P$	Polarization
$P^{(j)}$	Polarization of jth order
ψ	Time-independent wavefunction in quantum mechanics
Ψ	Time-dependent wavefunction in quantum mechanics
q	Charge
R	Radius
R	Reflection coefficient
$\mathbf{r}$	Position vector with $\mathbf{r} = (x, y, z)^\mathrm{T}$
$\mathbf{R}$	Vector indicating the position of the mass center
Ry	Rydberg energy
R_∞	Ry/(hc)
ρ	Mass density
ρ	Specific resistance
σ	Electrical conductivity
σ	Standard deviation in a Gaussian distribution
σ_stat	Static value for the electric conductivity
T	Absolute temperature
T	Transmission coefficient
ϑ	Step function
ϑ	Absolute temperature "in cm^{-1}": $\vartheta \equiv \frac{k_B T}{hc}$
θ	Angle
T_kin	Kinetic energy

t	Time
τ	Time constant, relaxation time
u	Spectral density
U, u	Potential energy
$u_{\mathbf{k},j}(\mathbf{r})$	Lattice periodic term in the Bloch function
UV	Ultraviolet spectral region
$\hat{V}$	Perturbation operator
V	Volume
V_{ij}	Matrix element of the perturbation operator
v_{phase}	Phase velocity
v_{group}	Group velocity
VIS	Visible spectral region
VP	Cauchy's principal value of the integral
W	Energy in classics
W	Occupation probability of a quantum state
w	Probability
w	Energy density
ω	Angular frequency
ω_0	Eigenfrequency, resonance frequency
ω_p	Plasma frequency
$\tilde{\omega}_0$	Shifted with respect to local field effects resonance frequency
ω_{nm}	Transition frequency, resonance frequency in quantum mechanics
$\Delta\omega$	Spectral bandwidth
x	Cartesian coordinate, position
χ	Linear dielectric susceptibility
$\chi^{(j)}$	Susceptibility of jth order
y	Cartesian coordinate
Z	Number of quantum states
Z	Atomic order
z	Cartesian coordinate

Survey of Constants

$$\varepsilon_0 = 8.86 * 10^{-12} \text{ F/m}$$

$$\mu_0 = 1.256 * 10^{-6} \text{ Vs/(Am)}$$

Planck's constant: $\quad h = 6.625 * 10^{-34} \text{ Ws}^2 = 4.136 * 10^{-15} \text{ eVs}$

$$\hbar = \frac{h}{2\pi} \approx 1.054 * 10^{-34} \text{Ws}^2$$

Elementary charge: $\quad e = 1.602 * 10^{-19} \text{ As}$

Electron rest mass: $\quad m_e = 9.108 * 10^{-31} \text{ kg}$

Proton rest mass: $\quad m_p = 1.672 * 10^{-27} \text{ kg}$

Boltzmann's constant: $\quad k_B = 1.38 * 10^{-23} \text{ Ws/K}$

Gravitational constant: $\quad G = 6.674 * 10^{-11} \text{ m}^3\text{kg}^{-1}\text{s}^{-2}$

Part I
Selected Facets of the Light–Matter Interaction in Classical Physics

"Sonnenaufgang im Tatio Tal" (Sunrise in El Tatio, Atacama Desert, Chile). Painting and Photograph by Astrid Leiterer, Jena, Germany (www. astrid-art.de). Photograph reproduced with permission

In many cases, the interaction of light with matter is rather unspectacular. However, there appear situations where dramatic phenomena are observed. You do not need lasers for that. One of the world's largest geyser fields, the El Tatio geyser field, is located in the Atacama Desert at an altitude of 4320 m. Because of the large intraday temperature differences, the geysers show the most spectacular activity immediately after sunrise.

Introduction

1

Abstract

Relevant characteristic dimensions of matter are introduced in order to provide a feeling for what we will later call the micro- and macroworlds. Selected basic physical and mathematical skills are repeated to facilitate sophisticated solution of the problems that are offered at the end of each chapter.

1.1 Some Characteristic Dimensions

This short introduction is to make you familiar with some characteristic dimensions that are specific for different material systems. Our focus will be on orders of magnitude rather than on accurate numbers.

Everybody has some intuitive feeling on what a length is, what a time interval is, or what a mass is. And we have certain personal experience and skills in handling these quantities. This daily experience works over many orders of magnitude: We have a feeling for length dimensions ranging from less than 1 mm up to hundreds of kilometers (8 orders of magnitude), masses from approximately 1 g to 100 kg (5 orders of magnitude), time scales from approximately 0.1 s (humans reaction time) up to 100 years (more is seldom)—10 orders of magnitude. And we are familiar with velocities ranging from that of a propagating snail up to that of an airplane. The classical Newtonian mechanics provide us with a quantitative theory that excellently reproduces all mechanical phenomena observed at these scales.

Why to develop further physical concepts? In order to understand this need, it makes sense to have a look at the corresponding characteristic dimensions occurring in the nature around us. Table 1.1 gives a short overview on the dynamic range in spatial extensions, characteristic times, and masses that are observed for different macroscopic and microscopic objects.

We note that the dynamic ranges are immense: Although the electron is by far not the lightest particle, we note that the universe is for more than 80 orders of

© The Author(s), under exclusive license to Springer Nature Switzerland AG 2022
O. Stenzel, *Light–Matter Interaction*, UNITEXT for Physics,
https://doi.org/10.1007/978-3-030-87144-4_1

Table 1.1 Selected orders of magnitude in nature

System	Characteristic spatial dimension/m	Characteristic time/s	Mass at rest/kg
Universe	10^{26}	Age: 10^{18}	10^{53}
Sun	10^{9}	Age: 10^{17}	10^{30}
Earth	10^{7}	Age: 10^{17}	10^{25}
Human	10^{0}	Life expectation: 10^{9}	10^{2}
Molecule	$>10^{-10}$	"Vibration period" 10^{-13} (see Chap. 2)	$>10^{-27}$
Atom	$\approx 10^{-10}$	"Vibration period" 10^{-15} (see Chap. 2)	10^{-25} to 10^{-27}
Electron	Classics: $r_e \approx 3 \times 10^{-15}$ m		10^{-30}

magnitude heavier than a single electron. A similar situation is observed when comparing the length and time scales. All these dimensions vary over much more orders of magnitude than our experience provides. This is exemplified in Fig. 1.1, which combines some of the data from Table 1.1 with the mentioned orders of magnitude normally accessible to our experience. It would therefore be a methodological mistake to automatically extend the range of validity of our experience

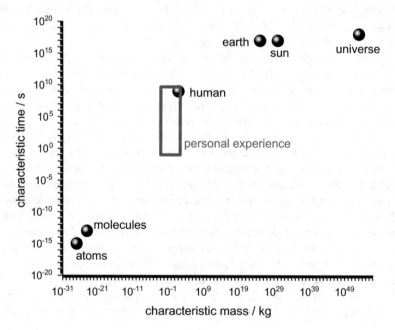

Fig. 1.1 Characteristic orders of magnitude in nature from Table 1.1 and of personal experience of a human

to phenomena which are observed at much larger or smaller dimensions. Instead, we even have to expect, that at very large or small dimensions, things will happen that contradict our daily experience. Unfortunately, this is something that we have to accept. So we need additional theoretical tools that describe the properties of matter at dimensions not accessible to our daily experience.

Taking into account the information given in Table 1.1, it appears not so astonishing that so far there exists no unique physical theory that describes the properties of matter at all length, time, and mass scales. The one extreme—the universe—is rather a topic to be described in terms of general relativity. Small objects, however, like molecules or atoms are accurately described by means of quantum mechanics.

1.2 The Organization of This Book

In this book, an approach to the structure of matter and its interaction with light is proposed that is designed to the needs of students of photonic disciplines. Our focus will be on degrees of freedom in matter that efficiently interact with light, mainly covering spectral ranges like infrared, visible, or ultraviolet light. Note that throughout this book, we will for simplicity use the termini "light" and "electromagnetic radiation" as synonyms. As it will be shown in the next chapter, the relevant degrees of freedom in matter arise from the excitation of different internal "vibrations" of electrons and nuclei in atoms, molecules or condensed systems. Therefore, the focus of our approach will be

- On atoms, molecules as well as the dynamics of charge carriers in solids
- Their interaction with electromagnetic radiation.

When consulting Table 1.1 (atoms, molecules), we see that we will arrive at dimensions which are very much smaller than those where our experience works. Therefore, instead of our experience, we will have to use formal theoretical tools in order to develop a quantitative description of what is going on when light is interacting with an atom or molecule. Fortunately, an adequate theory is available, it is the already mentioned apparatus of quantum mechanics.

One remark concerning the organization of the book: Starting from Chap. 2, all chapters will be organized in an identical manner, including the Sections:

1. Starting point (accepted experimental data or theoretical concepts)
2. Physical idea(s) (how we want to proceed...)
3. Theoretical considerations (new material)
4. Consistency considerations
5. Application to practical problems
6. Advanced material
7. Tasks for self-check (solutions will be provided at the end of the book)

8. Literature. Note that general literature shall be easily accessible. For special examples, I sometimes make use of specific (maybe older) references that are not so easily accessible. Nevertheless I feel some obligation to indicate the original source.

Generally, in each of the chapters, the reader is supposed to be familiar with the technical material presented in the corresponding first section. If not, it is recommended to consult other literature sources, such as collected in the section: General Literature, and/or to repeat the previous chapter(s) of this book.

For quick reading, it might be, in principle, sufficient to read the sections 2–5 in each chapter. These sections contain the basic learning stuff of the given chapter. Nevertheless, the remaining 6th section may also provide new ideas or concepts, and may make the following chapters more understandable.

Note that the main body of the text is written on a normal white background.

A grey background indicates rules of thumb or mnemonic tricks for memorizing. Also, verbal illustrations for a better "visualization" of theoretical results may be offered here. Note that these verbal illustrations are never strong in a scientific sense.

The remark providing some additional material. The remarks are not necessary to read for understanding the further material.

1.3   A Remark on Calculations

Just for repetition, it makes sense to emphasize the correct treatment of measurement units. A physical quantity Q is usually written as: $Q = \{Q\}[Q]$; where $\{Q\}$ is the numerical value of Q, while $[Q]$ is the measurement unit of Q. Thus, if we indicate the length of 15 m, we shall write:

$l = 15$ m. Here, "m" is the measurement unit (a meter), i.e., $[l] = $ m, while "15" is the numerical value, i.e., $\{l\} = 15$. **_Both_** information must be indicated in order to provide an unambiguous quantification of l.

In the following chapters, a lot of calculations will be provided in order to make the reader familiar with characteristic values of the parameters we are working with. Not all of these calculations are shown in full detail, but the reader is rather invited to check the calculations himself. But here we will provide one characteristic example and a corresponding recommendation how to deal with this type of calculation. This example is to emphasize the calculation principle and therefore included into this classical part of the book, although it will contain a "quantum constant", namely $\hbar$. Our example is to calculate what is hidden behind the following expression:

$$\frac{4\pi\varepsilon_0\hbar^2}{m_e e^2} =?$$

The corresponding constants are given in the section: Survey of constants. From there we have:

$$\frac{4\pi\varepsilon_0\hbar^2}{m_e e^2} = \frac{4\pi * 8.86 \times 10^{-12}\frac{F}{m}\left(1.054 \times 10^{-34}Ws^2\right)^2}{9.108 \times 10^{-31}\text{kg}\left(1.602 \times 10^{-19}As\right)^2}$$

The recommendation is now to separate measurement units and orders of magnitude from the rest. That gives:

$$\frac{4\pi\varepsilon_0\hbar^2}{m_e e^2} = \underbrace{\frac{4\pi * 8.86 * (1.054)^2}{9.108 * (1.602)^2}}_{\approx 5.29} * \underbrace{10^{-12-34-34+31+19+19}}_{10^{-11}} * \underbrace{\frac{FW^2s^4}{\text{mkg}A^2s^2}}_{?}$$

We obtain:

$$\frac{4\pi\varepsilon_0\hbar^2}{m_e e^2} \approx 0.53 \times 10^{-10} * \underbrace{\frac{FW^2s^4}{\text{mkg}A^2s^2}}_{?}$$

It remains to calculate the resulting measurement unit. We have:

$$\frac{FW^2s^4}{\text{mkg}A^2s^2} = \frac{\frac{As}{V}V^2A^2s^4}{\text{mkg}A^2s^2} = \frac{AsVs^2}{\text{mkg}} = \frac{VAs}{\frac{\text{mkg}}{s^2}} = \frac{Ws}{\frac{\text{kgm}^2}{s^2}}m = m$$

Consequently,

$$\frac{4\pi\varepsilon_0\hbar^2}{m_e e^2} \approx 0.53 \times 10^{-10}m = 0.053\,\text{nm}$$

What we have calculated is some characteristic length, as it is evident from the obtained measurement unit. The physical sense of that length will become clear in Chap. 8.

1.4   Two Useful Integrals

There are two types of integrals of significance for this course. We will provide
their solution here. The first integral is:

$$\int_0^\infty x^n e^{-px} \, dx = ?? \quad (n - \text{integer}; \ p > 0)$$

Here, two ways of solving the integral are provided. The first way makes use
of partial integrating:

$$\int_0^\infty x^n e^{-px} \, dx = \frac{1}{p^{n+1}} \int_0^\infty \xi^n e^{-\xi} \, d\xi \quad \text{with } \xi = px$$

$$\int_0^\infty \xi^n e^{-\xi} \, d\xi = n \int_0^\infty \xi^{n-1} e^{-\xi} \, d\xi = n(n-1) \int_0^\infty \xi^{n-2} e^{-\xi} \, d\xi$$

$$= \cdots = n! \int_0^\infty e^{-\xi} \, d\xi$$

$$\Rightarrow \int_0^\infty x^n e^{-px} \, dx = \frac{n!}{p^{n+1}}$$

The second way makes use of differentiating by a parameter:

$$\int_0^\infty x^n e^{-px} \, dx = (-1)^n \frac{d^n}{dp^n} \int_0^\infty e^{-px} \, dx$$

$$= (-1)^n \frac{d^n}{dp^n} \frac{1}{p} \int_0^\infty e^{-\xi} \, d\xi = (-1)^n \frac{d^n}{dp^n} \frac{1}{p} e^{-\xi} \Big|_\infty^0$$

$$= (-1)^n \frac{d^n}{dp^n} \frac{1}{p} = (-1)^n (-1)^n \frac{n!}{p^{n+1}} = \frac{n!}{p^{n+1}}$$

$$\Rightarrow \int_0^\infty x^n e^{-px} \, dx = \frac{n!}{p^{n+1}}$$

Of course, both approaches lead to the same result:

$$\int_0^\infty x^n e^{-px} \, dx = \frac{n!}{p^{n+1}} \tag{1.1}$$

Let us now come to the second integral:

$$\frac{2}{L}\int_0^L x\sin\left(\frac{n\pi x}{L}\right)\sin\left(\frac{m\pi x}{L}\right)\mathrm{d}x = ??; \quad n \neq m \quad n, m \text{ integer}$$

Here, we again make use of partial integrating:

$$\xi = \frac{\pi x}{L} \Rightarrow \frac{2}{L}\int_0^L x\sin\left(\frac{n\pi x}{L}\right)\sin\left(\frac{m\pi x}{L}\right)\mathrm{d}x$$

$$= \frac{2L}{\pi^2}\int_0^\pi \xi\sin(n\xi)\sin(m\xi)\mathrm{d}\xi$$

$$= \frac{L}{\pi^2}\int_0^\pi \xi[\cos((n-m)\xi) - \cos((n+m)\xi)]\mathrm{d}\xi$$

$$= \frac{L}{\pi^2}\left\{0 - \int_0^\pi\left[\frac{\sin((n-m)\xi)}{n-m} - \frac{\sin((n+m)\xi)}{n+m}\right]\mathrm{d}\xi\right\}$$

$$= \frac{L}{\pi^2}\left[\frac{\cos((n-m)\xi)}{(n-m)^2} - \frac{\cos((n+m)\xi)}{(n+m)^2}\right]_0^\pi$$

$$= \frac{L}{\pi^2}\left[\frac{\cos((n-m)\pi - 1)}{(n-m)^2} - \frac{\cos((n+m)\pi - 1)}{(n+m)^2}\right]$$

Hence, we finally obtain:

$$n - m \text{ odd} \Rightarrow \frac{2}{L}\int_0^L x\sin\left(\frac{n\pi x}{L}\right)\sin\left(\frac{m\pi x}{L}\right)\mathrm{d}x$$

$$= \frac{2L}{\pi^2}\left[\frac{1}{(n+m)^2} - \frac{1}{(n-m)^2}\right]$$

$$n - m \text{ even} \Rightarrow \frac{2}{L}\int_0^L x\sin\left(\frac{n\pi x}{L}\right)\sin\left(\frac{m\pi x}{L}\right)\mathrm{d}x = 0 \qquad (1.2)$$

1.5 Tasks for Self-check

1.5.1 measurement units:

Simplify the following combinations of SI-units:

Example: $1\frac{Ws^3}{\mathrm{kgm}} = 1Ws\frac{s^2}{\mathrm{kgm}} = 1\,\mathrm{Nm}\frac{s^2}{\mathrm{kgm}} = 1N\frac{s^2}{\mathrm{kg}} = 1\frac{\mathrm{kgm}}{s^2}\frac{s^2}{\mathrm{kg}} = 1m$

Expression	Solution
$1\frac{V\,As}{m^2\,Pa} =?$	
$1\frac{kg m^2}{s^2 V^2} =?$	
$1\frac{JF}{As} =?$	
$1\frac{A^2 s^5}{m^2 kg F^2} =?$	
$1\frac{V\,As^2}{JF} =?$	
$1 s^2 MPa =?$	
$1\frac{V\,As^3}{m^2} =?$	
$1\frac{W\,s^2}{\Omega^2 F A^2} =?$	

Even if you are confronted with an unknown equation, you may be able to identify the correct measurement units of some of the individual variables or parameters. Try this by completing the following table: (Example: $V = l^3$ and $[V] = m^3 \rightarrow [l] = m$.

Information:	Task	Your answer:
$\int_V |\psi|^2 dV = 1$ with V—volume	$[dV] =?$	
	$[\psi] =?$	
$\psi = \sqrt{\frac{2}{L}}\sin kx$ with $[L] = m$ and $[x] = m$	$[\psi] =?$	
	$[k] =?$	
$I = I_0 e^{-\alpha x}$ with $[I] = Wm^{-2}$ and $[x] = m$	$[I_0] =?$	
	$[\alpha] =?$	
$\psi = \frac{1}{\sqrt{\pi}}\left(\frac{Z}{a_0}\right)^{\frac{3}{2}} e^{-Zr/a_0}$ with $[Z] = 1$ and $[r] = m$	$[a_0] =?$	
	$[\psi] =?$	
$Re\varepsilon(\omega) = 1 + \frac{1}{\pi} V P \int_{-\infty}^{\infty} \frac{Im\varepsilon(\xi)d\xi}{\xi-\omega}$ with $[Im\varepsilon] = 1$ and $[\xi] = s^{-1}$	$[\omega] =?$	
	$[Re\varepsilon(\omega)] =?$	

1.5.2 Calculate the following expressions:
- $\sqrt{\varepsilon_0\mu_0} =?$ (ε_0 is the permittivity of free space, μ_0 is the permeability of free space)
- $\frac{e^2}{4\pi\varepsilon_0\hbar c} =?$
- $\sqrt{\frac{\hbar G}{c^3}} =?$ (G—gravitational constant)
- $\frac{2e}{h} =?$
- $\frac{h}{e^2} =?$

1.5.3 From the Maclaurin series, derive the following expansion rules:
- $(1+x)^\mu = 1 + \frac{\mu}{1!}x + \frac{\mu(\mu-1)}{2!}x^2 + \frac{\mu(\mu-1)(\mu-2)}{3!}x^3 + \cdots$ (μ real; $|x| < 1$)

- $e^x = 1 + \frac{x}{1!} + \frac{x^2}{2!} + \frac{x^3}{3!} + \cdots$
- $\sin x = \frac{x}{1!} - \frac{x^3}{3!} + \frac{x^5}{5!} - + \cdots$
- $\cos x = 1 - \frac{x^2}{2!} + \frac{x^4}{4!} - + \cdots$

1.5.4 Basing on the last three relations, proof the validity of Euler's formula:

$$e^{ix} = \cos x + i \sin x$$

1.5.5 Imagine one liter of water uniformly spread over the surface of a sphere with the radius of the earth. Estimate the number of water molecules observed in 1 cm^2 of the surface of the sphere! (after [1])

1.5.6 Liquid helium has a density of $\rho = 0.13$ gcm^{-3}. Estimate the radius of a helium atom, assuming that each atom has a weight of 4 nucleons, and that the atoms are packed in the densest possible configuration, which fills 74% of the space (after [2]).

1.5.7 Calculate and compare momentum and kinetic energy for the following objects:

- Earth on its orbit around the sun $\left(m \approx 6 \times 10^{24} \text{kg}; \ v \approx 30 \frac{\text{km}}{\text{s}}\right)$
- A tectonic plate with $m \approx 3 \times 10^{20}$ kg; $v \approx 1 \frac{\text{cm}}{\text{year}}$
- A train with $m \approx 10^6$ kg; $v \approx 100 \frac{\text{km}}{\text{h}}$
- A runner with $m \approx 80$ kg; $v \approx 20 \frac{\text{km}}{\text{h}}$
- An electron moving with $v = 0.1c$. Consider a nonrelativistic case!

References

Specific References

1. U. Fano, L. Fano, *Physics of Atoms and Molecules. An Introduction to the Structure of Matter* (University of Chicago Press, 1973), Task 1.1
2. H. Haken, H.C. Wolf, *The Physics of Atoms and Quanta* (Springer, 2005), task 2.2

General Literature

3. P.A. Tipler, G. Mosca, *Physik, 7* (Springer, Auflage, 2015)

Simplest Model Treatment of the Classical Interaction of Light with Matter

Abstract

A short review of selected aspects of light–matter interactions in classical physics is provided. The treatment is restricted to electric dipole interactions. After a short introduction to the two-body problem, a classical approach to light emission and absorption is presented. Characteristic absorption frequencies or wavelength is estimated in terms of simple model assumptions.

2.1 Starting Point

Let us start from recalling some basic knowledge on electromagnetic radiation. Imagine the simplest situation: a monochromatic electromagnetic plane wave that is propagating in vacuum (Fig. 2.1).

In complex notation, its electric field strength vector $\mathbf{E}$ may be given as:

$$\mathbf{E}(\mathbf{r}, t) = \mathbf{E}_0 e^{-i(\omega t - \mathbf{kr})} \tag{2.1}$$

Here, ω is the angular frequency ($\omega = \frac{2\pi}{T}$); T is the period (i.e., the time necessary for performing one oscillation). Moreover, $\mathbf{k}$ is the wavevector ($|\mathbf{k}| = k = \frac{2\pi}{\lambda}$; $\mathbf{k} = k\mathbf{e}$); λ is the wavelength in vacuum, and $\nu = \frac{1}{\lambda}$ is called the wavenumber. $\mathbf{e}$ is the so-called unit propagation vector. We have the obvious relations:
$\omega = \frac{2\pi}{T} = 2\pi f = \frac{2\pi c}{\lambda} = kc$, with f—ordinary frequency and c—velocity of light in vacuum. Some selected spectral regions of electromagnetic radiation are summarized in Table 2.1.

In a plane electromagnetic wave propagating in vacuum, the magnetic field strength $\mathbf{H}$ follows from (2.1a):

$$\frac{1}{\mu_0 c}\mathbf{e} \times \mathbf{E} = \mathbf{H} \tag{2.1a}$$

O. Stenzel, *Light–Matter Interaction*, UNITEXT for Physics,
https://doi.org/10.1007/978-3-030-87144-4_2

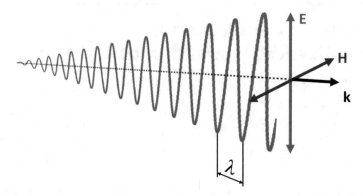

Fig. 2.1 Illustration of a plane electromagnetic wave, arriving from a remote light source. The picture shall be impressive, but it must not be misinterpreted: Of course, field strength vectors are not measured in length units, and therefore, they are not subject to perspective distortions

Table 2.1 Typical frequencies and wavelength (orders of magnitude)

Spectral range		Wavelength	Frequency/s^{-1}
Microwave	MW	mm	10^{11}
Infrared	IR	μm	10^{13}
Visible	VIS	400–700 nm	10^{14}–10^{15}
Ultraviolet	UV	$\approx$10–400 nm	10^{16}–10^{17}
X-ray	X	Between UV and γ	

A more detailed classification of spectral ranges will be given later in Table 14.1

   Of course, our goal is to describe the interaction of light with *matter*. Therefore, our focus is not on the propagation of electromagnetic waves in vacuum; instead, we shall bring the light wave into interaction with a material system. Depending on the specific kind of matter, the latter may be characterized by specific characteristic internal time and length constants. On the other hand, the monochromatic plane wave introduced by (2.1) has only one characteristic time (the period, corresponding to a certain frequency) and only one characteristic length (the wavelength).

As an intuitive rule of thumb, we have to expect rather strong (resonant) interaction between light and matter, when:

- ω comes close to a resonance angular frequency (eigenfrequency) of the material system, or
- λ comes close to a characteristic spatial dimension of the material system.

If neither the frequency nor the wavelength match any internal eigenfrequency of characteristic length of the material system, the interaction is nonresonant and, as a rule, much weaker than in the resonant case.

Remark Let us make a remark concerning a convention implicitly made when writing down (2.1). Of course, the natural writing of the electric field in a monochromatic plane wave would operate with real functions and coefficients only. For such real fields, we could use a description of the type:

$$E_{\text{real}}(\mathbf{r}, t) = E_{0,\text{real}} \cos(\omega t - \mathbf{kr} + \delta) \tag{2.2}$$

However, the cosine function appears to be quite inconvenient with respect to our further mathematical treatment. On the other hand, it can be written as:

$$E_{\text{real}}(\mathbf{r}, t) = \frac{1}{2}\left[E_{0,\text{real}} e^{-i(\omega t - \mathbf{kr})} e^{-i\delta} + E_{0,\text{real}} e^{i(\omega t - \mathbf{kr})} e^{i\delta} \right]$$
$$\equiv E_0 e^{-i(\omega t - \mathbf{kr})} + c.c \tag{2.3}$$

Here, "c.c." denotes the conjugate complex to the preceding expression. It turns out that the assumed monochromatic real electric field may be expressed as the sum of a complex field and its conjugate complex counterpart, while the latter does not contain any new physical information. Hereby, we have introduced the complex field amplitude E_0 as:

$$E_0 \equiv \frac{E_{0,\text{real}} e^{-i\delta}}{2} \tag{2.4}$$

In practice, it appears much more convenient to build the further theory using complex electric fields $E(\mathbf{r},t)$ according to (2.1) and (2.4) instead of working with the real version (2.2). Therefore, in our treatment we make use of the complex field defined by (2.1), keeping in mind that the real field will be obtained when adding the complex conjugate to (2.1). Or, in other words:

$$E_{\text{real}}(\mathbf{r}, t) = 2\text{Re}E((\mathbf{r}, t) \tag{2.5}$$

Here, $E(\mathbf{r},t)$ is given by (2.1) and (2.4).

But I would like to direct your attention onto a specific detail. The choice of (2.1) for the complex writing of the electric field defines a particular convention, which is used throughout this course. When looking at (2.3), it becomes evident that we could have used the writing:

$$E_{\text{real}}(\mathbf{r}, t) = \frac{1}{2}\left[E_{0,\text{real}}e^{+i(\omega t - \mathbf{kr})}e^{+i\delta} + E_{0,\text{real}}e^{-i(\omega t - \mathbf{kr})}e^{-i\delta} \right]$$

$$\equiv E_0 e^{+i(\omega t - \mathbf{kr})} + c.c;$$

$$E_0 \equiv \frac{E_{0,\text{real}}e^{+i\delta}}{2}$$

as well. In our further treatment, we will use the minus sign as fixed in (2.1). In other sources, the other convention (with the plus sign) may be used, which may result in differences in some of the equations to be derived in this course.

The intensity (=energy per time and area, averaged over a characteristic time interval) of the wave propagating in vacuum is then:

$$I = \frac{1}{2\mu_0 c}\left| E_{0,\text{real}} \right|^2 = \frac{2}{\mu_0 c}|E_0|^2 \tag{2.6}$$

2.2   Physical Idea

Let us now start discussing the interaction between a propagating light wave and a material system. We have already introduced the plane monochromatic wave as our model for describing the impact of the electromagnetic field. What we still need is a corresponding model for the material system.

We will choose the two-body system as the simplest possible model system for understanding the general principles of the interaction of light with matter. Indeed, basic optical properties arise from so-called electric dipole interactions between the electric field of the electromagnetic wave and the charge carriers in the material system. But an electric dipole is formed by a minimum of *two* charges, a positive and a negative one. So in our basic model, we will have to regard two bodies (here material points), which will later be assumed to be electrically charged and interacting with a light wave (Fig. 2.2).

If so, the interaction mechanism is rather clear. The oscillating electric field of the wave gives rise to Coulomb forces that result in an oscillatory movement of the charge carriers in the medium (here the two-body system(s)). This way, oscillating electric dipoles are excited, which provide the basic mechanism of the interaction of light with matter in the so-called electric dipole approximation. In a classical language, the excitation of the oscillating dipoles can be described in terms of Newton's equation of motion.

Note that in this approximation, magnetic interactions are neglected. We will not consider any Lorentz forces acting on the charge carriers. Therefore, in the

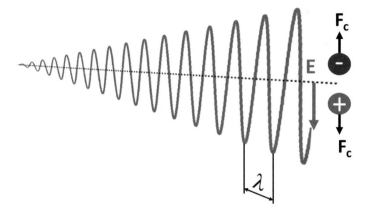

Fig. 2.2 Illustration of the formation of an oscillating electric dipole by the electric field of the electromagnetic wave

electromagnetic wave, we focus on the electric field strength vector and (at least at the moment) forget the magnetic one.

Why can we do so? Let us have a look at the expressions for the Coulomb and Lorentz forces $\mathbf{F}_C$ and $\mathbf{F}_L$. We have:

$$\mathbf{F}_C = q\mathbf{E}$$
$$\mathbf{F}_L = q\mathbf{v} \times \mathbf{B}$$

Obviously, a resting charge feels a Coulomb force, but no Lorentz force. In order to give rise to a relevant Lorentz force, the charge carrier should have a corresponding velocity $\mathbf{v}$. Hence, if we start from "resting" charges, the electric field must first bring the charge carriers into motion, in order to allow the Lorentz force coming into the game. But if the oscillating electric field is not too strong, the charge carriers will never achieve the necessary velocity to make magnetic interactions comparable to the electric ones.

Remark Note that in a plane monochromatic electromagnetic wave propagating in vacuum, the relation $B = \frac{E}{c}$ holds (compare (2.1a) and consider $B = \mu_0 H$). Here, $B = |\mathbf{B}|$ and so on. Therefore, making use of a rectangular geometry, we have: $F_L = qvB = q\frac{v}{c}E = \frac{v}{c}F_C$. In the nonrelativistic case, $v \ll c$, and therefore, $F_L \ll F_C$.

2.3   Theoretical Considerations

2.3.1   On the Two-Body System

Imagine now two point masses m_1 and m_2 in a Cartesian coordinate system as shown in Fig. 2.3.

Their distance from each other is characterized by the vector $\mathbf{r}$.

Obviously:

$$\mathbf{r_1} + \mathbf{r} = \mathbf{r_2}$$

Their kinetic energy is given by:

$$T_{\text{kin}} = \frac{m_1}{2}\dot{\mathbf{r}}_1^2 + \frac{m_2}{2}\dot{\mathbf{r}}_2^2$$

Let us change from the coordinates $\mathbf{r_1}, \mathbf{r_2}$ to the coordinates

$$\mathbf{r} = \mathbf{r_2} - \mathbf{r_1}$$
$$\mathbf{R} = \frac{m_1\mathbf{r_1} + m_2\mathbf{r_2}}{M} \tag{2.7}$$

where $M = m_1 + m_2$. Then, the vector $\mathbf{R}$ marks the position of the mass center of the two point masses. We then find:

$$\mathbf{r_1} = \mathbf{R} - \frac{m_2}{M}\mathbf{r}$$
$$\mathbf{r_2} = \mathbf{R} + \frac{m_1}{M}\mathbf{r}$$

Fig. 2.3 Assumed geometry in a two-body system

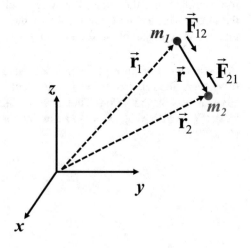

This way, we obtain the kinetic energy:

$$T_{kin} = \frac{M}{2}\dot{\mathbf{R}}^2 + \frac{\mu}{2}\dot{\mathbf{r}}^2 \qquad (2.8)$$

where $\mu = \frac{m_1 m_2}{m_1 + m_2}$ is called the *reduced mass* of the system of the two point masses.

The full kinetic energy of the system is thus composed from the translational movement of the mass center and the relative movement of the point masses with respect to each other.

When placing the origin of coordinates into the center of mass, we set:

$$\mathbf{R} = 0$$
$$\dot{\mathbf{R}} = 0. \qquad (2.8a)$$

Then,

$$T_{kin} = \frac{\mu}{2}\dot{\mathbf{r}}^2 \qquad (2.9)$$

The full derivative of $\mathbf{r}$ with respect to time may be composed from two contributions: the change in the *distance* between the point masses (absolute value of $\mathbf{r}$, we simply write r for that) and a change in *orientation* of $\mathbf{r}$, which can be expressed in terms of the polar angle φ. Hence, in terms of the Pythagoras theorem we have:

$$\dot{\mathbf{r}}^2 = \dot{r}^2 + r^2\dot{\phi}^2 \qquad (2.10)$$

Let us regard two special cases:

(a) Pure rotation ($\dot{r} = 0$; $\dot{\phi} \neq 0$)
    Then,

$$T_{kin} = \frac{\mu}{2}r^2\dot{\phi}^2 = \frac{I}{2}\Omega^2 \qquad (2.11)$$

with $I = \mu r^2$—mass moment of inertia, and $\dot{\phi} = \Omega$—angular frequency of rotation. We obtain the well-known expression for the rotational energy of a rigid body in classical physics.

(b) Pure distance change (e.g., pure vibration): ($\dot{\phi} = 0$; $\dot{r} \neq 0$).

Let us assume that both material points interact with an interaction potential $U = U(|\mathbf{r}_2 - \mathbf{r}_1|) = U(r)$. Then, each of the material points will experience an interaction force from the other one, namely $\mathbf{F}_{12}$ and $\mathbf{F}_{21} = -\mathbf{F}_{12}$ (Fig. 2.3). On the other hand, we have Newton's law of motion:

$$\mathbf{F} = \left(-\frac{dU}{d\mathbf{r}}\right) = m\mathbf{a} \qquad (2.12)$$

We obtain
$$\ddot{\mathbf{r}} = \ddot{\mathbf{r}}_2 - \ddot{\mathbf{r}}_1 = \frac{\mathbf{F}_{21}}{m_2} - \frac{\mathbf{F}_{12}}{m_1} = \frac{\mathbf{F}_{21}}{\mu} \equiv \frac{\mathbf{F}}{\mu} \ \text{or}$$

$$\mu\ddot{\mathbf{r}} = \mathbf{F} \tag{2.12a}$$

which is again identical to the classical Newton's law of motion provided that the "normal" mass is replaced by the reduced mass. In order to describe a vibration, the introduced interaction forces must be of a restoring nature; i.e., they should push the system back to an equilibrium position when the latter has been perturbed.

From (2.8) to (2.10), we learn that besides of translation, the two-body system may have "vibrational" and "rotational" degrees of freedom.

What is the use of the presented theory? Provided that our interest to the dynamics of a two-body system is restricted to its rotational and vibrational degrees of freedom, the introduction of the reduced mass concept allows a significant facilitation of the task. Indeed, the system of two (possibly interacting by a potential $U(|\mathbf{r}_2 - \mathbf{r}_1|)$) particles with masses m_1 and m_2, as well as coordinates $\mathbf{r}_1$ and $\mathbf{r}_2$, is practically replaced by a single particle with the reduced mass $\mu = \frac{m_1 m_2}{m_1 + m_2}$ and the coordinates $\mathbf{r}$ in an external potential $U(r)$. In the special case of $r = \text{const.}$, we will call it a rigid rotor.

So far, our two-body system is a rather abstract model configuration. In Sect. 2.5, we will use it to perform simple estimations of characteristic eigenfrequencies of different types of material systems.

2.3.2 The Classical Picture of Light Absorption

Let us for simplicity assume some microscopic object (e.g., an atom), which has no permanent electric dipole moment. That means, the dipole moment should be zero when the applied electric field is zero for a sufficiently long time.

When being illuminated by light, the oscillating electric field strength in the wave $\mathbf{E}$ causes a Coulomb force acting on the charge carriers, thus resulting in the formation of an oscillating induced dipole moment. This is the situation that has been sketched in Fig. 2.2.

We should mention in this connection that we deal with a microscopic object that is much smaller than the wavelength of the light. Therefore, when assuming an electric field as given by (2.1):

$$\mathbf{E}(\mathbf{r}, t) = \mathbf{E}_0 e^{-i(\omega t - \mathbf{kr})}$$

It is reasonable neglecting the spatial variation of the electric field over the extension of the object. Instead, in our model, we replace (2.1) by an oscillating field according to:

$$\mathbf{E} = \mathbf{E}(t) = \mathbf{E}_0 e^{-i\omega t}$$

Let us now regard the motion of a charge carrier, which is bound to its equilibrium position ($x = 0$) by an elastic restoring force (spring constant $\kappa = \mu\omega_0^2$). The oscillating field may lead to small ($x \ll \lambda$) movements of the charge carriers, thus inducing dipoles that interact with the field. While the atoms are composed from nuclei and electrons, the nuclei are much heavier than the electrons, so that they will be considered as fixed. Then, in our model only the electrons are in motion when a harmonic electric field is applied, and the reduced mass is almost identical to the electron mass. The equation of motion of a single charge carrier with mass m (usually an electron) around its equilibrium position is then:

$$qE = qE_0 e^{-i\omega t} = m\ddot{x} + 2\gamma m\dot{x} + m\omega_0^2 x \tag{2.13}$$

This is the equation for driven oscillations of a damped harmonic oscillator with the eigenfrequency ω_0. m and q are the mass and charge of the charge carriers, and γ is a damping constant necessary to consider the damping of the charge carrier (here electrons) movement. We assume that the electric field is polarized along the x-axis; hence, we consider only movements of the electrons along the x-axis. Assuming well-established oscillations according to $x(t) = x_o e^{-i\omega t}$, we obtain:

$$\frac{qE}{m} = \omega_0^2 x - \omega^2 x - 2i\gamma\omega x$$

This leads to the following expression for the elongation of the charge carriers:

$$x = \frac{qE}{m} \frac{1}{\omega_0^2 - \omega^2 - 2i\omega\gamma} \tag{2.14}$$

This relation describes a resonant behavior of the accumulated (absorbed) energy, when the angular frequency ω of the field approaches ω_0. In this resonance condition, the interaction between radiation and matter is expected to be most effective.

The energy W gained by the charge carrier must have been taken from the light wave, so we really describe a light absorption process. In a symbolic writing, we have:

$$\text{absorption} \quad \propto |x|^2 \propto \frac{1}{\left(\omega_0^2 - \omega^2\right)^2 + 4\omega^2\gamma^2}$$

If damping is weak, we have $\gamma^2 \ll \omega_0^2$. In the immediate vicinity of the resonance frequency, we assume $\omega \approx \omega_0$, and then we obtain:

$$\text{absorption} \quad \propto \frac{1}{(\omega_0 - \omega)^2 + \gamma^2} \tag{2.14a}$$

Fig. 2.4 Lorentzian line

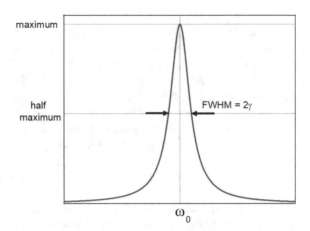

We find that absorption is centered in a symmetric manner in a certain frequency interval around the central absorption frequency. In this connection, we speak on an absorption line. The absorption line defined by (2.14a) has a symmetric shape and is called a *Lorentzian line* (Fig. 2.4). It describes the shape of an absorption line in terms of the so-called classical single oscillator model.

Apart from resonance, the absorption decreases and achieves 50% of the maximum value at the frequencies:

$$\omega - \omega_0 = \pm \gamma \tag{2.15}$$

The value 2γ therefore represents the so-called full width at half Maximum (FWHM) and is an important characteristic of any spectral line. In the present simplest version of our classical theory, the width of a spectral line is exclusively determined by the damping constant γ.

Note that such an absorption line is completely defined by three parameters: the central frequency, the linewidth, and the line intensity. We will provide a more detailed discussion of the classical oscillator model later in Chap. 14.

2.3.3 The Classical Picture of Light Emission

In order to deal with light emission, we will have to make use of a result from classical electrodynamics. From there, it is known that any oscillating electric dipole acts as an emitter of electromagnetic radiation. The full loss of energy per time caused by light emission is given by:

$$\frac{dW}{dt} = -\frac{q^2}{6\pi\varepsilon_0 c^3} \langle \ddot{x}^2 \rangle\big|_t \tag{2.16}$$

Here, the average is taken over a relevant time period, say a few oscillation periods, the latter being performed along the x-axis again. For the motion of a classical harmonic oscillator, we have:

$$x = x_0 \cos \omega_0 t \quad \text{with} \quad W = \frac{m\omega_0^2 x_0^2}{2}$$

Let us assume weak damping; i.e., the amplitude of the oscillation x_o is almost constant during one period. Substituting into (2.16) yields:

$$\frac{dW}{dt} = -\frac{q^2\omega_0^4}{6\pi\varepsilon_0 c^3}\frac{x_0^2}{2} = -\frac{q^2\omega_0^2}{6\pi\varepsilon_0 c^3 m}W \propto W \propto x_0^2 \tag{2.17}$$

In order to get access to the spectral shape of the emission line, let us have a look on the frequency spectrum of the emitted light. An oscillating dipole performing damped but free oscillations is described by:

$$m\ddot{x} + 2\gamma m\dot{x} + m\omega_0^2 x = 0$$

This equation may be solved by means of the approach:

$$x = x_{00}e^{\xi t}$$

That leads to:

$$\xi = -\gamma \pm \sqrt{\gamma^2 - \omega_0^2} = -\gamma \pm i\sqrt{\omega_0^2 - \gamma^2}$$

We further assume weak damping again:

$$\omega_0^2 >> \gamma^2$$

and obtain after having chosen the minus sign

$$x(t) \approx x_{00}e^{-\gamma t}e^{-i\omega_0 t} = x_0(t)e^{-i\omega_0 t} \tag{2.18}$$

This is no more a pure harmonic oscillation, and therefore, the emitted spectrum is not monochromatic. When performing a Fourier transformation, we get information on the spectral composition $F(\omega)$ of $x(t)$. We find:

$$F(\omega) \propto \int_{-\infty}^{\infty} xe^{i\omega t}dt$$

Let the emission process start at $t = 0$. Then,

$$F(\omega) \propto \int_{0}^{\infty} xe^{i\omega t}dt = x_{00}\int_{0}^{\infty} e^{i\omega t}e^{-\gamma t}e^{-i\omega_0 t}dt \propto \frac{1}{\omega_0 - \omega - i\gamma}$$

The intensity spectrum $I(\omega)$ may be approximated by [1]:

$$I(\omega) \propto F(\omega)F^*(\omega) \propto \left(\frac{1}{\omega_0 - \omega - i\gamma}\right)\left(\frac{1}{\omega_0 - \omega + i\gamma}\right) = \frac{1}{(\omega_0 - \omega)^2 + \gamma^2}$$

The conclusion is that the emission is centered around the central frequency, such that we have to observe an emission line. It also has to be a Lorentzian lineshape.

Remark From $x \approx x_{00}e^{-\gamma t}e^{-i\omega_0 t}$, we find the expected damping of the oscillation amplitude, with a decay time of $\tau_{amplitude} = \gamma^{-1}$. As the energy is proportional to the square of the amplitude, it will dissipate with half the decay time, so that we get:

$$\tau_W = (2\gamma)^{-1} \tag{2.19}$$

Hence, the decay time for the energy τ_W equals the reciprocal value of the FWHM of the corresponding Lorentzian, when the latter is given in angular frequency units. The longer the energy remains in the system, the narrower is the corresponding absorption line. The linewidth defined through τ_W is called the natural linewidth of the oscillator. If one is able to measure the natural linewidth experimentally, the decay time may be calculated.

We further note in this context that the Lorentzian lineshape appears as a necessary result of exponential damping of the dipole oscillations in the time domain.

Let us finally estimate the emission time, when damping is caused by light irradiation only.

From (2.17), we have:

$$\frac{dW}{dt} = -\frac{q^2\omega^2}{6\pi\varepsilon_0 c^3 m}W \propto W$$

we find that

$$W \propto e^{-\frac{t}{\tau_W}};$$
$$\tau_W = \frac{6\pi\varepsilon_0 c^3 m}{q^2\omega^2} \tag{2.20}$$

For an electron oscillating with an angular frequency of 10^{16} s^{-1}, the emission time τ_W will be around $2 * 10^{-9}$ s. That corresponds to an FWHM of $5 * 10^8$ s^{-1}.

2.4   Consistency Considerations

Although the theoretical considerations presented in this chapter are based on classical physics only, we have already arrived at an extremely important result: Single resonances in a piece of matter may give rise to spectroscopic features such as absorption lines and emission lines. We have found that absorption and emission lineshapes are identical. This is by the way necessary in order to keep consistency with Kirchhoff's radiation laws (see later Sect. 7.5.2). Absorption and emission resonance frequencies are specific to different materials and are widely used in applied spectroscopy for identification purposes.

However, it appears that the physical picture developed so far is not consistent with our experience on the stability of atoms and molecules. Generally, the emission of electromagnetic radiation by accelerated charge carriers is nothing specific to optics; instead, it is of greatest importance, for example, in the radio frequency (rf) technology. But in application to microscopic objects like atoms, this leads to rather peculiar consequences. Thus, if in a classical picture, an atom is thought to consist of a positively charged nucleus, surrounded by electrons which move around the nucleus on circular orbits, their movement should permanently give rise to electromagnetic radiation. This must result in a loss of kinetic energy of the electrons, such that after a certain time, the atom collapses because the electrons fall into the nucleus. Let us estimate the "lifetime" of a classical atom.

For that estimation, let us choose a hydrogen atom, consisting of a proton surrounded by an electron, the latter moving with a velocity v on a (nearly) circular orbit of radius r. This is a two-body system with a reduced mass close to that of a single electron. On a circular orbit, the Coulomb attraction force should be equal to the centrifugal force, such that

$$\frac{1}{4\pi\varepsilon_0}\frac{e^2}{r^2} \approx \mu\frac{v^2}{r} = \mu\omega_0^2 r \Rightarrow \omega_0^2 \approx \frac{1}{4\pi m_e\varepsilon_0}\frac{e^2}{r^3}$$

When choosing (Table 1.1) $r \approx 0.08$ nm, we can estimate $\omega_0 \approx 2.2 * 10^{16}\,\mathrm{s^{-1}}$. This angular frequency falls deep into the UV spectral region, which is nevertheless a qualitatively acceptable result, because many of the hydrogen atom emission lines are really observed in the UV. However—according to (2.20)—this angular frequency corresponds to a classical "lifetime" that is smaller than 10^{-9} s. Thus, a classical hydrogen atom should collapse within 10^{-9} s (Fig. 2.5)—a really peculiar result that is in complete contradiction to reality. In general, for higher order atoms, the assumed orbit radius of the outermost electrons (the *valence* electrons) should be larger, such that the angular frequency becomes smaller and rather falls into the near-UV or into the -VIS spectral regions. Accordingly, classical lifetimes of the order of 10^{-9}–10^{-8} s may be predicted, but this does not eliminate the obvious paradoxon. We come to the conclusion that something becomes wrong with classical physics of matter, at least when applying it to microscopic objects like atoms. But that means that whenever we want to understand the origin of optical material properties on a microscopic level, we have to make use of another theory

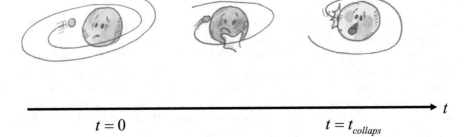

$t = 0$ $t = t_{collaps}$

Fig. 2.5 Collaps of a classical atom. Cartoon by Dr. Alexander Stendal. Printed with permission

that is consistent with the practical observations. The corresponding theoretical apparatus has been developed approximately 100 years ago and forms what we call today quantum mechanics.

In finishing this section, let us note that it was not only the description of the dynamics of microscopic matter that revealed inconsistencies between the classical predictions and experimental results. At the beginning of the twentieth century, it became more and more clear that the classical wave picture of light was also insufficient to explain the growing amount of new experimental results. As a prominent example, let us regard the photoelectric effect: Under the action of incident light, electrons (so-called photoelectrons) with a certain kinetic energy T_{kin} may be emitted from metallic surfaces (Fig. 2.6).

In terms of the classical wave picture, one would expect that a growing light intensity should result in an increase of the kinetic energy of the emitted electrons. But this was not observed. Instead, the maximum kinetic energy of the individual electrons turned out to be proportional to the light frequency. Moreover, a threshold frequency was observed: Light with a smaller frequency did not produce photoelectrons at all. Albert Einstein was able to explain the effect in terms of a quantum hypothesis: He assumed that light may transfer energy to matter only in discrete portions, which are proportional to the frequency. That "photon energy" E_{phot} should be large enough to release the electron from the metal plate, and the rest of the photon energy may be transferred to kinetic energy. Hence, the energy balance requires that:

$$E_{phot} \geq W_A + T_{kin}$$

where W_A is the material-specific work function, i.e., the minimum energy necessary to release an electron. Hence, it follows that

$$T_{kin} \leq E_{phot} - W_A$$

If now the photon energy is assumed to be proportional to the light frequency f:

$$E_{phot} = hf$$

Fig. 2.6 Photoelectric effect. Cartoon by Dr. Alexander Stendal. Printed with permission

with h—Planck's constant, we find:

$$T_{\text{kin}} \leq hf - W_A$$

The threshold frequency $f_{\min}$ is given from the condition:
$0 \leq T_{\text{kin}} \Rightarrow hf \geq W_A \Rightarrow f_{\min} = \frac{W_A}{h}$, and consequently:

$$T_{\text{kin}} \leq h(f - f_{\min})$$

which was later shown to reproduce the experimental findings and allowed estimating Planck's constant. The important conclusion for us is that in contrast to classical predictions, certain experimental findings indicate that light energy is only transferred in discrete portions, given by $E_{\text{phot}} = hf$.

2.5 Application to Practical Problems

2.5.1 Simple Classical Estimation of the Frequency of Nuclei Vibration in a Diatomic Molecule

In Sects. 2.3.2 and 2.3.3, we have used some "extreme" version of the two-body system, where one of the participating masses was considered to be practically

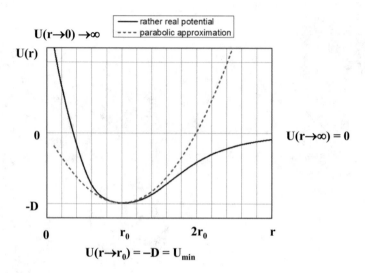

Fig. 2.7 Assumed potential of the interatomic interaction

infinitely large. Then, the reduced mass of the two-body system is simply identical to the mass of the other particle.

We will now identify our two masses m_1 and m_2 with the nuclei in a diatomic molecule; i.e., assume that they may be of the same order of magnitude. In order to consider a suitable restoring force, we will start from a potential $U(r)$ as shown in Fig. 2.7 in black.

At large distances, the two atoms do not interact, and therefore, U should approach zero. At $r = r_0$, we postulate the existence of a local minimum in U, which marks the equilibrium interatomic distance in the molecule. Obviously, a dissociation energy D is necessary to excite the molecule until dissociation. From the well-known relation:

$$\mathbf{F} = -\nabla U \tag{2.21}$$

We further find:

If $r > r_0$, $F = -\frac{dU}{dr} < 0$ and

$$r < r_0, \quad F = -\frac{dU}{dr} > 0$$

Indeed, when being elongated, the restoring force resulting from the potential curve shown in Fig. 2.7 will push the system back to the equilibrium distance r_0.

For small elongations $|x| = |r - r_0| \ll r_0$, the real potential may be approximated by a parabola (red curve) $U(r) \rightarrow U(x) = \frac{\kappa}{2}x^2 - D$.

From here, the restoring force could be estimated if κ would be known. As a rough estimation (Fig. 2.7), we assume a shape of the parabola such that $U(x =$

Fig. 2.8 Classical
illustration of diatomic
molecule vibrations

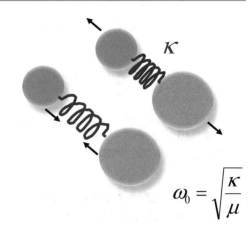

$$\omega_0 = \sqrt{\frac{\kappa}{\mu}}$$

$r_0) = 0$. Assuming further the reasonable values $D = 5$ eV and $r_0 = 0.1$ nm (Table 1.1), we find:

$$0 = \frac{\kappa}{2}r_0^2 - D \Rightarrow \kappa = \frac{2D}{r_0^2} = \frac{10\,\mathrm{eV}}{(0.1\,\mathrm{nm})^2} = 10^{21}\,\mathrm{eVm^{-2}} \approx 160\,\frac{\mathrm{N}}{\mathrm{m}}$$

The restoring force is

$$F = -\frac{\mathrm{d}U(x)}{\mathrm{d}x} = -\kappa x$$

For small elongations, the system is now equivalent to two masses connected by a spring, while κ is the spring constant (Fig. 2.8).

The resonance frequency is well known from mechanics:

$$\omega_0 = \sqrt{\frac{\kappa}{\mu}}$$

For estimation, we set μ equal to 10 proton masses. We then find:

$$\omega_0 = \sqrt{\frac{10^{21}}{10 * 1.672 * 10^{-27}}\,\frac{\mathrm{eV}}{\mathrm{kgm^2}}} = \sqrt{\frac{1}{1.672}10^{47}\,\frac{\mathrm{eV}}{\mathrm{kgm^2}}}$$

Using:

$$1\,\mathrm{eV} = \frac{1.602}{10^{19}}\,\mathrm{Ws} = 1.602 * 10^{-19}\,\frac{\mathrm{kgm^2}}{\mathrm{s^2}}$$

We find

$$\omega_0 = \sqrt{\frac{1.602}{1.672}10^{47-19}\,\frac{\mathrm{kgm^2}}{\mathrm{kgm^2 s^2}}} \approx 10^{14}\,\mathrm{s^{-1}}$$

This vibration frequency becomes resonant with infrared electromagnetic radiation of a wavelength of approximately

$$\lambda_0 = \frac{2\pi c}{\omega_0} \approx \frac{1.9 * 10^9}{10^{14}} \, \text{m} \approx 2 * 10^{-5} \, \text{m} = 20 \, \mu m$$

Vibrations of nuclei in molecules are really observed in the so-called middle infrared (MIR) spectral region and are widely used for analytical purposes in the infrared spectroscopy (fingerprint spectra).

2.5.2 Classical Estimation of the Vibration Frequency of a Valence Electron in an Atom or Molecule

Let us now have a look at "electronic" excitations. In our model, one of the bodies is now identical to an electron, and the other to a nucleus. In classics, the only difference to Sect. 2.5.1 is in the value of the reduced mass, which is now about 10^4 times smaller, while the "spring constants" are of the same order. Thus, the angular frequency will be about 100 times higher (around $10^{16} \, \text{s}^{-1}$) and the wavelength about 100 times smaller (around 200 nm). Practically, we arrive in the same spectral region as we have already estimated in Sect. 2.4. The wavelength of 200 nm corresponds to UV radiation (Table 2.1), in practice those transitions are really observed in the UV or in the VIS.

2.5.3 Classical Estimation of the Characteristic Angular Frequency of the Rotation of a Diatomic Molecule

In Sect. 2.3.1, we have found the relations (2.9) and (2.10):

$$T_{\text{kin}} = \frac{\mu}{2}\dot{\mathbf{r}}^2 \ \text{ and } \ \dot{\mathbf{r}}^2 = \dot{r}^2 + r^2\dot{\phi}^2$$

Let us now regard the case of pure rotation, assuming:

$$\dot{r} = 0 \ \text{ and } \ T_{\text{kin}} = \frac{\mu}{2}r^2\dot{\phi}^2 = \frac{I}{2}\Omega^2$$

Let us again identify our two masses with the nuclei in a diatomic molecule. Once their distance is fixed, we obtain a pure rotation of a rigid molecule as shown in Fig. 2.9:

Let us assume a moment of inertia $I = 2.6 * 10^{-47} \, \text{kgm}^2$. Can we estimate the rotation angular frequency Ω of such a molecule in a gas that is held at room temperature ($T = 300$ K) in a classical manner?

Yes, we can. Let us assume that in equilibrium conditions, T_{kin} is of the order of $k_B T$. For our estimation, we set:

Fig. 2.9 Illustration of diatomic molecule rotation

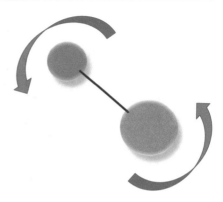

Table 2.2 Overview on classical estimations

Phenomenon	Responds in spectral range
Free molecular rotation	FIR
Vibration of nuclei in a molecule	MIR
Excitations of valence electrons	VIS, UV

$T_{kin} = \frac{1}{2}\Omega^2 \approx \frac{k_B T}{2}$ (equipartition theorem) with $k_B = 1.38*10^{-23} WsK^{-1}$—Boltzmann's constant. Then,

$\Omega \approx \sqrt{\frac{k_B T}{I}} \approx 1.3 * 10^{13} s^{-1}$. An electromagnetic wave with such a frequency would have a wavelength

$$\lambda_{rot} = \frac{2\pi c}{\Omega} \approx \frac{1.9 * 10^9}{1.3 * 10^{13}} \, m \approx 1.5 * 10^{-4} \, m = 150 \, \mu m$$

That frequency corresponds to the far infrared (FIR).

In conclusion, our estimations have suggested that the discussed degrees of freedom in atoms or molecules respond in the following spectral regions of electromagnetic radiation (see Table 2.2).

Core electrons are "closer" to the nucleus than valence electrons and therefore feel stronger restoring forces. In a classical picture, this may be tackled assuming substantially larger spring constants κ. According to $\omega_0 = \sqrt{\frac{\kappa}{\mu}}$, this will correspond to higher resonance frequencies, deeper in the UV or in the X-ray spectral region.

2.6 Advanced Material: Doppler Broadening in Gases

As in equilibrium conditions, the distribution of gas particles (let them be molecules) with respect to their velocities is isotropic, some of them fly in direction to the light source, and some of them away from the source. Due to the Doppler effect, a resting observer will detect that the molecules propagating in direction to

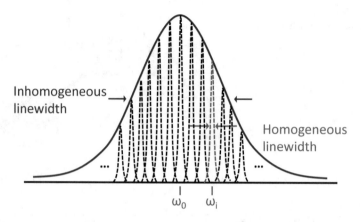

Fig. 2.10 Mechanism of inhomogeneous line broadening

the source may absorb at a slightly lower light frequency than those propagating away from the source. Hence, the molecules differ with respect to the physical condition essential for the process of light absorption. This is a typical situation for an inhomogeneous line broadening mechanism.

Inhomogeneous line broadening occurs when many individual narrow spectral lines with different individual resonance angular frequencies ω_i superimpose to a broader absorption line. This broader structure may also be characterized by an central absorption angular frequency ω_0 and a total FWHM, the latter called inhomogeneous linewidth (Fig. 2.10). On the contrary, the individual linewidth of the constituents of that superimposed absorption feature is called the homogeneous linewidth. Natural line broadening as well as collision broadening may influence the homogeneous linewidth. Thus, broadening mechanisms which are common to all of the oscillators in the ensemble contribute to the homogenous linewidth, while a spread in physical conditions essential for light–matter interaction of the individual oscillators results in contributions to the inhomogeneous linewidth.

Doppler broadening is a particular case of inhomogeneous line broadening; in the case of Doppler broadening, the full absorption line will be composed from a large number of narrow lines shifted with respect to each other due to the Doppler effect. This gives rise to a Gaussian resulting lineshape.

Indeed, this particular case may be mathematically treated in an exact manner. Let us assume that the light wave propagates along the z-axis. Due to Maxwell's distribution, the number of molecules with a given z-component of their velocity is:

$$N(v_z)\mathrm{d}v_z \propto e^{-\frac{mv_z^2}{2k_BT}}\,\mathrm{d}v_z$$

m is the mass of a molecule, k_B Boltzmann's constant, and T the absolute temperature. Let ω_0 be the resonance frequency of the molecule in rest. Due to the movement along z, the molecule absorbs no longer at ω_0, but at a shifted frequency ω_D:

$$\omega_D = \omega_0\left(1 + \frac{v_Z}{c}\right)$$

The number of molecules absorbing at ω_D is then:

$$N(v_Z)\mathrm{d}(v_Z) = N(v_Z)\frac{\mathrm{d}v_Z}{\mathrm{d}\omega_D}d\omega_D \equiv N(\omega_D)\mathrm{d}\omega_D$$

so that we finally get:

$$N(\omega_D) = \left[N(v_Z)\frac{\mathrm{d}v_Z}{\mathrm{d}\omega_D}\right]_{v_Z = f(\omega_D)}$$

This reveals the probability density distribution for the Doppler-shifted absorption frequency ω_D in an assembly of gas molecules or atoms. In the case that this distribution is considerably broader than the homogeneous linewidth, it will dominate the absorption lineshape of the assembly, and we find a Gaussian spectral shape with an FWHM given as:

$$\Delta\omega_D = \frac{2\omega_0}{c}\sqrt{\frac{2\ln 2k_B T}{m}} \tag{2.22}$$

which is, of course, dependent on the temperature.

For transferring linewidth data into wavelength units, it is convenient making use of implicit differentiating:

$$\omega = 2\pi f = \frac{2\pi c}{\lambda} \Rightarrow \ln\omega = \ln 2\pi c - \ln\lambda \Rightarrow \frac{\mathrm{d}\omega}{\omega} = -\frac{\mathrm{d}\lambda}{\lambda} \Rightarrow \left|\frac{\Delta\omega}{\omega}\right| \approx \left|\frac{\Delta\lambda}{\lambda}\right|$$

2.7   Tasks for Self-check

2.7.1 Multiple choice test: Mark all answers which seem to you correct!

The light wavelength 1064 nm belongs to the	Ultraviolet spectral range
	Visible spectral range
	Infrared spectral range
The wavelength of 500 nm belongs to the	Microwave spectral range
	Middle infrared spectral range
	X-ray spectral range

Vibrations of nuclei in molecules and solids cause light absorption primarily in the	Ultraviolet spectral range	
	Visible spectral range	
	Infrared spectral range	
The emission line of an assembly of resting light emitters appears to be naturally broadened. Its spectral shape may therefore be described by a	Lorentzian lineshape	
	Gaussian lineshape	
	Triangular lineshape	
A Doppler-broadened emission line as emitted from a hot gas has a	Gaussian lineshape	
	Lorentzian Lineshape	
	Rectangular lineshape	
The mass moment of inertia of a diatomic molecule is typically of the order	10^{-27} kgm^2	
	10^{-47} kgm^2	
	10^{-67} kgm^2	

2.7.2 True or wrong? Make your decision!

Assertion	True	Wrong
The reduced mass of a system of two bodies is always in between the masses of the individual two bodies		
Molecular fingerprint spectra are typically recorded in the middle infrared spectral range		

2.7.3 The interatomic potential between two neutral atoms may be approximated by:
$$U(r) = 4D\left[\left(\tfrac{\sigma}{r}\right)^{12} - \left(\tfrac{\sigma}{r}\right)^{6}\right]$$ (D, σ are constants). Sketch that potential as a function of r in a graph. Find an expression for the distance r_0 where $U(r)$ has a local minimum.

2.7.4 In a HCl molecule, assume an interatomic distance $r_0 = 0.127$ nm. Calculate full mass and reduced mass of the molecule and compare these data with the assumed masses of the H and Cl single atoms. Calculate the mass moment of inertia.

2.7.5 A photon releases a photoelectron with $T_{kin} = 2$ eV from a metal which has a work function $W_A = 2$ eV. Indicate the smallest possible value for the energy of this photon (after [2]).

2.7.6 Basing on (2.6), derive a formula that relates light intensities given in W/cm^2 to the real electric field amplitude in V/m!

References

Specific References

1. А.Н. Матвеев, Оптика, Москва „Высшая школа" (1985), pp. 56–68
2. H. Haken, H.C. Wolf, *The Physics of Atoms and Quanta* (Springer 2005), task 5.13

General Literature

3. P.A. Tipler, G. Mosca, *Physik, 7* (Springer, Auflage, 2015)
4. W. Demtröder, *Atoms, Molecules, and Photons* (Springer, 2010)
5. L.D. Landau, E.M. Lifshitz, *Mechanics* (Elsevier 1976)

Part II
A Pedestrian's Guide to Quantum Mechanics

"Jena movit". Painting and Photograph by Astrid Leiterer, Jena, Germany (www.astrid-art.de). Photograph reproduced with permission

Wave and particle properties seem to be complementary. But with some fantasy, they may be united, particularly by a pedestrian walking through Jena old town after having had a drink (preferably a Long Island Ice Tea*) in the "Stilbruch" restaurant.*

Waves as Particles and Particles as Waves

3

Abstract

Particle and wave properties of traditional particles and electromagnetic waves are discussed in the context of simple light absorption, emission, and scattering phenomena. Attention is paid to the propagation of wave packets, including the introduction of the group velocity as well as pulse broadening phenomena.

3.1 Starting Point

Today, it is commonly accepted that light may exhibit both wave-like and corpuscular properties, depending on the concrete experimental conditions. The corresponding Janus-like construction that combines wave and particle features is called a photon.

Let us start with a short survey of typical wave and particle parameters. What we will need in the following is summarized in Table 3.1.

In fact, the debate on the nature of light is very old: Christiaan Huygens may be regarded as the father of the wave picture of light, while Isaac Newton developed an early corpuscular theory. Each of them was able to explain certain facets of the experimentally observed phenomena. Thus, Huygen's construction led to a principal understanding of refraction and birefringence phenomena, while his idea of longitudinal light waves did not allow for introducing the important concept of light polarization. Newton developed a color theory and introduced the concept of light polarization, assuming the existence of light particles with somewhat anisotropic internal properties.

Somewhat later, the exploration of light interference phenomena and the development of the Maxwell's electromagnetic theory resulted in a seemingly final victory of the wave concept over the corpuscular approach. But again, new theoretical and experimental results required a rethinking on the nature of light. We will mention three corresponding milestones:

Table 3.1 Wave versus particle characteristics

Waves: $\Psi(\mathbf{r}, t) \propto e^{-i(\omega t - \mathbf{k r})}$	Freely propagating point particles: rest mass m, coordinates $\mathbf{r}$	
Phase velocity $\mathbf{v}_{\mathrm{ph}}$	Particle velocity $\mathbf{v}$	$\mathbf{v} \equiv \frac{\mathrm{d}\mathbf{r}}{\mathrm{d}t} = \dot{\mathbf{r}}$
Group velocity $\mathbf{v}_{\mathrm{gr}}$		
Wavevector $\mathbf{k}$	Momentum $\mathbf{p}$	Relativistic: $\mathbf{p} = \dfrac{m\mathbf{v}}{\sqrt{1 - \frac{v^2}{c^2}}}$
		Nonrelativistic: $\mathbf{p} = m\,\mathbf{v}$
Angular frequency ω	Energy E	Relativistic: $E = \sqrt{p^2 c^2 + m^2 c^4}$
		rest energy: $E = mc^2$
		Non-relativistic: $E \approx mc^2 + T_{\mathrm{kin}}$; $T_{\mathrm{kin}} = \frac{p^2}{2m} = \frac{mv^2}{2}$
$v_{\mathrm{ph}} \equiv \frac{\omega}{k}$	$v = \frac{\mathrm{d}E}{\mathrm{d}p}$	
$v_{\mathrm{gr}} \equiv \frac{\mathrm{d}\omega}{\mathrm{d}k}$		

- The spectral behavior of the so-called blackbody irradiation could be theoretically explained when accepting the ad hoc hypothesis that the light energy is transferred in small portions proportional to the frequency of light [Planck's formula—see later Chap. 7)].
- The quantitative exploration of the photoelectric effect also suggested that the light energy is transferred in corresponding energy portions (Albert Einstein's interpretation).
- The experimental results on scattering of X-ray radiation at electrons (Compton effect) are consistent with the idea that light may behave like a particle with well-defined energy and momentum.

3.2   Physical Idea

3.2.1   Energy and Momentum of „Light Particles"

The idea of the following treatment is to establish a quantitative correspondence between the "classical" wave and particle parameters summarized in Table 3.1. That will allow us to attribute a combination of both wave and particle properties to traditional waves as well as to traditional particles. From the experimental findings mentioned in 3.1, we will start with the assumption that the smallest portions of energy transferred by light of frequency f are given by:

$$E_{\text{phot}} = hf = \hbar\omega \tag{3.1}$$

Let us call this the energy of a photon. On the other hand, when regarding light as an assembly of particles propagating with c, they surely have to be regarded as relativistic particles with a rest mass $m = 0$. For their total energy, we therefore make use of the relativistic expression, following from Einsteins theory of special relativity:

$$E = \sqrt{p^2 c^2 + m^2 c^4} \tag{3.2}$$

Hence for a photon with vanishing rest mass, we have:

$$E_{\text{phot}} = \sqrt{p^2 c^2 + m^2 c^4}\Big|_{m=0} = p_{\text{phot}} c \tag{3.3}$$

With p_{phot}—momentum of a photon. This results in (assuming propagation in vacuum).

$$p_{\text{phot}} = \frac{E_{\text{phot}}}{c} = \hbar\frac{\omega}{c} = \hbar\frac{2\pi}{\lambda} = \frac{h}{\lambda} \quad \text{or}$$

$$\mathbf{p}_{\text{phot}} = \hbar\mathbf{k} \tag{3.4}$$

Equations (3.1) and (3.4) represent the relations between photon energy and momentum as particle characteristics with (angular) frequency, wavelength, and wavevector as typical wave characteristics.

3.2.2 Phase and Group Velocities

When returning to Table 3.1, we recognize that we have not yet spoken about the concepts of a wave's phase and group velocities. Instead of a general treatment, we will illustrate these concepts in terms of a simple example that is more easy to handle (Example after [1]).

The distinction between group and phase velocities becomes interesting when light waves with a modulated amplitude are taken into consideration. Note that this situation is already beyond the phenomena described by our simplest ansatz (2.1). Instead, we will now focus on a special case of a propagating wave packet.

Consider a wave packet propagating in space along the z-axis without damping. Let us assume for simplicity that it has a rectangularly shaped Fourier spectrum centered at $k = k_0$ with the width Δk (k is the absolute value of the wavevector). Hence, the wave packet is obtained as a superposition of plane monochromatic waves, each with the real amplitude $A/\Delta k$. In our model calculation, the wave packet may thus be described by the wavefunction:

$$\psi(z,t) = \frac{A}{\Delta k} \int_{k_0 - \frac{\Delta k}{2}}^{k_0 + \frac{\Delta k}{2}} \mathrm{e}^{-i(\omega t - kz)} \mathrm{d}k$$

Let us further assume an arbitrary dispersion law $\omega = \omega(k)$ (we will identify the term "dispersion law" with any relationship that establishes a relation between k and ω). When expanding it into a Taylor's series, we have:

$$\omega = \omega(k_0) + (k - k_0) \left.\frac{\mathrm{d}\omega}{\mathrm{d}k}\right|_{k=k_0} + \frac{(k - k_0)^2}{2} \left.\frac{\mathrm{d}^2\omega}{\mathrm{d}k^2}\right|_{k=k_0} + \cdots \tag{3.5}$$

In first-order dispersion theory, we restrict on the first and second terms in the above written equation. That leads to the simplified dispersion law:

$$\omega = \omega(k_0) + (k - k_0) \left.\frac{\mathrm{d}\omega}{\mathrm{d}k}\right|_{k=k_0} \tag{3.6}$$

We then obtain:

$$\psi(z,t) = \mathrm{e}^{-i(\omega_0 t - k_0 z)} \frac{A}{\Delta k} \int_{k_0 - \frac{\Delta k}{2}}^{k_0 + \frac{\Delta k}{2}} \mathrm{e}^{-i(k-k_0)\left(t\left.\frac{\mathrm{d}\omega}{\mathrm{d}k}\right|_{k=k_0} - z\right)} \mathrm{d}k = A\mathrm{e}^{-i(\omega_0 t - k_0 z)} \frac{\sin \xi}{\xi} \tag{3.7}$$

Here we have introduced the new variable:

$$\xi = \frac{\Delta k}{2}\left[z - t\left.\frac{\mathrm{d}\omega}{\mathrm{d}k}\right|_{k=k_0}\right] \tag{3.8}$$

The obtained expression (3.7) for $\psi(z,t)$ obviously describes a propagating wave packet, analytically composed from two factors: a term $\mathrm{e}^{-i(\omega_0 t - k_0 z)}$ that corresponds to a plane wave propagating along z with the phase $\omega_0 t - k_0 z$, and a term $A\frac{\sin \xi}{\xi}$, which describes a modulation of the wave amplitude. From here we conclude, that a point of constant phase may propagate with a velocity different from that of a point of constant amplitude. Indeed, for a point of constant phase traveling along z, we find the phase velocity according to:

$$\omega_0 t - k_0 z = \mathrm{const} \Rightarrow \frac{\mathrm{d}z}{\mathrm{d}t} = v_{\mathrm{ph}} = \frac{\omega_0}{k_0} \tag{3.9}$$

Remark: In practice, points of constant phase cannot be used for signal transfer. Therefore, it may happen that the phase velocity exceeds c.

On the other hand, a point of constant amplitude corresponds to

$$\xi = \text{const.}$$

Correspondingly, a point of constant amplitude propagates along the z-axis with the group velocity defined by:

$$\xi = \text{const.} = \frac{\Delta k}{2} \left[z - t \left. \frac{d\omega}{dk} \right|_{k=k_0} \right] \Rightarrow \frac{dz}{dt} = v_{gr} = \left. \frac{d\omega}{dk} \right|_{k=k_0} \tag{3.10}$$

Note that this expression has been obtained neglecting any damping as well as any higher order terms in the expansion of the presumed dispersion law (3.6). In other words: if there is no damping mechanism and no higher order dispersion term relevant, than any of the points with given amplitude propagates with the group velocity given by (3.10). That means that all points with given amplitude propagate with the *same* velocity. Consequently, in first-order dispersion theory, a given pulse propagates without any distortion of its shape. Once a point of constant amplitude can be used for signal transfer, it shall never move faster than c.

3.3 Theoretical Considerations

3.3.1 Waves as Particles

Provided that the angular frequency and the wavevector of a light wave are given, the photon energy and momentum are immediately given by (3.1) and (3.4). The phase velocity $v_{ph} \equiv \frac{\omega}{k}$ of a photon propagating in vacuum is then (compare Sect. 3.2.2):

$$v_{ph} = \frac{\omega}{k} = \frac{\hbar\omega}{\hbar k} = \frac{E_{phot}}{p_{phot}} = c$$

Moreover, we obtain the group velocity $v_{gr} \equiv \frac{d\omega}{dk}$:

$$v_{gr} = \frac{d\omega}{dk} = \frac{dE_{phot}}{dp_{phot}} = c$$

Remark: According to (3.1), the photon energy is unambiguously related to a certain light frequency. Then, according to Sect. 3.2.1, we have $\omega = \frac{2\pi}{T} = 2\pi f = \frac{2\pi c}{\lambda}$, so that a given photon energy is also associated with a well-defined vacuum wavelength or a wavenumber. Thus, $E_{phot} = hc\nu = \frac{hc}{\lambda}$, while $hc \approx 1240\,\text{eVnm}$.

Therefore, as a rule of thumb, the photon energy in eV may be converted into wavelength in nm or wavenumber in cm^{-1} by the following relations:

$$\lambda \approx \frac{1240}{E_{phot}} nmeV; \quad \nu \approx 8065 E_{phot}(eV)^{-1} cm^{-1}$$

3.3.2 Particles as Waves

One of the most striking features of quantum mechanics is now that wave properties may also be attributed to typical particles with a rest mass $m \neq 0$. Again, we assume (3.2):

$$E = \sqrt{p^2c^2 + m^2c^4}$$

For a *resting* particle, $p = 0$, and the above equation results in Einstein's famous formula:

$$E = mc^2$$

Thus, for a resting electron, we have $E \approx 0.511$ MeV, and for a resting proton, $E \approx 0.938$ GeV.

In the nonrelativistic case with $p \neq 0$ but $m^2c^2 >> p^2$ we have instead:

$$E = mc^2\sqrt{1 + \frac{p^2}{m^2c^2}} \approx mc^2\left(1 + \frac{p^2}{2m^2c^2}\right) = mc^2 + \frac{p^2}{2m} = mc^2 + T_{kin}$$

In the relativistic case however, we have

$$E^2 = p^2c^2 + m^2c^4 \Rightarrow 2EdE = 2pc^2dp \Rightarrow \frac{dE}{dp} = c^2\frac{p}{E}$$

Let us now attribute a so-called de Broglie wavelength to a propagating particle according to (compare (3.4) again):

$$\lambda = \frac{h}{p} \tag{3.11}$$

The corresponding matter wave (or de Broglie wave) is again characterized by a phase and a group velocity. But these velocities differ from those of a photon. Keeping in mind the relativistic mass increase, we have:

$$p^2 = \frac{m^2v^2}{1 - \frac{v^2}{c^2}} \Rightarrow E^2 = p^2c^2 + m^2c^4 = \frac{m^2c^4}{1 - \frac{v^2}{c^2}}$$

From here:

$$p^2 = E^2 \frac{v^2}{c^4} \text{ or } p = \frac{v}{c^2} E$$

That gives us the phase velocity:

$$v_{\text{ph}} = \left(\frac{\omega}{k} = \frac{\hbar\omega}{\hbar k} \right) = \frac{E}{p} = \frac{c^2}{v} > c$$

And the group velocity:

$$v_{\text{gr}} = \left(\frac{\mathrm{d}\omega}{\mathrm{d}k} = \frac{\mathrm{d}(\hbar\omega)}{\mathrm{d}(\hbar k)} \right) = \frac{\mathrm{d}E}{\mathrm{d}p} = c^2 \frac{p}{E} = c^2 \frac{v}{c^2} = v$$

We come to the conclusion that a de Broglie wave moves with a group velocity that exactly coincides with the velocity of the particle—a quite reasonable result.

3.4 Consistency Considerations

We have established relations between energy, momentum, frequency, wavelength, phase, and group velocities such that particle and wave properties may be ascribed to photons as well as traditional particles. In particular, the group velocity of propagating particles coincides with their "usual" velocity, which is a quite encouraging result. The phase velocity of propagating relativistic particles turns out to be larger than c, which does not contradict relativity, because the phase velocity is not relevant for signal transfer. All in all, so far no inconsistency has been observed.

But what about other wave phenomena? Can particles cause interference and diffraction pattern? Can they be used in microscopes to produce magnified images of an object? Yes, they can, as demonstrated by double slit experiments or electron microscopes. Wave properties of particles are in fact in the basis of all kind of modern technology today.

Once accelerated electrons may behave like matter waves, they should suffer diffraction at the atoms of a solid. Figure 3.1 shows the electron diffraction pattern obtained from an amorphous (see later Chap. 17) thin solid film. From the observed diffraction fringes, one may conclude on the interatomic spacing within the material. A Fourier transformation of the diffraction picture provides access to the so-called radial distribution function RDF.

Electron diffraction has thus become an important tool in the characterization of the atomic structure of solid matter.

From here, we make the following primary conclusions:

- Once we have experimental evidence that electrons suffer diffraction, it makes sense to describe their dynamics in a wave picture. We will develop this picture starting from Chap. 4.

Fig. 3.1 Electron diffraction fringes obtained from an amorphous thin film coating [2]. Courtesy of Johannes Ebert, Laseroptik GmbH

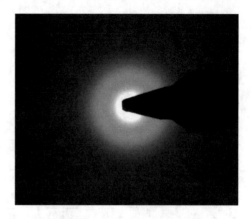

- The diffraction picture shown in Fig. 3.1 is at the same time a manifestation of a more fundamental principle that is known as an uncertainty relation. Indeed, in complete accordance to wave optics, the smaller the interatomic distance in the film is, the larger the radius of the diffraction fringes will be. Thus, a stronger confinement in the coordinate of the particle will result in a larger scatter of possible momenta along that direction. Therefore, the coordinates that may be measured in future become less predictable. That makes it conceptionally challenging to ascribe a *path* to a moving quantum particle.

3.5   Application to Practical Problems

3.5.1   The Compton Effect

The scattering of an X-ray photon at an electron can be regarded as an elastic collision event, where total energy and momentum of the photon–electron system must be preserved. Assuming an X-ray (or γ-photon), which is incident on an initially resting electron, we find from energy conservation:

$$p_1 c + m_e c^2 = p_2 c + \sqrt{p_e^2 c^2 + m_e^2 c^4}$$

On the other hand, from momentum conservation, we find:

$$\mathbf{p}_1 = \mathbf{p}_2 + \mathbf{p}_e$$

Here, $\mathbf{p}_e$ is the momentum of the electron after collision, and $\mathbf{p}_1$ and $\mathbf{p}_2$—the momenta of the photon before and after collision (compare Fig. 3.2).
Momentum conservation results in:

$$\mathbf{p}_1 - \mathbf{p}_2 = \mathbf{p}_e \Rightarrow p_e^2 = p_1^2 + p_2^2 - 2 p_1 p_2 \cos\theta$$

Fig. 3.2 The Compton effect as a collision event: on right: geometry as presumed in the theoretical model. Cartoon by Dr. Alexander Stendal. Printed with permission

The scattering angle θ is formed between $\mathbf{p}_1$ and $\mathbf{p}_2$. Then, from energy conservation, we find:

$$(p_1 - p_2)c + m_e c^2 = \sqrt{(p_1^2 + p_2^2 - 2p_1 p_2 \cos\theta)c^2 + m_e^2 c^4}$$

From here we obtain:

$$m_e c(p_1 - p_2) = p_1 p_2 (1 - \cos\theta) \text{ or}$$
$$\frac{1}{p_2} - \frac{1}{p_1} = \frac{1}{m_e c}(1 - \cos\theta).$$

Therefore,

$$\lambda_2 - \lambda_1 = \frac{h}{m_e c}(1 - \cos\theta) \equiv \Lambda(1 - \cos\theta)$$

Thus, the scattering of a photon at the initially resting electron results in a gain in kinetic energy of the electron. Hence, the photon must have lost energy, which may be recorded as a smaller frequency or a higher wavelength of the scattered X-rays. Our simple theory predicts that the wavelength shift depends of the scattering angle. This predicted behavior may be verified experimentally and was found to be in excellent agreement with experimental data. Hence, our assumptions on the energy and momentum of a photon are consistent with the experimental results from scattering experiments. The so-called Compton wavelength Λ for scattering at electrons is given by:

$$\frac{h}{m_e c} \equiv \Lambda \approx 2.4 \times 10^{-12}\,\text{m}$$

Note that the Compton wavelength is explicitly dependent on the mass. If, instead of an electron, X-ray scattering occurs at a heavier particle with mass m, the wavelength change becomes smaller by modulus in a reciprocal manner. In the limiting case of $m \to \infty$, we have $\lambda_2 \to \lambda_1$, and we arrive at the result known from daily experience that incident light may be diffusely scattered by macroscopic objects without wavelength change (elastic scattering of light at macroscopic objects).

3.5.2   Absorption of a Photon by an Initially Resting Atom

Let us apply our knowledge to the following situation: Let a resting atom with mass m absorb a photon with angular frequency ω. As a result of energy conservation, the atom is assumed to be transferred from a state of low internal energy (let us call this here the ground state) to a state of higher internal energy (let us call this the excited state). Moreover, as a result of momentum conservation, after absorption the atom will no more be at rest. The task is to estimate the ratio of the kinetic energy of the atom after absorption and the photon energy for $\omega = 10^{15}\,\text{s}^{-1}$ and $m = 10^{-26}$ kg [3].

In a nonrelativistic approach, it is sufficient to consider nonrelativistic momentum conservation. The momentum of the incident photon must be equal to the momentum of the atom after absorption, which results in:

$$\hbar \frac{\omega}{c} = mv \Rightarrow v = \hbar \frac{\omega}{mc}$$

A simple consistency check reveals that the resulting velocity after absorption is indeed much smaller than c: $\frac{v}{c} = \frac{\hbar\omega}{mc^2} \approx 10^{-10} \ll 1$ as required. Therefore

$$T_{\text{kin}} = \frac{m}{2} v^2 = \frac{\hbar^2 \omega^2}{2mc^2} \Rightarrow \frac{T_{\text{kin}}}{\hbar\omega} = \frac{\hbar\omega}{2mc^2} \approx 5 \times 10^{-11}$$

In the relativistic calculation, we have to distinguish between the rest masses of the atom in the ground state (before photon absorption) and in the excited state (after photon absorption). We set:

m: rest mass of the atom in the ground state

m': rest mass of the atom atom in the excited state

p: momentum of the atom after absorption of the photon

Relativistic energy conservation yields:

$$E^2 = \left(\hbar\omega + mc^2\right)^2 = p^2c^2 + m'^2c^4 \tag{3.12}$$

Momentum conservation yields:

$$p = \hbar k = \hbar\frac{\omega}{c} = \frac{m'v}{\sqrt{1 - \frac{v^2}{c^2}}} \quad \Rightarrow pc = \hbar\omega \tag{3.13}$$

Let us substitute this into (3.12). We obtain:

$$\left(\hbar\omega + mc^2\right)^2 = \hbar^2\omega^2 + m'^2c^4 \tag{3.14}$$

From (3.13), we obtain an expression for the velocity of the atom after absorption as: $v^2 = \frac{\hbar^2\omega^2}{m'^2c^2}\left(1 - \frac{v^2}{c^2}\right) \Rightarrow v^2 = \frac{c^2\hbar^2\omega^2}{\hbar^2\omega^2 + m'^2c^4}$.

When substituting the denominator in this expression by the left side of (3.14), we get:

$$v^2 = \frac{c^2\hbar^2\omega^2}{\left(\hbar\omega + mc^2\right)^2} \Rightarrow v = \frac{c\hbar\omega}{\hbar\omega + mc^2}$$

Once $\hbar\omega \ll mc^2$ is certainly fulfilled for any light quanta in the visible or ultraviolet spectral range, we practically arrive at the nonrelativistic result again.

Remark: Note that $\hbar\omega \ll mc^2$ is formally fulfilled for any electronic excitation in a real atom, supposed that the latter is excited but not ionized by resonant absorption of a photon. Thus, when assuming the Rydberg theory (see later Chap. 8), we have $\hbar\omega \leq Z^2 Ry$ for the excitation of a single electron, and $mc^2 \approx 2Zm_pc^2$. Here, Z is the atomic order, and Ry the Rydberg energy: $1Ry \approx 13.6$ eV. Then, for any realistic Z, $\hbar\omega \ll mc^2$ will be fulfilled (see Fig. 3.3). Even if we assume the hypothetical case that the photon absorption results in an instantaneous excitations of all electrons in the atom, we still have $\hbar\omega < Z^3 Ry \ll mc^2$ (Fig. 3.3).

Fig. 3.3 $Z^2 Ry$, $Z^3 Ry$ and $2Zm_pc^2$ versus atomic order Z

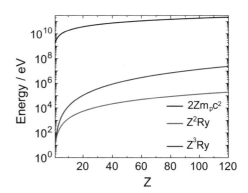

3.5.3 Emission of a Photon by an Initially Resting Atom

Let us now regard the opposite process: A resting atom with mass m emits a photon
with angular frequency ω. As a result of momentum conservation, after absorption,
the atom will no more be at rest. Again, let us judge velocity and kinetic energy
of the atom after photon emission.

When using the same symbols as before, we have for the nonrelativistic
situation:

$$\text{momentum conservation (projection)}: \Rightarrow \hbar\frac{\omega}{c} + mv = 0 \Rightarrow v = -\hbar\frac{\omega}{cm}$$

$$\Rightarrow T_{\text{kin}} = \frac{m}{2}v^2 = \frac{(\hbar\omega)^2}{2mc^2}.$$

i.e., the same as in the absorption process, except the "–" sign. The relativistic
case yields:

$$\text{momentum conservation (projection)}: \Rightarrow \hbar\frac{\omega}{c} + p = 0 \Rightarrow pc = -\hbar\omega = p = \frac{mv}{\sqrt{1 - \frac{v^2}{c^2}}}.$$

$$\text{energy conservation}: m'c^2 = \hbar\omega + \sqrt{p^2c^2 + m^2c^4}$$

$$\Rightarrow \left(m'c^2 - \hbar\omega\right)^2 = p^2c^2 + m^2c^4 = (\hbar\omega)^2 + m^2c^4$$

These equations transform to (3.13) and (3.14) when substituting:

$$\omega \leftrightarrow -\omega$$
$$m' \leftrightarrow m$$

Consequently, we have: $v^2 = \dfrac{c^2\hbar^2\omega^2}{\hbar^2\omega^2 + m^2c^4} = \dfrac{c^2\hbar^2\omega^2}{\left(m'c^2 - \hbar\omega\right)^2}.$

3.5.4 Radiation Pressure

Remark: In Sect. 3.5.2 (photon absorption), momentum conservation resulted in
$p_{\text{atom after absorption}} = \hbar k = \hbar\frac{\omega}{c}$, which means that the momentum of the photon is
completely transferred to the atom. This differs from what would be observed for back
reflection of light from an object. This is obvious from our treatment of the Compton
effect (Sect. 3.5.1). In that case momentum conservation yielded: $\mathbf{p}_1 = \mathbf{p}_2 + \mathbf{p}$. In
the case of back reflection of light, $\mathbf{p}_1$ and $\mathbf{p}_2$ have opposite directions. Moreover,
if the object is macroscopic (i.e., has a large mass), $\lambda_2 \rightarrow \lambda_1$, such that $\mathbf{p}_1$ and $\mathbf{p}_2$
must have the same absolute value. Therefore, $\mathbf{p}_1 = -\mathbf{p}_2 \Rightarrow \mathbf{p} = 2\mathbf{p}_1$. Thus, in back
reflection, the momentum transfer is twice as effective as in absorption, each back
reflected photon transfers a momentum that is twice as large as the incident photon
momentum. The momentum transfer from absorbed or back reflected photons gives
rise to the macroscopic phenomenon of radiation pressure.

Let us now assume that a mirror is illuminated (normal incidence) with a light intensity $I = 1500$ W/m^2. What is the radiation pressure p_l applied to the mirror?

In Sect. 3.2.1, we introduced the light intensity as the average amount of light energy arriving per time interval per unit area. Hence, $I = \frac{dE}{A dt}$. The pressure is $p_l = \frac{F}{A}$. Eliminating A, we have: $p_l = \frac{F I dt}{dE}$.

From Newtons law, we have $F = \frac{dp}{dt}$, where p is the total momentum transferred to the mirror. Hence, $dp = 2p_{\mathrm{phot}}dN_{\mathrm{phot}}$, while $dE = E_{\mathrm{phot}}dN_{\mathrm{phot}}$; N_{phot} is the number of photons incident to the mirror. Hence, $p_l = \frac{F I dt}{dE} = \frac{I dp}{dE} = 2I \frac{p_{\mathrm{phot}}}{E_{\mathrm{phot}}}$. From $\frac{E_{\mathrm{phot}}}{p_{\mathrm{phot}}} = c$ we finally have: $p_l = 2I \frac{p_{\mathrm{phot}}}{E_{\mathrm{phot}}} = \frac{2I}{c}$.

For $I = 1500$ W/m^2, we obtain a radiation pressure of approximately 10^{-5} Pa, independent on the light frequency. The obtained radiation pressure is in rough correspondence to the solar radiation pressure in the earth orbit. If the mirror would be replaced by an absorber surface, the radiation pressure would be $p_l = I \frac{p_{\mathrm{phot}}}{E_{\mathrm{phot}}} = \frac{I}{c}$ for reasons explained in the remark.

3.6   Advanced Material: Wave Packet Distortion

In order to obtain an illustrative picture of the particle-wave dualism, it seems prospective to understand the wave character of particles in a naïve picture where the particles are represented by localized wave packets, propagating with a given group velocity (=particle velocity, as we have already seen). In order to have a reasonable correspondence between our classical picture of a propagating particle (which docs not change its shape when propagating with a constant speed) and the mentioned wave packet, a minimum requirement to this wavepacket is that it also propagates without distortion. As we have seen in 3.2.2, such propagation is principally possible. The goal of this section is to understand in how far this picture is self-consistent when being applied to propagating macroscopic and microscopic particles.

As we have shown in Sect. 3.2.2, in first-order dispersion theory and absence of damping, a wave packet propagates without any distortion of its shape. So let us return to our special wave packet. In particular, we wrote down an expansion of the dispersion law $\omega = \omega(k)$ into a Taylors series according to (3.5):

$$\omega = \omega(k_0) + (k - k_0) \left.\frac{d\omega}{dk}\right|_{k=k_0} + \frac{(k - k_0)^2}{2} \left.\frac{d^2\omega}{dk^2}\right|_{k=k_0} + \cdots$$

In order to introduce the concepts of phase and group velocities, we truncated that series after the linear term (=first-order dispersion theory) and obtained (3.6):

$$\omega \approx \omega(k_0) + (k - k_0) \left.\frac{d\omega}{dk}\right|_{k=k_0}$$

But the approximation (3.6) makes sense only when the higher order terms in (3.5) give only marginal contributions. In particular, (3.6) lacks any sense in the case that

$$\frac{(k-k_0)^2}{2} \left.\frac{\mathrm{d}^2\omega}{\mathrm{d}k^2}\right|_{k=k_0} t \geq 2\pi$$

is fulfilled [1]. At least after this time t, the conclusions obtained from first-order dispersion theory are no more valid. In particular, wave packet distortion will occur. Let us estimate this characteristic time and call it τ.

Then, if $\left.\frac{\mathrm{d}^2\omega}{\mathrm{d}k^2}\right|_{k=k_0} \neq 0$, wave packet distortion has to be expected after:

$$\tau \approx \frac{4\pi}{\left(\frac{\Delta k}{2}\right)^2 \left.\frac{\mathrm{d}^2\omega}{\mathrm{d}k^2}\right|_{k=k_0}}$$

Generally, the more localized along the z-axis the wave packet is, the broader its Fourier spectrum will be. This leads to some kind of uncertainty relation like:

$$\Delta k \Delta z \geq 2\pi \tag{3.15}$$

Or

$$\Delta p \Delta z \geq 2\pi \hbar = h \tag{3.15a}$$

For the sinc-like wave packet discussed in 3.2.2, localization along the z-direction corresponds directly to ξ-values confined between $-\pi$ and $+\pi$. Hence, $\Delta \xi \approx 2\pi$, and at $t = 0$, we have:

$\Delta \xi \approx 2\pi \approx \frac{\Delta k}{2}\Delta z$ or $\Delta k \Delta z \approx 4\pi$ [which is consistent with the mentioned uncertainty relation (3.15)]. Thus, Δk can be expressed in terms of the spatial extension of the wave packet Δz at $t = 0$. Hence,

$$\tau \approx \frac{(\Delta z)^2}{\pi \left.\frac{\mathrm{d}^2\omega}{\mathrm{d}k^2}\right|_{k=k_0}} = \frac{(\Delta z)^2}{\pi \hbar \left.\frac{\mathrm{d}^2E}{\mathrm{d}p^2}\right|_{p=p_0}}$$

In a nonrelativistic situation,

$$E = \frac{p^2}{2m} \Rightarrow \frac{\mathrm{d}^2E}{\mathrm{d}p^2} = m \Rightarrow \tau \approx \frac{2m(\Delta z)^2}{h} \tag{3.16}$$

This enables us estimating the characteristic time τ. Let us associate the spatial extension of the wave packet with the characteristic dimension of the propagating particle. We recognize that the time τ is mass and size dependent: the larger m and Δz are, the larger τ will be. Hence, quite different results are expected when comparing the behavior of "microscopic" and "macroscopic" particles.

For a macroscopic particle with $\Delta z = 10$ cm and $m = 1$ kg, we find $\tau \approx 3 \times 10^{31}$ s. Hence, when comparing with the age of the universe (Table 1.1), such a wavepacket remains undistorted, i.e., localized at all practically relevant time scales. The particle can thus be associated with a localized wavepacket that propagates without distortion with a group velocity that coincides with the usual particle velocity. Here, our illustrative picture works well.

In the case of a nonrelativistic electron localized in an atom (Table 1.1), $\Delta z \approx 10^{-8}$ cm and $m \approx 9 \times 10^{-31}$ kg, wavepacket distortion (In most cases the wavepacket will simply broaden) is expected within $\tau \approx 1.5 \times 10^{-17}$ s. Practically, the wavepacket broadens instantaneously. Such a picture is not consistent with the mechanical movement of a classical particle, i.e., such a microscopic particle should not be described as a well localized wave packet associated with a corpuscular picture. Instead, its wave character must be taken into account explicitly.

This result is qualitatively consistent with (3.15a). Indeed, the localization (at $t = 0$) at a very small Δz results in a large uncertainty of the momentum, and therefore our illustrative picture leads to the result, that at $t > 0$, the position of the particle can no more accurately be predicted. The only chance we might have is to indicate a probability to observe the electron in a certain space region. In the next chapter, we will therefore develop (not derive!) a picture of the particle dynamics that is based on the solution of a wave equation (the Schrödinger equation). We will then use that solution (the wavefunction of the electron) to predict the probability to observe the electron somewhere in space.

3.7 Tasks for Self-check

3.7.1 Estimate the so-called classical electron radius r_e by setting its classical electrostatic self-energy ($\approx \frac{e^2}{4\pi\varepsilon_0 r_e}$) equal to its relativistic total energy at rest ($m_e c^2$)! (after [4]).

3.7.2 (after [5]) Imagine a photon with an energy of 2 MeV. In the vicinity of a heavy nucleus, this photon is converted into a positron–electron pair. Please calculate:
   (a) The photon wavelength
   (b) The velocity of the electron or positron provided that as a result of the convertion process, each of them obtains a relativistic energy of 1 MeV, and electrostatic interaction between them may be ignored.

3.7.3 Consider the scattering of a γ-quant at an electron. When assuming a Compton wavelength of 2.4 pm, indicate the maximum wavelength shift possible as the result of a single scattering event. What would be the corresponding scattering angle?

References

Specific References

1. А.А. Соколов, И.М. Тернов, В.Ч. Жуковский: Квантовая механика, Москва "Наука" (1979), pp. 22–28
2. O. Stenzel, *Optical Coatings. Material Aspects in Theory and Practice* (Springer 2014), p. 112
3. А.Н.Матвеев, Механика и теориа относительности, Москва „Высшая школа" pp. 283–285
4. R.H. Good, *Classical Electromagnetism* (Saunders College Publishing, 1999), pp. 156–158
5. H. Haken, H.C. Wolf, *The Physics of Atoms and Quanta* (Springer, 2005), task 5.4

General Literature

6. P.A. Tipler, G. Mosca, *Physik, 7* (Springer, Auflage, 2015)
7. J. Bricmont, *Quantum Sense and Nonsense* (Springer, 2017)

The Schrödinger Equation and Model System I

4

Abstract

We introduce the wavefunction of a single-particle quantum system as well as the corresponding Schrödinger equation. As a first model system, we consider the behavior of a single quantum particle in a potential box with infinitely high potential walls. The results are used to introduce the concept of densities of states.

4.1 Starting Point

In the previous chapter, we came to the conclusion that it should be reasonable to ascribe wave properties to typical microscopic particles—such as electrons.

In a conventional particle picture (if we associate the electron with a point particle), the electrons position at a given time would be completely defined by a single vector $\mathbf{r} = \mathbf{r}(t)$, the time dependence being defined by Newtons equation of motion. In Chap. 2, we used that concept for developing a classical theory of light–matter interactions, where the actual positions of the electrons defined the actual values of the (induced) oscillating dipole moments.

A wave picture is something essentially different. A wave is more or less a delocalized phenomenon; typical wave features appear to be smeared over the space. Therefore, instead of ascribing a single radius vector to the electron, a new concept has to be introduced—the wavefunction Ψ. The wavefunction of a single electron depends on the electrons coordinates and the time, so we write $\Psi = \Psi(\mathbf{r}, t)$. Clearly, its evolution with time can no more be described by Newton's law. So what we need is a new equation that describes the time evolution of Ψ. This wave equation is called the Schrödingers equation. How such a wave equation could look like?

© The Author(s), under exclusive license to Springer Nature Switzerland AG 2022
O. Stenzel, *Light–Matter Interaction*, UNITEXT for Physics,
https://doi.org/10.1007/978-3-030-87144-4_4

4.2   Physical Idea

In order to illustrate the structure of Schrödingers equation, let us regard a plane wave (without any amplitude modulation) propagating along the x-axis according to:

$$\Psi(x, t) \propto e^{-i(\omega t - kx)} \tag{4.1}$$

This delocalized plane wave propagating in free space will further be associated with the wavefunction of a single freely propagating particle, i.e., a particle propagating in a potential that is zero in any point x. In view of our discussion from Chap. 3, we further make use of the relations

$$E = \hbar\omega$$

and

$$p = \hbar k.$$

Let us start by differentiating the wavefunction (4.1). Then we obviously find:

$$\frac{\partial}{\partial t}\Psi(x, t) = -i\omega\Psi(x, t) = -i\frac{E}{\hbar}\Psi(x, t)$$

$$\frac{\partial}{\partial x}\Psi(x, t) = ik\Psi(x, t) = i\frac{p}{\hbar}\Psi(x, t)$$

When acting on such a wavefunction Ψ, the operation $i\hbar\frac{\partial}{\partial t}$ "extracts" the particle energy from the wavefunction, while $-i\hbar\frac{\partial}{\partial x}$ "extracts" the x-component of a particle momentum (in a 3D case, the momentum vector will be associated with $-i\hbar\nabla$). Then, for a freely propagating nonrelativistic particle (neglecting rest mass contributions, which provide a constant offset only), replacing energy and momentum by the mentioned differential operations leads to:

$$E = T_{\text{kin}} \Rightarrow E = \frac{p^2}{2m} \Rightarrow E\Psi = \frac{p^2}{2m}\Psi \Rightarrow$$

$$i\hbar\frac{\partial}{\partial t}\Psi(x, t) = \left[-\frac{\hbar^2}{2m}\frac{\partial^2}{\partial x^2}\right]\Psi(x, t) \quad 1D - \text{case}$$

$$i\hbar\frac{\partial}{\partial t}\Psi(\mathbf{r}, t) = \left[-\frac{\hbar^2}{2m}\Delta\right]\Psi(\mathbf{r}, t) \quad 3D - \text{case}$$

where Δ stands for the Laplace operator:

$$\Delta \equiv \frac{\partial^2}{\partial x^2} + \frac{\partial^2}{\partial y^2} + \frac{\partial^2}{\partial z^2}$$

We will now proceed by postulating the nonrelativistic Schrödinger equation of a single particle in an arbitrary potential $U(\mathbf{r})$. We proceed in exactly the same manner as with the free particle, but instead of $E = T_{kin}$, we now write $E = T_{kin} + U(\mathbf{r})$. That leads to:

$$
\begin{aligned}
E = \frac{p^2}{2m} + U(\mathbf{r}) &\Rightarrow i\hbar\frac{\partial}{\partial t}\Psi(\mathbf{r}, t) \\
&= \left[-\frac{\hbar^2}{2m}\Delta + U(\mathbf{r})\right]\Psi(\mathbf{r}, t) \\
&= \left[\hat{T}_{kin} + U(\mathbf{r})\right]\Psi(\mathbf{r}, t) = \hat{H}\Psi(\mathbf{r}, t)
\end{aligned}
\tag{4.2}
$$

This is the nonrelativistic Schrödinger equation for a single particle in the potential $U(\mathbf{r})$ (in the following, we will use the terminology that this particle and the corresponding potential form a quantum system). The term $\hat{H} = \left[-\frac{\hbar^2}{2m}\Delta + U(\mathbf{r})\right]$ is called the Hamilton operator or *Hamiltonian* $\hat{H}$ of the quantum system. Solving (4.2) shall supply the wavefunction $\Psi(\mathbf{r}, t)$, but it should be noted that there may be several wavefunctions that suffice (4.2). The idea is now to associate the set of wavefunctions which suffice (4.2) with the set of all possible physical states of the quantum system. Note that (4.2) is a linear differential equation, such that the superposition principle is valid. If any two wavefunctions $\Psi_1(\mathbf{r}, t)$ and $\Psi_2(\mathbf{r}, t)$ are solutions of (4.2), then any linear superposition of them also describes a possible physical state.

4.3 Theoretical Considerations

4.3.1 The Time-Independent Schrödinger Equation

Let us now assume a stationary case, defined by

$$
\frac{\partial}{\partial t}\hat{H} = \frac{\partial}{\partial t}U = 0
$$

In the classical particle picture, the energy of the nonrelativistic particle would then be given by

$$
E = T_{kin} + U = \frac{m}{2}\mathbf{v}^2 + U
$$

Let us regard the energy as a function of the particles coordinates, momenta, and time. We then obtain:

$$
E = T_{kin} + U = \frac{\mathbf{p}^2}{2m} + U
$$

$$\Rightarrow dE = \frac{\partial E}{\partial \mathbf{p}} d\mathbf{p} + \frac{\partial E}{\partial \mathbf{r}} d\mathbf{r} + \frac{\partial E}{\partial t} dt = m\mathbf{v} d\mathbf{v} + \frac{\partial U}{\partial \mathbf{r}} d\mathbf{r} + \frac{\partial U}{\partial t} dt$$

$$\Rightarrow \frac{dE}{dt} = \overbrace{\frac{\mathbf{p}}{m}}^{=\mathbf{v}} \underbrace{\frac{d\mathbf{p}}{dt}}_{=m\mathbf{a}} + \frac{\partial U}{\partial \mathbf{r}} \underbrace{\frac{d\mathbf{r}}{dt}}_{=\mathbf{v}} + \frac{\partial U}{\partial t} = \mathbf{v}\left(m\mathbf{a} + \frac{\partial U}{\partial \mathbf{r}}\right) + \frac{\partial U}{\partial t}$$

Because of: $\mathbf{F} = -\frac{\partial U}{\partial \mathbf{r}}$ and $\mathbf{F} = m\mathbf{a}$ we have (compare (2.21) and (2.12)):

$$\left.\begin{array}{l} \mathbf{F} = -\dfrac{\partial U}{\partial \mathbf{r}} \\[2mm] \mathbf{F} = m\mathbf{a} \end{array}\right\} \Rightarrow m\mathbf{a} + \frac{\partial U}{\partial \mathbf{r}} = 0 \Rightarrow \frac{dE}{dt} = \frac{\partial U}{\partial t}$$

Hence, if $\frac{\partial U}{\partial t} = 0 \Rightarrow E = $ const.

We come to the result that the mechanical energy of the classical particle in the potential U is a constant when U is not explicitly time-dependent. This is by the way a conclusion from more general symmetry considerations, as formulated in the Noether theorems.

What we are now doing is to combine our knowledge on the particle and wave approaches. Let us search special solutions of (4.2) according to:

$$\Psi(\mathbf{r}, t) \equiv \psi(\mathbf{r}) f(t) \tag{4.3}$$

If $\frac{\partial}{\partial t}\hat{H} = \frac{\partial}{\partial t}U = 0$, that leads to

$$i\hbar\frac{\partial}{\partial t}\Psi(\mathbf{r}, t) = \hat{H}\Psi(\mathbf{r}, t) \Rightarrow i\hbar\psi\dot{f} = f\hat{H}\psi \Rightarrow i\hbar\frac{\dot{f}}{f} = \frac{\hat{H}\psi}{\psi} \tag{4.3a}$$

Let us have a closer look at the equation

$$i\hbar\frac{\dot{f}}{f} = \frac{\hat{H}\psi}{\psi}$$

Obviously, $i\hbar\underbrace{\dfrac{\dot{f}}{f}}_{\text{independent on } \mathbf{r}}$ $=$ $\underbrace{\dfrac{\hat{H}\psi}{\psi}}_{\text{independent on } t}$. Then, $i\hbar\underbrace{\dfrac{\dot{f}}{f}}_{\text{independent on } \mathbf{r}}$ is also indepen-

dent on t; hence, it is a constant. Therefore, we obtain:

$$i\hbar\frac{\dot{f}}{f} = \frac{\hat{H}\psi}{\psi} = \text{const} \tag{4.3b}$$

From there, we have $i\hbar\dot{f} - \text{const} * f = 0 \Rightarrow f = e^{-i\frac{\text{const}}{\hbar}t}$.

Although our discussion is no more about a free particle, the time-dependent term f of the wavefunction coincides with that from (4.1), which was associated with a freely propagating particle. When setting const$/\hbar = \omega$, from $\hbar\omega = E$, we find const $= E$. This is in complete agreement with the above-mentioned classical considerations, provided that we associate the constant in (4.3b) with the energy E of the system again. Note that this is rather a postulate, our arguments can by no means be accepted as a strong derivation. But at least for classical objects, this interpretation makes sense; on the other hand, there is no rigorous criterion which would forbid applying (4.2) to a classical (*vulgo* "macroscopic") material system and allow it for a quantum mechanical (*vulgo* "microscopic") one. Moreover, what we have postulated here in an ad hoc manner is consistent with a more general formulation of basic postulates in quantum mechanics, and we will return to this question in Chap. 5. At the moment, let us fix the conclusion that if the Hamiltonian of the system is not explicitly time-dependent, wavefunctions described by (4.3) correspond to physical states with a constant energy. How large is that energy?

Let us mention that from (4.3b), we get two equations:

$$i\hbar \dot{f} - Ef = 0 \Rightarrow f = \mathrm{e}^{-i\frac{E}{\hbar}t}$$
$$\hat{H}\psi = E\psi \tag{4.4}$$

While the first equation immediately leads us to the function f, the second equation (the time-independent Schrödinger equation) poses an eigenvalue problem with solutions (further numbered by the index n) that are specific to the chosen potential $U(x)$. Having solved that eigenvalue problem, we obtain a set of energies $\{E_n\}$ that are consistent with the Schrödinger equation, while the wavefunctions of the system (which describe its eigenstates) are given by:

$$\Psi_n(\mathbf{r}, t) = \mathrm{e}^{-i\frac{E_n}{\hbar}t}\psi_n(\mathbf{r}) \text{ if } \frac{\partial U}{\partial t} = 0 \tag{4.4a}$$

A quantum state as given by (4.4a) is called a stationary quantum state. It is characterized by a fixed energy value. The set of eigenvalues $\{E_n\}$ will further be associated with the set of "allowed" energy levels of the quantum system.

Note that because of the linearity of (4.2), linear superpositions of different wavefunctions according to (4.4a) will also describe possible physical states of the system.

4.3.2 The Free Particle

For a free particle ($U = 0$), the time-independent Schrödinger equation in one dimension may be written as:

$$\left[-\frac{\hbar^2}{2m}\frac{\mathrm{d}^2}{\mathrm{d}x^2}\right]\psi(x) = E\psi(x)$$

Table 4.1 Wavefunction of a single particle in a constant potential U_0. (1D case)

Time-independent Schrödinger equation	Solution	Case	Result
$\left[-\frac{\hbar^2}{2m}\frac{d^2}{dx^2} + U_0\right]\psi(x) = E\psi(x)$	$\psi(x) \propto e^{\pm ikx}$ $k = \frac{\sqrt{2m(E-U_0)}}{\hbar}$	$U_0 = 0$	$\psi(x) \propto e^{\pm i \frac{\sqrt{2mE}}{\hbar} x}$
		$E > U_0 \neq 0$	$\psi(x) \propto e^{\pm i \frac{\sqrt{2m(E-U_0)}}{\hbar} x}$
		$E < U_0 \neq 0$	$\psi(x) \propto e^{\pm \frac{\sqrt{2m(U_0-E)}}{\hbar} x}$

Let us search the solution as:

$$\psi(x) \propto e^{\pm ikx} \tag{4.5}$$

Substituting into Schrödingers equation leads to: $E = \frac{\hbar^2 k^2}{2m} = \frac{p^2}{2m}$, which is quite a reasonable result. In this case, the energy coincides with the kinetic energy of the freely propagating particle. The wavefunction becomes: $\Psi(x,t) = e^{-i\frac{E}{\hbar}t}\psi(x) \propto e^{-i\frac{E}{\hbar}t}e^{\pm ikx}$ which is nothing else than the initially assumed plane wave propagating along the x-axis.

For completeness, Table 4.1 shows the corresponding solutions for the time-independent Schrödinger equation when assuming a constant potential U_0. Note that we obtain a continuous spectrum of eigenvalues E in this case.

4.3.3 Model System I: Particle in a Box

In order to get a feeling on the mathematical apparatus which is hidden behind the equations written down in Sect. 4.3.1, let us regard a simplest case, namely the one-dimensional movement of a quantum particle in a box with infinitely high potential walls. We assume the time-independent Schrödinger equation according to:

$$\hat{H}\psi = E\psi;$$

$$\hat{H} = -\frac{\hbar^2}{2m}\frac{d^2}{dx^2} + U(x)$$

$$U = 0; \quad 0 \leq x \leq L$$

$$U \to \infty; \quad x < 0 \text{ or } x > L$$

We will start with the reasonable assumption that behind the infinitely high walls, the wavefunction must be zero (compare Table 4.1, red-marked field with $U_0 \to \infty$). Although we have not yet discussed the sense of the introduced wavefunction (this will be done in Chap. 5), let us postulate that whenever the wavefunction is zero in a certain space region, the particle cannot be observed there. Indeed, in a 3D situation, the term

$$dw = \Psi^* \Psi dx dy dz = \Psi^* \Psi dV \tag{4.6}$$

may be associated with the probability to observe the particle in the volume element dV. This is the essence of Max Born's statistical interpretation of the wavefunction (Born's rule). Then, the term $\Psi^* \Psi$ is the corresponding *probability density*. Once the probability to observe the particle elsewhere in the volume V occupied by the system must be equal to 1, we find the normalization rule:

$$\int_V \Psi^* \Psi dV = 1 \tag{4.6a}$$

So our assumption means that the captured particle may never be observed behind the infinitely high (und therefore impermeable) potential walls. We will also assume that the wavefunction is continuous. Then we should use the boundary conditions:

$$\psi(0) = \psi(L) = 0.$$

Within the box, the time-independent Schrödinger equation is:

$$-\frac{\hbar^2}{2m} \frac{d^2}{dx^2} \psi(x) = E\psi(x); \ 0 \le x \le L$$

Let us search the solution as $\psi(x) \propto \sin kx$. In order to have the boundary conditions fulfilled, we must require:
$\sin kL = 0 \Rightarrow kL = n\pi; \ n = 0, 1, 2, \ldots$ Hence, we get a set of wavefunctions numbered by n: $\psi(x) = \psi_n(x) \propto \sin \frac{n\pi}{L} x$.

Note that for $n = 0$, the wavefunction is identical to zero. In terms of (4.6) that means, that the probability density to find the particle anywhere in the box is zero, which corresponds to an empty box without any particle inside. In this connection, n is called a *quantum number*. Consequently, we obtain a set of allowed k-values such that $k = k_n = \frac{n\pi}{L}$ holds. On the other hand, we have $k_n = \frac{n\pi}{L} = \frac{2\pi}{\lambda_n} \Rightarrow \lambda_n = \frac{2L}{n}$, so that the associated eigenfunctions $\psi_n(x)$ represent nothing else than the set of possible standing wave modes within the box (Fig. 4.1).

For the corresponding energy values (*energy levels*), we find:

$$-\frac{\hbar^2}{2m} \frac{d^2}{dx^2} \psi_n(x) = \frac{\hbar^2}{2m} \frac{n^2\pi^2}{L^2}$$
$$\psi_n(x) = E_n \psi_n(x); \ 0 \le x \le L$$
$$\Rightarrow E = E_n = \frac{h^2}{8mL^2} n^2 = \frac{\pi^2 \hbar^2}{2mL^2} n^2 \tag{4.7}$$

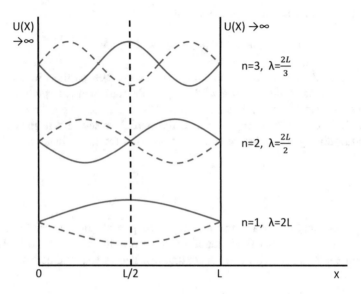

Fig. 4.1 Eigenfunctions of the particle in the box, see (4.7a)

In contrast to what has been obtained in Sect. 4.3.2 for a freely propagating particle, we now find a discrete spectrum of eigenvalues E_n. Note that the thus allowed energy values and their separation become larger when the length of the box L becomes smaller.

There is a simple mnemonic trick for quick reproduction of (4.7) without memorizing. Once inside the box, the potential energy is zero, the energy is purely kinetic, i.e., $E_n = \frac{p_n^2}{2m}$. But $p_n = \pm\hbar k_n = \pm\frac{h}{\lambda_n}$. From $\lambda_n = \frac{2L}{n}$ (nevertheless you must memorize Fig. 4.1), we have $p_n^2 = \frac{h^2 n^2}{4L^2}$ and therefore $E = E_n = \frac{h^2}{8mL^2}n^2$. Note that as a rule, we derive from here that the larger the spatial extension of a quantum system is, the smaller the kinetic ground-state energy tends to be.

Remark For reasons which will become clear later, we do not consider negative quantum numbers n. Also, in the 1D case discussed here, instead of (4.6a), it makes sense to normalize the wavefunctions according to: $\int_0^L \psi_n^2(x)\mathrm{d}x = 1 \ \forall n$. This requirement leads to the normalized final wavefunctions:

$$\psi_n(x) = \sqrt{\frac{2}{L}} \sin \frac{n\pi}{L}x; n = 1, 2, 3, \ldots \tag{4.7a}$$

The three-dimensional generalization of our results (4.7), (4.7a) is:

$$E_{n_x,n_y,n_z} = \frac{h^2}{8m}\left(\frac{n_x^2}{L_x^2} + \frac{n_y^2}{L_y^2} + \frac{n_z^2}{L_z^2}\right) \tag{4.8}$$

$$\psi_{n_x,n_y,n_z}(x, y, z) = \sqrt{\frac{8}{L_x L_y L_z}}\,\sin\left(\frac{n_x\pi}{L_x}x\right)\sin\left(\frac{n_y\pi}{L_y}y\right)\sin\left(\frac{n_z\pi}{L_z}z\right) \tag{4.8a}$$

L_x, L_y, and L_z mark the spatial extensions of the box in x-, y-, and z-directions, respectively.

4.4 Consistency Considerations

I suppose that you, dear reader, are now feeling some frustration, because so far in this chapter, practically nothing was sophistically derived, but everything was rather postulated or illustrated in terms of analogies. This is indeed true, because such a general concept like expressed by Schrödingers equation cannot be derived in a mathematically strong manner.

In this section, let us nevertheless shortly discuss the consistency of the results from Sects. 4.3.2 and 4.3.3 with the wave packet picture developed in Chap. 3, and the resulting preliminary formulation of the uncertainty relation (3.15).

Clearly, the solution (4.5) of the time-independent Schrödinger equation for a free particle (Sect. 4.3.2) suffices the uncertainty relation (3.15), because the fixed k-value corresponds to complete delocalization of the particles wavefunction in space. But once the Schrödinger equation (4.2) is linear, the superposition principle holds, and therefore, linear superpositions of solutions like (4.1) also represent physically possible states of the particle. This way we can construct wavepackets like introduced in Sect. 3.2.2, and according to (3.15), the resulting variance in k-values now allows for certain localization of the particles wavefunction in space. But according to (3.16), this localization may be destroyed as a result of wave packet broadening as discussed in Sect. 3.6: For small particles with a small mass, in free space, the wave packet broadens practically instantaneously. Thus localization in free space is quickly lost.

Therefore, in order to keep the wavefunction of a quantum particle localized, the latter should interact "with somewhat." As an example, in Sect. 3.3.3, we introduced what we called a box potential and thus introduced localization of the quantum particle within the length of the box. On the other hand, the momentum of the quantum particle is now no more fixed; we may observe it moving to the left as well as moving to the right. This leads to a certain variance in the momentum, and as it will be shown later in task 5.7.7, the variances in coordinate and momentum together are again consistent with the uncertainty relation. So all in all, the considerations from Chap. 3 are consistent with what we have obtained so far from the Schrödinger equation in this chapter.

But more generally, let us emphasize that Schrödingers equation has proven correct because it gives accurate and correct predictions of what goes on in nature. Or alternatively, of what can be observed or measured. That depends on the particular interpretation of quantum mechanics, and this is still a topic of debates. You, dear reader, have now two basic choices. You can—as many physicists do quite successfully—accept the mathematical apparatus of quantum mechanics as an outstandingly reliable recipe for rigorous quantitative treatment of all phenomena we will discuss throughout this book. And you can, in addition to becoming familiar with the mathematical apparatus—try to develop your idea on what quantum mechanics is about, i.e., what are the "objects" investigated by quantum mechanics, and what can be learned about them by performing quantum mechanical calculations. And finally, on what quantum mechanics tells us about nature. In this case, you have to go deeper into the existing quantum theories (there are several of them), but this will not be topic of this book. Why? The answer is simple: Such books exist, but in Sect. 5.1, some extra recommendations on further reading will be provided.

Thus, instead of a formal consistency consideration, I would it like to refer to the outstanding performance of quantum mechanics in making quantitative predictions here. The following practical problems provide at least some examples for that.

4.5   Application to Practical Problems

4.5.1   Size Effects

A trivial example of the application of the theory developed so far is in the use of quantum confinement for designing material systems with tailored frequencies of optical transitions. Indeed, let us assume that the energy gap between two adjacent energy levels in (4.7) or (4.8) may be bridged by absorption or emission of a single light quantum (photon). Such a situation is sketched in Fig. 4.2. A more detailed treatment of this situation is topic of Chap. 6, but our knowledge developed so far is sufficient to build at least a qualitative picture. According to Fig. 4.2, if a photon initiates a light absorption process between the levels n and $n + 1$, as a result of energy conservation, we assume its angular frequency according to:

$$\hbar\omega = E_{n+1} - E_n = \frac{\hbar^2\pi^2}{2mL^2}\left[(n+1)^2 - n^2\right]$$

$$\Rightarrow \omega = \frac{\hbar\pi^2}{2mL^2}(2n+1) = \frac{2\pi c}{\lambda} \tag{4.9}$$

Obviously, the absorption (or emission) frequency may be tuned by the choice of a suitable length L. In technological practice, the usual situation is that an electron is confined in a nanometer-sized crystal (a "nanocrystal") that we call a quantum dot. In fact, the potential inside the dot is not really constant. Instead, the

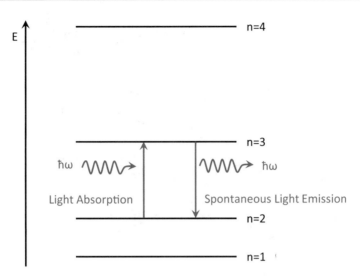

Fig. 4.2 Visualization of light absorption and light emission between the energy levels $n = 2$ and $n = 3$

electrons "feel" the periodic potential of the atomic cores that build the crystal. Therefore, (4.9) provides a very crude estimation only, but it nevertheless demonstrates that by changing the box length in the sub-nanometer range, the wavelength may be tuned over quite a broad spectral range.

4.5.2 Color Centers in Ionic Crystals

Crystals are characterized by a nearly perfect periodic arrangement of their atoms, which we call a crystalline lattice. In solid crystals built from positively and negatively charged ions (such as for example LiF, NaCl or KCl), electrons may be trapped in a vacancy defect formed by a vacant halogen atom. The void left by the vacancy acts as a 3D potential box, with a spatial extension defined by the lattice period a. An electron confined in that box forms what is called a color center; the name arises from its characteristic light absorption in the VIS. Although the potential within the box is surely not constant (the electrons will preferably be found in the vicinity of the positive ions), the absorption wavelength should depend on the lattice period a in the crystal in a qualitative manner as described by (4.9) (Fig. 4.3).

Such a dependence is experimentally observed, while the wavelength at maximum absorption λ_{max} satisfies the relation:

$\lambda_{max} = \text{const} * a^2$. From the experiment, the constant turns out to be equal to:

$$\text{const} = 6 * 10^{12}\,\text{m}^{-1}$$

(compare [1]).

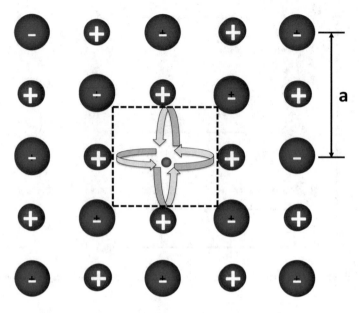

Fig. 4.3 Color center in a crystal composed from oppositely charged ions

Note that the corresponding proportionality constant as predicted from (4.9) is approximately $\frac{4m_e c}{3\hbar\pi} \approx 1.2 * 10^{12}\,\mathrm{m}^{-1}$ and therefore, regardless on the crudeness of the model, still of the correct order of magnitude.

4.5.3 Organic Chain Molecules with Conjugated Double Bonds

Let us regard a hydrocarbon chain molecule built according to the recipe:

$$... - C = C - C = C - C = ...$$

from a total of N atoms. The so-called π-electrons which are responsible for the double bounds are delocalized, i.e., rather mobile within the molecule, so that the mentioned chain may be modeled as a 1D potential box with approximately N electrons captured. The length of the box corresponds to $(N - 1)$ times the distance between two adjacent C-atoms, which is approximately $L_0 = 0.14$ nm, multiplied with a phenomenological correction factor C, which takes the nonlinear geometry of the real molecule into account.

We are now confronted with the new circumstance, that there is not only one electron trapped in the assumed box. Instead, we have N electrons (for simplicity, let that number be even). We have not yet introduced Paulis exclusion principle (see later Sect. 9.3.1) and therefore just postulate, that each of the energy levels may be populated by a maximum of two electrons (with different spin orientations, see later Sect. 8.6.1). Hence, the energy levels up to $n = N/2$ are expected to be

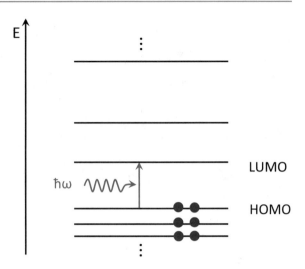

Fig. 4.4 Visualization of an optical HOMO–LUMO transition

populated, and absorption from the highest occupied quantum state to the lowest unoccupied quantum state will correspond to excitations from the energy level $n = N/2$ to $n = N/2 + 1$. In molecular spectroscopy, it is common to call this transition the HOMO–LUMO transition, where HOMO means "Highest Occupied Molecular Orbital" and LUMO "Lowest Unoccupied Molecular Orbital" (Fig. 4.4).

Then, when neglecting any interaction between the π-electrons, from (4.9), we find

$$\lambda = \frac{4 m_e c (N-1)^2 L_0^2 C^2}{\hbar \pi (N+1)}$$

For a long chain (N is large), that gives

$$\lambda \approx \frac{4 m_e c N L_0^2 C^2}{\hbar \pi}$$

as an estimation for the wavelength that corresponds to the HOMO–LUMO transition, we obtain:

$$\frac{4 m_e c L_0^2}{\hbar \pi} \approx 65 \, \text{nm}$$

So the estimated wavelength becomes:

$$\lambda \approx 65 \, \text{nm} N C^2$$

Already for $C = 1$, we find that an increasing chain length (increased N) is accompanied by an increase in the absorption wavelength, which is the quantitative expression of an empirical rule in organic dye chemistry [2]. With a moderate chain

Fig. 4.5 Colors in nature: Alpine flora

length (moderate value of N), their absorption falls into the visible spectral region, which explains the beautifully colored appearance of our biological environment (see Fig. 4.5 for examples).

4.6   Advanced Material

4.6.1   Densities of States

Let us now assume that we deal with a cubic 3D box with $L_x = L_y = L_z \equiv L$. Let us further assume, that $L \to \infty$, so we turn to macroscopic systems. Then, according to (4.8) or (4.9), the energy levels will practically form a continuum of allowed energy values, because the spacing between different allowed energy values becomes close to zero. That allows us introducing a function called the density of states, i.e., the number of quantum states (described by the eigenfunctions of $\hat{H}$) with energies that fall into a given energy interval. In our 3D example, quantum states are defined by given values of the quantum numbers n_x, n_y, and n_z.

From (4.8) we find:

$$
\begin{aligned}
E_{n_x,n_y,n_z} &= \frac{h^2}{8m}\left(\frac{n_x^2}{L_x^2} + \frac{n_y^2}{L_y^2} + \frac{n_z^2}{L_z^2}\right) \\
&= \frac{h^2}{8mL^2}\left(n_x^2 + n_y^2 + n_z^2\right) \equiv \frac{h^2}{8mL^2}n^2 = E_n
\end{aligned}
\tag{4.10}
$$

with $n^2 \equiv n_x^2 + n_y^2 + n_z^2$.

If $L \to \infty$, the energy should approach that of a free particle, hence

$$
E_n = \frac{h^2}{8mL^2}n^2 \to \frac{p^2}{2m} = \frac{\hbar^2 k^2}{2m} = \frac{h^2 k^2}{8\pi^2 m}
$$

Fig. 4.6 To the calculation of the number of quantum states

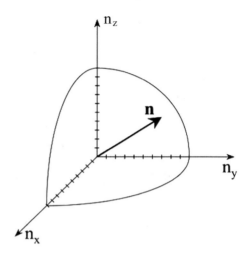

For large L, n may be tackled as a continuous variable. It is connected to the wavevector $\mathbf{k}$ via:

$$k^2 = k_x^2 + k_y^2 + k_z^2 = \left(\frac{\pi}{L}\right)^2 \left(n_x^2 + n_y^2 + n_z^2\right) \equiv \left(\frac{\pi}{L}n\right)^2 \tag{4.11}$$

Note that the thus defined value n is not a quantum number.

Let us now calculate the number of states that fall into a given n-interval. Each state corresponds to a certain triple of n_x, n_y, and n_z and therefore occupies a cube of the volume 1 in the n-space [3] (see Fig. 4.6).

In particular, the figure visualizes the n-space occupied by the states corresponding to n-values between zero and a given maximum value of n. The volume of the sphere with the radius n in the figure corresponds to the full number of states, however, due to the circumstance that n_x, n_y, and n_z shall not be negative, we should only count the states in the first octant. Hence, we find for the full number of states Z:

$$Z = \frac{1}{8} \cdot \frac{4\pi}{3} n^3$$

We therefore have for the full number of quantum states Z:

$$Z = \frac{\pi}{6} n^3 \Rightarrow \frac{dZ}{dn} = \frac{\pi n^2}{2} \tag{4.12}$$

The function $\frac{dZ}{dn} \equiv D(n)$ represents the full number of quantum states per n-interval. In order to find the mentioned density of states as the number of states per energy interval, we set:

$$D(E) \equiv \frac{dZ}{dE} = \frac{dZ}{dn}\frac{dn}{dE} = \pi n^2 \frac{1}{2}\frac{dn}{dE} \tag{4.12a}$$

From (4.10), we have

$$E = \frac{h^2}{8mL^2}n^2 \Rightarrow \frac{\mathrm{d}E}{\mathrm{d}n} = \frac{nh^2}{4mL^2}$$

That yields:

$$D(E) \equiv \frac{\mathrm{d}Z}{\mathrm{d}E} = \frac{\mathrm{d}Z}{\mathrm{d}n}\frac{\mathrm{d}n}{\mathrm{d}E} = \pi n^2 \frac{1}{2}\frac{4mL^2}{nh^2} = \frac{2\pi mL^2}{h^2}n$$

Or, when using $n = \frac{L}{h}\sqrt{8mE}$ (compare (4.10))

$$D(E) = \frac{2\pi mL^2}{h^2}\frac{L}{h}\sqrt{8mE} = 2\pi L^3\left(\frac{2m}{h^2}\right)^{\frac{3}{2}}\sqrt{E} \tag{4.13}$$

For completeness, let us have a look on the 2D and 1D cases.
In the 2D case, $n^2 \equiv n_x^2 + n_y^2$, and for Z we have::

$$Z = \frac{1}{4}\cdot\pi n^2 \Rightarrow \frac{\mathrm{d}Z}{\mathrm{d}n} = D(n) = \frac{\pi n}{2} \tag{4.14}$$

$$D(E) = \frac{\mathrm{d}Z}{\mathrm{d}n}\frac{\mathrm{d}n}{\mathrm{d}E} = \pi n\frac{1}{2}\frac{4mL^2}{nh^2} = \frac{2\pi mL^2}{h^2} = \text{const.} \tag{4.15}$$

In the 1D case, we obtain:

$$n^2 \equiv n_x^2$$

$$Z = n \Rightarrow \frac{\mathrm{d}Z}{\mathrm{d}n} = D(n) = 1 \tag{4.16}$$

$$D(E)\frac{\mathrm{d}Z}{\mathrm{d}n}\frac{\mathrm{d}n}{\mathrm{d}E} = \frac{4mL^2}{nh^2} = \frac{4mL^2}{h^2 n}$$

Or, when using $n = \frac{L}{h}\sqrt{8mE}$

$$D(E) = \frac{4mL^2}{h^2}\frac{h}{L\sqrt{8mE}} = \frac{L}{h}\sqrt{\frac{2m}{E}} \tag{4.17}$$

The obtained densities of states will become important later in Chap. 17, when we will turn to the discussion of the optical properties of crystalline solids.

Remark We will introduce the terminus *degeneration* later in Sect. 5.3.3. Nevertheless, it is worth mentioning already here that in reality, even when n_x, n_y, and n_z are specified, more than one physically different quantum state may be possible. In such a situation, we will say that the quantum state specified by n_x, n_y, and n_z is *degenerated*. In our case, if we speak on electrons as the quantum particles confined in the box potential, each state defined by n_x, n_y, and n_z may be populated with *two* electrons which are in different states with respect to their *spin orientation* (compare Sect. 4.5.3). The electron spin will be introduced later in Chap. 8. In (4.12)–(4.17), degeneration may be taken into account by multiplying with an additional degeneration factor g. In the case of electron spin degeneration, that factor is $g = 2$.

4.6.2 Quantum Tunneling

Let us now come to a typical quantum effect that has no classical analog, namely quantum tunneling through a barrier. Again we will restrict ourselves to the 1D case. For the present model calculation, we choose a rectangular potential given by:

$$x < 0 \text{ and } x > a : \ U = 0$$
$$0 \leq x \leq a : \ U = U_0 = \text{const}$$

According to Table 4.1, we use the following ansatz for the particles wavefunction:

$$x < 0 : \ U = 0 : \ \psi_1(x) = Ae^{i\alpha x} + Be^{-i\alpha x}; \ \alpha = \frac{\sqrt{2mE}}{\hbar}$$
$$0 \leq x \leq a : \ U = U_0 : \ \psi_2(x)$$
$$= Ce^{\beta x} + De^{-\beta x}; \ \beta = \frac{\sqrt{2m(U_0 - E)}}{\hbar}$$
$$x > a : \ U = 0 : \ \psi_3(x) = A\xi e^{i\alpha x}; \ \alpha = \frac{\sqrt{2mE}}{\hbar}$$

Here, we supposed that the particle arrives from negative x-values and moves toward positive x-values. We naturally expect that the barrier may cause reflection of the particle, hence at $x < 0$, we consider both forward and back traveling waves. Once the particle has "penetrated" the wall, it may move further into the positive x-direction and will never come back. Hence, for $x > a$, no back traveling wave will be considered. Obviously, ξ may be understood as the ratio of the amplitudes of the forward-traveling part of the wavefunctions behind and in front of the barrier.

Note further that we will assume $E < U_0$. In classical physics, in this case, a particle with energy E would not be able to pass the barrier. However, in quantum physics, it turns out that there is a possibility to observe the particle behind the wall.

Once the potential is zero on both sides of the barrier, we define the transmission coefficient T through:

$$T = |\xi|^2$$

In order to find ξ, we have to consider the boundary conditions of the wavefunctions at $x = 0$ and $x = a$.

When requiring continuity of the wavefunction and its first derivative, we obtain at $x = 0$:

$$\psi_1(0) = \psi_2(0) \Rightarrow A + B = C + D$$

$$\frac{d}{dx}\psi_1(x)\bigg|_{x=0} = \frac{d}{dx}\psi_2(x)\bigg|_{x=0} \Rightarrow i\alpha(A - B) = \beta(C - D)$$

Remark The requirement on the continuity of the first derivative of the wavefunction is consistent to the requirement that the so-called probability current density should be continuous (it is even constant in a situation like here). The probability current density is dominated by terms like $\psi \frac{d\psi^*}{dx}$. For continuity of this term, it is sufficient that both ψ and $\frac{d\psi}{dx}$ are continuous. Note that if $\psi = 0$, continuity of $\frac{d\psi}{dx}$ does not matter (compare Sect. 4.3.3).

The probability current density will be introduced in Sect. 5.3.1. Therefore, dear reader, for the moment you are asked to accept the above-mentioned boundary conditions as a mathematical model.

From here, we find immediately:

$$2C = A + B + i\frac{\alpha}{\beta}(A - B)$$

$$2D = A + B - i\frac{\alpha}{\beta}(A - B)$$

At $x = a$, we observe:

$$\psi_3(a) = \psi_2(a) \Rightarrow A\xi e^{i\alpha a} = Ce^{\beta a} + De^{-\beta a}$$

$$\frac{d}{dx}\psi_3(x)\bigg|_{x=a} = \frac{d}{dx}\psi_2(x)\bigg|_{x=a} \Rightarrow i\frac{\alpha}{\beta}A\xi e^{i\alpha a} = \left(Ce^{\beta a} - De^{-\beta a}\right)$$

From here it follows that

$$2Ce^{\beta a} = \left(1 + i\frac{\alpha}{\beta}\right)e^{i\alpha a}A\xi$$

$$2De^{-\beta a} = \left(1 - i\frac{\alpha}{\beta}\right)e^{i\alpha a}A\xi$$

When substituting now $2C$ and $2D$ by the previously obtained expressions, we find two equations in A,B and ξ. Resolving the first one with respect to B, we arrive at:

$$B = \frac{1 + i\frac{\alpha}{\beta}}{1 - i\frac{\alpha}{\beta}}\left(\xi e^{(i\alpha - \beta)a} - 1\right)A$$

Substituting now B in the second equation, for ξ, we find:

$$\xi = \frac{-4i\frac{\alpha}{\beta}}{\left(1 - i\frac{\alpha}{\beta}\right)^2 e^{(i\alpha + \beta)a} - \left(1 + i\frac{\alpha}{\beta}\right)^2 e^{(i\alpha - \beta)a}}$$

In order to come to a more compact expression, let us regard the special case of a very high or/and broad barrier. Hence, we require $\beta a \gg 1$. Then, we easily find [4, pp. 857–858]:

$$T = |\xi|^2 \approx \frac{16\alpha^2\beta^2}{\left(\beta^2 + \alpha^2\right)^2}e^{-2\beta a} = \frac{16E(U_0 - E)}{U_0^2}e^{-2\frac{\sqrt{2m(U_0 - E)}}{\hbar}a} \qquad (4.18)$$

Obviously, even with particle energies $E \ll U_0$, the transmission coefficient is different from zero. Hence, the wavefunction of the particle is different from zero at both sides of the barrier. So we have a chance to observe the particle at the other side of the wall, which is the essence of quantum tunneling. For macroscopic objects, with the mass of a soccer ball, the transmission coefficient will be vanishingly small, such that please do not expect to observe the situation visualized in Fig. 4.7 on a real soccer field.

4.6.3 Model Potential for Describing the α-decay in Nuclear Physics

The α-decay, i.e., the emission of α-particles by radioactive nuclei, cannot be understood in terms of classical physics. Natural radioactive materials emit α-particles with energies in the range of some MeV, with half-life times varying between 10^{-5} s and 10^{11} years. There is no classical model that allows reproducing that dynamic range of decay times.

On the contrary, the process has early been successfully modeled in terms of a model potential, which combines features of the particle-in-a-box potential and a repulsive Coulombs potential according to

$$U(r) = \frac{1}{4\pi\varepsilon_0}(2e)\frac{Ze}{r}$$

Here, $2e$ is the electric charge of an α-particle (it is a helium nucleus composed from two protons and two neutrons), and Z the atomic order (=number of protons

Fig. 4.7 Tunneling through a barrier. Cartoon by Dr. Alexander Stendal. Printed with permission

in the nucleus) after emission of the α-particle. The assumption was that prior to its emission, the α-particle can be modeled as being confined in a potential box with the radius of the nucleus r_0 (see Fig. 4.8 on left), with a potential wall of finite height. This box potential is to model the attractive action of the nuclear forces. The latter are only efficient at shortest distances, of the order of r_0. Outside the nucleus, the α-particle only feels the repulsive Coulomb potential of the nucleus. As seen in the figure, the superposition of these model potentials defines a potential barrier of finite height. But in this case, there is no reason to assume that the wavefunction of the α-particle is zero outside the nucleus. Instead, one would expect then within the barrier, the wavefunction is damped with increasing distance from the nucleus as long as the energy of the α-particle $E < U(r) = \frac{1}{4\pi\varepsilon_0}(2e)\frac{Ze}{r}$

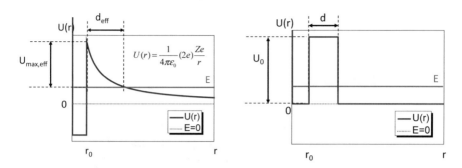

Fig. 4.8 Model potentials for the α-decay. On left: box potential combined with Coulomb potential. On right: simplified model with rectangular potential barrier [4, p. 858]

holds (compare Table 4.1 and Sect. 4.6.2). At a certain distance from the nucleus, we find $E > U(r)$, and the wavefunction would correspond to that of a propagating particle. At $r \rightarrow \infty$, we have $U \rightarrow 0$, and the particles wavefunction should approach that of a freely propagating particle with kinetic energy E (Fig. 4.8 on left). Thus, there is a chance that the α-particle may be found outside the nucleus, which means that an α-particle has been emitted.

As we have seen in Sect. 4.6.2, a higher and/or broader potential barrier should result in a smaller tunneling probability—in our example to a larger decay time. But as it is illustrated in the figure, the effective barrier height and width d_{eff} depend on the particles energy. So that a higher kinetic energy of the α-particle after decay should correspond to a smaller decay time. This dependence can be measured, and the results on measurements of the characteristic decay time of different radioactive elements and the kinetic energy of the corresponding α-particles were in best agreement to calculations performed on the basis of such a model potential (Georg Gamow 1928).

Let us estimate such a decay time in terms of a simplified model, namely a 1D rectangular potential barrier (Fig. 4.8 on right). We choose the model potential:

$$U(x) = \begin{cases} \infty; \ x < 0 \\ 0; \ 0 \leq x \leq L \ \text{and} \ x > L + d \\ U_0; \ L < x \leq L + d \end{cases}$$

where L corresponds to r_0. Let us perform an estimation of the decay time of the α-decay process, assuming $L = 10^{-12}$ cm; $d = 2*10^{-12}$ cm; $U_0 = 22$ MeV, while the kinetic energy E of the α-particle is 2 MeV.

The following treatment cannot be considered as a stringent quantum mechanical approach; instead, it is rather an illustrative classical picture that makes use of a transmission coefficient estimated in terms of the concept of quantum tunneling. First of all, let us estimate the tunneling probability through the rectangular barrier with width d. From (4.18), we have:

$$T \approx \frac{16E(U_0 - E)}{U_0^2} e^{-2\frac{\sqrt{2m_{\text{alpha}}(U_0-E)}}{\hbar}d}.$$

$$\Rightarrow T \approx 1.32 * e^{-78.5} \approx 1 * 10^{-34}$$

Let us now assume that the emission of the α-particle may be understood as the result of the tunneling of an α-particle through said barrier. If before tunneling, the α-particle is confined in the box potential between $x = 0$ and $x = L$, then it propagates between the walls of the box with a velocity $v = \sqrt{\frac{2E}{m_{\text{alpha}}}}$. Therefore, the time span Δt between two bounces on the barrier is

$$\Delta t = \frac{2L}{v} \approx 3 * 10^{-21} s$$

Per bounce the escaping probability is T. From the rate equation valid for a large number N of emitters $\frac{dN}{dt} = -\frac{1}{\Delta t} T N$, we find that the decay time τ is

$$\tau = \frac{\Delta t}{T} \approx 8.3 * 10^{13} s \approx 9.5 * 10^5 \text{ years}$$

The half-life time is equal to $\tau \ln 2 \approx 6.58 * 10^5$ years. Note that the transmission coefficient T is extremely sensitive to d, which, in practice, is related to the radius of the nucleus. Smallest variation in d or E cause tremendous changes in T, and therefore, half-live times of real nuclei vary between submilliseconds up to more than 10^{11} years [5]. This dynamic range of half-life times is absolutely impossible to be reproduced in terms of classical physics.

Note further that the exponent in (4.18) results in drastic changes of the transmission coefficient, when small changes in U_0, E, d, or m occur. Thus, it has been shown that for the hypothetical emission of single protons, with realistic parameters of the potential wall, one would obtain decay times less than 10^{-12} s. That means that such a decay process would occur practically instantaneously and not on the time scales typically observed for radioactivity.

Note finally that the idea of an α-particle propagating through the nucleus and frequently bouncing on the barrier walls is of course a model idealization; in fact, there is no α-particle migrating through the nucleus prior to its emission. The good reproduction of the characteristic half-times by this model is the more astonishing.

4.7   Tasks for Self-check

4.7.1. Calculate phase and group velocities of a nonrelativistic particle propagating in a constant potential U_0. Hint: make use of (4.2) with the ansatz (4.1) for the wavefunction (after [6])!

4.7.2. Assume a 3D cubic potential box with edge length $L = 2$ nm. How many electrons with energy $E \leq 1$ eV may be placed into this box? (after [7])

4.7.3. A color center in LiF shows maximum absorption at a photon energy of 5 eV, while the corresponding absorption maximum in KBr is observed at approximately 2 eV. Which of the corresponding crystals is expected to have the larger lattice period?

4.7.4. Consider the 1D potential $U(x) = -A\delta(x)$ and determine the eigenenergies of a particle with mass m trapped in such a potential $(A > 0)$! $\delta(x)$ is Diracs delta-function ([4, pp. 862–863], compare also [8]).

References

Specific References

1. Gross/Marx, *Festkörperphysik*, 2nd edn. (de Gruyter, 2014), p.41
2. N. Treitz, Von der Unbestimmtheitsrelation zur Farbe der Tomate, in *Spektrum der Wissenschaft* (Nov. 2010), p. 30–32
3. C. Kittel, *Thermal Physics* (Wiley, 1969), chapter 10
4. M. Bartelmann, B. Feuerbacher, T. Krüger, D. Lüst, A. Rebhan, A. Wipf, *Theoretische Physik*, (Springer Spektrum, 2015)
5. E.W. Schpolski, *Atomphysik II*, (VEB Deutscher Verlag der Wissenschaften, 1956), pp. 453–456,
6. W. Macke, *Quanten. Ein Lehrbuch der theoretischen Physik*, 3. (Auflage, Akademische Verlagsgesellschaft, 1959), p.466
7. W. Demtröder, *Experimentalphysik 3, Atome, Moleküle und Festkörper* (Springer, 2016), task 13.1
8. H.Haken, H.C. Wolf, *The Physics of Atoms and Quanta* (Springer, 2005), task 9.3

General Literature

9. P.A. Tipler, G. Mosca, *Physik, 7* (Springer, Auflage, 2015)
10. J. Bricmont, *Quantum Sense and Nonsense* (Springer, 2017)
11. M. Bartelmann, B. Feuerbacher, T. Krüger, D. Lüst, A. Rebhan, A. Wipf, *Theoretische Physik*, (Springer Spektrum, 2015)
12. L.D. Landau, E.M. Lifshitz, *Quantum Mechanics* (Volume 3 of A Course of Theoretical Physics) (Pergamon Press, 1965)

Operators in Quantum Mechanics and Model System II

5

Abstract

We introduce basic concepts of quantum mechanics. Thereby, the focus is placed on the introduction of skills that are necessary for our further discussion of light–matter interaction phenomena. As a second model system, the harmonic oscillator is discussed in terms of Heisenbergs matrix method as well as in terms of ladder operators. The results are used to illustrate the idea of electromagnetic field quantization.

5.1 Starting Point

What comes now is a chapter crammed with rather abstract mathematical expressions, which will be introduced in order to provide the necessary tools for a quantitative description of the material systems which are in the focus of this book. Whenever you, dear reader, will find it boring or tedious to familiarize with the mathematical apparatus of quantum mechanics, please remember that this is a theory that allows predicting experimental results (among them those concerning the interaction of light with matter) with an accuracy that has never been observed before. This alone makes it worth for every student of photonics obtaining at least basic skills in quantum mechanical calculations.

But what quantum mechanics is about? Newtonian mechanics allow predicting the dynamic behavior of systems of point masses, so we know that it is about systems of point masses. On the contrary, the Schrödingers equation allows calculating a wavefunction, and knowledge of this wavefunction will enable us to predict certain properties of the system that is in a physical state described by this wavefunction. But what is the wavefunction? And has quantum mechanics to be regarded as a theory on wavefunctions?

Once the physical meaning of the complex wavefunction itself is not so obvious, it might seem unsatisfactory to define quantum mechanics as a theory on

© The Author(s), under exclusive license to Springer Nature Switzerland AG 2022
O. Stenzel, *Light–Matter Interaction*, UNITEXT for Physics,
https://doi.org/10.1007/978-3-030-87144-4_5

wavefunctions only. In fact, as a result of this "uncertainty," it is not only one unique quantum theory that has been developed in the past; in fact at present, there are several competing quantum theories [1, pp. 31–45]. A common feature of them is that all they describe most fundamental building units of matter, such as atoms or electrons, as well as electromagnetic radiation, in excellent agreement with experimental findings. But therefore, no experiment available today can verify the correct one or falsify the others, because all these theories provide (at least almost) identical predictions on the result of an experiment.

Let us mention some of these competing quantum theories:

- The orthodox Copenhagen school
- The de Broglie—Bohmian theory
- Collaps theories
- Many-world theories
- And, of course, several others…

I am physicist by education, so that during my studies, I had to listen to a regular one-year (30 weeks) course of quantum mechanics based on the Copenhagen school, consisting of two lectures and one seminar per week. I still remember how fascinated I was about the obtained recipes for exactly calculating optical spectra of atoms, molecules, and solids. Nevertheless I am honest enough to disclose that it became not finally clear to me, what quantum mechanics is basically about (wavefunctions? waves?, particles?, something hybrid?). At the end of the course, I had arrived at the conclusion that quantum mechanics (=wave mechanics) is a theory on all microscopic objects that do never move along a definite path. It is also characteristic that in orthodox quantum mechanics, the concept of the **observable** (something that may *be observed*) has a key function. Note that John Stuart Bell later opposed the concept of the **Beable** (something that may *exist*) to that observable concept.

Among the books I read on the Copenhagen version of quantum mechanics, I would it like to recommend the books written by Landau and Lifshitz as well as that by Bartelmann et al. to the reader (see Sect. 5.8.1).

The de Broglie–Bohmian theory is a nonrelativistic quantum theory of point particles. In this quantum theory (and in contrast to the Copenhagen school), particles do have well-defined coordinates and paths. Here, the point particles are the "Beables," while the Schrödinger equation is necessary to provide a wavefunction that acts as a "pilot wave" guiding the movement of the particle. Thus, the Schrödinger equation is in the basis of both the orthodox and the de Broglie–Bohmian formulations of quantum mechanics. Therefore, their predictions are the same (in the nonrelativistic case).

In collaps theories, the Schrödinger equation is modified; it contains stochastic nonlinear terms. In fact there are a couple of different collaps theories with different fundamental "Beables." We note here that the mentioned modifications of the Schrödinger equation result in small differences between the predictions of collaps theories and the previously mentioned orthodox or de Broglie–Bohmian approaches. When sufficiently accurate experiments will be available, a verification or falsification of the collaps theories should therefore be possible.

In many-world theories, a rather universal wavefunction of the whole universe is the basic object of investigations. It is again the typical linear Schrödinger equation that describes the evolution of that universal wavefunction, and therefore, the predictions will again be the same.

For our practical purposes, the important point is that all these theories make almost identical quantitative predictions when being applied to concrete problems. We will therefore focus our attention to those elements of their mathematical apparatus, which are common to all these theories, namely the solution of the Schrödinger equation (which supplies the wavefunction) and the way how to calculate the system properties when the wavefunction is known. But we will not introduce the mathematical apparatus of quantum mechanics in the formal manner as it is done in every standard textbook for students of physical disciplines; instead, we will restrict on the introduction of a minimum of concepts necessary for our narrow purposes.

As a consequence, the presentation of the material will not adhere to any of the standard textbooks. This may be unsatisfactory for ongoing physicists, but this textbook does not provide a course on quantum mechanics for physicists. And why should I do so? There are enough excellent textbooks on quantum mechanics. The basic material presented in this chapter can already be found in rather old-fashioned textbooks like those written by Landau and Lifshitz, or by Davydov. But I also would like to recommend three more books for further reading:

- J. Bricmont: Quantum sense and nonsense. Read it if you want to know what quantum mechanics is about, practically without referring to mathematics
- S. Flügge: Practical quantum mechanics. Read it if you want to learn solving practical problems
- M. Haug, S.W.Koch, Quantum Theory of the Optical and Electronic Properties of Semiconductors, (World Scientific Singapore, 1990). I recommend it to everybody who wants to get a deeper insight into the quantum picture of light–matter interactions.

The starting point for our following treatment is:

Schrödinger Equation (4.2): Let us regard a simple quantum system (a single-particle propagating in the potential U) which is described by the Hamiltonian $\hat{H}$. Let $\Psi(\mathbf{r}, t)$ be the solution of the corresponding Schrödinger equation:

$$i\hbar \frac{\partial}{\partial t} \Psi(\mathbf{r}, t) = \left[-\frac{\hbar^2}{2m} \Delta + U(\mathbf{r}) \right] \Psi(\mathbf{r}, t) \equiv \hat{H} \Psi(\mathbf{r}, t)$$

We will assume that $\Psi(\mathbf{r}, t)$, which describes a certain quantum state of the system, contains valuable information about the dynamics of the considered quantum system.

Born's rule (4.6): In a 3D situation ($\mathbf{r} = \begin{pmatrix} x \\ y \\ z \end{pmatrix}$), the term

$$\mathrm{d}w = \Psi^*\Psi\mathrm{d}x\mathrm{d}y\mathrm{d}z = \Psi^*\Psi\mathrm{d}V$$

is the probability to observe the single particle in the volume element $\mathrm{d}V$.

We will be working exclusively in the so-called coordinate representation, i.e., all functions will depend on the coordinates that describe the system configuration in space. Through (4.6), the wavefunction in coordinate representation gives a direct idea on the space region where we have a real chance to observe the quantum particle, and we will call that space region an orbital.

5.2   Physical Idea

If we accept the idea that

- a certain physical state of a quantum system is characterized by a certain wavefunction, and
- that wavefunction contains valuable information about the quantum system in that state,

it should be the task to develop a recipe on how to extract that information from the wavefunction. However, the wavefunction obtained from solving (4.2) does not provide knowledge on the exact coordinates of the particle (except the wavefunction is a δ-function). So, the result of a measurement of a coordinate cannot be exactly predicted, but according to (4.6), we can indicate a statistical probability to find the particle in a certain volume element. It is therefore reasonable assuming that the information about other system characteristics may have a statistical character as well.

If so, we can become confronted with two situations:

i. In a given quantum state, it may happen that a single measurement of a certain characteristic gives a well-predictable unambiguous measurement result.
ii. In a given quantum state, it may happen that a single measurement of a certain characteristic has an unpredictable result. That means, that different measurement results are possible, but as a result of the Born's rule, each of them will be observed with a well-defined probability.

If you, dear reader, find this somewhat irritating, then maybe a classical analogy provides a helpful illustration: Imagine the case of free oscillations of a classical (nonrelativistic) 1D harmonic oscillator (e.g., a mass on a spring with a small oscillation amplitude). Its kinetic and potential energies sum up to its total mechanical energy, which is constant in the case that there is <u>no damping</u>. Thus, we have:

$$E = \text{const} = T_{\text{kin}} + U = \frac{m}{2}\dot{x}^2 + \frac{m}{2}\omega_0^2 x^2 = \frac{m}{2}(\dot{x}^2 + \omega_0^2 x^2) \quad (5.1)$$

Let us assume now that we have no indication on the actual value of x. Then, we cannot predict the actual values of U and T_{kin} (as corresponding to situation ii). But nevertheless, we surely know the value of E (as corresponding to situation i), which does not depend on the actual value of x.

In this chapter, we will now turn to the development of a mathematical recipe that allows predicting possible results of a measurement of system characteristics as well as their statistical probability, provided the quantum state of the system (in terms of its wavefunction) is known. We will try to keep this section as short as possible, by writing down the main equations without superfluous prosa.

5.3 Theoretical Considerations

5.3.1 General Properties of the Wavefunction

<u>Normalization</u>: In the given coordinate representation, the wavefunction is normalized—if possible—according to (4.6) and (4.6a).

<u>Superposition principle</u>: Let Ψ_1 and Ψ_2 describe physical states of the regarded system. Then, the wavefunction:

$$\Psi = a_1\Psi_1 + a_2\Psi_2 \quad (5.2)$$

(a_1, a_2-constants) also describes a possible physical state of the system.

Let Ψ_1 describe a physical state of the regarded system. Then, the wavefunction:

$$\Psi = \text{const.} * \Psi_1$$

describes the same state.

Note that in (4.7a), the substitution of a positive quantum number n by its negative counterpart would be equivalent to the multiplication of the wavefunction with the constant value -1, which describes exactly the same quantum state. Therefore, we restricted on positive quantum numbers only when dealing with the particle-in-a-box system.

<u>Construction of the Hamiltonian</u>: The recipe is: Start with the classical Hamilton function, and replace all observables (dynamic variables that can be measured) by

corresponding quantum mechanical operators according to:

coordinate: $x \to \hat{x}$

momentum: $p_x \to \hat{p}_x = -i\hbar \dfrac{\partial}{\partial x}$

angular momentum: $L_x = [\mathbf{r} \times \mathbf{p}]_x \to \hat{L}_x = \hat{y}\hat{p}_z - \hat{z}\hat{p}_y$ (5.3)

5.3.2 Continuity Equation and Probability Current Density

From the Schrödinger equation (4.2)

$$i\hbar \frac{\partial}{\partial t}\Psi = \left[-\frac{\hbar^2}{2m}\nabla^2 + U\right]\Psi$$

we have:

$$\frac{\partial}{\partial t}\Psi = \frac{i\hbar}{2m}\nabla^2\Psi - i\frac{U}{\hbar}\Psi \Rightarrow \Psi^*\frac{\partial}{\partial t}\Psi = \frac{i\hbar}{2m}\Psi^*\nabla^2\Psi - i\frac{U}{\hbar}\Psi^*\Psi$$

$$\frac{\partial}{\partial t}\Psi^* = -\frac{i\hbar}{2m}\nabla^2\Psi^* + i\frac{U}{\hbar}\Psi^* \Rightarrow \Psi\frac{\partial}{\partial t}\Psi^* = -\frac{i\hbar}{2m}\Psi\nabla^2\Psi^* + i\frac{U}{\hbar}\Psi\Psi^*$$

$$\Rightarrow \frac{\partial}{\partial t}(\Psi^*\Psi) + \frac{i\hbar}{2m}(\Psi\nabla^2\Psi^* - \Psi^*\nabla^2\Psi) = 0$$

$$\Rightarrow \frac{\partial}{\partial t}(\Psi^*\Psi) + \mathrm{div}\left[\frac{i\hbar}{2m}(\Psi\nabla\Psi^* - \Psi^*\nabla\Psi)\right] = 0$$

This is a continuity equation like known from the electromagnetic theory. Therefore, when $\Psi^*\Psi$ is associated with a probability density, the term:

$$\mathbf{j} = \frac{i\hbar}{2m}(\Psi\nabla\Psi^* - \Psi^*\nabla\Psi)$$ (5.4)

has to be interpreted as the *probability current density*. Alternative writings are:

$$\mathbf{j} = \frac{i\hbar}{2m}(\Psi\nabla\Psi^* - \Psi^*\nabla\Psi) = \frac{\hbar}{m}\mathrm{Im}(\Psi^*\nabla\Psi) = \frac{1}{m}\mathrm{Re}(\Psi^*\hat{\mathbf{p}}\Psi)$$

Note that for the solution of any one-dimensional time-independent Schrödinger equation $\left[-\frac{\hbar^2}{2m}\frac{d^2}{dx^2} + U(x)\right]\psi = E\psi$, the probability current density turns out to be a constant.

In order to obtain a feeling for $\mathbf{j}$, let us calculate it for a freely propagating particle. Assuming (4.1), we have:

$$\hat{\mathbf{p}}\Psi = \hbar\mathbf{k}\Psi = \mathbf{p}\Psi = m\mathbf{v}\Psi$$

From here:

$$\mathbf{j} = \frac{1}{m}\text{Re}(\Psi^*\hat{\mathbf{p}}\Psi) = \mathbf{v}\text{Re}(\Psi^*\Psi) = \mathbf{v}\Psi^*\Psi = \mathbf{v}|\Psi|^2 \tag{5.4a}$$

Thus, the probability current density of a freely propagating particle appears to be the product of its formally calculated probability density with its velocity and is, of course, constant.

Remark As we have mentioned in Sect. 5.1, in the de Broglie–Bohmian quantum theory, particles are assumed to have coordinates and to move on paths. The latter is determined by the wavefunction; the wave "guides" the particles on their way and thus acts as a "pilot wave." Indeed, for the discussed single particle, we have: $\mathbf{v} = \dot{\mathbf{r}} = \frac{\mathbf{j}}{|\Psi|^2} = \frac{\hbar}{m}\text{Im}(\frac{\nabla\Psi}{\Psi})$, while Ψ follows from the Schrödinger equation $i\hbar\frac{\partial}{\partial t}\Psi = \hat{H}\Psi$. Thus, in the de Broglie–Bohmian quantum theory, the Schrödinger equation delivers the wavefunction, which guides the particle on its path. Note that this interpretation is contrary to what is in the essence of the Copenhagen school; there (5.4a) must not be interpreted as defining a particle's path.

5.3.3 Relevant Properties of Linear Operators

Let us now consider a mathematical operation $\hat{F}$ that transforms a function $u(x)$ into a function $v(x)$:

$$u(x) \underbrace{\rightarrow}_{\text{operation } \hat{F}} v(x)$$

We will then use the writing

$$\hat{F}u(x) = v(x)$$

Which means that the operator $\hat{F}$ is defined through the result of its action on a function $u(x)$, while $v(x)$ is another function of the same coordinate(s) as $u(x)$. (Here and in the following, x may be understood as a single coordinate or as a couple of coordinates describing the system.) The operator $\hat{F}$ is linear, if and only if:

$$\hat{F}(u_1 + u_2) = \hat{F}u_1 + \hat{F}u_2$$
$$\hat{F}(\text{const.} * u_1) = \text{const.} * \hat{F}u_1$$

is fulfilled for any functions $u_1(x)$ and $u_2(x)$. Examples of such operators are:

Example 1 $\hat{F} = \hat{x} \Rightarrow \hat{F}u(x) = v(x) = xu(x)$.

Example 2 $\hat{F} = \hat{p}_x \Rightarrow \hat{F}u(x) = v(x) = -i\hbar\frac{du(x)}{dx}$.

One may also define new operators by the combined application of other operators:

Example 3 $\hat{F} = \hat{x}\hat{p}_x \Rightarrow \hat{F}u(x) = v(x) = -i\hbar x\frac{du(x)}{dx}$.

$$\hat{F} = \hat{p}_x\hat{x} \Rightarrow \hat{F}u(x) = v(x) = -i\hbar\frac{d[xu(x)]}{dx}$$

Example 4

$$= -i\hbar u(x) - i\hbar x\frac{du(x)}{dx} = -i\hbar\left(1 + x\frac{d}{dx}\right)u(x)$$

Example 5 $\hat{F} = \hat{x}\hat{p}_x - \hat{p}_x\hat{x} \Rightarrow \hat{F}u(x) = v(x) = +i\hbar u(x)$.

Example 5 represents a particular case of what we call a *commutation rule* (or simply a commutator):

$$\left[\hat{F}, \hat{G}\right] \equiv \hat{F}\hat{G} - \hat{G}\hat{F}$$

The terminology is:

$$\left[\hat{F}, \hat{G}\right] \equiv \hat{F}\hat{G} - \hat{G}\hat{F} \begin{cases} = 0 : \text{ The operators } \hat{F} \text{ and } \hat{G} \text{ commute} \\ \neq 0 : \text{ The operators } \hat{F} \text{ and } \hat{G} \text{ do not commute} \end{cases}$$

Obviously, because of example 5:

$$\left[\hat{x}, \hat{p}_x\right] \equiv \hat{x}\hat{p}_x - \hat{p}_x\hat{x} = +i\hbar \neq 0 \tag{5.5}$$

The operators $\hat{x}$ and $\hat{p}_x$ do not commute.

Self-adjoint operators:

The operator $\hat{F}$ is self-adjoint, if and only if

$$\int_X u_1(x)^* \hat{F}u_2(x)dx = \int_X u_2(x)\left[\hat{F}u_1(x)\right]^* dx \tag{5.6}$$

holds for any functions $u_1(x)$ and $u_2(x)$. X is the domain of definition of $u_1(x)$ and $u_2(x)$, which are supposed to be square-integrable functions over the full domain of definition now.

Eigenfunctions and eigenvalues:

Let $u(x)$ be a continuous and square-integrable function that suffices:

$$\hat{F}u(x) = fu(x) \tag{5.7}$$

With f—some constant. Then $u(x)$ represents an *eigenfunction* of the operator $\hat{F}$ and f the corresponding *eigenvalue*.

Note that more than one linearly independent eigenfunction may belong to the same eigenvalue. In this case, the eigenvalue is called to be *degenerated*.

Eigenvalues of self-adjoint operators:

Let u be an eigenfunction of $\hat{F}$. Let us regard ((5.6) and set $u_1 = u_2 = u$. Then

$$\int_X u(x)^* \hat{F} u(x) dx = \int_X u(x) \left[\hat{F} u(x) \right]^* dx$$

From (5.7), we find:

$$\int_X u(x)^* \hat{F} u(x) dx = f \int_X u(x)^* u(x) dx$$

$$= \int_X u(x) \left[\hat{F} u(x) \right]^* dx = f^* \int_X u(x) u(x)^* dx$$

Or

$$\left(f - f^* \right) \int_X u(x)^* u(x) dx = 0$$

From here, we find $f = f^*$, i.e., the eigenvalues of a self-adjoint operator are real. The set of eigenvalues of an operator forms the *spectrum of eigenvalues* of that operator. Note that the spectrum of eigenvalues may be discrete or continuous.

Orthogonality of eigenfunctions of self-adjoint operators (discrete spectrum only!).

Let $\hat{F}$ be a self-adjoint operator with eigenfunctions u_1, u_2, u_3, ..., and corresponding discrete eigenvalues $f_1, f_2, f_3,...$ Let us regard two eigenfunctions u_n, u_m with $f_n \neq f_m$. From (5.6) we find:

$$\int_X u_n(x)^* \hat{F} u_m(x) dx = \int_X u_m(x) \left[\hat{F} u_n(x) \right]^* dx$$

$$\Rightarrow f_m \int_X u_n(x)^* u_m(x) dx = f_n \int_X u_m(x) u_n(x)^* dx$$

$$\Rightarrow (f_m - f_n) \int_X u_n(x)^* u_m(x) dx = 0$$

However, we have required $f_n \neq f_m$. Then, it follows that

$$\int_X u_n(x)^* u_m(x) dx = 0 \text{ if } (n \neq m) \tag{5.8}$$

must be fulfilled. Equation (5.8) is an expression of the fact that the eigenfunctions of a self-adjoint operator corresponding to different eigenvalues are mutually orthogonal. If the eigenfunctions are additionally normalized according to (4.6a), we can write:

$$\int_X u_n(x)^* u_m(x) \mathrm{d}x = \delta_{nm} = \begin{cases} 0; & n \neq m \\ 1; & n = m \end{cases} \tag{5.9}$$

Eigenfunctions $u(x)$ that suffice (5.9) are called *orthonormalized*.

Note that the eigenfunctions that belong to degenerated eigenvalues are not automatically orthogonal to each other. But in this case, proper linear combination of the eigenfunctions may be constructed such that the resulting eigenfunctions are again mutually orthogonal.

Let us now recapitulate our treatment from Chap. 4 in view of the results of the operator approach discussed in this section. Obviously, the Hamiltonian as introduced in (4.2) is a linear self-adjoint operator. Hence, it has real eigenvalues, and a set of othonormalized eigenfunctions may always be constructed, as corresponding to the solutions of (4.4) in the stationary case. The Hamiltonian (4.2) itself is a sum of the operators of kinetic and potential energies, and what we have done in Sect. 4.3.1 is to associate its eigenvalues with the allowed energy levels of the quantum system. And the same procedure will now be carried out with other operators representing physical (measurable) quantities: We will represent each of them by a self-adjoint linear operator and will assume that their eigenvalues correspond to the set of possible ("allowed") results of a measurement of that quantity.

5.3.4 Operators with Joint Eigenfunctions

Imagine now that we are able to prepare a quantum system in a well-specified state, described by the wavefunction Ψ. Let us further assume that we want to perform a measurement of the quantity F (coordinate, momentum or the like) in that quantum system. We will call a physical quantity that can be measured an *observable*. The observable F should be associated with the self-adjoint operator $\hat{F}$, with corresponding eigenfunctions Ψ_n and real eigenvalues f_n.

According to Sect. 5.2, it may happen that although the system is prepared in a well-defined manner (with respect to a given wavefunction Ψ), the result of a particular measurement cannot be reliably predicted. Instead, when performing the cycle of system preparation and measurement often enough, it may turn out that different measurement results are recorded but each with a certain individual probability.

However, in special cases, it may be observed that *always* the same measurement result is obtained, i.e., it has the probability 100%. As a postulate, let us now assume that whenever a particular measurement result F is obtained with 100% probability, then the systems quantum state Ψ is identical to an eigenstate Ψ_n of the operator $\hat{F}$, and the measurement result F is the corresponding eigenvalue f_n.

And vice versa, if we prepared an eigenstate Ψ_n of $\hat{F}$, the result of a measurement of F will be the eigenvalue f_n with absolute certainty.

It should be emphasized that the above discussion does not concern questions of imperfections or inaccuracies of the real measurement device and principle. It has to be understood in the sense that the observable F can adopt only such values, which coincide with one of the eigenvalues of $\hat{F}$. As a consequence, only such values are measured.

For that reason, eigenfunction and eigenvalue problems play an exceptional role in quantum mechanics. In order to guarantee real eigenvalues that correspond to real measurement results, the operators associated to the observables should be self-adjoint.

As a consequence, the spectrum of eigenvalues $\{f_n\}$ of the operator $\hat{F}$ is further associated with the full set of possible measurement results of the observable F. From that, we will later conclude that the set of eigenfunctions of $\hat{F}$ forms a basis for the wavefunction Ψ in any quantum state.

Let the eigenfunctions Ψ_n of $\hat{F}$ suffice:

$$\hat{F}\Psi_n = f_n\Psi_n \tag{5.10}$$

With $\{f_n\}$—spectrum of eigenvalues. Let us assume that the spectrum is discrete, and that each eigenvalue corresponds to only one eigenfunction (non-degenerated case). Provided that the Ψ_n form a complete set of eigenfunctions (we will call it a basis), we may write:

$$\Psi = \sum_n a_n\Psi_n \tag{5.11}$$

(It is here assumed that Ψ and Ψ_n are square-integrable functions of the same variables and suffice the same boundary conditions). The coefficients in (5.11) are easily found provided that the Ψ_n are orthonormalized according to (5.9):

$$\Psi_m^*\Psi = \sum_n a_n\Psi_m^*\Psi_n$$

$$\int_V \Psi_m^*\Psi dV = \sum_n a_n \int_V \Psi_m^*\Psi_n dV = \sum_n a_n\delta_{mn} = a_m \ \forall m \tag{5.11a}$$

In that connection, it is interesting to know, whether or not two different observables may deliver well-defined measurement values in a given quantum state. In the context of the above argumentation, in this case, the state of the system should correspond to a common eigenfunction of both operators. Is this possible?

Let us assume that the mentioned two observables correspond to the operators $\hat{F}$ and $\hat{G}$. We restrict on the case of discrete eigenvalue spectra without degeneration. Let us assume, that

$\hat{F}$ and $\hat{G}$ have a complete set of joint eigenfunctions $\{\Psi_n\}$.

We find:

$$\hat{F}\Psi_n = f_n\Psi_n$$

$$\hat{G}\Psi_n = g_n\Psi_n$$

$$\left.\begin{aligned}\Rightarrow \hat{G}\hat{F}\Psi_n = \hat{G}f_n\Psi_n = f_n g_n\Psi_n \\ \hat{F}\hat{G}\Psi_n = \hat{F}g_n\Psi_n = f_n g_n\Psi_n\end{aligned}\right\} \Rightarrow \left(\hat{G}\hat{F} - \hat{F}\hat{G}\right)\Psi_n = 0$$

Once we assume a complete set of common eigenfunctions, any function Ψ may be expanded into a series of Ψ_n. Then

$$\left(\hat{F}\hat{G} - \hat{G}\hat{F}\right)\Psi = \left(\hat{F}\hat{G} - \hat{G}\hat{F}\right)\sum_n a_n\Psi_n = \sum_n a_n\left(\hat{F}\hat{G} - \hat{G}\hat{F}\right)\Psi_n = 0 \Rightarrow \left[\hat{F}, \hat{G}\right] = 0$$

We come to the result that if the operators have a complete set of joint eigenfunctions, they must commute. Let us note in this context, that according to (5.5), the operators of the coordinate and the momentum do not commute. In fact they have no joint eigenfunctions. Consequently, there is no state where well-defined values of both coordinate and momentum may be measured. This results in the formulation of Heisenbergs uncertainty relation (for a short derivation, see Sect. 5.6.1):

$$\langle(\Delta x)^2\rangle\langle(\Delta p_x)^2\rangle \geq \frac{\hbar^2}{4} \tag{5.12}$$

Let us now prove, that if $\left[\hat{F}, \hat{G}\right] = 0$, the operators will have a complete set of joint eigenfunctions. Let Ψ_n be an eigenfunction of $\hat{F}$. Then, from

$$\hat{F}\Psi_n = f_n\Psi_n$$

We have

$$\hat{F}\hat{G}\Psi_n = \hat{G}\hat{F}\Psi_n = f_n\hat{G}\Psi_n$$

That means that $\hat{G}\Psi_n$ is an eigenfunction of $\hat{F}$ corresponding the eigenvalue f_n. But the same is true for Ψ_n. Once we have excluded degeneration, $\hat{G}\Psi_n$ can differ from Ψ_n only by a constant factor. But then we find that

$$\hat{G}\Psi_n = \text{const.} * \Psi_n$$

which means that Ψ_n is also an eigenfunction of $\hat{G}$ independently of the chosen value n.

We come to the result that $\hat{F}$ and $\hat{G}$ have a complete set of joint eigenfunctions if and only if they commute.

Remark In the case of degenerated eigenvalues, the discussion should be modified. Let us assume that the eigenvalue f_n is degenerated. Then, in the general case, $\hat{G}\Psi_n \neq$ const. $* \Psi_n$. In this case, however it turns out that a suitable linear superposition of the wavefunctions Ψ_n that belong to the degenerated eigenvalue f_n allows constructing an eigenfunction of $\hat{G}$. Once the linear superposition of eigenfunctions belonging to the same f_n is per se at the same time an eigenfunction of $\hat{F}$, the conclusion on the existence of joint eigenfunctions of $\hat{F}$ and $\hat{G}$ remains true even in the case of degeneration.

5.3.5 Quantum Mechanical Expectation Values

In quantum mechanics, the expectation value of the observable F in a system described by the wavefunction Ψ is given by (5.13):

$$\langle F \rangle = \int_V \Psi^* \hat{F} \Psi \, dV \tag{5.13}$$

This equation corresponds to the 3D case; in our often-used 1D model, it reads like:

$$\langle F \rangle = \int_X \Psi^* \hat{F} \Psi \, dx \tag{5.13a}$$

In the case that Ψ is an eigenfunction of $\hat{F}$ with the eigenvalue f, we have $\hat{F}\Psi = f\Psi$, and from (5.13), it follows that $\langle F \rangle = \int_V \Psi^* \hat{F} \Psi \, dV = f \int_V \Psi^* \Psi \, dV = f$. Also, we have $\langle F^2 \rangle = \int_V \Psi^* \hat{F} \hat{F} \Psi \, dV = f \int_V \Psi^* \hat{F} \Psi \, dV = f^2$.

Therefore, if Ψ is an eigenfunction of $\hat{F}$, for the variance of the observable F, we obtain:

$$\langle (\Delta F)^2 \rangle = \langle F^2 \rangle - \langle F \rangle^2 = f^2 - f^2 = 0$$

If the wavefunction Ψ in (5.13) does not belong to the set of eigenfunctions of $\hat{F}$, it can be expanded into a series of eigenfunctions of $\hat{F}$. Thus, let us use the expansion (5.11) together with (5.10). That leads to:

$$\langle F \rangle = \int_V \Psi^* \hat{F} \Psi \, dV = \sum_n \sum_m a_n^* a_m \int_V \Psi_n^* \hat{F} \Psi_m \, dV$$
$$= \sum_n \sum_m a_n^* a_m f_m \int_V \Psi_n^* \Psi_m \, dV$$

$$= \sum_n \sum_m a_n^* a_m f_m \delta_{nm} = \sum_n |a_n|^2 f_n \qquad (5.14)$$

Here, we have made use of the orthonormalization condition (5.9). Obviously, the expectation value is composed from all eigenvalues of $\hat{F}$ (i.e., all possible measurement results) entering into (5.14) with specific weights, which correspond to the square of the absolute values of the expansion coefficients in (5.11). When comparing with classical expressions for the expectation value, it becomes clear that the $|a_m|^2$ have the sense of individual probabilities to observe a certain eigenvalue f_m in the given quantum state. Once the set of $\{f_m\}$ contains all principally possible values of F, the probabilities must sum up to 1. Hence, we have:

$$1 = \sum_n |a_n|^2 \qquad (5.15)$$

But this is the condition that in (5.11), the functions Ψ_n form a basis for the expansion of Ψ. Therefore, the expansion (5.11) is consistent within our theoretical treatment.

Again, provided that Ψ is identical with the m-th eigenfunction of $\hat{F}$, then from (5.11) and (5.14) we find:

$$\Psi = \Psi_m \Rightarrow |a_m| = 1; \ a_{n \neq m} = 0 \Rightarrow \langle F \rangle = \sum_n |a_n|^2 f_n = f_m$$

Hence, in an eigenstate, the observable F adopts its eigenvalue with 100% probability. It follows that the recipe (5.13) is fully consistent with the discussion performed at the beginning of Sect. 5.3.4.

Remark What happens to the wavefunction of a quantum system when a measurement is performed? And even more generally: What is a measurement?

In fact, these questions are still debated, and the provided answers may be used to distinguish between different quantum theories. Although this topic is far beyond the scope of this textbook, I will try to give a short idea on what is the problem.

Let us have a closer look at (5.11).

$$\Psi = \sum_n a_n \Psi_n$$

Let us assume that prior to the measurement, the system was in a state described by (5.11). Let the measurement of F result in the value f_m. What is the state of the system after the measurement?

The more or less generally accepted answer is:

$$\Psi|_{\text{after measurement}} = \Psi_m$$

This equation clearly asserts that the measurement performed on a quantum state is invasive, i.e., it changes the state of the quantum system. The transformation

$\Psi \to \Psi_m$ is called a collapse of the initial wavefunction Ψ. But what is the concrete physical mechanism behind the transformation:

$$\Psi = \sum_n a_n \Psi_n \underset{???}{\longrightarrow} \Psi = \Psi_m ???$$

It turns out that this collapse of the wavefunction cannot be explained as a result of the time evolution defined by the linear Schrödinger (4.2) only. Therefore, that wavefunction collapse is introduced as an independent postulate of the Copenhagen school, i.e., as an independent addendum to the Schrödinger equation. Thus, when a measurement is performed, the Schrödinger time evolution is replaced by the postulated spontaneous collapse of the wavefunction, which formally solves the problem. Provided that we have a consistent idea on which physical processes may be regarded as a measurement.

Alternative quantum theories provide alternative suggestions. Thus, in the collapse theories, the collapse of the wavefunction $\Psi = \sum_n a_n \Psi_n \underset{\text{collapse}}{\longrightarrow} \Psi_m$ occurs as a result of additional *nonlinear* terms in the Schrödinger equation.

In many-world theories, no fundamental collapse of the wavefunction occurs. It is instead assumed that whenever we measure the result "f_m"; so this measurement result is obtained only in "our world," while the other possible results $f_{n \neq m}$ are observed in "other" worlds. So the wavefunction does not collapse but splits into different branches, each of them corresponding to another world. Of course, from the viewpoint of one selected world, say the m-th one, a wavefunction collaps according to $\Psi \to \Psi_m$ has occurred.

In the de Broglie–Bohmian theory, it is however assumed that the wavefunction alone does *not* contain the full information on the system but is complemented by an equation describing the real coordinates of the quantum particles, i.e., the real configuration of the system. During and after the measurement, the wavefunction of the whole system (i.e., quantum system + measurement apparatus) naturally evolves according to the linear Schrödinger (4.2), i.e., it does not suffer any spontaneous collapse. Whenever the value "f_m" is measured, it has the meaning that there is a real configuration of the system as defined by its coordinates which corresponds to that measurement result. In this sense, one could say that it is only the "branch" Ψ_m of the full wavefunction (5.11) that "guides" the dynamics of the system, while the other branches $\Psi_{n \neq m}$ are "empty" and represent what is sometimes called "ghost wavefunctions." But these ghost wavefunctions are of no practical relevance. This can be shown in terms of a stringent Bohmian approach to the theory of measurement processes in quantum mechanics, which leads to the concept of an *effective* collapse of the wavefunction.

Note that all these theories represent refined theoretical buildings and are much more complex as it might seem from the few rows written here. Interested readers are kindly referred to the recommended books for getting more information.

5.3.6 Ehrenfests Theorems

In quantum mechanics, the derivative of an observable with respect to time is defined through its expectation value. It is defined according to:

$$\langle \dot{F} \rangle = \frac{\mathrm{d}}{\mathrm{d}t} \langle F \rangle \tag{5.16}$$

Hence, whenever we can write:

$$\frac{\mathrm{d}}{\mathrm{d}t} \langle F \rangle = \int_V \Psi^* \hat{K} \Psi \mathrm{d}V \equiv \langle K \rangle \tag{5.17}$$

From (5.16), it follows that

$$\frac{\mathrm{d}}{\mathrm{d}t} \hat{F} = \hat{K} \tag{5.17a}$$

Let us now assume that the operator $\hat{F}$ does not explicitly depend on time. Once the state Ψ in (5.17) must suffice the Schrödinger equation, from (5.16) and (5.13), we have:

$$\frac{\mathrm{d}}{\mathrm{d}t} \langle F \rangle = \frac{\mathrm{d}}{\mathrm{d}t} \int_V \Psi^* \hat{F} \Psi \mathrm{d}V$$

$$= \int_V \left(\frac{\partial}{\partial t} \Psi^* \right) \hat{F} \Psi \mathrm{d}V + \int_V \Psi^* \hat{F} \left(\frac{\partial}{\partial t} \Psi \right) \mathrm{d}V$$

The derivatives of the wavefunction with respect to time may be substituted according to (4.2). We find:

$$\frac{\mathrm{d}}{\mathrm{d}t} \langle F \rangle = \frac{i}{\hbar} \left\{ \int_V \left(\hat{H} \Psi \right)^* \hat{F} \Psi \mathrm{d}V - \int_V \Psi^* \hat{F} \left(\hat{H} \Psi \right) \mathrm{d}V \right\}$$

Let us now remember that the Hamiltonian is self-adjoint (condition (5.6)). We then obtain:

$$\int_V \left(\hat{H} \Psi \right)^* \hat{F} \Psi \mathrm{d}V = \int_V \Psi^* \hat{H} \hat{F} \Psi \mathrm{d}V$$

$$\Rightarrow \frac{\mathrm{d}}{\mathrm{d}t} \langle F \rangle = \frac{i}{\hbar} \int_V \Psi^* \left(\hat{H} \hat{F} - \hat{F} \hat{H} \right) \Psi \mathrm{d}V = \frac{i}{\hbar} \int_V \Psi^* \left[\hat{H}, \hat{F} \right] \Psi \mathrm{d}V$$

Hence,

$$\frac{d}{dt}\langle F\rangle = \frac{i}{\hbar}\left\langle\left[\hat{H}, \hat{F}\right]\right\rangle \tag{5.18}$$

Thus, whenever the operators $\hat{H}$, $\hat{F}$ commute, the expectation value of F does not change with time (provided that $\hat{F}$ does not explicitly depend on the time).

Compared to (5.17), we can also write:

$$\frac{d}{dt}\hat{F} = \frac{i}{\hbar}\left[\hat{H}, \hat{F}\right] \tag{5.18a}$$

When setting $F = x$, from (5.18), we immediately find:

$$\frac{d}{dt}\langle x\rangle = \frac{i}{\hbar}\left\langle\left[\hat{H}, \hat{x}\right]\right\rangle$$

When substituting the Hamiltonian (5.1) into this equation, we obtain:

$$\frac{d}{dt}\langle x\rangle = \frac{i}{\hbar}\left\langle\left[\hat{H}, \hat{x}\right]\right\rangle = -\frac{i\hbar}{m}\left\langle\frac{\partial}{\partial x}\right\rangle = \frac{\langle p_x\rangle}{m} \tag{5.19}$$

Hence, the time derivative of the expectation value of the coordinate, multiplied with the particles mass, is equal to the expectation value of the momentum. In an analogous way, from (5.18), we find for the momentum:

$$\frac{d}{dt}\langle p_x\rangle = \frac{i}{\hbar}\left\langle\left[\hat{H}, \hat{p}_x\right]\right\rangle = -\left\langle\frac{\partial U}{\partial x}\right\rangle \tag{5.20}$$

Equations (5.19) and (5.20) form the essence of Ehrenfests theorems.

Remark When in (5.20) replacing the momentum by $m dx/dt$ and using (5.16), we obtain:

$$m\frac{d^2}{dt^2}\langle x\rangle = -\left\langle\frac{\partial U}{\partial x}\right\rangle \tag{5.20a}$$

This is similar to the classical Newtons equation of motion:

$$m\frac{d^2}{dt^2}x = -\frac{\partial U}{\partial x} \tag{5.20b}$$

In this sense, one might assume that (5.20a) describes some converging of quantum mechanics to classical physics, where instead of the coordinate x itself, its expectation value $<x>$ is used as the variable. However, this would only be true when the relation

$$\left\langle\frac{\partial U(x)}{\partial x}\right\rangle \approx \frac{\partial U}{\partial x}(\langle x\rangle) \tag{5.20c}$$

would be fulfilled [1, p. 19]. According to (5.13), this requires the wavefunction to be strongly localized in the vicinity of <x>. Here, the picture of the wavepacket is illustrative. Indeed, when identifying a propagating particle with a well localized wave packet, (5.20c) would be fulfilled, and hence

$$m \frac{d^2}{dt^2} \langle x \rangle \approx -\frac{\partial U}{\partial x}(\langle x \rangle) \qquad (5.20d)$$

would be valid, which is a direct analog on to the classical relation (5.20b). As we have seen at the end of Sect. 3.6, for a macroscopic particle, the wave packet remains localized during propagation, such that a classical description makes sense. For an electron, however, the wave packet tends to broaden at a very short time scale, such that (5.20c) and (5.20d) become invalid. Hence, it is the wave description that should be used for such kind of microscopic particle, and (5.20) or (5.20a) do not describe any convergence to classical equations of motion.

Without proof, let us mention a further important property of quantum mechanical expectation values. Whenever the movement of a single quantum particle is spatially confined in a potential $U \propto r^k$, as a result of the quantum mechanical version of the virial theorem, the following relation holds in any possible state:

$$2 \langle T_{kin} \rangle = k \langle U \rangle \qquad (5.20e)$$

Let us close this section by mentioning the Hellmann–Feynman theorem: Provided that the time-independent Hamiltonian of the regarded system depends on a parameter Λ, then the following relationship is valid:

$$\frac{\partial E_n}{\partial \Lambda} = \left\langle \frac{\partial \hat{H}}{\partial \Lambda} \right\rangle_n \qquad (5.20f)$$

Here, the subscript n indicates that the expectation value is obtained in the n th eigenstate. The proof of (5.20e) is proposed as task 5.7.17.

5.3.7 Matrix Representation

From (4.3) and (4.4), we have (1D; $\frac{\partial U}{\partial t} = 0$):

$$\Psi_n(x, t) = e^{-i \frac{E_n}{\hbar} t} \psi_n(x)$$
$$\hat{H} \psi_n(x) = E_n \psi_n(x) \qquad (5.21)$$

The wavefunction (5.21) is called the wavefunction of a *stationary state*. The stationary state with the lowest energy is called the *ground state*.

When expanding Ψ into the basis provided by (5.21), we have:

$$\Psi = \sum_n a_n \Psi_n(x, t) = \sum_n a_n e^{-i\frac{E_n}{\hbar}t} \psi_n(x) \tag{5.22}$$

And generally (3D)

$$\langle F \rangle = \int_V \Psi^* \hat{F} \Psi dV = \sum_n \sum_m a_n^* a_m \int_V \Psi_n^* \hat{F} \Psi_m dV$$

$$= \sum_n \sum_m a_n^* a_m F_{nm}(t)$$

$$\int_V \Psi_n^* \hat{F} \Psi_m dV = F_{nm}(t)$$

$$= e^{i\frac{E_n - E_m}{\hbar}t} \int_V \psi_n^* \hat{F} \psi_m dV = e^{i\omega_{nm}t} F_{nm}$$

$$\omega_{nm} = \frac{E_n - E_m}{\hbar} \tag{5.23}$$

Here, we have introduced:
$\{F_{nm}(t)\}$—matrix of the operator $\hat{F}$.
$F_{nm}(t)$—the *matrix elements*.
F_{nm}—the time-independent *matrix elements*.
For $n \neq m$, those matrix elements are also called *transition matrix elements*.
$\omega_{nm} = \frac{E_n - E_m}{\hbar}$—*transition angular frequency*
When introducing the derivative of $\hat{F}$ with respect to time according to:

$$\langle \dot{F} \rangle = \frac{d}{dt} \langle F \rangle = \sum_n \sum_m a_n^* a_m \dot{F}_{nm}(t) \tag{5.24}$$

We find

$$\dot{F}_{nm}(t) = i\omega_{nm} F_{nm}(t)$$

On the other hand, the time-dependent matrix element of $\dot{F}$ may also be written as the product of the time-independent matrix element and the corresponding oscillating exponential term according to:

$$\dot{F}_{nm}(t) = \left(\dot{F}\right)_{nm} e^{i\omega_{nm}t}$$

where $\left(\dot{F}\right)_{nm}$ denotes the time-independent matrix element of the operator $\dot{F}$. Thus, we have

$$\dot{F}_{nm}(t) = i\omega_{nm} F_{nm}(t) = i\omega_{nm} F_{nm} e^{i\omega_{nm}t} = \left(\dot{F}\right)_{nm} e^{i\omega_{nm}t}$$

or

$$\left(\dot{F}\right)_{nm} = i\omega_{nm} F_{nm} \tag{5.25}$$

Remark Note that the matrix elements of a product of two operators are calculated according to the usual recipe for matrix multiplication. Indeed, let us expand the function $\hat{F}\psi_n$ into a series of eigenfunctions ψ_m. From (5.11), we have:

$$\hat{F}\psi_n = \sum_m F_{mn}\psi_m$$

Then:

$$\hat{F}\hat{G}\psi_n = \sum_m (FG)_{mn}\psi_m = \hat{F}\left(\hat{G}\psi_n\right)$$

$$= \hat{F}\sum_k G_{kn}\psi_k = \sum_k G_{kn}\hat{F}\psi_k$$

$$= \sum_k \sum_m G_{kn}F_{mk}\psi_m = \sum_m \sum_k F_{mk}G_{kn}\psi_m = \sum_m \left(\sum_k F_{mk}G_{kn}\right)\psi_m$$

$$\Rightarrow (FG)_{mn} = \sum_k (F_{mk}G_{kn})$$

The latter expression coincides with the well-known rule for matrix multiplication.

5.3.8 Time-Independent First-Order Perturbation Theory (Without Degeneration)

So far we have implicitly assumed that the Schrödinger (4.2) can be solved analytically in an exact manner. This is, however, by far not always the case. Often enough an analytical solution of the Schrödinger equation cannot be indicated. The perturbation theory provides a tool for an approximate solution of the Schrödinger equation in such a case.

Let us assume the time-independent Hamiltonian:

$$\hat{H} = \hat{H}^{(0)} + \hat{V} \tag{5.26}$$

Let us further assume that the time-independent Schrödinger equation:

$$\hat{H}^{(0)}\psi^{(0)} = E^{(0)}\psi^{(0)}$$

may be solved exactly. We will now consider $\hat{H}^{(0)}$ as the Hamiltonian of what we call the unperturbed system, while $\hat{V}$ represents what we call the perturbation.

In practice, it will often happen that the Schrödinger equation of the perturbed system:

$$\hat{H}\psi = E\psi$$

cannot be solved exactly. We will therefore search an approximate solution.

First of all, let us represent the unknown wavefunction ψ as a superposition of the wavefunctions of the unperturbed system:

$$\psi = \sum_m a_m \psi_m^{(0)}$$

(a_m –constant coefficients). Substitution into the Schrödinger equation leads to:

$$\hat{H}\psi = E\psi = \sum_m a_m \left(\hat{H}^{(0)} \psi_m^{(0)} + \hat{V} \psi_m^{(0)} \right)$$

$$= \sum_m a_m \left(E_m^{(0)} \psi_m^{(0)} + \hat{V} \psi_m^{(0)} \right) = E \sum_m a_m \psi_m^{(0)}$$

When multiplying this equation from the left with $\left[\psi_k^{(0)} \right]^*$ and integrating over the full volume, as the result of the orthonormalization of the unperturbed wavefunctions, we obtain:

$$a_k \left[E - E_k^{(0)} \right] = \sum_m a_m V_{km}^{(0)}$$

Here, the perturbation matrix element is calculated with respect to the unperturbed wavefunctions, i.e.,

$$V_{km}^{(0)} = \int \left[\psi_k^{(0)} \right]^* \hat{V} \psi_m^{(0)} \, \mathrm{d}V$$

With V-volume in the case of a single-particle system. Do not confuse with the perturbation operator $\hat{V}$.

Let us now start with approximations. We assume that the perturbation is "small." Therefore we postulate that the perturbation results in a small change $E^{(1)}$ in the energy states of the system, compared with the unperturbed one. So we postulate:

$$E \approx E^{(0)} + E^{(1)}$$

Let us further assume, that in a zeroth approximation, ($\hat{V} = 0$), the system is in the n-th unperturbed eigenstate. Hence, in zeroth order,

$$E = E_n^{(0)}; \quad a_n^{(0)} = 1; \quad a_{m \neq n}^{(0)} = 0$$

In first approximation, we set accordingly:

$$a_m = a_m^{(0)} + a_m^{(1)}$$

Setting now $k = n$, we find:

$$a_n\left[E - E_n^{(0)}\right] \approx \sum_m a_m V_{nm}^{(0)}$$

$$\Rightarrow \left[a_n^{(0)} + a_n^{(1)}\right]\left[E_n^{(0)} + E_n^{(1)} - E_n^{(0)}\right] \approx \sum_m \left[a_m^{(0)} + a_m^{(1)}\right] V_{nm}^{(0)}$$

$$\Rightarrow \left[1 + a_n^{(1)}\right]\left[E_n^{(1)}\right] \approx \sum_m \left[\delta_{nm} + a_m^{(1)}\right] V_{nm}^{(0)}$$

When keeping only terms which are small to the first order, we have immediately:

$$E_n^{(1)} \approx V_{nn}^{(0)} \tag{5.27}$$

So that the first-order correction to the nth eigenenergy value is identical to the matrix element $V_{nn}^{(0)}$.

Setting now $k \neq n$, we find:

$$a_k\left[E - E_k^{(0)}\right] \approx \sum_m a_m V_{km}^{(0)}$$

$$\Rightarrow \left[a_k^{(0)} + a_k^{(1)}\right]\left[E_n^{(0)} + E_n^{(1)} - E_k^{(0)}\right] \approx \sum_m \left[a_m^{(0)} + a_m^{(1)}\right] V_{km}^{(0)}$$

$$\Rightarrow a_k^{(1)}\left[E_n^{(0)} + E_n^{(1)} - E_k^{(0)}\right] \approx \sum_m \left[\delta_{nm} + a_m^{(1)}\right] V_{km}^{(0)}$$

When keeping only terms which are small to the first order, we have immediately:

$$a_k^{(1)}\left[E_n^{(0)} - E_k^{(0)}\right] \approx V_{kn}^{(0)} \Rightarrow a_k^{(1)} \approx \frac{V_{kn}^{(0)}}{E_n^{(0)} - E_k^{(0)}}$$

Note that this is a "small" correction only when the condition $\left|\frac{V_{kn}^{(0)}}{E_n^{(0)} - E_k^{(0)}}\right| \ll 1$ is satisfied. This condition must be fulfilled in order to apply the derived first order perturbation approach. Note that the presented approach cannot be applied to systems with degenerated energy levels!

Note further that the expansion coefficient $a_n^{(1)}$ remains undefined this way. For normalization reasons, it should be set equal to zero. Indeed, when setting as before:

$\psi = \sum_m a_m \psi_m^{(0)} \equiv \psi_n^{(0)} + \psi_n^{(1)}$, we find:

$$\int |\psi|^2 dV = \int \left(\psi_n^{(0)} + \psi_n^{(1)} \right)^* \left(\psi_n^{(0)} + \psi_n^{(1)} \right) dV$$

$$= \underbrace{\int \left| \psi_n^{(0)} \right|^2 dV}_{=1} + \int \left(\psi_n^{(0)*} \psi_n^{(1)} + \psi_n^{(0)} \psi_n^{(1)*} \right) dV + \underbrace{\int \left| \psi_n^{(1)} \right|^2 dV}_{\text{neglected}}$$

Thus, normalization seems no more guaranteed. But if $\psi_n^{(1)}$ is assumed to be small in first-order, second-order terms may be neglected, and normalization may be preserved in first-order perturbation theory when $\int \left(\psi_n^{(0)*} \psi_n^{(1)} \right) dV = 0$ is postulated. From here, as the result of the orthonormalization of the unperturbed wavefunctions, it follows that $a_n^{(1)} = 0$.

5.4 Consistency Considerations

Let us start with a short resume here. In Sect. 2.4, we have clearly demonstrated that a purely classical approach cannot explain basic experimental data concerning the interaction of electromagnetic radiation with matter. We have claimed that we need another theory for that.

In Sect. 3.4, we have collected indications on the relevance of wave characteristics for the description of the dynamics of matter. Therefore, the idea was to develop a theory that is based on a wave equation.

In Sect. 4.4, it was emphasized that this wave equation cannot be derived in a stringent manner but is rather guessed. The solution of the Schrödinger equation for a simple box potential enabled us later to understand the characteristics of several real physical phenomena, among them the α-decay. We have emphasized that it is the consistency of the theoretical predictions of quantum mechanics with experimental data that certifies its application.

In this sense, what we have done in this chapter is the consistent continuation of this logical path. We have formulated the Schrödinger equation, we have worked out mathematical tools that shall help us to find its solutions, and we have formulated recipes to gain further information on the physical system characteristics from the solution of the Schrödinger equation.

We will proceed now with two concrete examples, namely the harmonic and anharmonic oscillators. We will demonstrate that the harmonic oscillator may be treated in terms of the theoretical material introduced in Sect. 5.3.7. In order to tackle the anharmonic oscillator, the perturbation approach form Sect. 5.3.8 will find application.

5.5   Application to Practical Problems

5.5.1   Model System II: Harmonic Oscillator (1D)

For the 1D harmonic oscillator, we have the Hamiltonian (compare Sect. 2.5.1):

$$\hat{H} = \left[-\frac{\hbar^2}{2m}\frac{d^2}{dx^2} + \frac{m\omega_0^2}{2}x^2 \right] \tag{5.28}$$

Fortunately, energy levels and coordinate matrix elements of the harmonic oscillator may be obtained without explicitly solving Schrödingers equation by applying the matrix technique (Heisenberg 1925). Let us start from differentiating (5.1) with respect to time. That leads to:

$$\ddot{x} + \omega_0^2 x = 0$$

For the matrix elements, we correspondingly have:

$$(\ddot{x})_{nm} + \omega_0^2 x_{nm} = 0$$

From (5.25), we find:

$$(\ddot{x})_{nm} = i\omega_{nm}(\dot{x})_{nm} = -\omega_{nm}^2 x_{nm} \Rightarrow x_{nm}\left(\omega_0^2 - \omega_{nm}^2\right) = 0 \Rightarrow$$
$$x_{nm} = 0 \text{ if } \omega_{nm} \neq \pm\omega_0$$

Let us introduce a counting rule according to:

$$\omega_{nm} = \pm\omega_0 \Leftrightarrow m = n \mp 1 \tag{5.29}$$

As a consequence, for a harmonic oscillator, only those matrix elements x_{nm} may be different from zero where $m = n \pm 1$ holds.

According to the definition (5.6), for any self-adjoint operator $\hat{F}$, we find: $F_{nm} = F_{mn}^*$. When assuming real wavefunctions in our case, as a direct consequence of the self-adjoint nature of the participating operators, we have $x_{nm} = x_{mn}$.

Let us now make use of the Ehrenfest theorem (5.19)

$$\frac{d}{dt}\langle x \rangle = \frac{\langle p \rangle}{m}$$

Then, from (5.5), (5.17), (5.17a), we find the commutation rule:

$$\hat{x}\hat{\dot{x}} - \hat{\dot{x}}\hat{x} = i\frac{\hbar}{m} \Rightarrow$$
$$(x\dot{x})_{nn} - (\dot{x}x)_{nn} = i\frac{\hbar}{m}$$

From here, considering (5.25), we find:

$$i \sum_l (x_{nl} \omega_{l\,n} x_{l\,n} - \omega_{nl} x_{nl} x_{l\,n}) = i \frac{\hbar}{m} = 2i \sum_l \omega_{l\,n} x_{nl}^2 \qquad (5.30)$$

$$\Rightarrow (x_{n+1,n})^2 - (x_{n,n-1})^2 = \frac{\hbar}{2m\omega_0}$$

This way we obtained a recursive recipe for calculating the matrix elements. What we need is an initial value. As a postulate, we set: $x_{0,-1} = 0$ and obtain:

$$(x_{n,n-1})^2 = \frac{n\hbar}{2m\omega_0} \qquad (5.31)$$

Remark When introducing the so-called *oscillator strength* $f_{l\,n}$ according to:

$$f_{l\,n} = \frac{2m}{\hbar} \omega_{l\,n} |x_{nl}|^2 \qquad (5.32)$$

Note that according to (5.23), we have $f_{l\,n} = -f_{nl}$, and $f_{ll} = 0$. Then, from (5.30), the so-called sum rule for the oscillator strength is immediately obtained:

$$\sum_l f_{l\,n} = \frac{2m}{\hbar} \sum_l \omega_{l\,n} |x_{nl}|^2 = 1 \qquad (5.33)$$

The energy level E_n corresponds to the expectation value of the Hamiltonian in the stationary state described by ψ_n, i.e., by its own n-th eigenfunction. We find:

$$\begin{aligned}
E_n = H_{nn} &= \frac{m}{2}\left[(\dot{x}^2)_{nn} + \omega_0^2 (x^2)_{nn}\right] \\
&= \frac{m}{2} \sum_l \left(i\omega_{n\,l} x_{nl} i\omega_{l\,n} x_{l\,n} + \omega_0^2 x_{nl} x_{l\,n}\right) \\
&= \frac{m}{2} \sum_{l=n\pm 1} \left(\omega_{nl}^2 + \omega_0^2\right) x_{nl}^2 \\
&= m\omega_0^2 \left[\frac{n\hbar}{2m\omega_0} + \frac{(n+1)\hbar}{2m\omega_0}\right] = \frac{\omega_0}{2}\hbar(2n+1) \\
&= \hbar\omega_0\left(n + \frac{1}{2}\right) \qquad (5.34)
\end{aligned}$$

We find that the Hamiltonian of the harmonic oscillator (5.28) has a discrete and equidistant eigenspectrum with ground-state energy $E_0 = \frac{\hbar\omega_0}{2}$ (see Fig. 5.1). Note that obviously, the parameter $x_0 \equiv \sqrt{\frac{\hbar}{m\omega_0}}$ is given in length units and corresponds

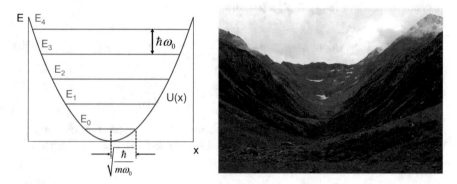

Fig. 5.1 Left: Potential curve and energy levels of the harmonic oscillator. Right: Example of a nearly parabolic potential wall with macroscopic dimensions in nature: The Wimmertal in the Alps, Austria

to the elongation from the potential minimum such that $U(x_0) = E_0 = \frac{\hbar\omega_0}{2}$ is fulfilled.

Without derivation, let us write down the eigenfunctions of (5.28). They are given by Hermitian polynomials H_n as:

$$\psi_n(x) = \left(\frac{m\omega_0}{\pi\hbar}\right)^{\frac{1}{4}} \frac{1}{\sqrt{2^n n!}} e^{-\frac{m\omega_0}{2\hbar}x^2} H_n\left(x\sqrt{\frac{m\omega_0}{\hbar}}\right);$$

with

$$H_n(\xi) = (-1)^n e^{\xi^2} \frac{d^n e^{-\xi^2}}{d\xi^n} \tag{5.35}$$

Remark Although we will not too often make explicit use of it, let us shortly remark the Dirac notation used in many books on quantum mechanics. When introducing the Ket-Vector as:

$$\Psi \rightarrow |\Psi\rangle$$

Table 5.1 Presents lowest-order Hermitian polynomials in an explicit manner

n	E_n	$H_n(\xi)$
0	$\frac{1}{2}\hbar\omega$	1
1	$\frac{3}{2}\hbar\omega$	2ξ
2	$\frac{5}{2}\hbar\omega$	$(4\xi^2 - 2)$
3	$\frac{7}{2}\hbar\omega$	$(8\xi^3 - 12\xi)$

And the Bra-vector:

$$\Psi^* \rightarrow \langle\Psi|$$

The eigenvalue (5.10) $\hat{F}\Psi_n = f_n\Psi_n$.
Can be written as:
$\hat{F}|\Psi_n\rangle = f_n|\Psi_n\rangle$, or simply as:
$\hat{F}|n\rangle = f_n|n\rangle$. Instead of (5.11), we have:

$$|\Psi\rangle = \sum_n a_n|n\rangle \Rightarrow$$

$$\langle m \mid \Psi\rangle = \int_V \Psi_m^* \Psi \, dV$$

$$= \sum_n a_n \int_V \Psi_m^* \Psi_n \, dV = \sum_n a_n\langle m \mid n\rangle = \sum_n a_n\delta_{nm}$$

$$\Rightarrow a_n = \int_V \Psi_n^* \Psi \, dV = \langle n \mid \Psi\rangle$$

Note that $\langle m \mid n\rangle = \langle n \mid m\rangle^*$ holds. For matrix elements, we have:

$$F_{nm}(t) = \int_V \Psi_n^* \hat{F}\Psi_m \, dV = \langle n \left| \hat{F} \right| m\rangle$$

Note further, that in the Dirac notation, an operator $\hat{F}$ may act "to the right side," but also "to the left side." Hence,
$\langle n \left| \hat{F} \right| m\rangle = \langle a \mid m\rangle = \langle n \mid b\rangle$ with $\langle a| = \langle n|\hat{F}$ and $|b\rangle = \hat{F}|m\rangle$.
As an example, let us rederive (5.27) in the Dirac notation. In first order, we set

$$\psi = \psi_n^{(0)} + \psi_n^{(1)}$$
$$\hat{H} = \hat{H}_0 + \hat{V}$$

where $\hat{V}$ and $\psi_n^{(1)}$ are supposed to be small. Thereby, according to Sect. 5.3.8,
$\int\left(\psi_n^{(0)*}\psi_n^{(1)}\right) dV = 0 = \left\langle\psi_n^{(0)} \mid \psi_n^{(1)}\right\rangle$. Then, keeping only terms small in first order, we have:

$$E \equiv E^{(0)} + E^{(1)} = \left\langle\psi\left|\hat{H}\right|\psi\right\rangle = \left\langle\psi_n^{(0)} + \psi_n^{(1)}\left|\hat{H}_0 + \hat{V}\right|\psi_n^{(0)} + \psi_n^{(1)}\right\rangle$$

$$= \underbrace{\left\langle\psi_n^{(0)}\left|\hat{H}_0\right|\psi_n^{(0)}\right\rangle}_{=E^{(0)}} + \underbrace{\left\langle\psi_n^{(1)}\left|\hat{H}_0\right|\psi_n^{(0)}\right\rangle}_{=E^{(0)}\left\langle\psi_n^{(1)}\middle|\psi_n^{(0)}\right\rangle=0} + \underbrace{\left\langle\psi_n^{(0)}\left|\hat{H}_0\right|\psi_n^{(1)}\right\rangle}_{=0} + \underbrace{\left\langle\psi_n^{(1)}\left|\hat{H}_0\right|\psi_n^{(1)}\right\rangle}_{\text{neglected}}$$

$$+ \left\langle \psi_n^{(0)} \left| \hat{V} \right| \psi_n^{(0)} \right\rangle + \underbrace{\left\langle \psi_n^{(0)} \left| \hat{V} \right| \psi_n^{(1)} \right\rangle + \left\langle \psi_n^{(1)} \left| \hat{V} \right| \psi_n^{(0)} \right\rangle + \left\langle \psi_n^{(1)} \left| \hat{V} \right| \psi_n^{(1)} \right\rangle}_{\text{neglected}}$$

$$\Rightarrow E \equiv E^{(0)} + E^{(1)} \approx E^{(0)} + \left\langle \psi_n^{(0)} \left| \hat{V} \right| \psi_n^{(0)} \right\rangle$$

Which coincides with (5.27).

5.5.2   Anharmonic Oscillator

As the next example, let us regard the model case of an anharmonic oscillator described by the model potential:

$$\hat{H}_{\text{anharm}} = \left[-\frac{\hbar^2}{2m} \frac{d^2}{dx^2} + \frac{m\omega_0^2}{2} x^2 + \beta x^4 \right] \tag{5.36}$$

Here, β is some small parameter. The Hamiltonian is thus composed from that of a harmonic oscillator ($= \hat{H}^{(0)}$) and a small perturbation, while the perturbation $\hat{V}$ is given by:

$$\hat{V} = \beta x^4$$

According to (5.27), the first-order correction to the energy levels of a harmonic oscillator is now given by:

$E_n^{(1)} \approx V_{nn} = \beta \left(x^4 \right)_{nn}$, where $|n\rangle$ is an eigenfunction of the harmonic oscillator according to (5.35).

The calculation of the matrix element is straightforward. We have:

$$\left(x^2 \right)_{n,n} = \sum_j x_{nj} x_{jn} = \left(x_{n,n-1} \right)^2 + \left(x_{n,n+1} \right)^2 = \frac{n\hbar}{2m\omega_0} + \frac{(n+1)\hbar}{2m\omega_0} = \frac{(2n+1)\hbar}{2m\omega_0}$$

And $\left(x^4 \right)_{nn} = \left(x^2 x^2 \right)_{nn} = \sum_j \left(x^2 \right)_{nj} \left(x^2 \right)_{jn} = \sum_l \left[\left(x^2 \right)_{nl} \right]^2$

$$\left(x^2 \right)_{nl} = \sum_{j=n\pm1} x_{nj} x_{jl} = x_{n,n-1} x_{n-1,l} + x_{n,n+1} x_{n+1,l} \Rightarrow l = n, n \mp 2$$

$$\left(x^4 \right)_{nn} = \sum_{l=n,n\mp2} \left[\left(x^2 \right)_{nl} \right]^2 = \left[\left(x^2 \right)_{n,n-2} \right]^2$$

$$+ \left[\left(x^2 \right)_{n,n} \right]^2 + \left[\left(x^2 \right)_{n,n+2} \right]^2$$

$$\left[\left(x^2 \right)_{n,n-2} \right]^2 = \left[x_{n,n-1} x_{n-1,n-2} \right]^2 = \frac{n(n-1)\hbar^2}{4m^2\omega_0^2}$$

$$\left[\left(x^2\right)_{n,n+2}\right]^2 = \left[x_{n,n+1}x_{n+1,n+2}\right]^2 = \frac{(n+2)(n+1)\hbar^2}{4m^2\omega_0^2}$$

$$\left[\left(x^2\right)_{n,n}\right]^2 = \frac{(2n+1)^2\hbar^2}{4m^2\omega_0^2}$$

$$\left(x^4\right)_{nn} = \frac{\hbar^2}{4m^2\omega_0^2}\left[n(n-1) + (2n+1)^2(n+2)(n+1)\right]$$

$$= \frac{3\hbar^2}{4m^2\omega_0^2}\left(2n^2 + 2n + 1\right)$$

So we can write down the final result:

$$E_n^{(1)} \approx \beta\left(x^4\right)_{nn} = \beta\frac{3\hbar^2}{4m^2\omega_0^2}\left(2n^2 + 2n + 1\right) \tag{5.37}$$

Hence, the energy levels of the anharmonic oscillator described by (5.36) are given by:

$$E_n = E_n^{(0)} + E_n^{(1)} \approx \hbar\omega_0\left(n + \frac{1}{2}\right) + \beta\frac{3\hbar^2}{4m^2\omega_0^2}\left(2n^2 + 2n + 1\right) \tag{5.38}$$

Note that in contrast to the harmonic oscillator, the energy levels of an anharmonic oscillator as defined by (5.37) are no more equidistant!

5.6 Advanced Material

5.6.1 Derivation of Heisenbergs Uncertainty Relation (1D-Version)

Let us introduce the variances in coordinate and momentum according to:

$$\left\langle(\Delta x)^2\right\rangle = \left\langle x^2\right\rangle - \langle x\rangle^2; \left\langle(\Delta p_x)^2\right\rangle = \left\langle p_x^2\right\rangle - \langle p_x\rangle^2$$

We chose a coordinate system where $<x> = 0$ and $<p_x> = 0$. Then,

$$\left\langle(\Delta x)^2\right\rangle = \left\langle x^2\right\rangle = \int \Psi^* x^2 \Psi dx$$

$$\left\langle(\Delta p_x)^2\right\rangle = \left\langle p_x^2\right\rangle = \int \Psi^* p_x^2 \Psi dx = -\hbar^2 \int \Psi^* \frac{d^2}{dx^2}\Psi dx$$

Introducing now some real numbers α and β, we may state:

$$\int_X \left|\alpha x\Psi + \beta\frac{d\Psi}{dx}\right|^2 dx \geq 0$$

The integration is performed over the full domain of definition X. The integral may be rewritten as:

$$\int_X \left| \alpha x \Psi + \beta \frac{d\Psi}{dx} \right|^2 dx = A\alpha^2 - B\alpha\beta + C\beta^2 \geq 0 \qquad (5.39)$$

where the coefficients A, B, and C are given by:

$$A = \int_X x^2 \Psi^* \Psi dx = \langle x^2 \rangle$$

$$B = -\int_X x \frac{d}{dx}(\Psi^* \Psi) dx = 1$$

$$C = \int_X \left(\frac{d}{dx}\Psi^* \right) \left(\frac{d}{dx}\Psi \right) dx = -\int_V \Psi^* \left(\frac{d^2}{dx^2}\Psi \right) dx = \hbar^{-2}\langle p_x^2 \rangle$$

Let us now introduce the variable.
$\xi = -\frac{\alpha}{\beta}$. Once α and β are real, dividing (5.39) by β^2 yields:

$$A\xi^2 + B\xi + C \geq 0 \qquad (5.39a)$$

Let us now look at the function $y = A\xi^2 + B\xi + C$. It describes a parabola with a vertex ordinate $C - B^2/(4A)$. In order to have (5.39a) fulfilled for every ξ, it must be required that $A \geq 0$ and $C - B^2/(4A) \geq 0$. From here, we have

$$4CA \geq B^2 = 1 \Rightarrow \langle x^2 \rangle \langle p_x^2 \rangle \geq \frac{\hbar^2}{4}$$

Which coincides with (5.12) if $<x> = 0$ and $<p_x> = 0$ is fulfilled.

A more general writing of the uncertainty relation (compare Davydov) is ($\hat{F}, \hat{G}, \hat{M}$– self-adjoint):

If

$$\left[\hat{F}, \hat{G} \right] = i\hat{M} \Rightarrow \langle (\Delta F)^2 \rangle \langle (\Delta G)^2 \rangle \geq \frac{\langle M \rangle^2}{4} \qquad (5.40)$$

.

From here, in particular we find again (5.12):

$$\langle (\Delta x)^2 \rangle \langle (\Delta p_x)^2 \rangle \geq \frac{\hbar^2}{4}$$

5.6.2 Harmonic Oscillator and Ladder Operators

Let us return to the harmonic oscillator. When introducing the length $x_0 = \sqrt{\frac{\hbar}{m\omega_0}}$ (compare Fig. 5.1), the Hamiltonian of the 1D harmonic oscillator (5.28) may be rewritten as:

$$\hat{H} = \frac{\hbar\omega_0}{2}\left[\frac{\hat{p}_x^2 x_0^2}{\hbar^2} + \frac{\hat{x}^2}{x_0^2}\right]$$

Let us now define the ladder operators according to:

$$\hat{a} = \frac{1}{\sqrt{2}}\left[\frac{\hat{x}}{x_0} + i\frac{x_0}{\hbar}\hat{p}_x\right]; \quad \hat{a}^+ = \frac{1}{\sqrt{2}}\left[\frac{\hat{x}}{x_0} - i\frac{x_0}{\hbar}\hat{p}_x\right] \quad (5.41)$$

At this point, we need some extension to the material presented so far. Let us mention, that definition (5.6) for the self-adjoint operator may be written as: $\langle m|\hat{F}|n\rangle = \langle n|\hat{F}|m\rangle^*$. Note that not all operators are self-adjoint. Let us define the adjoint to $\hat{F}$ operator $\hat{F}^+$ through the relationship: $\langle m|\hat{F}|n\rangle = \langle n|\hat{F}^+|m\rangle^*$. Then, the operator $\hat{F}$ is self-adjoint if and only if $\hat{F} = \hat{F}^+$ holds. For the matrix elements of the adjoint operator, we obviously have the relation: $\left(\hat{F}\right)^*_{mn} = \left(\hat{F}^+\right)_{nm}$.

The following derivations adhere to [2]. Let us now have a look on the matrix elements of $\hat{a}^+$ and $\hat{a}$. We have (in the momentum operator matrix elements, we skip the subscript x for simplicity):

$$\left(\hat{a}\right)_{nm} = \frac{1}{\sqrt{2}}\left[\frac{x_{nm}}{x_0} + i\frac{x_0}{\hbar}p_{nm}\right]; \quad \left(\hat{a}^+\right)_{nm} = \frac{1}{\sqrt{2}}\left[\frac{x_{nm}}{x_0} - i\frac{x_0}{\hbar}p_{nm}\right]$$

Obviously
$\left(\hat{a}\right)^*_{mn} = \frac{1}{\sqrt{2}}\left[\frac{x^*_{mn}}{x_0} - i\frac{x_0}{\hbar}p^*_{mn}\right] = \frac{1}{\sqrt{2}}\left[\frac{x_{nm}}{x_0} - i\frac{x_0}{\hbar}p_{nm}\right]$ because $\hat{x}$ and $\hat{p}$ are self-adjoint. Therefore, we have $\left(\hat{a}\right)^*_{mn} = \left(\hat{a}^+\right)_{nm}$ so that $\hat{a}^+$ is adjoint to $\hat{a}$ and vice versa, but neither of $\hat{a}^+$ and $\hat{a}$ is self-adjoint. Note further that

$$x_{nm} = \frac{\left(\hat{a}\right)_{nm} + \left(\hat{a}^+\right)_{nm}}{\sqrt{2}}x_0 \quad (5.42)$$

From (5.41) and (5.5), we immediately find the important commutation rule:

$$\left[\hat{a}, \hat{a}^+\right] = \hat{a}\hat{a}^+ - \hat{a}^+\hat{a} = 1 \quad (5.43)$$

As well as:

$$\hat{a}\hat{a}^+ + \hat{a}^+\hat{a} = \frac{\hat{p}_x^2 x_0^2}{\hbar^2} + \frac{\hat{x}^2}{x_0^2} \quad (5.44)$$

For the harmonic oscillator, we can therefore write:

$$\hat{H} = \frac{\hbar\omega_0}{2}\left[\frac{\hat{p}_x^2 x_0^2}{\hbar^2} + \frac{\hat{x}^2}{x_0^2}\right] = \frac{\hbar\omega_0}{2}\left[\hat{a}\hat{a}^+ + \hat{a}^+\hat{a}\right]$$

$$= \frac{\hbar\omega_0}{2}\left[1 + \hat{a}^+\hat{a} + \hat{a}^+\hat{a}\right] = \hbar\omega_0\left[\hat{a}^+\hat{a} + \frac{1}{2}\right] \tag{5.45}$$

From here, we obtain

$$\hat{H}\hat{a}^+ = \hbar\omega_0\left[\hat{a}^+\hat{a} + \frac{1}{2}\right]\hat{a}^+ = \hbar\omega_0\left[\hat{a}^+\hat{a}\hat{a}^+ + \frac{1}{2}\hat{a}^+\right]$$

$$= \hat{a}^+\hbar\omega_0\left[\hat{a}\hat{a}^+ + \frac{1}{2}\right] = \hat{a}^+\hbar\omega_0\left[1 + \hat{a}^+\hat{a} + \frac{1}{2}\right]$$

$$= \hat{a}^+\left\{\hbar\omega_0\left[\hat{a}^+\hat{a} + \frac{1}{2}\right] + \hbar\omega_0\right\} = \hat{a}^+\left[\hat{H} + \hbar\omega_0\right]$$

Subsequently applying this procedure results in:

$$\hat{H}(\hat{a}^+)^n = (\hat{a}^+)^n\left[\hat{H} + n\hbar\omega_0\right] \tag{5.46}$$

Let us now postulate that there exists a state $|0\rangle$ such that $\hat{a}|0\rangle = 0$ is fulfilled. For this state, we obviously find:

$$\hat{H}|0\rangle = \hbar\omega_0\left[\hat{a}^+\hat{a} + \frac{1}{2}\right]|0\rangle = \hbar\omega_0\hat{a}^+\underbrace{\hat{a}|0\rangle}_{=0} + \frac{\hbar\omega_0}{2}|0\rangle = \frac{\hbar\omega_0}{2}|0\rangle$$

We conclude from here that $|0\rangle$ defines an eigenstate of the Hamiltonian of the harmonic oscillator with the eigenvalue $\frac{\hbar\omega_0}{2}$. The state $|0\rangle$ thus corresponds to the ground state of the harmonic oscillator. Then, the state defined by $(\hat{a}^+)^n|0\rangle$ is also an eigenstate of $\hat{H}$ corresponding to: $\hat{H}(\hat{a}^+)^n|0\rangle = \hbar\omega_0(n + \frac{1}{2})(\hat{a}^+)^n|0\rangle$.
Indeed,

$$\hat{H}(\hat{a}^+)^n|0\rangle = (\hat{a}^+)^n\left[\hat{H} + n\hbar\omega_0\right]|0\rangle$$

$$= (\hat{a}^+)^n\left[\underbrace{\hat{H}|0\rangle}_{=\frac{\hbar\omega_0}{2}|0\rangle} + n\hbar\omega_0|0\rangle\right] = \hbar\omega_0\left(n + \frac{1}{2}\right)(\hat{a}^+)^n|0\rangle$$

Therefore, the state $(\hat{a}^+)^n|0\rangle$ is an eigenstate of $\hat{H}$ and corresponds to the eigenvalue $\hbar\omega_0(n + \frac{1}{2})$. Obviously, the excited eigenstates of the harmonic oscillator $|n\rangle$ may be constructed by multiple application of $\hat{a}^+$ to the ground state $|0\rangle$. Thereby, the normalization requirement $\langle n \mid n\rangle = 1$ results in:

$$|n\rangle = \frac{1}{\sqrt{n!}}(\hat{a}^+)^n|0\rangle. \tag{5.47}$$

Remark Let us recall that in the Dirac notation, an operator $\hat{F}$ may act "to the right side" but also "to the left side." Thereby, the ket-vector $\hat{F}|\psi\rangle$ corresponds to a bra-vector $\langle\psi|\hat{F}^+$ and vice versa. Abstractly, this is a direct conclusion from the definition $\langle m|\hat{F}|n\rangle = \langle n|\hat{F}^+|m\rangle^*$.

Indeed, from $\langle m|\hat{F}|n\rangle = \langle n|\hat{F}^+|m\rangle^*$ defining $\langle\psi| = \langle m|\hat{F}$, we have:

$$\langle m|\hat{F}|n\rangle = \langle\psi \mid n\rangle = \langle n \mid \psi\rangle^* = \langle n|\hat{F}^+|m\rangle^*$$

Hence, from $\langle\psi| = \langle m|\hat{F} \Leftrightarrow |\psi\rangle = \hat{F}^+|m\rangle$ or $\langle\psi| = \langle m|\hat{F}^+ \Leftrightarrow |\psi\rangle = \hat{F}|m\rangle$.

A more illustrative but nevertheless strong derivation is obtained when using the matrix representation of the operator $\hat{F}$ as well as of the corresponding states, identifying the ket with a single column matrix, while the bra corresponds to a single row (compare task 5.7.14).

Indeed, when assuming $|n\rangle = C(\hat{a}^+)^n|0\rangle$, we have $\langle n| = C^*\langle 0|(\hat{a})^n$. Therefore,

$$\langle n \mid n\rangle = CC^*\langle 0|(\hat{a})^n(\hat{a}^+)^n|0\rangle$$

Multiple application of $\hat{a}\hat{a}^+ = \hat{a}^+\hat{a} + 1$ together with $\hat{a}|0\rangle = 0$ results in:

$$1 = \langle n \mid n\rangle = CC^*\langle 0|(\hat{a})^n(\hat{a}^+)^n|0\rangle = CC^*\langle 0|(\hat{a})^{n-1}(\hat{a}^+\hat{a} + 1)(\hat{a}^+)^{n-1}|0\rangle = \ldots = CC^*n!\langle 0 \mid 0\rangle = CC^*n!$$

Which is fulfilled when setting $C = \frac{1}{\sqrt{n!}}$.

Thus, from (5.47), we have:

$$|n\rangle = \frac{1}{\sqrt{n!}}(\hat{a}^+)^n|0\rangle$$

$$\Rightarrow |n+1\rangle = \frac{1}{\sqrt{(n+1)!}}(\hat{a}^+)^{n+1}|0\rangle = \frac{\hat{a}^+}{\sqrt{n+1}}\frac{1}{\sqrt{n!}}(\hat{a}^+)^n|0\rangle = \frac{1}{\sqrt{n+1}}\hat{a}^+|n\rangle$$

$$|n-1\rangle = \frac{1}{\sqrt{(n-1)!}}(\hat{a}^+)^{n-1}|0\rangle = \frac{1}{\sqrt{(n-1)!}}(\hat{a}^+)^{n-1}(\hat{a}\hat{a}^+ - \hat{a}^+\hat{a})|0\rangle$$

$$= \frac{1}{\sqrt{(n-1)!}}(\hat{a}^+)^{n-1}\hat{a}\hat{a}^+|0\rangle = \frac{1}{\sqrt{(n-1)!}}(\hat{a}^+)^{n-2}(\hat{a}\hat{a}^+ - 1)\hat{a}^+|0\rangle$$

$$= \frac{1}{\sqrt{(n-1)!}}(\hat{a}^+)^{n-2}\hat{a}(\hat{a}^+)^2|0\rangle - |n-1\rangle$$

$$\Rightarrow 2|n-1\rangle = \frac{1}{\sqrt{(n-1)!}}(\hat{a}^+)^{n-2}\hat{a}(\hat{a}^+)^2|0\rangle \ldots n|n-1\rangle = \frac{1}{\sqrt{(n-1)!}}\hat{a}(\hat{a}^+)^n|0\rangle$$

$$\Rightarrow n|n-1\rangle = \frac{1}{\sqrt{(n-1)!}}\hat{a}(\hat{a}^+)^n|0\rangle = \sqrt{n}\hat{a}\frac{1}{\sqrt{n!}}(\hat{a}^+)^n|0\rangle = \sqrt{n}\hat{a}|n\rangle$$

Therefore, we finally obtain (compare also tak 5.7.14):

$$\hat{a}^+|n\rangle = \sqrt{n+1}|n+1\rangle$$
$$\hat{a}|n\rangle = \sqrt{n}|n-1\rangle$$

(5.48)

Thereby, $\hat{a}|0\rangle = 0$ per definition. The effect of applying $\hat{a}$ to an eigenstate of the harmonic oscillator with quantum number n is to turn it into the eigenstate with quantum number $n-1$. $\hat{a}^+$ has the opposite function. This is visualized in Fig. 5.2; from there, the name "ladder operators" should become obvious.

From (5.48), one can immediately calculate the matrix elements $\langle n|\hat{a}|m\rangle$ and $\langle n|\hat{a}^+|m\rangle$ again.

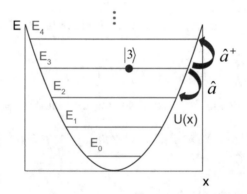

Fig. 5.2 Illustration of the action of the ladder operators on the harmonic oscillator state $|3\rangle$

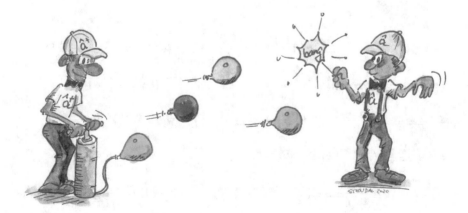

Fig. 5.3 Photon creation (left) and annihilation (right) operators. Cartoon by Dr. Alexander Stendal. Printed with permission

It turns out that the only different from zero matrix elements are of the type:

$$\langle n - 1|\hat{a}|n\rangle = \langle n|\hat{a}^+|n - 1\rangle = \sqrt{n} \tag{5.49}$$

Of course, from (5.49), we reobserve the well-known matrix elements:

$$x_{nm} = \frac{\left(\hat{a}\right)_{nm} + \left(\hat{a}^+\right)_{nm}}{\sqrt{2}} x_0 \Rightarrow x_{n,n-1} = \sqrt{\frac{n}{2}} x_0$$

5.6.3   Remarks on the Quantization of the Free Electromagnetic Field

Idea.
Let us start from (2.1) and write the classical expression for the real electric field (for simplicity, we will not take its vector character into explicit account here) of a plane wave propagating in vacuum according to:

$$E = E_0 e^{-i(\omega t - \mathbf{k r})} + E_0^* e^{+i(\omega t - \mathbf{k r})} \tag{5.50}$$

Each excitation of type (5.50) with a given ω, $\mathbf{k}$, (and polarization state) will be associated with what we call a field mode.

In Chap. 3, we have already discussed, however, that the energy of an electromagnetic field oscillating with the angular frequency ω_0 can only change in discrete portions $\hbar\omega_0$, which we have called photons. Hence, each photon that is added to or removed from the field changes its energy by a fixed value of $\hbar\omega_0$. Once the number of photons is always integer, the possible energy values of the monochromatic electromagnetic field form an infinite set of equidistant energy values, separated by $\hbar\omega_0$. Thus, these energy values exactly coincide with those of the harmonic oscillator provided that we accept a ground-state energy of the electromagnetic field of $\hbar\omega_0/2$, which means that even in the absence of any photons, there is still some energy in the corresponding field mode (or now photon mode).

Hence, we associate the field in a given field mode with a harmonic oscillator and make immediate use of the material developed in Sect. 5.6.2. This way we will introduce the quantization of a single mode field. In a multiple mode situation, we have to sum up the corresponding single mode expressions to the complete field.

Photon annihilation and creation operators
Particularly, according to (5.48), we introduce the photon annihilation operator $\hat{a}$ according to

$$\hat{a}|n\rangle = \sqrt{n}|n - 1\rangle$$

Here, $|n\rangle$ is associated with a quantum state with a well-defined number of photons n excited in the corresponding photon mode. We will call such states Fock states.

We further introduce the photon creation operator $\hat{a}^+$ according to:

$$\hat{a}^+|n\rangle = \sqrt{n+1}|n+1\rangle$$

In agreement with (5.45) we can now write:

$$\hat{H}|n\rangle = \hbar\omega_0\left[\hat{a}^+\hat{a}|n\rangle + \frac{1}{2}|n\rangle\right] = \hbar\omega_0\left[\hat{a}^+\sqrt{n}|n-1\rangle + \frac{1}{2}|n\rangle\right]$$

$$= \hbar\omega_0\left[n|n\rangle + \frac{1}{2}|n\rangle\right] = \hbar\omega_0\left[n + \frac{1}{2}\right]|n\rangle$$

We therefore introduce the self-adjoint operator of the photon number $\hat{N}$

$$\hat{N} \equiv \hat{a}^+\hat{a} \Rightarrow \hat{N}|n\rangle = n|n\rangle \tag{5.51}$$

Note that $\hat{N}$ is self-adjoint; hence, it has orthogonal eigenfunctions.

Mnemonic trick: There is no need to confuse $\hat{a}$ and $\hat{a}^+$. The one with the " $+$ " has the function to "add" a photon, so it is the photon creation operator. In (5.51), the order is:

$$\hat{a}^+\hat{a} \equiv \hat{N}$$

For memorizing: read from left to right and realize that a photon must first be created, before it may be annihilated (see Fig. 5.3).

If you wish to change the order of $\hat{a}$ and $\hat{a}^+$(first annihilate and then create), so you need to have an extra photon first. Therefore, change the order and add one photon (i.e., add 1), and find.

$$\hat{a}\hat{a}^+ = \hat{N} + 1,$$

which coincides with the commutation rule (5.43).

Electric field of a freely propagating plane wave in a Fock state

Clearly, all Fock states are eigenstates of both $\hat{N}$ and $\hat{H}$, such that the latter commute. We now perform the quantization of the electromagnetic field by replacing the classical field (5.50) by the self-adjoint operator of the electric field (compare [3, 4])

$$\hat{\mathbf{E}} = \mathbf{A}_0 e^{-i(\omega t - \mathbf{kr})}\hat{a} + \mathbf{A}_0^* e^{+i(\omega t - \mathbf{kr})}\hat{a}^+ \tag{5.52}$$

Remark The representation of the creation and annihilation operators in terms of the matrix elements (5.49) defines the so-called occupation number representation. It corresponds to a description of the field quantum state in terms of the quantum numbers n, i.e., the "occupation numbers." In this representation, the field operator (5.52) acts on the fields wavefunction expressed in terms of the occupation numbers n. The earlier introduced internal structure of $\hat{a}$ and $\hat{a}^+$ is here no more of interest; $\hat{a}$ and $\hat{a}^+$ are defined through the relations (5.43) and (5.48) only.

Then, because of $\langle n|\hat{a}|n\rangle = \sqrt{n}\langle n \mid n-1\rangle = 0 = \langle n|\hat{a}^+|n\rangle \Rightarrow \langle n|\hat{E}|n\rangle = 0 \forall n$.

The expectation value of the electric field is thus zero in each Fock state (5.47). We will use the following writing for that: $\langle E\rangle_n = 0$. It has the meaning that the expectation value of E in the state $|n\rangle$ is zero. Note that because of (5.48), Fock states cannot be eigenfunctions of $\hat{E}$. Therefore, $\left[\hat{E}, \hat{N}\right] \neq \mathbf{0}$.

This is an important result. In a photon state with a well-defined number of photons, the measurement result of the electric field is uncertain, and vice versa. This is contrary to the classical idea of the electric field, where both energy density and field strength are well defined.

Even in the ground state with $n = 0$, the field is therefore different from zero. Once the expectation value of the field is nevertheless equal to zero, we have to expect fluctuations of the electromagnetic field ("vacuum oscillations"), which obviously provide the reason for the finite ground-state energy of the radiation field mode that is also different from zero.

Let us quantify these field oscillations in terms of the variance of the electric field. When rewriting (5.52) for simplicity as:

$$\hat{E} = \mathbf{A}\hat{a} + \mathbf{A}^*\hat{a}^+; \ \mathbf{A} \equiv \mathbf{A}_0 e^{-i(\omega t - \mathbf{kr})}$$

We find the following general expression (regardless of the concrete quantum state) for the expectation value of the electric field according to:

$\langle E\rangle = \langle A\hat{a} + A^*\hat{a}^+\rangle = A\langle\hat{a}\rangle + A^*\langle\hat{a}^+\rangle$

$\Rightarrow \langle E\rangle^2 = A^2\langle\hat{a}\rangle^2 + (A^*)^2\langle\hat{a}^+\rangle^2 + 2\underbrace{AA^*}_{=|A|^2}\langle\hat{a}\rangle\langle\hat{a}^+\rangle$ Furthermore,

$$\langle E^2\rangle = \left\langle\left(A\hat{a} + A^*\hat{a}^+\right)^2\right\rangle = A^2\langle\hat{a}^2\rangle + (A^*)^2\left\langle\left(\hat{a}^+\right)^2\right\rangle + |A|^2\left\langle\underbrace{\hat{a}\hat{a}^+}_{\hat{a}^+\hat{a}+1} + \hat{a}^+\hat{a}\right\rangle$$

$$= A^2\langle\hat{a}^2\rangle + (A^*)^2\left\langle\left(\hat{a}^+\right)^2\right\rangle + 2|A|^2\left(\langle\hat{a}^+\hat{a}\rangle + \frac{1}{2}\right) \Rightarrow$$

$$\langle(\Delta E)^2\rangle = \langle E^2\rangle - \langle E\rangle^2$$

$$= A^2\left(\langle\hat{a}^2\rangle - \langle\hat{a}\rangle^2\right) + (A^*)^2\left(\left\langle\left(\hat{a}^+\right)^2\right\rangle - \langle\hat{a}^+\rangle^2\right)$$

$$+ 2|A|^2\left(\langle\hat{a}^+\hat{a}\rangle - \langle\hat{a}^+\rangle\langle\hat{a}\rangle + \frac{1}{2}\right) \tag{5.53}$$

Averaging over time eliminates the quickly oscillating terms in (5.53), which leads to the more familiar expression:

$$\overline{\left\langle (\Delta E)^2 \right\rangle}^t = 2|A|^2 \left(\left\langle \hat{a}^+ \hat{a} \right\rangle - \left\langle \hat{a}^+ \right\rangle \left\langle \hat{a} \right\rangle + \frac{1}{2} \right) \tag{5.53a}$$

In a Fock state, we have again:
$\left\langle \hat{a}^2 \right\rangle_n = \left\langle \hat{a} \right\rangle_n = \left\langle \left(\hat{a}^+ \right)^2 \right\rangle_n = \left\langle \hat{a}^+ \right\rangle_n = 0$, and therefore:

$$\left\langle (\Delta E)^2 \right\rangle_n = 2|A|^2 \left(\left\langle \hat{a}^+ \hat{a} \right\rangle_n + \frac{1}{2} \right) = 2|A|^2 \left(n + \frac{1}{2} \right) \tag{5.54}$$

Thus, for the photon field ground state with $n = 0$, the quantum mechanical treatment results in a variance of the electric field according to:

$$\left\langle (\Delta E)^2 \right\rangle_{n=0} = \left\langle E^2 \right\rangle_{n=0} - \left\langle E \right\rangle_{n=0}^2 = |A|^2 \neq 0 \tag{5.55}$$

Such that even in the absence of any photon, field strength oscillations as calculated in terms of the variance are different from zero. With an increasing number of photons, the variance (uncertainty) in the electric field increases drastically.

Glauber states

The question is now whether it is possible to construct photon states such that the field variance remains equal to that of the vacuum oscillations (5.55), although a considerable number of photons is excited?

Yes, it is possible. The searched photon states $|\alpha\rangle$ are eigenstates of the photon annihilation operator according to:

$$\begin{aligned} \hat{a}|\alpha\rangle &= \alpha|\alpha\rangle \\ \langle \alpha \mid \alpha \rangle &= 1 \end{aligned} \tag{5.56}$$

Note that $\hat{a}$ is not self-adjoint, such that its eigenvalues α may be complex, and its eigenfunctions do not need to be orthogonal.

From the definition $\langle m|\hat{F}|n \rangle = \langle n|\hat{F}^+|m \rangle^*$ we find:

$$\underbrace{\langle \alpha|\hat{a}|\alpha\rangle^*}_{=\alpha^*} = \langle \alpha|\hat{a}^+|\alpha\rangle \Rightarrow \langle \alpha|\hat{a}^+ = \langle \alpha|\alpha^*$$

The eigenfunctions $|\alpha\rangle$ describe the so-called Glauber states. In a Glauber state, we obviously have:

$$\begin{aligned} \langle \alpha|\hat{a}|\alpha\rangle &= \alpha, \ \langle \alpha|\hat{a}^+|\alpha\rangle = \alpha^* \\ \Rightarrow \left\langle \hat{a}^2 \right\rangle_\alpha &= \left\langle \hat{a} \right\rangle_\alpha^2; \left\langle \left(\hat{a}^+ \right)^2 \right\rangle_\alpha = \left\langle \hat{a}^+ \right\rangle_\alpha^2; \left\langle \hat{a}^+ \hat{a} \right\rangle_\alpha \\ &= \langle \alpha|\hat{a}^+ \hat{a}|\alpha\rangle = |\alpha|^2 = \left\langle \hat{a}^+ \right\rangle_\alpha \left\langle \hat{a} \right\rangle_\alpha \end{aligned}$$

Therefore, from (5.53) we have

$$\langle (\Delta E)^2 \rangle_\alpha = |A|^2 \tag{5.57}$$

These so-called coherent states of the radiation field provide states where the variances in photon number and electric field are in a reasonable compromise. They can be expressed through Fock states as:

$$|\alpha\rangle = e^{-\frac{|\alpha|^2}{2}} \sum_{n=0}^{\infty} \frac{\alpha^n}{\sqrt{n!}} |n\rangle \tag{5.58}$$

This can easily checked by the reader himselves by substituting (5.58) into (5.56) and checking the correctness of normalization. Note that the field mode ground state $|n = 0\rangle$ is also a Glauber state, because

$$|\alpha \to 0\rangle = \lim_{\alpha \to 0} \left[e^{-\frac{|\alpha|^2}{2}} \sum_{n=0}^{\infty} \frac{\alpha^n}{\sqrt{n!}} |n\rangle \right] = |0\rangle$$

It is easy to obtain the following expectation value:

$$\langle N \rangle_\alpha = \langle \alpha | \hat{N} | \alpha \rangle = \langle \alpha | \hat{a}^+ \hat{a} | \alpha \rangle = |\alpha|^2 \tag{5.59}$$

In a Glauber state, the expectation value of the photon number is equal to $|\alpha|^2$, such that α should be proportional to the classical field amplitude. Indeed, we find:

$$\langle E \rangle_\alpha = \langle A\hat{a} + A^*\hat{a}^+ \rangle_\alpha = A\langle \alpha | \hat{a} | \alpha \rangle + A^* \langle \alpha | \hat{a}^+ | \alpha \rangle = A\alpha + A^*\alpha^* \tag{5.60}$$

This coincides—by structure—with (5.50). Note that $\alpha \to 0$ results in both $<N>_\alpha \to 0$ and $<E>_\alpha \to 0$.

General results for important expectation values are summarized in Table 5.2. In a Fock state, in general, variances in the photon number are vanishing, but those in the field strength are large. In Glauber states, there is certain variance in the photon number, but that in the electric field is smaller than in a Fock state. Therefore, Glauber states can be regarded as photon states where a reasonable

Table 5.2 Important expectation values

State	Formula	$\langle N \rangle$	$\langle (\Delta N)^2 \rangle =$ $\langle N^2 \rangle - \langle N \rangle^2$	$\langle E \rangle$	$\langle (\Delta E)^2 \rangle =$ $\langle E^2 \rangle - \langle E \rangle^2$
Fock	$|n\rangle$	n	0	0	$2|A|^2 \left(n + \frac{1}{2}\right)$
Glauber	$|\alpha\rangle = e^{-\frac{|\alpha|^2}{2}} \sum_{n=0}^{\infty} \frac{\alpha^n}{\sqrt{n!}} |n\rangle$	$|\alpha|^2$	$|\alpha|^2$	$2\mathrm{Re}(A\alpha)$	$|A|^2$

compromise between uncertainties in photon number and field strength is achieved. In particular, for $|\alpha| = \sqrt{\langle N \rangle} \to \infty$, the relative uncertainties in field strength and photon number approach zero:

$$|\alpha| \to \infty \Rightarrow \begin{cases} \dfrac{\sqrt{\langle (\Delta N)^2 \rangle}}{\langle N \rangle} = \dfrac{|\alpha|}{|\alpha|^2} \to 0 \\[4mm] \dfrac{\sqrt{\langle (\Delta E)^2 \rangle}}{\langle E \rangle} = \dfrac{|A|}{2 \mathrm{Re}(\alpha A)} \to 0 \end{cases}$$

Thus, the quantum mechanical Glauber states come closest to the classical idea of the electric field with well-defined energy density and field strength. Obviously, because of the finite value of the vacuum field oscillations, quantum states with a vanishing relative standard deviation in the electric field must correspond to states with an infinitely large energy.

Note again that for $n = \alpha = 0$, the results obtained for Fock and Glauber states are identical.

So everything seems fine with the expectation values, if only we would have an idea on the somewhat dubious amplitude A_0 in (5.52). Let us finally clarify this. Let us assume a Fock state. From (5.54) and $<E>_n = 0$, we have:

$$\langle E^2 \rangle_n = 2|A|^2 \left(n + \frac{1}{2} \right)$$

In a plane wave propagating in free space, $\mathbf{k}$ is real, and $|A| = |A_0|$. Let us now regard the energy density w (= energy per volume V). In free space, we have:

$$\langle w \rangle_n = \varepsilon_0 \langle E^2 \rangle_n = 2|A_0|^2 \varepsilon_0 \left(n + \frac{1}{2} \right) = \frac{\hbar \omega}{V} \left(n + \frac{1}{2} \right) \Rightarrow |A_0|^2 = \frac{\hbar \omega}{2 \varepsilon_0 V} \quad (5.61)$$

5.7 Tasks for Self-check

5.7.1 Prove the relationship: $\left[\hat{A}\hat{B}, \hat{C} \right] = \hat{A}\left[\hat{B}, \hat{C} \right] + \left[\hat{A}, \hat{C} \right]\hat{B}$.

5.7.2 Calculate the commutation relation $\left[\hat{x}, \hat{p}_x^2 \right] = ?$

5.7.3 Check whether or not the following operators are self-adjoint!
   (a) $-i \frac{d}{dx}$
   (b) $\hat{F}\hat{G}$, if both $\hat{F}$ and $\hat{G}$ are self-adjoint and commute with each other
   (c) $\left(\hat{F} + \hat{G} \right)$, if both $\hat{F}$ and $\hat{G}$ are self-adjoint

5.7.4 Consider a particle trapped in a one-dimensional box potential with length L and infinitely high potential walls, i.e.,

$$U = 0; \quad 0 \leq x \leq L$$
$$U \to \infty; \quad x < 0 \text{ or } x > L$$

Calculate the expectation value of the x-coordinate in any eigenstate!

5.7.5 Find the probability current density for a particle in the box with impermeable walls in an arbitrary eigenstate.

5.7.6 Assume a state described by a wavefunction $\Psi(\mathbf{r}, t) = e^{-i\frac{E_n}{\hbar}t}\psi(\mathbf{r})$ with $\psi(\mathbf{r})$—real. Show that in this state, the expectation value of the momentum is zero. Also, calculate the probability density current corresponding to that situation!

5.7.7 Return to the 1D particle-in-the-box system from task 5.7.4. Show that in any eigenstate, Heisenbergs uncertainty relation for the coordinate and the momentum is fulfilled!

5.7.8 From Heisenbergs uncertainty relation, find an estimation for the minimum energy of the ground state of a harmonic oscillator!

5.7.9 Estimate the kinetic energy of an electron, which is spatially confined to a region of
(a) 10^{-10} m
(b) 10^{-15} m (after [5])

5.7.10 Show that $\left(\hat{A}\hat{B}\right)^+ = \hat{B}^+\hat{A}^+$.

5.7.11 From (5.35), derive the recursive recipe for Hermitian polynomials:

$$H_{n+1}(x) = \left(2x - \frac{\mathrm{d}}{\mathrm{d}x}\right)H_n(x)$$

Setting further $H_0=1$, make clear which of the Hermitian polynomials are even functions of x, and which are uneven!

5.7.12 Calculate the expectation value of the potential energy of a 1D harmonic oscillator in any state $|n\rangle$!

5.7.13 Assume a harmonic oscillator in eigenstate $|n\rangle$. Calculate all oscillator strength values f_{jn}!

5.7.14 Rewrite (5.48) in an explicit matrix representation!

5.7.15 Make sure that (5.58) is a solution of (5.56) for every α!

5.7.16 In a Glauber state, calculate $\langle(\Delta N)^2\rangle$.

5.7.17 Proof relation (5.20f)

References

Specific References

1. D. Dürr, D. Lazarovici, *Verständliche Quantenmechanik* (Springer Spektrum, 2018)
2. А.А. Соколов, И.М. Тернов, В.Ч. Жуковский: Квантовая механика, Москва "Наука" (1979), pp. 125–126
3. H. Paul, Photonen. Eine Einführung in die Quantenoptik, Teubner Studienbücher Physik (1995) Anhang A

4. L.D. Landau, E.M. Lifshitz, *Relativistic Quantum Theory part 1* (Pergamon Press, 1971), pp. 5–9
5. W. Macke, *Quanten. Ein Lehrbuch der theoretischen Physik, 3.* Auflage, Akademische Verlagsgesellschaft (1959), p.474

General Literature

6. L. Landau, E.M. Lifshitz, *Quantum Mechanics* (Volume 3 of A Course of Theoretical Physics) (Pergamon Press, 1965)
7. A.S. Davydov, *Quantum Mechanics* (Pergamon, 1965)
8. M. Bartelmann, B. Feuerbacher,T. Krüger, D. Lüst, A. Rebhan, A. Wipf, *Theoretische Physik* (Springer Spektrum, 2015)
9. J. Bricmont, *Quantum Sense and Nonsense* (Springer, 2017)
10. S. Flügge, *Practical Quantum Mechanics* (Springer 1999)
11. E.W. Schpolski, *Atomphysik II* (VEB Deutscher Verlag der Wissenschaften, Berlin, 1956)

Einstein Coefficients and Quantum Transitions

6

Abstract

The chapter is focused on the introduction of a perturbative treatment of quantum transitions. In order to apply this concept to light–matter interactions, the electric dipole interaction between the electric field of the incident light wave and the material system is considered as a time-dependent perturbation of the quantum state of material system within a semiclassical approach. The results are applied to a two-level system interacting with light in terms of the Einstein coefficients.

6.1   Starting Point

We have now collected enough knowledge to return to our basis topic, namely the interaction of a material (quantum) system with electromagnetic radiation. The very simplest possibility to deal with the radiation–matter topic in quantum mechanics is to construct a model that has two energy levels only. We will call this a two-level system (Fig. 6.1).

The idea is to neglect the multiplicity of energy levels a real material system might have and to concentrate on two selected energy levels. This makes sense, when the electromagnetic wave has a frequency close to the eigenfrequency of the subsystem described by the two energy levels.

A very popular and transparent treatment of the interaction of such a two-level system with electromagnetic irradiation may be performed in terms of the so-called Einstein coefficients. We will make use of a treatment published earlier in [1].

© The Author(s), under exclusive license to Springer Nature Switzerland AG 2022   121
O. Stenzel, *Light–Matter Interaction*, UNITEXT for Physics,
https://doi.org/10.1007/978-3-030-87144-4_6

Fig. 6.1 Illustration of a two-level system. Left: Ground state; right: first and only excited state. Cartoon by Dr. Alexander Stendal. Printed with permission

6.2   Physical Idea

Let us have a look at Fig. 6.2. It shows two discrete energy levels, E_1 and E_2. The level 2 corresponds to a higher energy state of the system than level 1. Simply for unambiguity, let us associate the first level with a ground state and the second one with an excited state. In order to describe the interaction between radiation and the two-level system in terms of Einstein coefficients, we have to consider three phenomena: light absorption, spontaneous emission of light, and stimulated emission of light by the two-level system.

Let us assume that the system is in the first, low energy state. When the radiation source is switched on, and the radiation frequency is close to the eigen-frequency of the system, an absorption process of light is expected to transfer

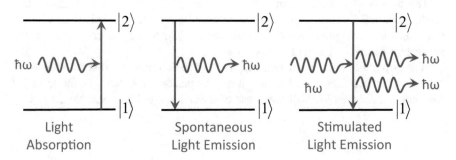

Fig. 6.2 Introduction of Einstein coefficients

the system from energy level 1 to the level 2 (the quantum system becomes excited—compare also Fig. 4.2). Due to energy conservation, this energy gain of the two-level system must be accompanied by an energy loss of the radiation field; hence, the energy is transferred from the electromagnetic field to the two-level system. This transfer of energy from the radiation field to internal degrees of freedom of the material system is what we will call further absorption of light. This absorption process becomes more probable when the electromagnetic radiation is more intense. The transition rate from level 1 to level 2 by absorption of light is therefore expected to be proportional to:

- The intensity of the incident radiation
- The statistical probability to find the system in state 1

Let us assume in this context that the electromagnetic field in fact interacts with a large amount of such two-level systems. Then, a considerably large part of the energy of the field may be transferred to the assembly of two-level systems. However, the energy loss of the field is always equal to an integer multiple of the excitation energy of the two-level systems. In our simple description, these single portions of light energy that may be absorbed have been called photons.

We will now consider the case that a single selected two-level system is in the second (excited) state. From our experience (or more accurate: from thermodynamics), we know that within a certain characteristic time any excited system tends to lose its energy, thus returning back into the ground state. In order to lose energy, our system has to perform a process that is reverse to the absorption of light, namely the emission of light (compare also Fig. 4.2). Let us postulate that an excited quantum system may lose energy without any stimulation arising from an external light source by emitting an energy portion of light that exactly corresponds to the energy difference between the two energy levels. In this case, we speak on the spontaneous emission of a photon. The transition rate from level 2 to level 1 by spontaneous emission will be proportional to:

- The statistical probability to find the system in state 2

There is a second mechanism to turn the system from the excited state into the ground state. We postulate that the system may also perform a so-called stimulated emission process. This has to be understood as an emission of light activated by the incident electromagnetic wave. The transition rate from state 2 to state 1 caused by this process should be proportional to:

- The intensity of the radiation
- The statistical probability to find the system in state 2

Any of these elementary processes enters into the resulting full transition rate with a specific proportionality coefficient, the so-called Einstein coefficients. It is a common practice to use the following symbols for Einstein coefficients:

- A_{21} for spontaneous emission $(2 \rightarrow 1)$
- B_{21} for stimulated emission $(2 \rightarrow 1)$
- B_{12} for absorption $(1 \rightarrow 2)$

The next section will deal with a mathematical treatment of Einstein coefficients, and it will be our purpose to derive—step by step—exact expression for Einstein coefficients in the dipole approximation.

Finally, Fig. 6.2 gives a schematic representation of all the mentioned elementary processes. Here, the vertical arrows correspond to the transitions between ground and excited states, while the sinusoidal structures demonstrate annihilation or creation of a photon.

6.3   Theoretical Considerations

6.3.1   The Rate Equation

Commonly, in the philosophy of Einstein's coefficients, the electromagnetic field is characterized by the so-called spectral density of the radiation field defined as:

$$u \equiv \frac{\mathrm{d}E}{V \, \mathrm{d}\omega} \tag{6.1}$$

This equation defines the spectral density as the field energy per angular frequency interval and per volume. E is again used for the energy. This may lead to confusion with the electric field strength, and we will try to avoid any misinterpretations using suitable subscripts when necessary.

Let us further assume that we have an assembly of N_0 two-level systems, interacting with the radiation field. Let N_1 be the number of systems in the ground state, and N_2 in the excited one. Obviously,

$$N_1 + N_2 = N_0 = \text{const.} \tag{6.2}$$

Due to the action of the radiation field, the population of the excited state may be changed. In terms of the mechanisms proposed in Sect. 6.2, the corresponding rate equation is:

$$\frac{\mathrm{d}N_2}{\mathrm{d}t} = N_1 B_{12} u - N_2 B_{21} u - N_2 A_{21} \tag{6.3}$$

Corresponding to Fig. 6.2, the first term in (6.3) describes absorption, which leads to an increase in the population of the excited state. The second term corresponds to stimulated emission, and the third one to spontaneous emission, both resulting in a decrease of the population of the excited state. Of course, here and throughout this section we only take the spectral density at a frequency corresponding to the eigenfrequency of the two-level system into consideration.

Interesting information may be obtained regarding the special case of thermodynamic equilibrium between radiation and matter. In this situation, $dN_2/dt = 0$, and from (6.3) it follows:

$$\text{equilibrium}: \frac{N_1}{N_2} = \frac{B_{21}u + A_{21}}{B_{12}u} \tag{6.4}$$

On the other hand, in equilibrium conditions with $N_0 = \text{const}$, Boltzmann's statistics hold, resulting in

$$\text{equilibrium}: \frac{N_1}{N_2} = e^{\frac{E_2 - E_1}{k_B T}} \tag{6.5}$$

where T is the absolute temperature. From (6.4) and (6.5) in combination, we obtain an expression for the spectral density of the radiation field in equilibrium conditions as:

$$\text{equilibrium}: u = \frac{A_{21}}{B_{12}\left(e^{\frac{E_2 - E_1}{k_B T}} - \frac{B_{21}}{B_{12}}\right)} \tag{6.6}$$

It is useful to discuss some special cases resulting from (6.6). Let us consider the case of $T \to 0$. Then, N_2 must be zero, and from (6.4) it follows that $u = 0$. In the other extreme case ($T \to \infty$), it makes sense to assume that the radiation density becomes infinitively large. If so, from (6.6) we must demand that:

$$B_{12} = B_{21} \tag{6.7}$$

Hence, the postulation of the stimulated emission appears to be absolutely necessary to suffice thermodynamics.

In fact, we do not need to rely on our feeling of an infinitively large spectral density at infinitively large temperatures. Condition (6.7) will be obtained independently as a result of the following perturbation theory treatment of quantum transitions.

6.3.2 Perturbation Theory of Quantum Transitions

In order to get information about the mathematical structure of Einstein coefficients, it becomes now necessary to apply the mathematical apparatus of quantum mechanics to the interaction of light with matter. The behavior of the system is described by a wavefunction Ψ, obtained as the solution of Schrödinger's (4.2).

The particular problem which will be considered now is sketched in Fig. 6.3. Imagine a time-independent Hamiltonian $\hat{H}_0$ with a set of discrete and nondegenerate eigenfunctions and corresponding eigenvalues $\{E_n\}$. Consider further that at

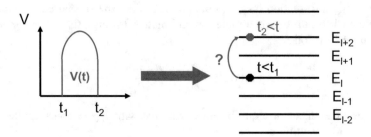

Fig. 6.3 Quantum transitions

a certain moment at $t < 0$, the system is definitely in the stationary lth quantum state and has the energy E_l.

In this case, the wavefunction $\Psi_l(\mathbf{r},t)$ suffices Schrödinger's equation

$$i\hbar \frac{\partial}{\partial t} \Psi_l(\mathbf{r}, t) = \hat{H}_0 \Psi_l(\mathbf{r}, t) \tag{6.8}$$

while the energy level E_l is a solution of the eigenvalue problem

$$\hat{H}_0 \psi_l(\mathbf{r}) = E_l \psi_l(\mathbf{r}) \tag{6.9}$$

and

$$\Psi_l(\mathbf{r}, t) = e^{-i\frac{E_l}{\hbar}t} \psi_l(\mathbf{r}) \tag{6.10}$$

We will now make the situation more complicated. Let us illuminate the regarded system with light. The light source will be switched on at the moment $t_1 = 0$. We will now have a completely different situation. The system is no more described by the time-independent Hamiltonian $\hat{H}_0$. Instead, the full Hamilton operator is now given by:

$$\hat{H} = \hat{H}(t) = \hat{H}_0 + \hat{V}(t) \tag{6.11}$$

where the certainly time-dependent perturbation operator $\hat{V}$ describes the interaction between the light and the system.

Finally, let us switch off the light at $t_2 = t_0$. Again, (6.8)–(6.10) are valid for the system. The question is: Is there any chance to find the system now in a quantum state m different from that which has been occupied at $t = 0$? If yes, then we will state that the perturbation $\hat{V}$ has caused a quantum transition between the states l and $m \neq l$. It is our task now to understand the conditions necessary for such a transition.

First of all, let us introduce the common terminology:

- If the probability of the transition $l \rightarrow m$ is equal to zero, then the transition is called *forbidden* with respect to the given perturbation $\hat{V}$.

- If the probability of the transition $l \to m$ is larger than zero, then the transition is called *allowed* with respect to the given perturbation $\hat{V}$.
- The recipe which classifies any transition as allowed or forbidden is called a *selection rule*.

Clearly, Einstein's coefficient B_{12} must be correlated to the transition probability. In particular, for a forbidden transition, the Einstein coefficient should be zero.

Let us now turn to the mathematics. We tackle the interesting time interval:

$$0 < t < t_0$$

Because the perturbation may be time-dependent, we have to regard the time-dependent Schrödinger's Equation (4.2) with the Hamiltonian (6.11). In order to find the solution, the unknown wavefunction is expanded into a series of eigenfunctions of the unperturbed Hamiltonian $\hat{H}_0$ (compare (5.22)):

$$\Psi(\mathbf{r}, t) = \sum_n a_n(t) \Psi_n(\mathbf{r}, t) \tag{6.12}$$

where the expansion coefficients a_n may now depend on time. Note that all these eigenfunctions should be ortho-normalized. According to the normalization condition, we have:

$$\sum_n |a_n(t)|^2 = 1 \tag{6.13}$$

The system as described by (6.12) is in a quantum superposition state. Following the usual interpretation of quantum mechanics, the value

$$|a_n(t)|^2$$

has to be understood as the probability to find the system in the nth quantum state, when a corresponding measurement is performed.

It is therefore the time evolution of the expansion coefficients $\{a_n\}$ that is utmost interesting for us. Substituting the wavefunction in (4.2) by (6.12) yields:

$$i\hbar \frac{\partial}{\partial t} \Psi(\mathbf{r}, t) = i\hbar \frac{\partial}{\partial t} \left[\sum_n a_n(t) \Psi_n(\mathbf{r}, t) \right]$$

$$= i\hbar \sum_n \Psi_n(\mathbf{r}, t) \frac{\partial}{\partial t} a_n(t) + i\hbar \sum_n a_n(t) \frac{\partial}{\partial t} \Psi_n(\mathbf{r}, t)$$

$$= \hat{H} \Psi(\mathbf{r}, t) = \hat{H} \sum_n a_n(t) \Psi_n(\mathbf{r}, t)$$

$$= \hat{H}_0 \sum_n a_n(t) \Psi_n(\mathbf{r}, t) + \hat{V} \sum_n a_n(t) \Psi_n(\mathbf{r}, t)$$

On the other hand, for each n we have:

$$i\hbar\frac{\partial}{\partial t}\Psi_n(\mathbf{r}, t) = \hat{H}_0\Psi_n(\mathbf{r}, t)$$

Therefore, we get the final equation:

$$i\hbar\sum_n \Psi_n(\mathbf{r}, t)\frac{\partial}{\partial t}a_n(t) = \hat{V}\sum_n a_n(t)\Psi_n(\mathbf{r}, t) \tag{6.14}$$

Let us now see whether or not the perturbation is able to transfer the system from state l to state m. We multiply (6.14) from the left side with the conjugate complex function $\Psi_m^*(\mathbf{r}, t)$ and integrate over all coordinates. Due to the normalization and orthogonality of wavefunctions, we have:

$$\int \Psi_m^*(\mathbf{r}, t)\Psi_n(\mathbf{r}, t)\mathrm{d}V = \delta_{mn}$$

and therefore, from (6.14) we obtain:

$$i\hbar\dot{a}_m = \sum_n a_n \int \Psi_m^*(\mathbf{r}, t)\hat{V}\Psi_n(\mathbf{r}, t)\mathrm{d}V \tag{6.15}$$

As following from (6.10), we may write:

$$\Psi_m^*(\mathbf{r}, t) = e^{i\frac{E_m}{\hbar}t}\psi_m^*(\mathbf{r})$$

$$\Psi_n(\mathbf{r}, t) = e^{-i\frac{E_n}{\hbar}t}\psi_n(\mathbf{r})$$

Let us remember the transition angular frequency according to:

$$\omega_{mn} \equiv \frac{E_m - E_n}{\hbar} \tag{6.16}$$

Equation (6.15) may then be rewritten as:

$$i\hbar\dot{a}_m = \sum_n a_n V_{mn}e^{i\omega_{mn}t} \tag{6.17}$$

where the (transition) matrix element V_{mn} is defined as:

$$V_{mn} \equiv \int \psi_m^*(\mathbf{r})\hat{V}\psi_n(\mathbf{r})\mathrm{d}V \tag{6.18}$$

Let us now formulate the initial conditions. At $t = 0$, we require:

$$|a_l| = 1; a_{n\neq l} = 0$$

Particularly, at $t = 0$ we have $a_m = 0$. As long as a_l is close to one by modulus, it may be regarded as constant, and a population of the mth state would then require:

$$i\hbar\dot{a}_m\big|_{t\geq0} = a_l V_{ml} e^{i\omega_{ml}t} \neq 0 \qquad (6.19)$$

To fulfill (6.19), it is absolutely necessary that the matrix element V_{ml} is different from zero. What we have found this way is the general formulation of a selection rule: *Given a perturbation $\hat{V}$, it can only cause a quantum transition between the states l and m when the corresponding matrix element of the perturbation operator V_{ml} is different from zero.*

Let us now regard the concrete case of the interaction of a microscopic quantum system with light. We will restrict ourselves to a semiclassical description; i.e., although the material system is described in terms of quantum mechanics, we use a classical description of the electric field of the wave. Therefore, we make use of the ansatz (5.50) while assuming a real field amplitude. When the spatial extension of the system is much smaller than the wavelength, we can neglect the spatial structure of the wave and regard a homogenous but oscillating electric field. In that dipole approximation, the perturbation operator may be written as:

$$\hat{V} = -\mathbf{dE} \qquad (6.20)$$

With

$$E = E_{0,\text{real}} \cos \omega t \qquad (6.20a)$$

Such that

$$\hat{V} = -\frac{\left(\mathbf{dE}_{0,\text{real}} e^{-i\omega t} + \mathbf{dE}_{0,\text{real}} e^{i\omega t}\right)}{2} \equiv -\left(\mathbf{dE}_0 e^{-i\omega t} + \mathbf{dE}_0 e^{i\omega t}\right)$$

with $\mathbf{d}$—dipole moment, $\mathbf{E}$ as the electric field vector (do not confuse with energy), and $\mathbf{E_0}$ its amplitude. From (6.19), we find:

$$a_m(t) \approx a_l \frac{\mathbf{d}_{ml}\mathbf{E}_0}{\hbar} \left\{ \frac{e^{i(\omega_{ml}-\omega)t} - 1}{\omega_{ml} - \omega} + \frac{e^{i(\omega_{ml}+\omega)t} - 1}{\omega_{ml} + \omega} \right\} \qquad (6.21)$$

which is valid as long as $|a_l|\approx1$. $\mathbf{d}_{ml}$ is the matrix element of the dipole operator. From (6.21), we see that in case of a dipole transition, it is the matrix element of the dipole operator that needs to be different from zero. Moreover, we recognize that the transition frequency ω_{ml} plays the role of the resonance frequency: The closer the frequency of the electric field to one of the transition frequencies ω_{ml} or ω_{lm} is, the more probable the transition becomes.

In order to compare this result to (6.3), we will again assume that we deal with an assembly of quantum systems (e.g. atoms or molecules), while N_l is the number of systems in the lth quantum state. Moreover, the transition rate between the states l and m is given by the expression:

$$\frac{dN_m}{dt} \propto \frac{d}{dt}|a_m|^2 \propto |\mathbf{d}_{ml}|^2|\mathbf{E}_0|^2 N_l \tag{6.22a}$$

Interchanging the indices, we obtain

$$\frac{dN_l}{dt} \propto \frac{d}{dt}|a_l|^2 \propto |\mathbf{d}_{lm}|^2|\mathbf{E}_0|^2 N_m \tag{6.22b}$$

Obviously, when comparing with (6.3), one sees that (6.22a) should correspond to absorption, and (6.22b) to stimulated emission of light, provided that $E_m > E_l$. But the proportionality factors are completely identical, because the dipole operator is self-adjoint. So we have:

$$|\mathbf{d}_{ml}|^2 = |\mathbf{d}_{lm}|^2$$

For that reason, Einstein's coefficients B_{12} and B_{21} must be identical. On the other hand, from (6.22a) and (6.22b) it turns out that

$$B_{12} \propto |\mathbf{d}_{12}|^2 \tag{6.23}$$

In fact, this is the most important conclusion for our further treatment of Einstein coefficients. The second conclusion is that, according to (6.16), resonance of the radiation with the two-level system is expected to occur when the condition:

$$\omega = \omega_{21} \equiv \frac{E_2 - E_1}{\hbar} \tag{6.24}$$

is fulfilled. Correspondingly, the energy of the "resonant" photon as defined before must be equal to $\hbar\omega_{21}$.

Finally, let us rewrite (6.6) taking our new findings into account. We shall write:

$$\text{equilibrium}: u(\omega_{21}) = \frac{A_{21}}{B_{12}\left(e^{\frac{\hbar\omega_{21}}{k_B T}} - 1\right)} \tag{6.25}$$

6.4   Consistency Considerations

Let us return to (6.19) and substitute (6.20) into (6.19). Then, dipole transitions are expected to occur whenever the condition:

$$i\hbar\dot{a}_m\big|_{t\geq 0} = a_l V_{ml}e^{i\omega_{ml}t} = -a_l\left(\mathbf{d}_{ml}e^{-i(\omega-\omega_{ml})t} + \mathbf{d}_{ml}e^{i(\omega+\omega_{ml})t}\right)\mathbf{E}_0 \neq 0$$

Moreover, the larger a_m by absolute value becomes, the more efficient the transition will be. Taking into account (6.21), and particularly the obvious resonant behavior of this expression, we see that for an efficient dipole transition, the following conditions should be fulfilled:

$$|a_l| \approx 1$$
$$|\mathbf{d}_{ml}| \to'' \text{large}''$$
$$|\mathbf{E}_0| \to'' \text{large}''$$
$$\omega \to \pm\omega_{ml} \tag{6.26}$$

When comparing with the classical expression such as (2.14), we see that the quantum mechanical transition frequency ω_{ml} plays the role of the classical resonance frequency ω_0 of the corresponding absorption line. The population probability of the mth level and the transition matrix element are obviously important characteristics for calculating the intensity of the absorption line. And, in the absence of an external electric field, no transition is expected. This is a consequence of the semiclassical ansatz (6.20).

At this point, we arrive at an obvious inconsistency. Let us remember that the process of spontaneous emission of light occurs regardless of the presence or absence of an incident light wave. However, in the philosophy of Sect. 6.3.2, (6.21), no quantum transitions are allowed to occur when the field amplitude E_0 is zero. Hence, our theory as developed so far does generally not allow for any spontaneous transition processes. Nevertheless, these transitions occur in real life. So what is the reason for the discrepancy between our theory and experiment?

It turns out that it is the ansatz for the perturbation operator (6.20) that is incompatible with the existence of spontaneous quantum transitions resulting in the emission of a photon. In classical electrodynamics, the energy of the electromagnetic field is zero in the case of a vanishing field strength. A complete quantum mechanical treatment (including the quantization of the field itself) will lead to a somewhat different result discussed in Sect. 5.6.3. In particular, it turns out that in the absence of photons, the field still has a ground-state energy of $\hbar\omega/2$ and thus a certain energy density that is different from zero. On the other hand, when expressing the energy density through the electric field, we obtain a nonzero field strength variance resulting in:

$$\frac{1}{V}\frac{\hbar\omega}{2} = \varepsilon_0\langle E^2\rangle_{n=0} \Rightarrow \langle E^2\rangle_{n=0} = \frac{\hbar\omega}{2\varepsilon_0 V}$$

(compare also (5.61)). These "zero oscillations" of the electromagnetic field (single field mode approximation) provide the perturbation necessary for the initiation of spontaneous effects in optical spectroscopy, among them the spontaneous emission of light. A corresponding calculation will be provided later in Sect. 7.6.2.

Hence, in our semiclassical treatment of Einstein's coefficients, we have to accept the existence of spontaneous transitions as a conclusion from thermodynamic principles, because without them, no relaxation process would be available to return the excited system back to equilibrium once the radiation field is zero at a certain time t (note that in Sect. 6.2, other possible relaxation channels have been excluded). Indeed, when returning to (6.3) and assuming $u = 0$, we have:

$$\frac{dN_2}{dt} = -N_2 A_{21} \Rightarrow N_2 = N_{20} e^{-A_{21} t} \equiv N_{20} e^{-\frac{t}{\tau}} \qquad (6.27)$$

Therefore, the reciprocal value of A_{21} may be interpreted as the lifetime of an excited quantum level that relaxes down to equilibrium by spontaneous emission of light only.

Note that (6.27) asserts that $N_2 \to 0$ when $t \to \infty$. This is a result from (6.3), where no thermal population mechanism of the excited quantum state is incorporated at $u = 0$. In reality, N_2 will approach the thermal population level as given by the corresponding Boltzmann factors.

Hence, the consideration of all three processes of absorption, stimulated, and spontaneous emission altogether allows for a thermodynamically consistent description of the interaction of the two-level system with radiation, provided that $B_{12} = B_{21}$, and provided that spontaneous light emission provides the only relevant relaxation channel. Thereby, our perturbation treatment resulting in (6.22a) and (6.22b) is consistent with the requirement $B_{12} = B_{21}$.

6.5 Application to Practical Problems

6.5.1 Light Absorption and Light Amplification

Let us now come to an utmost important practical application of the theoretical apparatus derived in this chapter so far. From (6.3) and (6.2), we find that

$$\frac{dN_2}{dt} = (N_1 - N_2)(B_{12}u + A_{21}) - N_1 A_{21} \qquad (6.28)$$

In the stationary case, when $dN_2/dt = 0$, we obtain the stationary solution:

$$(N_1 - N_2) = \frac{N_1 A_{21}}{B_{12}u + A_{21}} > 0 \ \forall u \qquad (6.29)$$

No matter how intense the field is, as long as we deal with a two-level system, the stationary population of the ground state will always be higher than that of the excited one. Of course, this conclusion is only true as long as the excited level may exclusively be populated by direct optical pumping from the ground state, as presumed in (6.3). In this case, it is impossible to achieve a stationary population inversion ($N_2 > N_1$). On the contrary, for a sufficiently high u, N_1 and N_2 become nearly equal to each other. In this case, the transition $1 \rightarrow 2$ is called to be saturated.

On the other hand, a population inversion (if it could be achieved anyway) would offer prospective physical effects. Let us for a moment assume that we prepared the system in a way that $N_2 > N_1$ is fulfilled (population inversion). From (6.28), we find that

$$N_2 > N_1 \Rightarrow \frac{dN_2}{dt} < 0$$

As long as we have population inversion in a two-level system, absorption and emission processes in sum tend to transfer the system from the excited to the ground state. This conclusion is particularly true when the spontaneous emission processes may be neglected. Consequently, the energy is transferred from the two-level systems to the radiation field. An incident light beam may therefore be amplified when traveling through a medium with population inversion. The preparation of population inversion is therefore essential for the construction of light amplifiers.

Let us assume a plane wave propagating through a medium along the z-axis. Both stimulated absorption and emission processes may change its amplitude. Phenomenologically, we can introduce a linear absorption coefficient α when assuming that the intensity change per path increment dz is proportional to the intensity I at the point z. Hence, we postulate:

$$\frac{dI}{dz} \propto -I \Rightarrow \frac{dI}{dz} = -\alpha I \Rightarrow I = I(z) = I_0 e^{-\alpha z} \tag{6.30}$$

Equation (6.30) is often called Beer's law.

The previous discussion (see particularly (6.28)) leads us to a very important conclusion on the structure of the absorption coefficient in quantum mechanics. We must assume that the absorption coefficient explicitly depends on the population difference $N_1 - N_2$. Indeed, when neglecting for a moment the spontaneous emission processes, for (6.28) we have:

$$\frac{dN_2}{dt} \approx B_{12} u (N_1 - N_2) \tag{6.30a}$$

When comparing (6.23), (6.30), and (6.30a), the following hypothesis may be obtained:

$$\alpha \propto |\mathbf{d}_{21}|^2 (N_1 - N_2) \tag{6.31}$$

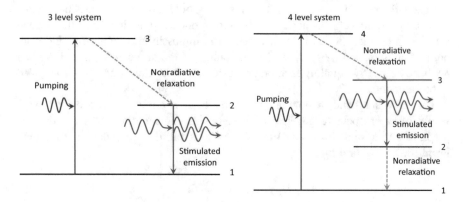

Fig. 6.4 Three- and four-level systems that allow to create population inversion by optical pumping. In the three-level system, population inversion is accomplished between the first and second states. This is what is done in the ruby laser. In the four-level system, one achieves population inversion between the second and third states (e.g., in the neodymium-YAG laser)

If so, it is the sign of the population difference that is crucial for the decision whether the material is absorbing (positive absorption coefficient) or amplifying (negative absorption coefficient). In saturation conditions, one should expect that absorption and stimulated emission processes compensate each other; hence in this case, a light beam would travel through the medium without any damping or amplification. Such a medium appears to be transparent, and the corresponding absorption coefficient is zero.

How to achieve population inversion? In practice, a population inversion may be achieved, for example, by collisions of electrons with atoms, which transfer the atoms into an excited state. For example, gas lasers such as the helium–neon laser work on this principle. Another way is to use three- or four-level systems with optical pumping, as shown in Fig. 6.4.

In the terminology of laser physics, a medium with population inversion is called an *active medium*. Keeping in mind that the absorption coefficient of an active medium is negative, from Beer's law (6.30) we find:

$$I = I_0 e^{-\alpha x} = I_0 e^{|\alpha|x} \tag{6.32}$$

Hence, in an active medium, the intensity of a propagating wave is expected to exponentially grow in intensity. What we obtain this way is an amplifier of electromagnetic waves.

6.5.2   Feedback

In our further discussion, we will simply assume that a population inversion has been achieved anyway. In complete analogy to electronics, we only need to add a positive feedback to an amplifier in order to build a generator of electromagnetic waves. That kind of light amplification due to stimulated emission, combined with a feedback mechanism, leads us to a specific kind of light source that is called a laser.

Let us have a look at Fig. 6.5.

Figure 6.5 sketches the idea of combining an amplifying element with a feedback mechanism. Let us start from the left side. We assume an external input, for example, an electromagnetic wave. The action of the amplifying element shall be to magnify the input by a complex factor B. After passing the amplifier for one time, we obtain an output according to:

$$\text{output} = B \cdot \text{input}, \quad |B| > 1 \tag{6.33}$$

Let us now discuss what happens when the mentioned feedback mechanism comes into play. A part of the output (say, the nominal output multiplied with a constant F, where F is again a possibly complex number with $|F| < 1$) is transferred back to the input side, again amplified, and so on. Then, instead of the simple output as given by (6.33) we get an effective output obtained as the result of an infinite number of loops through the amplifier caused by the feedback mechanism. Mathematically, this may be expressed in the following manner:

$$\text{effective output} = \left[(1 - F) + (1 - F)BF + (1 - F)B^2 F^2 + \ldots \right]$$

$$* \,\text{output} = B(1 - F) \sum_{j=0}^{\infty} (BF)^j \,\text{input} \equiv B_{\text{eff}} * \text{input}$$

$$B_{\text{eff}} = (1 - F)B \sum_{j=0}^{\infty} (FB)^j = \frac{(1 - F)B}{1 - FB} \tag{6.34}$$

Particularly, (6.34) leads to an infinitively large effective enhancement factor in the limit $BF \rightarrow 1$. Writing down $B \equiv |B|e^{i\phi_B}$ and $F \equiv |F|e^{i\phi_F}$, the condition

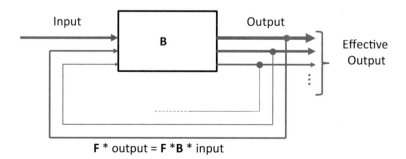

F * output = F *B * input

Fig. 6.5 An amplifier with feedback

$BF \to 1$ breaks down into two conditions for amplitude and phase according to:

$$|B| \to |F^{-1}|$$

$$\phi_B + \phi_F = 2j\pi, \text{ where } j \text{ is an integer} \tag{6.35}$$

In the case that the effective enhancement becomes infinitively large, an arbitrarily small input (e.g., a single photon, accidentally emitted as the result of a spontaneous emission process) may lead to a finite effective output of the system (in our case to electromagnetic irradiation). There is no contradiction to energy conservation, because the amplifying element (in our case the active medium) is pumped by an external energy source. Such a system works as a generator of light and is called a laser. Of course, it will also generate light when the light amplification defined by B is larger than the threshold value defined by (6.35).

Consequently, we have to fulfill two conditions in order to construct a generator of light. First of all, we have to take care that the light amplification is large enough to compensate any losses of light that leave the system. Technically, this is accomplished through a sufficiently high population inversion in the active medium, and the corresponding mathematical criterion is called the *laser condition*, but it will not be derived here.

The necessary feedback is usually achieved placing the active medium into a resonator (the cavity), which may be built up by two parallel mirrors. Let R_1 and R_2 be their reflectances (reflectance = ratio between the intensities of reflected and incident light—see later Sect. 13.5.2), while L is the cavity length. Usually, the active medium does not fill the complete cavity, but extends for a smaller length l. This situation is shown in Fig. 6.6.

Let us now rewrite (6.35) for the special case of the geometry from the figure. Imagine a light wave that performs one loop in the cavity. In performing one loop in the cavity, the wave has to cross the active medium for two times, and therefore,

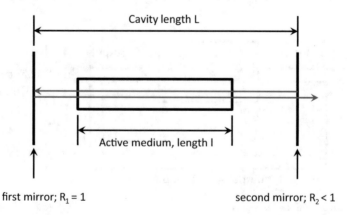

Fig. 6.6 Simple laser geometry

its intensity will grow by a factor $e^{2|\alpha|l}$. On the other hand, at the second mirror some light will escape from the cavity, so that after one loop, the intensity of the light wave becomes:

$$I \text{(after one loop)} = I_0 e^{2|\alpha|l} R_2 \tag{6.36}$$

Equation (6.36) describes an extremely simplified case, because no loss mechanisms inside the cavity have been taken into account. But the principle is nevertheless clear. The value $\sqrt{R_2 e^{2|\alpha|l}}$ is nothing else than the absolute value of the product BF as fixed in (10.58). Consequently, for light generation it is necessary that

$$\sqrt{R_2 e^{2|\alpha|l}} \geq 1 \Rightarrow e^{2|\alpha|l} \geq (R_2)^{-1} \tag{6.37}$$

is fulfilled.

The condition on the phases may be written in a similarly simple way. After one loop, according to (6.35), the phase of the wave is allowed to have changed only for an integer multiple of 2π. We thus have:

$$\phi \text{(after one loop)} = \phi_0 + 2j\pi \tag{6.38}$$

The phase gain is thus equal to $2j\pi$. On the other hand, the phase gain for a single loop of a light wave between two plane interfaces is equal to $2kL = 4\pi L/\lambda$ (provided that the refractive index is equal to one. The concept of the refractive index will be introduced formally in Chap. 13, but I will assume at this point that you, dear reader, already have an idea on what a refractive index is). This is clearly a rough simplification, but it still allows highlighting the main principles of the laser action. Moreover, possible phase shifts occurring upon reflection at the mirrors will also not be taken into account in our treatment, in order to keep the picture as simple as possible.

We therefore find the condition (remember Fig. 4.1 in this context):

$$2j\pi = \frac{4\pi L}{\lambda} \Rightarrow \lambda = \lambda_j = \frac{2L}{j} \tag{6.39}$$

The conditions (6.37) and (6.39) have to be fulfilled together in order to get the laser working. Let us therefore analyze their common solutions.

We will start from condition (6.37). It obviously defines a threshold value for the amplification coefficient which must be exceeded in order to get the laser work. As the amplification coefficient is wavelength-dependent (instead of the familiar absorption line, we now have an amplification line), one may expect that (6.37) defines one or several spectral ranges where the amplification coefficient is sufficiently large to achieve light generation. In order to handle this in a more convenient mathematical way, let us rewrite (6.37) in the more symbolic manner:

$$|\alpha| = |\alpha(\omega)| \geq \text{threshold} \tag{6.40}$$

where the generation threshold according to (6.37) is defined by the reflectivity of the second mirror and the length of the active medium. In general, other loss mechanisms may also be present in the cavity. In this case, the threshold will be enhanced, but the general formulation of criterion (6.40) remains the same. Of course, the threshold value itself may also depend on the frequency.

On the other hand, (6.39) defines a series of discrete wavelength values that are equidistant at the frequency scale as long as the refractive index is considered to be constant (in our case, it is equal to 1 regardless of the frequency). Indeed, from (6.39) the allowed frequency values may be calculated according to:

$$\omega_j = \frac{2c\pi}{\lambda_j} = j\frac{c\pi}{L} \qquad (6.41)$$

where $c\pi/L$ is the line spacing in angular frequency units. Equation (6.41) defines the set of allowed so-called longitudinal resonator modes.

The light frequencies that suffice both criteria (6.40) and (6.41) are given by a set of discrete lines confined to the frequency region defined by (6.40). Figure 6.7 sketches this situation in a simplified manner.

In practice, because of some Darwinian "survival of the fittest," a real laser will not generate light at all these principally allowed frequencies simultaneously; instead, usually a couple of modes is excited that completely "consume" the population inversion within the active medium. If you wish to prepare ultrashort laser pulses, a broad frequency spectrum is required, and you need to excite a large number of equidistant longitudinal modes for this purpose. On the contrary, if you need highest spectral resolution, it is favorable to let the laser generate at only one longitudinal mode. Both these extreme situations need special experimental precautions, and their discussion is beyond the scope of this book.

Fig. 6.7 Frequency spectrum of a laser as shown in Fig. 6.6. The vertical lines represent the longitudinal resonator modes. Generation may only occur when the amplification coefficient (red curve) exceeds the generation threshold (black curve). Hence, from the full set of longitudinal resonator modes only those in navy constitute the laser spectrum

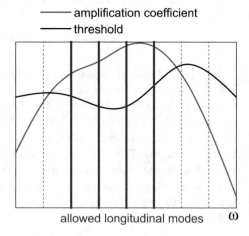

allowed longitudinal modes ω

6.6   Advanced Material

6.6.1   The Parity Selection Rule

The elaboration of selection rules in terms of (6.19) may be tedious in practice, because rather uncomfortable integrals may have to be calculated. Fortunately, in highly symmetric systems, simple selection rules may be obtained from symmetry considerations. An example for this is the parity selection rule, relevant in systems with an inversion centrum (centrosymmetric systems).

A formal inversion $\mathbf{r} \rightarrow -\mathbf{r}$ results in

$$\hat{H}(-\mathbf{r})\psi(-\mathbf{r}) = E\psi(-\mathbf{r})$$

Let us now regard the existence of an inversion center at $\mathbf{r} = \mathbf{0}$. In this case, we have:

$$\hat{H}(\mathbf{r}) = \hat{H}(-\mathbf{r})$$

The Hamiltonian of such a system commutes with the inversion operator $\hat{I}$, defined by:

$$\hat{I}f(\mathbf{r}) \equiv f(-\mathbf{r})$$

Indeed

$$\left.\begin{array}{l}\hat{I}\hat{H}(\mathbf{r})\psi(\mathbf{r}) = \hat{H}(-\mathbf{r})\psi(-\mathbf{r}) = E\psi(-\mathbf{r}) \\ \hat{H}(\mathbf{r})\hat{I}\psi(\mathbf{r}) = \hat{H}(\mathbf{r})\psi(-\mathbf{r}) = \hat{H}(-\mathbf{r})\psi(-\mathbf{r}) = E\psi(-\mathbf{r})\end{array}\right\} \Rightarrow \left[\hat{I}, \hat{H}\right] = 0$$

Therefore, $\hat{H}$ and $\hat{I}$ must have a complete system of common eigenfunctions. Let us have a look at these eigenfunctions and assume that $f(\mathbf{r})$ is an eigenfunction of $\hat{I}$ with the eigenvalue I. We find:

$$\left.\begin{array}{l}\hat{I}f(\mathbf{r}) \equiv f(-\mathbf{r}) = If(\mathbf{r}) \\ \hat{I}\hat{I}f(\mathbf{r}) \equiv \hat{I}f(-\mathbf{r}) \equiv f(\mathbf{r}) = \hat{I}If(\mathbf{r}) = I^2 f(\mathbf{r})\end{array}\right\} I^2 = 1 \Rightarrow I = \pm 1 \Rightarrow f(-\mathbf{r}) = \pm f(\mathbf{r})$$

Once the eigenfunctions of $\hat{H}$ must be eigenfunctions of the inversion operator in centrosymmetric systems, we have two types of wavefunctions:

$$\psi(-\mathbf{r}) = \begin{cases} +\psi(\mathbf{r}) \\ -\psi(\mathbf{r}) \end{cases}$$

In the case $\psi(-\mathbf{r}) = +\psi(\mathbf{r})$, we say that the wavefunction has even parity, while in the case $\psi(-\mathbf{r}) = -\psi(\mathbf{r})$, the parity is odd.

For an electric dipole-allowed transition, from Eq. (6.19) it follows that

$$\mathbf{d}_{nm} \neq 0$$

or

$$\mathbf{d}_{nm} = q\mathbf{r}_{nm} = q \int \psi_n^*(\mathbf{r})\,\mathbf{r}\psi_m(\mathbf{r})dV \neq 0$$

Note that $\mathbf{r}$ is an odd function itself. Therefore, if both Ψ_n and Ψ_m are even, the whole integrand is an odd function, and integration in symmetric integration limits results in a matrix element that is definitely equal to zero. The same is valid of course, if both wavefunctions are odd.

Therefore, in order to excite a dipole transition in a centrosymmetric system, it is necessary (but not sufficient) that the participating quantum states have a different parity. This is the essence of the *parity selection rule*.

6.7   Fermi's Golden Rule

Let us return to the stuff discussed in Sect. 6.3.2 and use this short section for deriving a useful expression for the transition rate between two states $|l\rangle$ and $|m\rangle$ in the particular case that the perturbation is given by [2]:

$$\hat{V}(t) = \theta(t)\hat{V}_0 \tag{6.42}$$

where $\hat{V}_0$ does not explicitly depend on time. Again, we assume that at $t < 0$, the system was definitely in the state $|l\rangle$. Then, when again making use of the ansatz (6.12), from (6.19) it follows that

$$a_m(t) \approx -\frac{\langle m|\hat{V}_0|l\rangle}{\hbar}\left\{\frac{e^{i\omega_{ml}t} - 1}{\omega_{ml}}\right\} \Rightarrow w_{l\to m} = |a_m(t)|^2 \approx \frac{4\left|\langle m|\hat{V}_0|l\rangle\right|^2}{\hbar^2\omega_{ml}^2}\sin^2\frac{\omega_{ml}t}{2}$$

Here, it has been assumed that $|a_l| \approx 1$ again (compare also (6.21)).
The transition rate $\frac{dw_{l\to m}}{dt}$ is obtained according to:

$$\frac{dw_{l\to m}}{dt} = \frac{2\left|\langle m|\hat{V}_0|l\rangle\right|^2}{\hbar^2\omega_{ml}}\sin\omega_{ml}t$$

Let us now regard the limit $t \to \infty$. We have:

$$\lim_{t\to\infty}\frac{dw_{l\to m}}{dt} = \frac{2\left|\langle m|\hat{V}_0|l\rangle\right|^2}{\hbar^2}\cdot\pi\underbrace{\lim_{t\to\infty}\left[\frac{1}{\pi}\frac{\sin\omega_{ml}t}{\omega_{ml}}\right]}_{=\delta(\omega_{ml})}$$

$$\Rightarrow \lim_{t\to\infty} \frac{\mathrm{d}w_{l\to m}}{\mathrm{d}t} = \frac{2\pi}{\hbar}\left|\langle m|\hat{V}_0|l\rangle\right|^2 \delta(E_m - E_l)$$

Provided that there is not only one potential final state $|m\rangle$, but rather a continuous distribution of potential final states, then it makes sense to introduce an energy density of states as $D(E) \equiv \frac{\mathrm{d}Z}{\mathrm{d}E} = \frac{\mathrm{d}Z}{\hbar\mathrm{d}\omega}$ (compare Sect. 4.6.1). The complete transition rate is then obtained as:

$$\frac{\mathrm{d}w}{\mathrm{d}t} \equiv \int \lim_{t\to\infty} \frac{\mathrm{d}w_{l\to m}}{\mathrm{d}t} D(E_m)\mathrm{d}E_m$$
$$= \frac{2\pi}{\hbar}\int \left|\langle m|\hat{V}_0|l\rangle\right|^2 D(E_m)\delta(E_m - E_l)\mathrm{d}E_m$$
$$= \frac{2\pi}{\hbar}\left|\langle m|\hat{V}_0|l\rangle\right|^2 D(E_m)|_{E_m=E_l}$$

The equation

$$\frac{\mathrm{d}w}{\mathrm{d}t} = \frac{2\pi}{\hbar}\left|\langle m|\hat{V}_0|l\rangle\right|^2 D(E_m)|_{E_m=E_l} \tag{6.42a}$$

is also called Fermi's golden rule.

Note further that the same derivation may be repeated with perturbations of the type:

$$\hat{V}(t) = \theta(t)\hat{V}_0\mathrm{e}^{\mp i\omega t} \tag{6.43}$$

Equation (6.21) corresponds to a particular case of such type of perturbation. Instead of (6.42a), we then find:

$$\frac{\mathrm{d}w}{\mathrm{d}t} = \frac{2\pi}{\hbar}\left|\langle m|\hat{V}_0|l\rangle\right|^2 D(E_m)|_{E_m=E_l\pm\hbar\omega} \tag{6.43a}$$

6.8 Tasks for Self-check

6.7.1. Assume a one-dimensional harmonic oscillator in its ground state (vibrational quantum number $n = 0$). What would be the oscillator strength for a transition $n = 0 \to n = 3$ in electric dipole approximation?

6.7.2. Assume an electron confined in a rectangular potential box of length $L = 1$ nm with impermeable walls (one-dimensional case). Obtain the transition matrix element x_{jk} and the corresponding oscillator strength for the transition $k = 1 \to j = 2,3,4$.

6.7.3. Obviously, both the box potential from Sect. 4.3.3 and the harmonic potential from Sect. 5.5.1 are centrosymmetric. Judge the parity of their eigenstates with respect to the inversion center! Interpret the results of the previous task in terms of the parity selection rule!

References

Specific References

1. O. Stenzel, *The Physics of Thin Film Optical Spectra. An Introduction* (Springer, 2016), pp. 229–254
2. M. Bartelmann, B. Feuerbacher, T. Krüger, D. Lüst, A. Rebhan, A. Wipf, *Theoretische Physik* (Springer Spektrum, 2015), pp. 988–989

General Literature

3. W. Demtröder, *Atoms, Molecules, and Photons* (Springer, 2010)
4. H. Haken, H.C. Wolf, *The Physics of Atoms and Quanta* (Springer, 2005)

Planck's Formula and Einstein Coefficients

7

Abstract

The discussion of Einstein coefficients is completed by a derivation of corresponding quantum mechanical expressions in the electric dipole approximation. As a side result, Planck's formula for blackbody radiation is obtained, and Bohr's correspondence principle is introduced. As applications, we discuss basic physics of lasers and introduce the concept of spectrally selective solar absorbers.

7.1   Starting Point

Let us now return to our two-level systems, interacting with electromagnetic radiation. In Chap. 6, we have introduced three Einstein's coefficients that described the efficiency of the three elementary processes of light absorption, spontaneous light emission, and stimulated light emission. For a quantitative treatment of the interaction of light with the two-level system, we need, of course, knowledge of these coefficients. So far, we have found that the Einstein coefficients B_{12} and B_{21} are identical (compare (6.7)). What we now need are expressions for the remaining two coefficients.

We will start with the derivation of a relation between the Einstein coefficients A_{21} and B_{12}. To do so, we will discuss the particular case of thermodynamic equilibrium between radiation and the two-level system. In this case, from (6.6), (6.7), and (6.25) we have:

$$\text{equilibrium} : u = \frac{A_{21}}{B_{12}\left(e^{\frac{E_2-E_1}{k_BT}} - 1\right)} = \frac{A_{21}}{B_{12}} \frac{1}{\left(e^{\frac{\hbar\omega}{k_BT}} - 1\right)} \tag{7.1}$$

© The Author(s), under exclusive license to Springer Nature Switzerland AG 2022   143
O. Stenzel, *Light–Matter Interaction*, UNITEXT for Physics,
https://doi.org/10.1007/978-3-030-87144-4_7

7.2   Physical Idea

In this chapter, we will make use of two different concepts. The first one is well established in physics, namely the basic principle that every macroscopic system— if it is not externally perturbed—tends to evolve toward a specific macroscopic state that is called thermodynamic equilibrium. In thermodynamic equilibrium, the average occupation of quantum states is controlled by Gibbs factors, and we will use this circumstance to derive an explicit expression for the spectral energy density of the electromagnetic radiation in equilibrium conditions. This way we will obtain what is called Planck's formula. A comparison with (7.1) will yield a relation between A_{21} and B_{12}.

Remark We have introduced the terminus of a macroscopic state here. A detailed definition of this concept is beyond the scopes of this book, but it might be intuitively clear that the description of a macroscopic system built up from some 10^{23} atoms in terms of a wavefunction depending on the coordinates of all these atoms would be a hopeless effort. Therefore, systems consisting of a huge number of particles are more conveniently described in terms of mean values of physical quantities characterizing the macroscopic system and its (still macroscopic) subsystems. The set of the mentioned mean values may then be used to define the macroscopic state of a system.

Then, there will be only one Einstein's coefficient left that needs to be determined. In the second part of this chapter, we will focus on an independent derivation of the coefficient A_{21}. As we have discussed in Sect. 6.4, our semiclassical perturbation treatment of the light–matter interaction is principally insufficient to deliver this coefficient. So, we need another concept for its derivation. In fact, we have two choices:

- Starting from Sect. 5.6.3, we can develop a completely quantum mechanical approach of the light–matter interaction. Then, Einstein's coefficients will be obtained practically as a side result. This will be the theoretically strong approach that is included into the Advanced Material Sect. 7.6.2.
- We can make use of an independent concept called the correspondence principle as postulated by Niels Bohr, which asserts that for sufficiently large energies, the quantum mechanical expressions shall converge to the classical ones.

In our basic treatment, we will use the correspondence principle, which is a useful and convenient heuristic tool in our semiclassical treatment of the interaction of light with matter. As mentioned by R.B. Laughlin (A different Universe: Reinventing Physics from the Bottom Down; Basic Books, New York (2005)), this principle is mathematically unprovable. Let us therefore note in this context that any result obtained in this book by means of the correspondence principle coincides with the corresponding result of a strong quantum mechanical treatment, while the latter is obtained with much more mathematical effort. This may be checked by the reader himself when comparing with corresponding textbooks on quantum mechanics. Thus, we use the correspondence principle as some shortcut in the derivation procedure only.

7.3 Theoretical Considerations

7.3.1 Planck's Formula

Ansatz

Equation (7.1) describes the spectral density of radiation in equilibrium with an assembly of two-level systems, held at a temperature T. The purpose of this section is to independently derive an alternative expression for this spectral density, which might be compared to Eq. (7.1), and will therefore give us an expression for the ratio between the coefficients A_{21} and B_{21}. The formula we will obtain is well known as the famous Planck's formula. We start from the definition (6.1):

$$u \equiv \frac{\mathrm{d}E}{V\,\mathrm{d}\omega}$$

The energy of the photon ensemble per angular frequency interval may be expressed as the energy per photon $E = \hbar\omega$ with an angular frequency ω, multiplied with the average number of photons $(\langle N_p \rangle)$ expected to be found in the corresponding quantum state, again multiplied with the number of those quantum states per angular frequency interval at the relevant frequency (the density of states $\frac{\mathrm{d}Z}{\mathrm{d}\omega}$). Hence, we make use of the approach:

$$\frac{\mathrm{d}E}{\mathrm{d}\omega} = \hbar\omega\langle N_p \rangle \frac{\mathrm{d}Z}{\mathrm{d}\omega} \tag{7.2}$$

It is now our task to calculate the single terms encountering into (7.2). Let us start with the average number of photons $\langle N_p \rangle$. Note that, in contrast to Sect. 5.6.3, we use here the symbol N_p for the number of photons, in order to avoid confusion with the quantum number n characterizing states in the box potential.

Planck's Distribution

In order to obtain an expression for $\langle N_p \rangle$ in the equilibrium case, let us calculate the energy accumulated in a quantum state when a number of N_p photons each with an energy $\hbar\omega$ are excited. Obviously, its energy will be $N_p\,\hbar\omega$. In equilibrium, the probability w to find the state with N_p photons excited is given by Boltzmann's factor (to be accurate: for the use of the Boltzmann factor instead of the Gibbs factor in this particular case, compare [1])

$$w\left(N_p\right) = \frac{\xi^{N_p}}{\sum_{N_p} \xi^{N_p}} \text{ with } \xi \equiv \mathrm{e}^{-\frac{\hbar\omega}{k_B T}}$$

Here, we did not consider the ground-state energy of the photon mode $\hbar\omega/2$, because it will immediately cancel out in the above expression. The average number of photons excited in the mode is now calculated in the usual way:

$$\langle N_p \rangle = \sum_{N_p} N_p w(N_p) = \frac{\sum_{N_p} N_p \xi^{N_p}}{\sum_{N_p} \xi^{N_p}} = \xi \frac{\mathrm{d}}{\mathrm{d}\xi} \ln \sum_{N_p} \xi^{N_p}$$

$$= \xi \frac{\mathrm{d}}{\mathrm{d}\xi} \ln(1-\xi)^{-1} = \frac{\xi}{1-\xi} \Rightarrow \langle N_p \rangle = \frac{1}{e^{\frac{\hbar\omega}{k_B T}} - 1} \qquad (7.3)$$

Equation (7.3) is known as the Planck distribution, applicable to systems of photons. It is a special case of the Bose–Einstein distribution.

Density of States and Planck's Formula

What remains to determine is the density of states. Let us remember a simple one-dimensional problem, namely the one-dimensional movement (e.g., along the x-axis) of a particle between two impermeable walls (Sect. 4.3.3). The corresponding wavefunctions have already been sketched in Fig. 4.1. We are now however facing macroscopic dimensions in order to discuss a thermodynamic system. Therefore, $L \to \infty$, and the standing wave picture rather resembles what is shown in Fig. 7.1. Of course, we will further assume that the propagating "particle" is nothing else than a photon.

When the separation between the two walls is L, from Sect. 4.3.3 it is known that the allowed eigenstates of the system correspond to standing waves as shown in Fig. 7.1, so that the allowed wavelength values become:

$$\lambda_n = \frac{2L}{n_x}; \; n_x = 1, 2, 3, \ldots \qquad (7.4)$$

That corresponds to allowed values of the wavevector:

$$k_x = \frac{2\pi}{\lambda} = \frac{\pi n_x}{L}$$

Fig. 7.1 Standing wave as solution of Schrödinger's equation for a particle between two walls

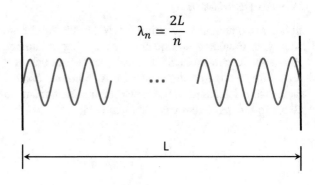

where n_x is now a quantum number.

Let us generalize our result to the three-dimensional case. Instead of the system like shown in Fig. 7.1, we should now imagine a hollow cube with a volume $V = L^3$ and count the allowed states inside the cube. Accordingly, we obtain for the wavevector:

$$k^2 = k_x^2 + k_y^2 + k_z^2 = \left(\frac{\pi}{L}\right)^2 \left(n_x^2 + n_y^2 + n_z^2\right) \equiv \left(\frac{\pi}{L}n\right)^2 \tag{7.5}$$

(compare (4.11)). From (4.12), we have:

$$Z = \frac{\pi}{3}n^3 \Rightarrow \frac{\mathrm{d}Z}{\mathrm{d}n} = \pi n^2 \tag{7.6}$$

Here, we assumed a degeneration factor $g = 2$, corresponding to two independent possible transversal polarization directions of the photon. From (7.5), it turns out that

$$k^2 = \frac{\omega^2}{c^2} = \left(\frac{\pi}{L}n\right)^2 \Rightarrow n^2 = \left(\frac{L\omega}{\pi c}\right)^2 \tag{7.7}$$

and

$$\frac{\mathrm{d}n}{\mathrm{d}\omega} = \frac{L}{\pi c} \tag{7.8}$$

Equations (7.6)–(7.8) in combination finally yield:

$$\frac{\mathrm{d}Z}{\mathrm{d}\omega} = \frac{\mathrm{d}Z}{\mathrm{d}n}\frac{\mathrm{d}n}{\mathrm{d}\omega} = \pi \left(\frac{L\omega}{\pi c}\right)^2 \frac{L}{\pi c} = \frac{L^3\omega^2}{c^3\pi^2} = V\frac{\omega^2}{c^3\pi^2} \tag{7.9}$$

We have now calculated all values that encounter into (7.2). In sum, we find:

$$\frac{\mathrm{d}E}{\mathrm{d}\omega} = \hbar\omega\langle N_p\rangle\frac{\mathrm{d}Z}{\mathrm{d}\omega} = \hbar\omega\frac{1}{e^{\frac{\hbar\omega}{k_BT}} - 1}V\frac{\omega^2}{c^3\pi^2} = V\frac{\hbar\omega^3}{c^3\pi^2}\frac{1}{e^{\frac{\hbar\omega}{k_BT}} - 1}$$

and

$$u(\omega, T) = \frac{\hbar\omega^3}{c^3\pi^2}\frac{1}{e^{\frac{\hbar\omega}{k_BT}} - 1} \tag{7.10}$$

Equation (7.10) represents Planck's famous formula for the so-called blackbody radiation.

Figure 7.2 shows the frequency dependence described by (7.10) for two different temperatures.

As it is obvious from Fig. 7.2, a body held at a higher temperature emits photons that have in average a higher frequency (and consequently a higher photon energy) than those emitted by a body held at a lower temperature. Thus, photons emitted from the surface of the sun have much higher average energies than those emitted by the surface of the earth.

Fig. 7.2 The
$u(\omega, T) = \frac{\hbar\omega^3}{c^3\pi^2} \frac{1}{e^{\frac{\hbar\omega}{k_B T}} - 1}$ for
two different temperatures,
assuming $T_1 < T_2$

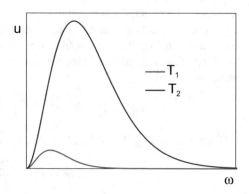

7.3.2   Final Expressions for Einstein Coefficients in the Dipole Approximation

Equation (7.10) describes the spectral density of electromagnetic irradiation in a hollow cube with walls held at a temperature T in the equilibrium case. Hence, for any selected angular frequency ω, the absorption of photons caused by the atoms or molecules of the wall is compensated by the processes of spontaneous and stimulated emission of photons by the same atoms or molecules. This is exactly the situation for which (7.1) has been derived. Therefore, comparing (7.1) and (7.10), we find an important relationship between Einstein coefficients, namely:

$$\frac{A_{21}}{B_{21}} = \frac{\hbar\omega^3}{\pi^2 c^3} \tag{7.11}$$

Therefore, only one of Einstein's coefficients remains to be determined. In the forthcoming, we will directly calculate the coefficient A_{21} making use of the correspondence principle. But before doing so, let us make one remark concerning (7.11):

Remark The relation between A_{21} (efficiency of spontaneous emission processes) and B_{21} (efficiency of stimulated processes) turns out to be strongly frequency-dependent. At low frequencies, the stimulated processes dominate, while at higher frequencies spontaneous processes become more efficient. The practical conclusion is that in infrared spectroscopy (low frequencies) one usually investigates stimulated processes (absorption spectroscopy), while in VIS and UV analytics (much higher frequencies), fluorescence spectroscopy has become an utmost important spectroscopic tool.

Let us now come to the derivation of A_{21} in the dipole approximation.

Recalling our knowledge on the structure of Einstein coefficients, from (6.23) and (7.11) and considering that $d = qx$ we may write:

$$A_{21} = C|x_{21}|^2 \tag{7.12}$$

C is a probably frequency-dependent coefficient, and it may be determined regarding any special case that is accessible to analytic calculations. Having found C, we may write down the final expression for Einstein coefficients. We choose the particular case of a harmonic oscillator and calculate the decay rate of the energy accumulated in the oscillator, when the latter is allowed to dissipate as a result of the spontaneous emission of photons. From the correspondence principle, we may write:

$$E \rightarrow \infty : \left. \frac{dE}{dt} \right|_{classics} = \left. \frac{dE}{dt} \right|_{quantum\ mechanics} \tag{7.13}$$

That means, for sufficiently high energies, the quantum mechanical expressions shall converge to the classical ones. From classical electrodynamics, we know that (compare (2.16)):

$$\left. \frac{dE}{dt} \right|_{classics} = -\frac{q^2}{6\pi \varepsilon_0 c^3} \langle \ddot{x}^2 \rangle \big|_t \tag{7.14}$$

The average is taken over a relevant time period, say the duration of one oscillation, the latter being performed along the x-axis. For the motion of a classical harmonic oscillator, we have:

$$x = x_0 \cos \omega_0 t \text{ with } E = \frac{m \omega_0^2 x_0^2}{2} \tag{7.15}$$

In the case of weak damping, after averaging from (7.14) and (7.15) we find:

$$\left. \frac{dE}{dt} \right|_{classics} = -\frac{q^2 \omega_0^2}{6\pi \varepsilon_0 c^3 m} E \tag{7.16}$$

Let us now turn to the quantum mechanical case. The energy of the harmonic oscillator is given by (5.34):

$$E_n = \hbar \omega_0 \left(n + \frac{1}{2} \right) \Rightarrow E_{n \rightarrow \infty} \approx \hbar \omega_0 n \tag{7.17}$$

Any quantum transition from level n to n-1 leads to an energy decay per time interval given by (compare (6.3) and (7.12)):

$$\left. \frac{dE}{dt} \right|_{quantum\ mechanics} = -\hbar \omega_0 A_{n,n-1} = -\hbar \omega_0 C \left| x_{n,n-1} \right|^2 \tag{7.18}$$

The matrix elements for the coordinate x of a harmonic oscillator are well known from (5.31):

$$\left| x_{n,n-1} \right|^2 = \frac{\hbar n}{2\omega_0 m} \approx -\frac{E}{2\omega_0^2 m} \tag{7.19}$$

That leads us to:

$$\frac{dE}{dt}\bigg|_{\text{quantum mechanics}} = -\frac{\hbar C E}{2\omega_0 m} \tag{7.20}$$

The constant C may now easily be found combining (7.13), (7.16), and (7.20). We obtain:

$$C = \frac{q^2 \omega_0^3}{3\varepsilon_0 \pi \hbar c^3}$$

So that from (7.12), Einstein's coefficient A_{21} is found as (we now replace the resonance frequency ω_0 by the transition frequency ω_{21}):

$$A_{21} = \frac{q^2 \omega_{21}^3 |x_{21}|^2}{3\varepsilon_0 \pi \hbar c^3} = \frac{\omega_{21}^3 |d_{21}|^2}{3\varepsilon_0 \pi \hbar c^3} \tag{7.21}$$

The other coefficients follow from (7.11) and (6.7):

$$B_{21} = B_{12} = \frac{\pi |d_{21}|^2}{3\varepsilon_0 \hbar^2} \tag{7.22}$$

Expressions (7.21) and (7.22) represent the final expressions for Einstein coefficients in the electric dipole approximation.

Let us make two final remarks:

Firstly, our derivation was based on the perturbation operator (6.20), which describes the electric dipole interaction. Therefore, the resulting Einstein coefficients are only valid in the (electric) dipole approximation. If, for any reason, that dipole transition is forbidden, nevertheless quantum transitions may occur as the result of other types of interaction—for example, magnetic dipole interaction. The corresponding Einstein coefficients may be derived analogously. However, when the electric dipole transition is allowed, in the nonrelativistic case it is usually much stronger than the other interaction terms, so that it is often sufficient to regard this first term in the multipole expansion of the interaction of an electromagnetic wave with matter.

Secondly, in the absence of incident irradiation ($u = 0$), the population of the excited quantum state decays according to (compare (6.27)):

$$N_2 = N_{20} e^{-A_{21} t} \equiv N_{20} e^{-\frac{t}{\tau}}$$

Therefore, the reciprocal value of A_{21} may be interpreted as the lifetime of an excited quantum level. Expression (7.23) enables us to estimate the energy decay time as introduced in Sect. 2.6 and consequently the natural linewidth as long as the latter is determined by radiative relaxation processes only.

Indeed, from (7.21) and (6.27), the radiative lifetime of the excited state in a two-level system in the dipole approximation turns out to be equal to:

$$\tau = \frac{3\varepsilon_0\pi\,\hbar c^3}{q^2\omega_{21}^3|x_{21}|^2} = \frac{3\varepsilon_0\pi\,\hbar c^3}{\omega_{21}^3|d_{21}|^2} \tag{7.23}$$

7.4 Consistency Considerations

For well-allowed dipole transitions in the VIS/UV, (7.24) yields lifetimes of the order of 10^{-8}–10^{-9} s. A measurement of these (and even shorter) relaxation processes requires the application of ultrafast spectroscopic tools, and it forms an individual branch of modern spectroscopy.

Note that the mentioned lifetime is of the order of what we have earlier estimated in terms of classical physics (Sect. 2.3.3). On the other hand, the lifetime becomes infinitively large, when the matrix element of the dipole operator vanishes. When we excite (by any means) a quantum level that cannot relax into the ground state via dipole radiation, then such a state may remain excited for a considerably long time. Of course, this time will be not infinitively large, because in fact there are still other relaxation channels than the electric dipole irradiation. But the lifetime may easily extend for minutes or hours, and this is the reason for the well-known phenomenon of phosphorescence observed in various materials.

It is now interesting to compare (7.24) with its classical counterpart (2.20). When setting:

$$\tau_{\mathrm{QM}} = \frac{3\varepsilon_0\pi\,\hbar c^3}{q^2\omega_{21}^3|x_{21}|^2} \text{ and } \tau_{\mathrm{classics}} = \frac{6\pi\varepsilon_0 c^3 m}{q^2\omega_0^2}; \text{ we obtain :}$$

$$\frac{\tau_{classics}}{\tau_{QM}} = \frac{\frac{6\pi\varepsilon_0 c^3 m}{q^2\omega_0^2}}{\frac{3\varepsilon_0\pi\,\hbar c^3}{q^2\omega_{21}^3|x_{21}|^2}} = \frac{2m\omega_{21}|x_{21}|^2}{\hbar} = f_{21}, \tag{7.24}$$

where f_{21} is the oscillator strength as defined in (5.32). Hence, for well allowed quantum transitions, the oscillator strength may be of the order of 1, and in this case the classical radiative lifetime and that calculated in terms of quantum mechanics are of the same order of magnitude. However, the classical lifetime does not "know" a concept like selection rules, and therefore, in the case of less allowed or forbidden quantum transitions, the two approaches will deliver quite different results.

Another interesting conclusion may be drawn from Fig. 7.2. This picture highlights the specific significance of the sun for the existence of life on our earth. It is sometimes believed that the major function of the sun is in delivering energy to the earth. But if the earth would gain net energy from the sun, the result would be a global warming, what is rather regarded as a threat for terrestrial (or at least human) life. In fact, the best living conditions are observed when the earth does *not* gain energy from the sun: What is gained at daylight should be lost per reradi-ation from the earth surface all over the day, particularly including the night. But in order to establish such a balance, the earth must compensate the energy gained from absorbing a certain number of high-energy photons by emitting a much larger number of low-energy photons into space. So, the earth emits much more photons than it absorbs, a process which results in an "export" of entropy [2]. This way the specific interaction between sun, earth, and space enables the earth to keep its entropy low, thus preventing a "thermal dead" on earth, which is absolutely essen-tial for the existence of life. Of course, the solar system as a whole moves toward equilibrium this way.

7.5   Application to Practical Problems

7.5.1   Number of Photons in a Cavity Held at Temperature T

Let it be the task to calculate the number of photons per unit volume in a cavity [3]. In equilibrium, we have (compare (6.1) and (7.10)):

$$\frac{dE}{d\omega} = V \frac{\hbar\omega^3}{c^3\pi^2} \frac{1}{e^{\frac{\hbar\omega}{k_BT}} - 1} \Rightarrow \frac{dN_p}{d\omega} = V \frac{\omega^2}{c^3\pi^2} \frac{1}{e^{\frac{\hbar\omega}{k_BT}} - 1}$$

$$\Rightarrow \frac{N_p}{V} = \frac{1}{c^3\pi^2} \int_0^\infty \frac{\omega^2}{e^{\frac{\hbar\omega}{k_BT}} - 1} d\omega = \frac{(k_BT)^3}{\hbar^3c^3\pi^2} \int_0^\infty \frac{x^2}{e^x - 1} dx$$

$$= \frac{(k_BT)^3}{\hbar^3c^3\pi^2} \int_0^\infty \frac{x^2 e^{-x}}{1 - e^{-x}} dx = 2\frac{(k_BT)^3}{\hbar^3c^3\pi^2} \left(1 + \frac{1}{2^3} + \frac{1}{3^3} + ...\right)$$

$$\approx 2.4 \frac{k_B^3}{\hbar^3c^3\pi^2} T^3 \approx \frac{20 * 10^6}{m^3 K^3} T^3$$

In this calculation, we made use of the expansion

$$\frac{1}{1-\xi} = 1 + \xi + \xi^2 + ...; \quad \xi = e^{-x}$$

and integral (1.1). Because of the assumed equilibrium, the shape of the cavity is of no significance; the radiation is completely isotropic. At room temperature, we thus find approximately a total of $5.4*10^8$ photons per cubic centimeter.

7.5.2   Black and Gray Bodies: Selective Absorbers

Let us regard a macroscopic object that is incident to electromagnetic radiation with a spectral flux density I_ω defined as:

$$I_\omega \equiv uc = \frac{c}{V} \frac{dE}{d\omega} = c \frac{dw}{d\omega} \tag{7.25}$$

Why the factor c? Let us try to illustrate this regarding the measurement units of the corresponding quantities.

We have introduced the energy spectral density in terms of (6.1) as:

$$u \equiv \frac{dE}{V\, d\omega}$$

Its rather exotic measurement unit may be written as $[u] = \text{Ws}/(\text{m}^3\text{s}^{-1})$. On the other hand, the energy density is defined as:

$$w = \frac{E}{V}, \text{ while } [w] = \frac{Ws}{m^3}.$$

It represents the energy in a unit volume. If multiplied with c, it represents the energy ΔE penetrating a unit surface normal to the propagation direction per unit time Δt. Hence, the flux density wc is obtained as:

$$wc = \frac{c}{V}E = \frac{1}{A}\frac{\Delta E}{\Delta t}; [wc] = \frac{W}{m^2}$$

i.e., what we call the flux density here has intensity units. So that generally, multiplication with c transforms the energy density into an intensity. The spectral flux density is then

$$I_\omega = \frac{d(cw)}{d\omega} = c\frac{dw}{d\omega}$$

and has the measurement unit:

$$[I_\omega] = \frac{W}{m^2 s^{-1}} = \frac{Ws}{m^2}$$

Let us further assume that a part of the incident flux is absorbed in the object. We define the absorptance as:

$$A(\omega, T) \equiv \frac{I_{\omega,\text{absorbed}}}{I_{\omega,\text{incident}}} \tag{7.26}$$

Clearly, the absorptance will also be a function of geometrical parameters, but we will not focus on this circle of problems. In the case that $A(\omega, T) = 1\ \forall\omega$, the corresponding body will be called a *black*body.

Note that in this case, all incident light is absorbed; hence, no light may be reflected from the object or transmitted through the system. Therefore, in any illumination conditions, such an object appears to be absolutely black.

In contrast, a *gray* body will be defined by an absorptance $0 < A < 1$ with a negligible frequency dependence. When restricting for simplicity on objects that are thick enough in order to prevent any light transmission, the amount of light that is not absorbed in the system (i.e., $1 - A > 0$) must be reflected. Obviously, $1 - A$ is frequency-independent as well. When being illuminated with white light, such an object appears to be gray. Correspondingly, the model system of a *white body* does not absorb light at all ($A = 0$) and appears to be white when being illuminated with white light.

What about objects with a strongly frequency-dependent absorptance? They form the class of spectrally selective absorbers. We will discuss some of their features in Sect. 7.6. In the case that the spectral selectivity concerns the VIS, spectrally selective absorbers often occur colored when being illuminated with white light. For illustration, Fig. 7.3 provides some examples. Clearly, the alpine flowers shown in Fig. 4.6 also belong to the class of spectrally selective absorbers. We will return to this topic later in Sect. 13.5.2.

For many practical purposes, it is essential to know the amount of energy irradiated by a heated surface. If we define the corresponding spectral flux density $I_{\omega,\text{em}}$ in analogy to (7.25), Kirchhoff's radiation law asserts that

$$I_{\omega,\text{em}}(\omega, T) = A(\omega, T) I_{\omega,\text{em,BB}}(\omega, T) \tag{7.27}$$

Here, $I_{\omega,\text{em,BB}}(\omega, T)$ represents the spectral flux density irradiated from a corresponding blackbody at the same temperature. Kirchhoff's radiation law is a direct conclusion from the second law of thermodynamics but will not be derived here.

According to (7.26), a blackbody absorbs all the incident light flux. If a blackbody is in thermal equilibrium with the radiation field, emission and absorption must exactly compensate each other, so that absorbed and emitted energies are identical. On the other hand, in thermal equilibrium conditions, Planck's formula (7.10) holds, such that the energy flux density as emitted from a heated blackbody will be defined by Planck's formula. Hence, $I_{\omega,\text{em,BB}}(\omega, T)$ may be calculated from Planck's formula. Therefore, (7.10) is sometimes called Planck's formula for blackbody irradiation.

Often, it is of interest to calculate the spectral flux density emitted from a plane unit surface of a heated body into all directions of the hemispace surrounding the body. This is done analogously to (7.25), setting:

$$I_{\omega,\text{em,BB}} \equiv u_{\text{Planck}} \frac{c}{4} = \frac{\hbar \omega^3}{4c^2 \pi^2} \frac{1}{e^{\frac{\hbar \omega}{k_B T}} - 1} \tag{7.28}$$

The factor 1/4 has two origins:

- 1/2 arises from the hemispherical emission into the solid angle 2π; emission into the full solid angle 4π would be twice as large. Note that u_{Planck} corresponds to equilibrium radiation, which is isotropic and therefore includes all directions in the full solid angle.
- Another 1/2 arises from integration over the cos-like directional dependence of the flux irradiated by a Lambertian emitter.

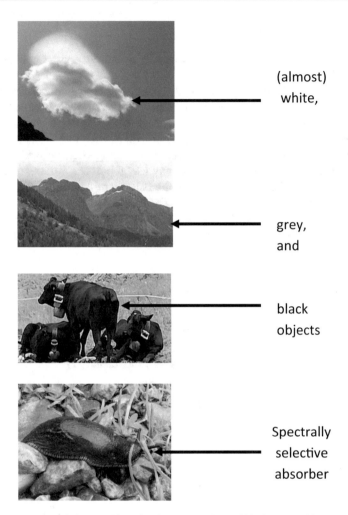

Fig. 7.3 Examples of almost white (cloud), gray (rock), and black (cow) objects, as well as of a spectrally selective absorber (the slug)

The blackbody is nevertheless a model idealization. For real objects, according to (7.27) we have:

$$I_{\omega,\text{em}} = \frac{\hbar\omega^3}{4c^2\pi^2}\frac{1}{e^{\frac{\hbar\omega}{k_B T}} - 1}A(\omega, T) \tag{7.28a}$$

7.6 Advanced Material

7.6.1 Spectrally Selective Absorbers in Solar Energy Conversion

Let us now return to the earlier introduced spectrally selective absorbers. We will regard a model situation of a selective absorber that is exposed to incident electromagnetic radiation with a spectral flux density $I_{\omega,\text{incident}}$. By light absorption, the absorber gains energy, and in our model we will assume that the latter is converted into heat. Of course, the hot absorber emits energy characterized by $I_{\omega,\text{em}}$. The rest of the absorbed energy is used for any useful purpose, for example, heating of water or the like. Let us assume that the balance of these three mechanisms results in a stationary absorber temperature T_A (Fig. 7.4).

You, dear reader, have already guessed that we are facing the absorber design for optimizing solar thermal collectors. How can we maximize their output?

We have the absorbed flux density I_A:

$$I_{\omega,\text{absorbed}} = A(\omega, T_A)I_{\omega,\text{incident}} \Rightarrow I_A \equiv \int_0^\infty I_{\omega,\text{absorbed}}d\omega$$
$$= \int_0^\infty A(\omega, T_A)I_{\omega,\text{incident}}d\omega$$

The energy lost by emission corresponds to a flux density I_E:

$$I_E \equiv \int_0^\infty I_{\omega,\text{em}}d\omega$$

Fig. 7.4 Simple light absorber for conversion into heat

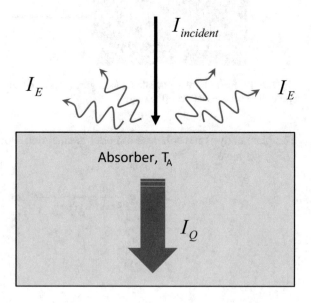

Then, the flux density I_Q remaining for conversion purposes is:

$$I_Q = I_A - I_E$$

Using further (7.28a), this may be written as:

$$I_Q = I_A - I_E = \int_0^\infty A(\omega, T_A) \left[I_{\omega,\text{incident}} - \frac{\hbar\omega^3}{4c^2\pi^2} \frac{1}{e^{\frac{\hbar\omega}{k_B T_A}} - 1} \right] d\omega \qquad (7.29)$$

Let it now be the task to maximize I_Q. How could we proceed?

Of course, one possibility would be to increase $I_{\omega,\text{incident}}$, for example, by focusing the incident light. However, if we speak on solar energy conversion, this is not so elementary in geographic regions with a rather cloudy weather, because the incident radiation is rather diffuse.

Let us therefore have a closer look at the expression in parentheses. Obviously, as depending on the absorber temperature T_A, it may be positive or negative. Let us identify two disjunct spectral regions, namely [4]:

$$I_{\omega,\text{incident}} - \frac{\hbar\omega^3}{4c^2\pi^2} \frac{1}{e^{\frac{\hbar\omega}{k_B T_A}} - 1} \geq 0 \Leftrightarrow \omega \in \{\omega_+\}$$

$$I_{\omega,\text{incident}} - \frac{\hbar\omega^3}{4c^2\pi^2} \frac{1}{e^{\frac{\hbar\omega}{k_B T_A}} - 1} < 0 \Leftrightarrow \omega \in \{\omega_-\}$$

Obviously, both these spectral regions in sum yield the full interval of frequencies considered in (7.29). Then, (7.29) may be rewritten as:

$$I_Q = \int_{\{\omega_+\}} A(\omega, T_A) \left[I_{\omega,incident} - \frac{\hbar\omega^3}{4c^2\pi^2} \frac{1}{e^{\frac{\hbar\omega}{k_B T_A}} - 1} \right] d\omega$$

$$- \int_{\{\omega_-\}} A(\omega, T_A) \left[\frac{\hbar\omega^3}{4c^2\pi^2} \frac{1}{e^{\frac{\hbar\omega}{k_B T_A}} - 1} - I_{\omega,incident} \right] d\omega \qquad (7.29a)$$

For a given absorber temperature and a given incident spectral flux density, I_Q can obviously be maximized by using a spectrally selective absorber that fulfills the requirement:

$$\begin{aligned} A(\omega \in \{\omega_+\}, T_A) &\to 1 \\ A(\omega \in \{\omega_-\}, T_A) &\to 0 \end{aligned} \qquad (7.30)$$

Note that $\{\omega_+\}$ and $\{\omega_+\}$ also depend on T_A. A selective absorber designed for a certain temperature may be far from being optimal at another temperature.

Equation (7.30) of course defines an idealized absorber specification, which cannot be fulfilled in practice. Therefore, in solar energy conversion technology, the quality of the absorber is characterized by two independent parameters, characterizing the degree of adherence to (7.30). Of course, $I_{\omega,\text{incident}}$ has then to be identified with the local terrestrial solar spectrum.

So, let us introduce the parameter $A_{\text{solar}}(T_A)$ as:

$$
A_{\text{solar}}(T_A) \equiv \frac{\int_{\{\omega_+\}} A(\omega, T_A) \left[I_{\omega,\text{incident}} - \frac{\hbar\omega^3}{4c^2\pi^2} \frac{1}{e^{\frac{\hbar\omega}{k_B T_A}} - 1} \right] d\omega}{\int_{\{\omega_+\}} \left[I_{\omega,\text{incident}} - \frac{\hbar\omega^3}{4c^2\pi^2} \frac{1}{e^{\frac{\hbar\omega}{k_B T_A}} - 1} \right] d\omega}
$$

And the parameter $E_{\text{therm}}(T_A)$ by

$$
E_{\text{therm}}(T_A) \equiv \frac{\int_{\{\omega_-\}} A(\omega, T_A) \left[\frac{\hbar\omega^3}{4c^2\pi^2} \frac{1}{e^{\frac{\hbar\omega}{k_B T_A}} - 1} - I_{\omega,\text{incident}} \right] d\omega}{\int_{\{\omega_-\}} \left[\frac{\hbar\omega^3}{4c^2\pi^2} \frac{1}{e^{\frac{\hbar\omega}{k_B T_A}} - 1} - I_{\omega,\text{incident}} \right] d\omega}
$$

In terms of these parameters, (7.29a) can be rewritten as:

$$
I_Q = A_{\text{solar}}(T_A) \int_{\{\omega_+\}} \left[I_{\omega,\text{incident}} - \frac{\hbar\omega^3}{4c^2\pi^2} \frac{1}{e^{\frac{\hbar\omega}{k_B T_A}} - 1} \right] d\omega
$$
$$
- E_{\text{therm}}(T_A) \int_{\{\omega_-\}} \left[\frac{\hbar\omega^3}{4c^2\pi^2} \frac{1}{e^{\frac{\hbar\omega}{k_B T_A}} - 1} - I_{\omega,\text{incident}} \right] d\omega
$$

It is maximized by requiring:

$$
A_{\text{solar}}(T_A) \to 1
$$
$$
E_{\text{therm}}(T_A) \to 0
$$

Depending on the absorber temperature, good selective solar absorbers have A-values well above 0.9, while good E-values at 100 °C are below 0.05.

Of course, solar thermal energy represents only one branch of the full solar energy conversion scene. Some more ideas on environmentally friendly energy conversion mechanisms are summarized in Fig. 7.5

7.6.2   Spontaneous Emission Revisited

Let us finish this chapter by returning to our derivation of the Einstein coefficient A_{21}, responsible for spontaneous emission of light. So far, we have provided a rather illustrative derivation based on the correspondence principle. But in fact,

Fig. 7.5   Solar energy conversion in a pupil's laboratory (= Schülerlabor). Cartoon by Dr. Alexander Stendal. Printed with permission

when drawing together our knowledge from Chaps. 5 to 7, we already have the necessary tools to provide a stronger derivation [5].

Let us regard a microscopic quantum system (e.g., an atom) interacting with electromagnetic radiation. For electric dipole interaction, we may write:

$$\hat{V} = -\hat{\mathbf{d}}\hat{\mathbf{E}} \tag{7.31}$$

In a strong quantum mechanical treatment, the operator of the electric field strength is given by (5.52):

$$\hat{\mathbf{E}} = \mathbf{A}_0 e^{-i(\omega t - \mathbf{kr})}\hat{a} + \mathbf{A}_0^* e^{-i(\omega t - \mathbf{kr})}\hat{a}^+ \tag{7.31a}$$

With (compare (5.61))

$$|A_0|^2 = \frac{\hbar\omega}{2\varepsilon_0 V}$$

When keeping in mind that the spatial dimensions of a microscopic quantum system are much smaller than the wavelength of light relevant for today's photonic applications, the spatial dependence in the term (7.31a) may be neglected when calculating the dipole transition matrix element $\mathbf{d}_{nl}$. Then, the perturbation potential "felt" by the atom corresponds to a superposition of terms like earlier discussed in (6.43).

Let us further assume that before emitting the photon, the material system is in state $|l\rangle$, while the photon field (photon system) is in state $|0\rangle$ for any possible photon mode. The emission of the photon turns the material system into state $|n\rangle$, while one photon mode becomes excited into state $|1\rangle$. The full system (materials system + photon system) is therefore in the states:

$$\text{before emission: } |l\rangle|0\rangle$$
$$\text{after emission: } \quad |n\rangle|1\rangle$$

From (5.49), we know the only different from zero matrix elements of the photon creation and annihilation operators. Therefore:

$$\langle 1|\hat{a}|0\rangle = 0$$
$$\langle 1|\hat{a}^+|0\rangle = 1$$

Consequently, the expression for $|V_{nl}|^2$ may be written as:

$$|V_{nl}|^2 = \frac{\hbar\omega}{2\varepsilon_0 V}|d_{nl}|^2 \tag{7.32}$$

This way our expression for the transition matrix element in the provided quantum mechanical picture is equivalent to the perturbation used in Sect. 6.3.2, when only accepting that the complex amplitude of the electric field is given by (5.61). This is consistent with our earlier obtained result that in a Fock state with no photon excited (i.e., the ground state of the regarded photon mode), the electric field is nevertheless different from zero. But having the transition matrix element (7.32) obtained, we can further proceed in terms of our usual approach earlier developed in Chap. 6.

The simple expression (7.32) is however only true when the electric field strength vector is parallel to the direction of dipole moment. In general, an averaging over the orientations must be carried out, which requires averaging $\cos^2\theta$ over the full solid angle. This results in an additional prefactor given by:

$$\frac{\int_0^{2\pi}\int_0^{\pi}\cos^2\theta\sin\theta\,d\varphi d\theta}{\int_0^{2\pi}\int_0^{\pi}\sin\theta\,d\varphi d\theta} = \frac{-2\pi\int_1^{-1}\xi^2 d\xi}{4\pi} = \frac{\int_{-1}^{1}\xi^2 d\xi}{2} = \frac{1}{6}\xi^3\Big|_{-1}^{+1} = \frac{1}{3}$$

Let us now return to Fermi's golden rule (6.43a). We have:

$$\frac{dw}{dt} = \frac{2\pi}{\hbar}\left|\langle n|\hat{V}|l\rangle\right|^2 D(E)\big|_{E_n=E_l-\hbar\omega} = \frac{\pi}{3}\frac{\omega}{\varepsilon_0 V}|d_{nl}|^2 D(E)\big|_{E_n=E_l-\hbar\omega} \tag{7.33}$$

Let us interpret $\frac{dw}{dt}$ now as the transition rate between the two discrete states of the material system $|l\rangle$ and $|n\rangle$. That might be an atom with the excited state $|l\rangle$ and the ground state $|n\rangle$.

Concerning the density of final states fixed in (7.33), so it might not be a priori clear what is meant by the density of atomic final states $|n\rangle$. But at this point, we must remember that our model treatment of the electric field quantization as provided in Sect. 5.6.3 was restricted to a single mode field. In reality, there are many different photon modes, and we have already counted them, namely in Sect. 7.3.1 (7.9). Then, the density of final states to be used in (7.33) turns out to be identical to the density of modes per photon energy interval the emitted photon "finds" in the environment of his place of birth. Therefore,

$$D(E) = \frac{dZ}{dE} = \frac{1}{\hbar} \frac{dZ}{d\omega}$$

We thus have:

$$\frac{dw}{dt} = \frac{\pi\omega}{3\hbar\varepsilon_0 V} |d_{nl}|^2 \frac{dZ}{d\omega}$$

Here, $\frac{dZ}{d\omega}$ represents the density of photon modes (states) at the emission angular frequency ω. From (7.9), we know:

$$\frac{dZ}{d\omega} = V \frac{\omega^2}{c^3 \pi^2}$$

Consequently,

$$\frac{dw}{dt} = \frac{\pi\omega}{3\hbar\varepsilon_0 V} |d_{nl}|^2 \frac{dZ}{d\omega} = \frac{\pi\omega}{3\hbar\varepsilon_0 V} |d_{nl}|^2 V \frac{\omega^2}{c^3 \pi^2} = \frac{\omega^3 |d_{nl}|^2}{3\hbar\pi \varepsilon_0 c^3}$$

which coincides with (7.21).

Is there a possibility to understand the identity of the density of the photon states with the density of atomic final states $|n\rangle$ in terms of an illustrative picture? I think there is. The full quantum state of the regarded atom is not only described by the excitation state of its electrons, but also by the propagation movement of its mass center. The latter contributes a term like (4.5) (of course in a 3D version) to the full wavefunction $|n\rangle$. According to Sect. 3.5.3, as a result of momentum conservation, any emitted photon provides a recoil momentum to the atom that is antiparallel to the wavevector of the photon. That means, all photon modes corresponding to a different wavevector direction will give rise to different recoil momentum directions of the atom that has emitted a photon, while the total energies of all these final states $|n\rangle$ are identical.

Note that the same type of discussion may be applied to photon modes that differ in light polarization, when taking angular momentum conservation into account.

This is by the way a good illustration of the general principle that energy conservation alone is not sufficient to make an optical transition allowed. Momentum and angular momentum conservation are important as well. In the forthcoming chapters, we will therefore sometimes make use of conservation laws in order to understand the physical essence of selection rules for optical transitions we will obtain in different physical systems. And this is a good time to change to the discussion of the quantum mechanics of single atoms, because this will give us the tools to under-stand the internal structure of atomic quantum states that we have simply called $|n\rangle$ so far.

7.7   Tasks for Self-check

7.7.1 Multiple-choice test: Mark all answers which seem to you correct!

A photon with a photon energy of 2000 eV corresponds to the	Visible spectral range
	Infrared spectral range
	X-ray spectral range
A photon with a photon energy of 0.5 eV corresponds to the	Visible spectral range
	Infrared spectral range
	X-ray spectral range
A resting particle has a de Broglie wavelength that is	Zero
	Infinitely large
	Equal to the particle dimension
$[\hat{x}, \hat{p}_y] =$	0
	$i\,\hbar$
	$-i\,\hbar$
The spectral energy density (u) of a radiation field may be given in	Js/m^3
	Ws^2/m^3
	kg/(ms)
The Einstein coefficient A_{21} responsible for spontaneous light emission is given in	s^{-1}
	W/m^2
	m/s
The ground state ($n = 1$) wavefunction of a particle in a rectangular potential box is	Even with respect to symmetry center
	Odd with respect to symmetry center
	Neither even or odd
The oscillator strength is	Dimensionless
	Given in cm$^{-3/2}$
	Given in s^{-1}

Which of the following quantum transitions in a 1D harmonic oscillator are dipole-allowed ($d_{nm} \neq 0$)?	$m = 2 \rightarrow n = 1$
	$m = 2 \rightarrow n = 3$
	$m = 2 \rightarrow n = 5$
	$m = 2 \rightarrow n = 7$
	$m = 2 \rightarrow n = 0$
Consider the second excited state ($E_2 = \frac{5}{2}\hbar\omega_0$) of a one-dimensional harmonic oscillator. In this state, the expectation value of the momentum will be	$+\sqrt{5\hbar\omega_0}\,\mathrm{m}$
	$-\sqrt{5\hbar\omega_0}\,\mathrm{m}$
	$+\sqrt{3\hbar\omega_0}\,\mathrm{m}$
	$-\sqrt{3\hbar\omega_0}\,\mathrm{m}$
	0
The parity selection rule allows dipole transitions	Between quantum states of even parity
	Between quantum states of odd parity
	Between states that have a different parity

7.7.2 True or wrong? Make your decision!

Assertion	True	Wrong
The phase velocity may never exceed c		
A rocket traveling with a velocity 1000 m/s has the same de Broglie wavelength like a proton traveling with the same velocity		
All linear operators are self-adjoint		
The operator $-i\frac{d}{dx}$ is self-adjoint		
The operator $-\frac{d^2}{dx^2}$ is self-adjoint		
All eigenvalues of a self-adjoint operator are real		
From $\left[\hat{A}, \hat{B}\right] = 0$ and $\left[\hat{A}, \hat{C}\right] = 0$ it follows that $\left[\hat{B}, \hat{C}\right] = 0$		
The Einstein coefficients are all dimensionless		
The Einstein coefficient A_{21} is given in s^{-1}		
The harmonic oscillator has equidistant energy levels		
All eigenfunctions of the harmonic oscillator are even functions of x		
In the ground state of a harmonic oscillator with $U \propto x^2$, $<x^2> = 0$		
In the ground state of a harmonic oscillator with $U \propto x^2$, $<x^3> = 0$		
In a harmonic oscillator, electric dipole-allowed quantum transitions occur only between adjacent energy levels		

7.7.3 From Planck's formula, obtain the low-frequency asymptote (Rayleigh–Jeans) and high-frequency (Wien) asymptote.

7.7.4 From (7.28), determine the temperature dependence of the full flux density irradiated from a unit area of a blackbody held at temperature T (Stefan–Boltzmann law)!

7.7.5 By making use of (6.1) and (7.10), find an expression for the energy density of radiation in equilibrium conditions *per wavelength interval*. From there, find the temperature dependence of the wavelength λ_{max} where that function has a maximum (Wien's displacement law).

7.7.6 Assume a two-level system in thermal equilibrium with a radiation field. Let the temperature be 300 K. What should be the transition wavelength between the two energy levels in order to observe equal intensities of the light generated by spontaneous emission and that generated by induced emission?

7.7.7 In electric dipole approximation, Einstein's coefficient A_{21} is given by: $A_{21} = \frac{q^2 \omega_{21}^3 |x_{21}|^2}{3\varepsilon_0 \pi \hbar c^3}$.

Calculate the relaxation time τ assuming that the excited state of the system relaxes to the ground state by spontaneous emission of light only. Calculate the bandwidth $\Delta\lambda$ (FWHM) of the emission line. Assume the following data: $q = -e$; $\lambda_{21} = 500$ nm; $|x_{21}| = 10^{-10}$ m.

References

Specific References

1. C. Kittel, *Thermal Physics* (Wiley, 1969, Chapter 15)
2. J. Bricmont, Science of chaos or chaos in science? The flight from Science and Reason. Ann. N.Y. Acad. Sci. **775**(1), 131–179 (1995)
3. Д.В. Сивухин, Общий курс физики. Оптика, Москва „Наука" 1980, p. 707
4. R. Sizmann, *Sonnenkollektoren mit Strahlungsselektivität, in Selektive Schichten in der Solartechnik* (Akademischer Verlag München, BMFT Statusbericht, 1992), pp. 25–44
5. M. Schubert, B. Wilhelmi, *Einführung in die nichtlineare Optik II* (BSB B. G. Teubner Verlagsgesellschaft Leipzig, 1971, Chapter 3)

General Literature

6. L.D. Landau, E.M. Lifshitz, *Statistische Physik* (Akademie-Verlag Berlin)
7. W. Demtröder, *Atoms, Molecules, and Photons* (Springer, 2010)

Part III

Energy Levels and Spectroscopic Properties of Simple Atoms

"Toconao" (Atacama Desert, Chile). Painting and Photograph by Astrid Leiterer, Jena, Germany (www.astrid-art.de). Photograph reproduced with permission

Every human consists of approximately $7*10^{27}$ atoms. It is statistically absolutely possible that some of the atoms that are currently hosting you, dear reader, have earlier been—at least for a short time—part of a pomegranate or even of a Tyrannosaurus rex.

The Hydrogen Atom

8

Abstract

We discuss the basic properties of the hydrogen atom. The treatment contains both the simple Bohr theory and a nonrelativistic quantum mechanical treatment, regarding the hydrogen atom as a particular case of a two-body system. In this context, basic aspects of the quantum mechanical treatment of the angular momentum are introduced. Finally, the electron spin is introduced.

8.1   Starting Point

So far, in the previous chapters we have introduced basic concepts of quantum mechanics and developed a quantitative semi-classical picture of the interaction of electromagnetic radiation with a material quantum system. Thereby, emphasis has been placed on the simplest quantum mechanical model, namely the two-level system.

In fact, this way we have already collected a lot of knowledge. We have been able to calculate the probability of quantum transitions in a quantum system as caused by its electric dipole interaction with light in terms of a perturbation approach. Particularly, we found out that the resonance frequencies depend on the position of the energy level in the discussed material system (6.16). The transition probability turned out to crucially depend on the numerical value of the matrix elements of the perturbation operator (6.18). For calculating the latter, the wavefunctions of the unperturbed material system must be known. The good news is that we have a recipe to calculate wavefunctions and energy levels: The "only" thing we have to do is to solve the Schrödinger equation.

So far, we have calculated energy levels and the relevant matrix elements for two model systems, namely the particle in the box potential and the harmonic oscillator. These are important model systems, but they are insufficient for a quantitative treatment of most of the real material systems that are surrounding us.

© The Author(s), under exclusive license to Springer Nature Switzerland AG 2022   167
O. Stenzel, *Light–Matter Interaction*, UNITEXT for Physics,
https://doi.org/10.1007/978-3-030-87144-4_8

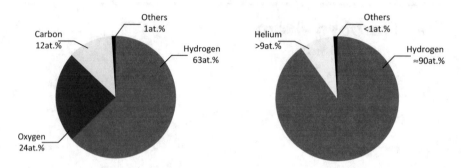

Fig. 8.1 On left: Abundance of different atoms in the human body. Data according to [1]. On right: same for the universe

Therefore, in the next chapters, we have to focus on solving the Schrödinger equation for more complicated model systems that are able to reflect important specifics of atoms, molecules, and solids. So that we are now focusing on what may be called the "structure of matter."

This is what we will be doing starting from this chapter. The bad news is that in the next chapters, we will basically be involved in rather complicated quantum mechanical calculations. But we have to obtain knowledge about energy levels and wavefunctions, before making use of them for understanding the specifics of the interaction of light with atoms, molecules, and solids. The goal is to calculate characteristic resonance frequencies and transition probabilities, or at least basic selection rules for electric dipole interaction.

We have now formulated the general roadmap. Let us now start from the simplest kind of atom, namely the hydrogen atom. Why?

First of all, the hydrogen atom is the only atom that is composed from two particles only, namely the nucleus (usually a proton) and one electron. It is thus an example of a two-body system (Sect. 2.3.1).

But it is not only the simplicity of the atom itself that makes it an important model system. To our knowledge, hydrogen is the atom with the highest abundance in the universe. Moreover, it is the atom with the highest abundance in the human body (Fig. 8.1).

8.2   Physical Idea

Thus, the hydrogen atom consists of one negatively charged electron ($q = -e$) and a positively charged nucleus ($q = +e$). The electron and the nucleus are interacting through an attractive Coulomb potential according to:

$$U(r) = -\frac{1}{4\pi\varepsilon_0}\frac{e^2}{r}$$

(8.1)

Fig. 8.2 Coulomb field of a proton as an example of a central field

In the case that the nucleus has the charge Ze, where Z is the atomic order, instead of (8.1) we have:

$$U(r) = -\frac{1}{4\pi\varepsilon_0}\frac{Ze^2}{r}\tag{8.2}$$

(8.2) corresponds to the potential energy of the remaining electron in a $(Z-1)$-times ionized atomic core of atomic order Z. Any two-body system built up from a single electron bound to a positively charged nucleus will further be called a hydrogen-like atom (or ion).

In a classical mechanical picture, r is the distance between the electron and the nucleus. The potential is only dependent on the distance between nucleus and electron; hence, it describes a central field (Fig. 8.2).

In such situations, instead of using Cartesian coordinates (x,y,z), the use of spherical coordinates (r,φ,θ) is favored. In a quantum mechanical picture, r will then correspond to the radial coordinate in the wavefunction of the electron, while the origin of the coordinate system is again in the mass center of the system.

It is reasonable assuming that the potential energy is zero in the case that nucleus and electron are separated by an infinitely large distance. Note further that $U < 0$ for any finite distance r. That means that quantum states corresponding to an electron that is bound to the nucleus must have negative energies. Positive electron energies correspond to electrons that can leave the nucleus with a finite kinetic energy.

Practically, we will use two approaches to calculate the energy levels corresponding to bound electron states. We will start with a short derivation of the Bohr's theory, before building a stringent (but nonrelativistic) quantum mechanical theory that provides the solution of Schrödinger's equation for a particle in the potential (8.1) or (8.2). This treatment is theoretically strong, but not very illustrative. The Bohr's theory, instead, is very illustrative: It merges the idea of the classical movement of a particle in a central field with a simple quantization condition like that illustrated in Figs. 4.1 or 7.1. The basic idea of the classical movement has already been explained in Sect. 2.4: We assume a motion on a nearly circular orbit, such that Coulomb and centrifugal force compensate each other.

8.3 Theoretical Considerations

8.3.1 Bohr's Theory

In this section, we will discuss the simplest version of that problem, namely the movement of the electron around the nucleus on a circular orbit. Then, the Coulomb force and the centrifugal force should be equal to each other by absolute value.

In fact, we are now dealing with a two-body system as introduced in Chap. 2. So we consider our two-body system built from one nucleus and one electron as a single rotor with reduced mass μ. Now, because of the uncertainty relation, we cannot fix both coordinates and momentum of the mass center as this was done in (2.8a), but for our further treatment it is sufficient to choose a coordinate system where $\dot{R} = 0$, such that (2.9) holds. From (8.2) we have:

$$U(r) = -\frac{1}{4\pi\varepsilon_0}\frac{Ze^2}{r} \Rightarrow F_{\text{Coulomb}} = -\frac{dU}{dr} = -\frac{1}{4\pi\varepsilon_0}\frac{Ze^2}{r^2} \Rightarrow \frac{1}{4\pi\varepsilon_0}\frac{Ze^2}{r^2} = \frac{\mu v^2}{r}$$

$$(8.3)$$

Here v is the velocity of the rotor on the circular orbit with radius r, while $\frac{\mu v^2}{r}$ is the centrifugal force acting on the electron. Because of $\dot{R} = 0$, from (8.3) and (2.9) we have:

$$\mu v^2 = 2T_{\text{kin}} = \frac{1}{4\pi\varepsilon_0}\frac{Ze^2}{r} = -U(r) \Rightarrow T_{\text{kin}} = -\frac{U(r)}{2} = \frac{1}{8\pi\varepsilon_0}\frac{Ze^2}{r} \Rightarrow$$

$$E = U + T_{\text{kin}} = \frac{1}{8\pi\varepsilon_0}\frac{Ze^2}{r} - \frac{1}{4\pi\varepsilon_0}\frac{Ze^2}{r} = -\frac{1}{8\pi\varepsilon_0}\frac{Ze^2}{r} \Rightarrow U = 2E \quad (8.4)$$

Thus, $2T_{kin} = -U$, which is a special realization of (5.20e). From (8.3) we find:

$$\frac{1}{4\pi\varepsilon_0}\frac{Ze^2}{r^2} = \frac{\mu v^2}{r} \Rightarrow \frac{1}{4\pi\varepsilon_0}\frac{Ze^2}{r^3} = \frac{\mu v^2}{r^2} = \mu\omega^2 = \frac{\mu^2 r^4\omega^2}{\mu r^4} = \frac{l^2\omega^2}{\mu r^4} = \frac{L^2}{\mu r^4}$$

With L—angular momentum of the rotor. When assuming that the latter is known, we find the radius as:

$$r = L^2\frac{4\pi\varepsilon_0}{Ze^2\mu} \qquad (8.5)$$

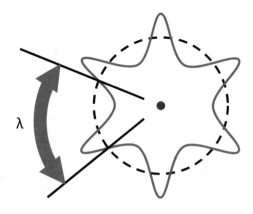

Fig. 8.3 Illustration of the quantization of circular orbits in Bohr's theory

Let us now return to (7.4). It defines a quantization condition, where an integer multiple of the wavelength fits exactly into the path corresponding to one loop in a cavity as visualized in Fig. 7.1. Hence, we had there:

$$\text{"path"} = n\lambda \quad (n = 1, 2, 3, \ldots)$$

Instead of the linear geometry postulated in (7.4), we now have a circular orbit. When generalizing (7.4) to this case by replacing the full path by the circumference of the orbit and using (3.11) (Fig. 8.3), we have:

$$2\pi r = n\lambda = n\frac{h}{p} \Rightarrow rp = L = n\frac{h}{2\pi} = n\hbar \tag{8.6}$$

Equations (8.5) and (8.6) together yield:

$$r = L^2 \frac{4\pi\varepsilon_0}{Ze^2\mu} = r_n = \frac{4\pi\varepsilon_0}{Ze^2\mu}n^2\hbar^2 = \frac{n^2}{Z}\frac{4\pi\varepsilon_0}{e^2\mu}\hbar^2 \equiv \frac{n^2}{Z}a_0 \tag{8.7}$$

With

$$a_0 = \frac{4\pi\varepsilon_0}{e^2\mu}\hbar^2 = \frac{\varepsilon_0}{e^2\pi\mu}h^2 \tag{8.8}$$

Bohr's radius. It is dependent on the relevant reduced mass and, as we have already calculated in Sect. 1.3, approximately equal to

$$a_0 \approx 0.53 * 10^{-10} \text{ m}$$

It is obviously related to the spatial extension of an atom (compare also Table 1.1). As it follows from (8.7), only discrete values of electron radii are allowed in Bohr's theory of the hydrogen atom. They define orbits that are postulated to be stable, i.e., the electrons may "move" on these orbits without loosing energy by radiation. Thus, Bohr's atoms are stable per postulate, in agreement with practical experience.

From (8.4), the energy of an electron moving on the nth orbit is immediately obtained as:

$$E = -\frac{1}{8\pi\varepsilon_0}\frac{Ze^2}{r} = E_n = -\frac{1}{8\pi\varepsilon_0}\frac{Ze^2}{r_n} = -\frac{1}{8\pi\varepsilon_0}\frac{Z^2e^2}{a_0n^2} = -Ry\frac{Z^2}{n^2} \qquad (8.9)$$

Equation (8.9) defines a set of discrete energy levels. The energy approaches zero if n approaches infinity. If the energy would become positive, the electron would no more be bound to the nucleus and move away, thus leaving the "ionized" nucleus alone. The positive energies of the electron would no more form a discrete spectrum of allowed energy values, but a continuous energy spectrum in accordance with what we know about a free electron (Sect. 4.3.2).

The abbreviation Ry denotes the Rydberg energy, given by:

$$Ry = \frac{1}{8\pi\varepsilon_0}\frac{e^2}{a_0} = \frac{e^4\mu}{8\varepsilon_0^2h^2} \approx 13.6\,\mathrm{eV} \qquad (8.10)$$

It corresponds to the ionization energy of a hydrogen atom.

8.3.2 Emission Spectrum of the Hydrogen Atom in the Bohr Theory

Equation (8.9) now allows calculating transition frequencies between different energy levels. From (5.23), we have:

$$\omega_{nm} \equiv \frac{E_n - E_m}{\hbar} = -\frac{Ry}{\hbar}\left(\frac{1}{n^2} - \frac{1}{m^2}\right) = \frac{Ry}{\hbar}\left(\frac{1}{m^2} - \frac{1}{n^2}\right) \qquad (8.11)$$

If instead of transition angular frequencies, we make use of wavenumbers, we get:

$$\nu_{nm} = \frac{\omega_{nm}}{2\pi c} = \frac{Ry}{ch}\left(\frac{1}{m^2} - \frac{1}{n^2}\right) \equiv R_\infty\left(\frac{1}{m^2} - \frac{1}{n^2}\right) \equiv G(n) - G(m)$$

The $G(n)$ are called *spectral terms.* $R_\infty \approx 1.097 * 10^5 \mathrm{cm}^{-1}$.

Traditionally, the emission spectrum of the hydrogen atom is separated in different series, depending on the final level m in the transition $n \to m$. We have (see Table 8.1, Fig. 8.4).

Thus, the Bohr theory of the hydrogen atom allows predicting the frequencies (or wavelength) of the emission spectrum of hydrogen atoms, which have been found to be in an excellent agreement with the experiment. This is a first utmost important consistency check of this theory. Note in particular that the emission frequencies are dependent on the reduced masses, i.e., are slightly different for

Table 8.1 Examples of hydrogen spectral series $(n > m)$

Series	m	ν_{nm}	Spectral range
Lyman	1	$R_\infty\left(1 - \frac{1}{n^2}\right)$	UV
Balmer	2	$R_\infty\left(\frac{1}{4} - \frac{1}{n^2}\right)$	VIS/UV
Paschen	3	$R_\infty\left(\frac{1}{9} - \frac{1}{n^2}\right)$	IR

Fig. 8.4 Illustration of the Lyman, Balmer, and Paschen series

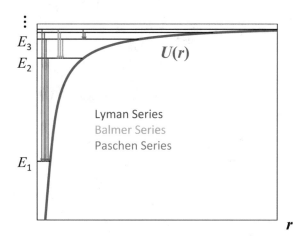

different isotopes. This gives rise to what is called the isotopic shift and historically resulted in the observation of different isotopes.

Concerning the intensities of the lines, so according to (6.3) and (7.21), the population of the energy levels as well as the transition matrix elements have to be known in order to predict them. However, so far no wavefunctions have been calculated, and consequently, we have no tool in hand to calculate the intensities. This provides the motivation for the development of a stronger quantum mechanical theory that should provide us with explicit expressions for the wavefunctions of the electrons in different eigenstates.

It is however possible to *guess* the relative number of electrons that may occupy a "shell" concerned with the quantum number n. This might be of significance in atoms with more than one electron. Let us develop the following illustration for that. Already from Fig. 8.3, it appears reasonable assuming that quantum states with a larger n correspond to orbits with a larger radius and correspondingly a larger circumference. In the 3D case that means that the surface area of the corresponding orbits is increasing, and consequently, orbits with a larger n should have the capacity to accept more electrons than those with a smaller n. Indeed, from (8.6) we have:

$$2\pi r = n\lambda \Rightarrow r_n = \frac{n\lambda}{2\pi} \Rightarrow r_1 = \frac{\lambda}{2\pi}$$

Here we made the unrealistic assumption that all orbits are circular, but this simplification does not destroy the qualitative picture. The assumed spherical surface A corresponding to the ground state ($n = 1$) orbital has an area A_1:

$$A_1 = 4\pi r_1^2 = \frac{\lambda^2}{\pi}$$

Let us assume that it may be occupied by a number of Z_1 electrons. Then, each electron occupies an area of

$$A_e = \frac{\lambda^2}{Z_1 \pi}$$

For the other orbitals $n > 1$, we have the area A_n:

$$r_n = \frac{n\lambda}{2\pi} \Rightarrow A_n = 4\pi r_n^2 = \frac{n^2 \lambda^2}{\pi} = n^2 A_1 = n^2 Z_1 A_e$$

It is now reasonable assuming that the number of electrons Z_n that may occupy the nth orbital is:

$$Z_n = \frac{A_n}{A_e} = n^2 Z_1 \Rightarrow \frac{Z_n}{Z_1} = n^2$$

This illustration leads to the "guess" that with increasing quantum number n, the number of electrons which fit into this orbital increases by a factor n^2.

8.3.3 Relativistic Corrections to the Rydberg Energy Levels

From (3.2), we know:

$$E = mc^2 \sqrt{1 + \frac{p^2}{m^2 c^2}}$$

When making use of:

$$(1+x)^\mu = 1 + \frac{\mu}{1!}x + \frac{\mu(\mu-1)}{2!}x^2 + \frac{\mu(\mu-1)(\mu-2)}{3!}x^3 + \dots \quad (\mu \text{ real}; |x| < 1)$$

We find:

$$E = mc^2\sqrt{1 + \frac{p^2}{m^2c^2}} \approx mc^2\left(1 + \frac{1}{2}\frac{p^2}{m^2c^2} - \frac{1}{8}\left(\frac{p^2}{m^2c^2}\right)^2 + \right)$$

$$= mc^2 + \frac{p^2}{2m} - \frac{1}{8}\frac{p^4}{m^3c^2} + \dots = mc^2 + T_{\text{kin}} - \frac{1}{8}\frac{p^4}{m^3c^2} + \dots$$

$$= mc^2 + T_{\text{kin}} - \frac{1}{4}\frac{p^2}{2m}\frac{p^2}{m^2c^2} + \dots = mc^2 + T_{\text{kin}}\left(1 - \frac{1}{4}\frac{p^2}{m^2c^2}\right)$$

$$+ \dots \approx mc^2 + T_{\text{kin}}\left(1 - \frac{1}{4}\frac{v^2}{c^2}\right) + \dots$$

Hence, the correction term to the kinetic energy is of the order of v^2/c^2. The latter term may already be estimated from the Bohr theory, namely from (8.3) and (8.7):

$$\frac{1}{4\pi\varepsilon_0}\frac{Ze^2}{r^2} = \frac{\mu v^2}{r} \Rightarrow \frac{v^2}{c^2} = \frac{1}{4\pi\varepsilon_0\mu c^2}\frac{Ze^2}{r} = \frac{1}{4\pi\varepsilon_0\mu c^2}\frac{Ze^2}{\frac{4\pi\varepsilon_0}{Ze^2\mu}n^2\hbar^2} = \frac{Z^2e^4}{16\pi^2c^2\varepsilon_0^2n^2\hbar^2}$$

$$= \frac{Z^2e^4}{4\varepsilon_0^2c^2n^2h^2} \Rightarrow \frac{v}{c} = \sqrt{\frac{Z^2e^4}{4\varepsilon_0^2c^2n^2h^2}} = \frac{e^2}{2\varepsilon_0ch}\frac{Z}{n} \equiv \alpha\frac{Z}{n} \qquad (8.12)$$

In (8.12), α is the *fine-structure constant*

$$\alpha = \frac{e^2}{2\varepsilon_0ch} = \frac{e^2}{4\pi\varepsilon_0c\hbar} \approx \frac{1}{137} \qquad (8.13)$$

Accordingly, the correction term to the energy of a Rydberg atomic energy level is of the order:

$$(\Delta E_n)_{\text{rel}} \approx -(E_n)_{\text{non-relativistic}}\frac{v^2}{c^2} \approx -(E_n)_{\text{non-relativistic}}\alpha^2\frac{Z^2}{n^2}$$

Note that when setting $\mu = m_e$, (8.10) and (8.8) may be rewritten as (Λ— Compton wavelength according to Sect. 3.5.1):

$$\alpha = \frac{\Lambda}{2\pi a_0}; \quad \frac{1}{2}\alpha^2 = \frac{Ry}{m_ec^2}$$

8.3.4 Quantum Mechanical Treatment of the Hydrogen-Like Atom

Schrödinger Equation

From the classical treatment of the relative movement of the constituents of two-body system (Sect. 2.3.1), we have:

$$T_{kin} = \frac{\mu}{2}\dot{\mathbf{r}}^2 = \frac{\mu}{2}(\dot{r}^2 + r^2\dot{\phi}^2) \equiv \frac{p_r^2}{2\mu} + \frac{L^2}{2\mu r^2}$$

We therefore assume for the Hamiltonian of the hydrogen atom:

$$\hat{H} = \frac{\hat{\mathbf{p}}^2}{2\mu} + U(r) = \frac{\hat{p}_r^2}{2\mu} + \frac{\hat{\mathbf{L}}^2}{2\mu r^2} + U(r) \tag{8.14}$$

where the potential energy is given according to (8.1) or (8.2). Thus, in our further treatment, the propagation of the mass center of the atom in space will not be taken into account.

We have the operator:

$$\hat{\mathbf{p}}^2 = -\hbar^2\Delta \tag{8.15}$$

where Δ denotes the Laplace operator (Laplacian) given by

$$\Delta = \frac{\partial^2}{\partial x^2} + \frac{\partial^2}{\partial y^2} + \frac{\partial^2}{\partial z^2} \tag{8.16}$$

When introducing spherical coordinates r, ϕ, θ according to:

$$x = r\sin\theta\cos\phi \, ; \quad y = r\sin\theta\sin\phi; \, z = r\cos\theta \tag{8.17}$$

We have the following expressions for the volume element

$$dV = dxdydz = r^2\sin\theta drd\phi d\theta \tag{8.18}$$

and for the Laplacian:

$$\begin{aligned}
\Delta &= \frac{\partial^2}{\partial x^2} + \frac{\partial^2}{\partial y^2} + \frac{\partial^2}{\partial z^2} = \frac{1}{r^2}\frac{\partial}{\partial r}\left(r^2\frac{\partial}{\partial r}\right) + \frac{1}{r^2\sin\theta}\frac{\partial}{\partial\theta}\left(\sin\theta\frac{\partial}{\partial\theta}\right) + \frac{1}{r^2\sin^2\theta}\frac{\partial^2}{\partial\phi^2} \\
&= \left(\frac{\partial^2}{\partial r^2} + \frac{2}{r}\frac{\partial}{\partial r}\right) + \frac{1}{r^2\sin\theta}\frac{\partial}{\partial\theta}\left(\sin\theta\frac{\partial}{\partial\theta}\right) + \frac{1}{r^2\sin^2\theta}\frac{\partial^2}{\partial\phi^2}
\end{aligned} \tag{8.19}$$

When combining (8.14), (8.15), and (8.19), we have:

$$\frac{\hat{\mathbf{p}}^2}{2\mu} = -\frac{\hbar^2}{2\mu}\Delta = -\frac{\hbar^2}{2\mu}\left(\frac{\partial^2}{\partial r^2} + \frac{2}{r}\frac{\partial}{\partial r}\right) - \frac{\hbar^2}{2\mu r^2}\left[\frac{1}{\sin\theta}\frac{\partial}{\partial\theta}\left(\sin\theta\frac{\partial}{\partial\theta}\right) + \frac{1}{\sin^2\theta}\frac{\partial^2}{\partial\phi^2}\right]$$

$$= -\frac{\hbar^2}{2\mu}\Delta_r - \frac{\hbar^2}{2\mu r^2}\Delta_{\theta,\phi} = \frac{\hat{p}_r^2}{2\mu} + \frac{\hat{L}^2}{2\mu r^2} \tag{8.20}$$

With

$$\hat{L}^2 = -\hbar^2\Delta_{\theta,\phi} = -\hbar^2\left[\frac{1}{\sin\theta}\frac{\partial}{\partial\theta}\left(\sin\theta\frac{\partial}{\partial\theta}\right) + \frac{1}{\sin^2\theta}\frac{\partial^2}{\partial\phi^2}\right] \tag{8.21}$$

Here $\hat{L}^2$ is the operator of the square of the angular momentum in spherical coordinates.

From (8.14), we find the time-independent Schrödinger equation:

$$\hat{H}\psi = \left[\frac{\hat{\mathbf{p}}^2}{2\mu} + U(r)\right]\psi = \left[\frac{\hat{p}_r^2}{2\mu} + \frac{\hat{L}^2}{2\mu r^2} + U(r)\right]\psi = E\psi \tag{8.22}$$

Let us set:

$$\psi = \psi(r,\theta,\phi) = R(r)\Phi(\theta,\phi) \tag{8.23}$$

$$\Rightarrow \hat{H}\psi = \Phi\frac{\hat{p}_r^2}{2\mu}R + R\frac{\hat{L}^2}{2\mu r^2}\Phi + U(r)R\Phi = ER\Phi$$

$$\Rightarrow \frac{1}{R}\frac{\hat{p}_r^2}{2\mu}R + \frac{1}{\Phi}\frac{\hat{L}^2}{2\mu r^2}\Phi + U(r) = E$$

$$\Rightarrow 2\mu r^2\left[E - U(r) - \frac{1}{R}\frac{\hat{p}_r^2}{2\mu}R\right] = \frac{1}{\Phi}\hat{L}^2\Phi = \text{const.} \tag{8.24}$$

We thus arrive at the eigenvalue problem:

$$\hat{L}^2\Phi = \text{const.}\Phi \tag{8.25}$$

Angular Momentum

We will now refer to standard textbooks on mathematics. It is well known that the eigenvalue problem (8.25) has the eigenfunctions:

$$\Phi(\theta,\phi) = Y_l^m(\theta,\phi) = \sqrt{\frac{2l+1}{4\pi}\frac{(l-m)!}{(l+m)!}}\,P_l^m(\cos\theta)e^{im\phi};$$

$$l = 0, 1, 2, \ldots; \quad m = 0, \pm 1, \pm 2, \ldots, \pm l \tag{8.26}$$

and the eigenvalues:

$$\text{const.} = \hbar^2 l(l+1) \tag{8.27}$$

(The Y_l^m denote spherical functions, and the P_l^m attributed Legendre polynomials). This way, the eigenvalues of $\hat{\mathbf{L}}^2$ are given in terms of the angular momentum quantum number l by:

$$\hat{\mathbf{L}}^2 \Phi = \hbar^2 l(l+1) \Phi \tag{8.28}$$

Let us now investigate the projection of the angular momentum on the z-axis. From (5.3), we have for the projection of the angular momentum:

$$L_z = \left[\mathbf{r} \times \mathbf{p}\right]_z \rightarrow \hat{L}_z = \hat{x}\hat{p}_y - \hat{y}\hat{p}_x = -i\hbar\left(x\frac{\partial}{\partial y} - y\frac{\partial}{\partial x}\right)$$

Note that:

$$\frac{\partial}{\partial \phi} = \frac{\partial}{\partial x}\frac{\partial x}{\partial \phi} + \frac{\partial}{\partial y}\frac{\partial y}{\partial \phi} + \frac{\partial}{\partial z}\frac{\partial z}{\partial \phi}$$

When considering (8.17), we find:

$$\frac{\partial}{\partial \phi} = -r\sin\theta\sin\phi\frac{\partial}{\partial x} + r\sin\theta\cos\phi\frac{\partial}{\partial y}$$

$$= -y\frac{\partial}{\partial x} + x\frac{\partial}{\partial y} \Rightarrow \hat{L}_z = -i\hbar\frac{\partial}{\partial \phi} \tag{8.29}$$

What are the eigenfunction and eigenvalues of $\hat{L}_z$? The eigenvalue problem is:

$$\hat{L}_z\psi = -i\hbar\frac{\partial}{\partial \phi}\psi = l_z\psi \Rightarrow \psi = e^{i\beta\phi} \Rightarrow \beta\hbar = l_z \tag{8.30}$$

The periodicity requirement yields: $\psi = e^{i\beta\phi} = e^{i\beta(\phi+2\pi)} \Rightarrow 2\beta\pi = 2m\pi$; $m = 0, \pm1, \pm2, ...$

$$\Rightarrow l_z = \beta\hbar = \hbar m \tag{8.30a}$$

Once the projection of L should not be larger that the full length of L, we have the constraint:

$$-l \leq m \leq l \tag{8.30b}$$

m is called the magnetic quantum number.

With an increasing value of the angular momentum quantum number l, the corresponding orbitals are named as s- ($l = 0$), p- ($l = 1$), d- ($l = 2$), f- ($l = 3$),... orbitals.

How to memorize these letters? Unfortunately, the recommended mnemonic trick is only efficient for german students: For

$$l = 0, 1, 2, 3. \text{ we have the orbitals :}$$
$$s, p, d, f \text{ Let us now focus on the first three letters}$$

From left to right this reads as "SPD", a german political party. In the reverse direction, we have the "FDP" which is another political party in Germany. I cannot guarantee that this trick will work in future.

Table 8.2 shows the first order spherical functions:

In an s-orbital, $Y_0^0 * Y_0^0 = \frac{1}{4\pi}$, i.e., the probability density to observe the electron in a unit area element on a spherical surface centered at the nucleus is of spherical symmetry. In a p-orbital, we have: $Y_1^{0*} * Y_1^0 \propto \cos^2 \theta$; $Y_1^{\pm 1*} * Y_1^{\pm 1} \propto \sin^2 \theta$. The probability density is no more of spherical symmetry, but of a type visualized in Fig. 8.5.

Let us try to develop a classical illustration on the physical origin of the orbital shapes shown in Fig. 8.5. Of course, in an s-orbital, the classical angular momentum of the electron would be zero, and thus there would be no centrifugal force counteracting the Coulomb attraction. Therefore, in classics there is no orbit corresponding to an s-orbital, instead, the existence of a stable 1 s-orbital

Table 8.2 First-order spherical functions

l	m	Y_l^m
0 (s-orbital)	0	$\frac{1}{2\sqrt{\pi}}$
1 (p-orbital)	± 1	$\mp \frac{1}{2}\sqrt{\frac{3}{2\pi}} \sin\theta e^{\pm i\phi}$
	0	$\frac{1}{2}\sqrt{\frac{3}{\pi}} \cos\theta$
2 (d-orbital)	± 2	$\frac{1}{4}\sqrt{\frac{15}{2\pi}} \sin^2\theta e^{\pm 2i\phi}$
	± 1	$\mp \frac{1}{2}\sqrt{\frac{15}{2\pi}} \cos\theta \sin\theta e^{\pm i\phi}$
	0	$\frac{1}{4}\sqrt{\frac{5}{\pi}}(2\cos^2\theta - \sin^2\theta)$
3 (f-orbital)	± 3	$\mp \frac{1}{8}\sqrt{\frac{35}{\pi}} \sin^3\theta e^{\pm 3i\phi}$
	± 2	$\frac{1}{4}\sqrt{\frac{105}{2\pi}} \cos\theta \sin^2\theta e^{\pm 2i\phi}$
	± 1	$\mp \frac{1}{8}\sqrt{\frac{21}{\pi}} \sin\theta(5\cos^2\theta - 1)e^{\pm i\phi}$
	0	$\frac{1}{4}\sqrt{\frac{7}{\pi}}(5\cos^3\theta - 3\cos\theta)$

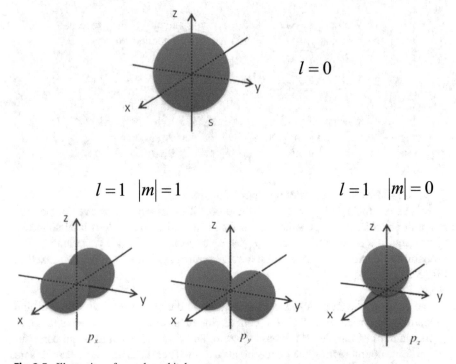

Fig. 8.5 Illustration of s- and p-orbitals

rather has to be understood as a consequence of Heisenberg's uncertainty relation (compare also Chap. 5 task 5.7.9 in this regard). The uncertainty relation
in application to point particles has no classical analog, and therefore, we have
to accept stable s-orbitals as a purely quantum- (or wave-) mechanical effect.
But if so, if it is the uncertainty relation that forces the s-electron to keep a
certain distance from the nucleus, then—keeping in mind the central symmetry
of the Coulomb potential—there is absolutely no reason to assume that there
is an angular dependence in the probability to detect the electron in a certain
space region. Therefore, the electron can be observed in any unit element of
the solid angle with the same probability, which is the reason for the spherical
symmetry of the s-orbital.

For p-states, the situation is somewhat different, and a classical illustration may
be provided. Let us look at the p_z orbital in Fig. 8.5. It corresponds to the
probability density $Y_1^{0*} * Y_1^0 \propto \cos^2 \theta$. Hence, the probability to observe the
electron in a unit element of the solid angle should be largest at $\theta = 0$ or π,
while it is smallest at $\theta = \pi/2$. This is visualized by the special dumbbell-like
shape of the p_z orbital in Fig. 8.5.

How to understand this behavior? Have a look at Fig. 8.6. It shows a classical
electron moving on a circular orbit in a clockwise direction such that the angular

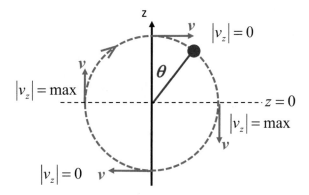

Fig. 8.6 Classical electron on a circular orbit with $L_z = 0$

momentum is directed perpendicular to the z-axis, hence $L_z = 0$. What we have illustrated here is therefore the classical analogon to an electron in a p_z-orbital. The rest should be clear from the figure. Random snapshots of the classical electron moving on a circular orbit with a constant angular velocity will observe the electron rather often nearby the extremal z values ($\theta = 0$ or π), but rather seldom around $z = 0$ ($\theta = \pi/2$). In principle, this simple consideration explains the shape of p_z in Fig. 8.5.

Note that $\hat{\mathbf{L}}^2$ commutes with each of its projections, but neither of the projections commutes with another one. Hence:

$$\left[\hat{L}_z, \hat{\mathbf{L}}^2\right] = 0 \tag{8.31}$$

$$\left[\hat{L}_x, \hat{L}_y\right] = i\hbar \hat{L}_z$$
$$\left[\hat{L}_y, \hat{L}_z\right] = i\hbar \hat{L}_x \tag{8.32}$$
$$\left[\hat{L}_z, \hat{L}_x\right] = i\hbar \hat{L}_y$$

The Radial Part of the Hydrogen Atoms Wavefunction
Let us start from (8.24) and use (8.20) and (8.27):

$$\left[\frac{\hat{p}_r^2}{2\mu} + \frac{\hbar^2 l(l+1)}{2\mu r^2} + U(r)\right]R = ER = \left[\frac{\hat{p}_r^2}{2\mu} + U_{\text{eff}}(r)\right]R$$
$$= \left[-\frac{\hbar^2}{2\mu}\left(\frac{d^2}{dr^2} + \frac{2}{r}\frac{d}{dr}\right) + U_{\text{eff}}(r)\right]R \tag{8.33}$$

Introducing the new function:

$$\chi = rR \tag{8.34}$$

We find:

$$r\left[-\frac{\hbar^2}{2\mu}\left(\frac{d^2}{dr^2}+\frac{2}{r}\frac{d}{dr}\right)\right]R + U_{\text{eff}}(r)\chi = E\chi \tag{8.35}$$

Let us investigate the behavior of R in the asymptotic case corresponding to $r \to \infty$. From:

$$\frac{d^2\chi}{dr^2} = \frac{d}{dr}\left[R + r\frac{dR}{dr}\right] = 2\frac{dR}{dr} + r\frac{d^2R}{dr^2}$$

$$= r\left[\frac{2}{r}\frac{dR}{dr} + \frac{d^2R}{dr^2}\right] \Rightarrow -\frac{\hbar^2}{2\mu}\frac{d^2}{dr^2}\chi + U_{\text{eff}}(r)\chi = E\chi \tag{8.36}$$

We find the asymptotic behavior of R:

$$r \to \infty \Rightarrow U_{eff}(r) \to 0 - \frac{\hbar^2}{2\mu}\frac{d^2}{dr^2}\chi = E\chi$$

$$\chi(r \to \infty) = Ae^{ikr} + Be^{-ikr};$$

$$k = \frac{\sqrt{2\mu E}}{\hbar} \Rightarrow R(r \to \infty) = \frac{A}{r}e^{ikr} + \frac{B}{r}e^{-ikr}$$

In bound states, we have $E < 0$. Therefore,

$$k = i|k| = i\kappa \Rightarrow R(r \to \infty)$$

$$= \frac{A}{r}e^{-\kappa r} + \frac{B}{r}e^{+\kappa r} \Rightarrow B = 0 \Rightarrow R(r \to \infty) \propto \frac{e^{-\kappa r}}{r}; \chi(r \to \infty) \propto e^{-\kappa r}$$

$$k = \frac{\sqrt{2\mu E}}{\hbar} \Rightarrow \kappa = \frac{\sqrt{-2\mu E}}{\hbar} \tag{8.37}$$

For $r \to 0$, from (8.33) to (8.36) we have:

$$-\frac{\hbar^2}{2\mu}\frac{d^2}{dr^2}\chi + U_{eff}(r)\chi = -\frac{\hbar^2}{2\mu}\frac{d^2}{dr^2}\chi + \frac{\hbar^2 l(l+1)}{2\mu r^2}\chi + U(r)\chi = E\chi$$

$$r \to 0 \Rightarrow \frac{\hbar^2}{2\mu}\frac{d^2}{dr^2}\chi = \frac{\hbar^2 l(l+1)}{2\mu r^2}\chi \; ; \; (l \neq 0)$$

With the ansatz:

$$\chi(r \to 0) \propto r^\alpha$$

We find:

$$\alpha(\alpha - 1)r^{\alpha-2} = l(l+1)r^{\alpha-2} \Rightarrow \alpha = l+1 \Rightarrow \chi(r \to 0) \propto r^{l+1}$$

$$\Rightarrow R(r \to 0) \propto r^l \tag{8.38}$$

An ansatz which combines both the dominant exponential decay at large r as well as the power-law dependence at small r could be of the type:

$$\chi \propto \chi(r \to 0)\chi(r \to \infty) \propto r^{l+1}e^{-\kappa r} \Rightarrow R \propto r^l e^{-\kappa r} \qquad (8.39)$$

Equation (8.39) provides a first guess on the mathematical structure of the solution R. We will therefore use the ansatz:

$$\text{Ansatz for each } r : R(r) = u(r)e^{-\kappa r} \qquad (8.40)$$

where $u(r)$ is a polynomial of r. From (8.33) and (8.2) we have:

$$E R = \left[-\frac{\hbar^2}{2\mu}\left(\frac{d^2}{dr^2} + \frac{2}{r}\frac{d}{dr} \right) + \frac{\hbar^2 l(l+1)}{2\mu r^2} - \frac{Ze^2}{4\pi\varepsilon_0 r} \right] R$$

Substituting (8.40), we find:

$$\frac{2}{r}\frac{d}{dr}R = \frac{2}{r}e^{-\kappa r}\frac{du}{dr} - \frac{2\kappa}{r}ue^{-\kappa r} = e^{-\kappa r}\left[\frac{2}{r}\frac{du}{dr} - \frac{2}{r}\kappa u \right]$$

$$\frac{d^2}{dr^2}R = \frac{d}{dr}\left[e^{-\kappa r}\frac{du}{dr} - \kappa u e^{-\kappa r} \right] = e^{-\kappa r}\left[-2\kappa\frac{du}{dr} + \frac{d^2 u}{dr^2} + \kappa^2 u \right]$$

$$\Rightarrow \left(\frac{d^2}{dr^2} + \frac{2}{r}\frac{d}{dr} \right) R = e^{-\kappa r}\left[\frac{d^2 u}{dr^2} + \frac{du}{dr}\left(\frac{2}{r} - 2\kappa \right) + u\left(\kappa^2 - \frac{2}{r}\kappa \right) \right]$$

$$\Rightarrow \frac{d^2 u}{dr^2} + \frac{du}{dr}\left(\frac{2}{r} - 2\kappa \right) + u\left(\kappa^2 - \frac{2}{r}\kappa \right) - u\frac{l(l+1)}{r^2} + \frac{2\mu Ze^2}{4\pi\varepsilon_0\hbar^2 r}u$$

$$= -\frac{2\mu E}{\hbar^2}u = \kappa^2 u$$

Let us set (compare also (8.8)):

$$a = \frac{\mu Ze^2}{4\pi\varepsilon_0\hbar^2} = \frac{Z}{a_0} : \qquad (8.41)$$

$$\Rightarrow \frac{d^2 u}{dr^2} + 2\frac{du}{dr}\left(\frac{1}{r} - \kappa \right) + u\left[\left(\frac{2a - 2\kappa}{r} \right) - \frac{l(l+1)}{r^2} \right] = 0 \qquad (8.42)$$

Remark From (8.42), it is rather easy to obtain the ground state wavefunction of the hydrogen atom directly. We start from analyzing the ansatz:

$$u = b_0 + b_1 r + b_2 r^2 + \ldots + b_{n-1}r^{n-1} \qquad (8.43)$$

Substituting (8.43) into (8.42), and performing a comparison of coefficients, results in a system of equations for the coefficients b_j. Particularly, for the term proportional to r^{n-2} we get:

$$2(n-1)b_{n-1}(-\kappa) + 2b_{n-1}(a - \kappa) = 0 \Rightarrow a = n\kappa \text{ if } b_{n-1} \neq 0.$$

For hydrogen, $Z = 1$, so that the use of (8.41) results in:

$$a = n\kappa = \frac{1}{a_0} = n\frac{\sqrt{-2\mu E}}{\hbar}$$

Therefore, when making use of (8.8):

$$a_0 = \frac{4\pi\varepsilon_0}{e^2\mu}\hbar^2, \text{ we find :}$$

$$E = -\frac{\hbar^2}{2\mu a_0^2}\frac{1}{n^2} = -\frac{1}{8\pi\varepsilon_0}\frac{e^2}{a_0}\frac{1}{n^2} = -\frac{Ry}{n^2}$$

Which obviously coincides with (8.9). The ground state corresponds to the lowest energy; hence, it corresponds to $n = 1$. So that the radial part of the hydrogen atoms ground state wavefunction according to (8.40) and (8.43) is given by:

$$u = b_0 = \text{const.}; \quad R(r) = b_0 e^{-\kappa r} = b_0\, e^{-\frac{r}{a_0}}$$

Moreover, from (8.38) and (8.43) it follows that for a given value of n, the angular momentum quantum number l must be smaller than n. Therefore, we have to set $l = 0$ in the hydrogen atoms ground state. According to $n = 1$ and $l = 0$, the ground state is also called a 1 s-state. The proportionality constant b_0 (compare Table 8.3) follows from the normalization requirement of the full wavefunction.

The method is applicable to the 2 s-state ($n = 2$ and $l = 0$) as well. In order to suffice (8.38) and (8.43), let us now set:

$$R(r) = u(r)e^{-\kappa r} \text{ with } u = (b_0 + b_1 r); \quad \frac{du}{dr} = b_1; \quad \frac{d^2 u}{dr^2} = 0$$

From (8.42) we have:

$$b_1\left(\frac{1}{r} - \kappa\right) + (b_0 + b_1 r)\left(\frac{a - \kappa}{r}\right)$$

$$= 0 \Rightarrow \begin{cases} -b_1\kappa + b_1(a - \kappa) = 0 \Rightarrow \kappa = \frac{a}{2} = \frac{1}{2a_0} \Rightarrow E = -\frac{Ry}{4} \\ \frac{1}{r}[b_1 + b_0(a - \kappa)] = \frac{1}{r}[b_1 + b_0\frac{a}{2}] = 0 \Rightarrow \frac{b_1}{b_0} = -\frac{a}{2} = -\frac{1}{2a_0} \end{cases}$$

$$\Rightarrow R = b_0\left(1 - \frac{r}{2a_0}\right)e^{-\frac{r}{2a_0}}$$

Which obviously coincides with the 2 s-wavefunction (Table 8.3) except the proportionality constant b_0 which again results from the normalization requirement.

According to (8.38) and (8.43), for the 2p-state ($n = 2$ and $l = 1$) one has to choose the ansatz: $R(r) = u(r)e^{-\kappa r}$ with $u = b_1 r$; $\frac{du}{dr} = b_1$; $\frac{d^2 u}{dr^2} = 0$ and $l = 1$. This quickly results in $E = -\frac{Ry}{4}$ and $\kappa = \frac{1}{2a_0}$, again. Hence $R = b_1 r e^{-\frac{r}{2a_0}}$.

Table 8.3 Electronic wavefunctions of the hydrogen-like atom

n	l	m	$\psi_{n,l,m}(r, \vartheta, \phi)$
1	0	0	$\frac{1}{\sqrt{\pi}}\left(\frac{Z}{a_0}\right)^{\frac{3}{2}} e^{-Zr/a_0}$
2	0	0	$\frac{1}{4\sqrt{2\pi}}\left(\frac{Z}{a_0}\right)^{\frac{3}{2}}\left(2 - \frac{Zr}{a_0}\right) e^{-Zr/2a_0}$
2	1	0	$\frac{1}{4\sqrt{2\pi}}\left(\frac{Z}{a_0}\right)^{\frac{3}{2}} \frac{Zr}{a_0} e^{-Zr/2a_0} \cos\theta$
2	1	± 1	$\frac{1}{8\sqrt{\pi}}\left(\frac{Z}{a_0}\right)^{\frac{3}{2}} \frac{Zr}{a_0} e^{-Zr/2a_0} \sin\theta\, e^{\pm i\phi}$
3	0	0	$\frac{1}{81\sqrt{3\pi}}\left(\frac{Z}{a_0}\right)^{\frac{3}{2}}\left(27 - 18\frac{Zr}{a_0} + 2\frac{Z^2 r^2}{a_0^2}\right) e^{-Zr/3a_0}$
3	1	0	$\frac{\sqrt{2}}{81\sqrt{\pi}}\left(\frac{Z}{a_0}\right)^{\frac{3}{2}}\left(6 - \frac{Zr}{a_0}\right)\frac{Zr}{a_0} e^{-Zr/3a_0} \cos\theta$
3	1	± 1	$\frac{\sqrt{2}}{81\sqrt{\pi}}\left(\frac{Z}{a_0}\right)^{\frac{3}{2}}\left(6 - \frac{Zr}{a_0}\right)\frac{Zr}{a_0} e^{-Zr/3a_0} \sin\theta\, e^{\pm i\phi}$
3	2	0	$\frac{1}{81\sqrt{6\pi}}\left(\frac{Z}{a_0}\right)^{\frac{3}{2}} \frac{Z^2 r^2}{a_0^2} e^{-Zr/3a_0} (3\cos^2\theta - 1)$
3	2	± 1	$\frac{1}{81\sqrt{\pi}}\left(\frac{Z}{a_0}\right)^{\frac{3}{2}} \frac{Z^2 r^2}{a_0^2} e^{-Zr/3a_0} \sin\theta \cos\theta\, e^{\pm i\phi}$
3	2	± 2	$\frac{1}{162\sqrt{\pi}}\left(\frac{Z}{a_0}\right)^{\frac{3}{2}} \frac{Z^2 r^2}{a_0^2} e^{-Zr/3a_0} \sin^2\theta\, e^{\pm 2i\phi}$

Analogously, the 3 s-state ($n = 3$ and $l = 0$) is obtained from:

$$R(r) = u(r)e^{-\kappa r} \text{ with } u = \left(b_0 + b_1 r + b_2 r^2\right); \quad \frac{du}{dr} = b_1 + 2b_2 r \; ; \quad \frac{d^2 u}{dr^2} = 2b_2$$

Substitution into (8.42) yields $E = -\frac{Ry}{9}$ and $\kappa = \frac{1}{3a_0}$. Correspondingly,

$$R \propto \left(1 - \frac{2r}{3a_0} + \frac{2r^2}{27a_0^2}\right)e^{-\frac{r}{3a_0}} \propto \left(27 - \frac{18r}{a_0} + \frac{2r^2}{a_0^2}\right)e^{-\frac{r}{3a_0}} \text{ and so on.}$$

Let us now come to the general solution. We choose the ansatz (compare (8.43)):

$$u(r) = \sum_{j=0}^{n} b_j r^j$$

$$\Rightarrow \frac{du}{dr} = \sum_{j=0}^{n} j b_j r^{j-1}$$

$$\Rightarrow \frac{d^2u}{dr^2} = \sum_{j=0}^{n} j(j-1)b_j r^{j-2}$$

With $b_n = 0$. Then, from (8.42) we have:

$$\Rightarrow \sum_{j=0}^{n} j(j-1)b_j r^{j-2} + 2\left(\frac{1}{r} - \kappa\right) \sum_{j=0}^{n} j b_j r^{j-1}$$

$$+ \left[\left(\frac{2a - 2\kappa}{r}\right) - \frac{l(l+1)}{r^2}\right] \sum_{j=1}^{n} b_j r^j = 0$$

$$\Rightarrow \sum_{j=0}^{n} \left\{ r^{j-2}[j(j-1)b_j + 2jb_j - l(l+1)b_j] + r^{j-1}[(2a - 2\kappa)b_j - 2\kappa j b_j] \right\} = 0$$

$$\Rightarrow \sum_{j=0}^{n} r^{j-2} b_j [j(j+1) - l(l+1)] + 2\sum_{j=0}^{n} r^{j-1} b_j [a - (j+1)\kappa] = 0$$

$$\Rightarrow \sum_{j=0}^{n} r^{j-2} b_j [j(j+1) - l(l+1)] = -2\sum_{j=0}^{n} r^{j-1} b_j [a - (j+1)\kappa]$$

In the sum on the right side, we change $j \to j' = j + 1 \to j = j'$-1

$$\sum_{j=0}^{n} r^{j-2} b_j [j(j+1) - l(l+1)] = 2\sum_{j'=1}^{n+1} r^{j'-2} b_{j'-1} [j'\kappa - a]$$

$$= 2\sum_{j=1}^{n+1} r^{j-2} b_{j-1} [j\kappa - a]$$

This results in the recursive recipe ($j > l$):

$$b_j = 2b_{j-1} \frac{j\kappa - a}{j(j+1) - l(l+1)}$$

while $b_{j<l} = 0$ because of (8.38). The truncation requirement is:

$$b_n = 0; b_{n-1} \neq 0 \Rightarrow n > l \wedge n\kappa = a \tag{8.43a}$$

When using (8.41), we find:

$$n\kappa = a \Rightarrow n^2 \kappa^2 = a^2 \Rightarrow -n^2 \frac{2\mu E}{\hbar^2}$$

$$= \left(\frac{\mu Z e^2}{4\pi \varepsilon_0 \hbar^2}\right)^2 \Rightarrow E = -\frac{\mu e^4}{8\varepsilon_0^2 h^2} \frac{Z^2}{n^2} = -Ry \frac{Z^2}{n^2} \tag{8.44}$$

i.e., the well-known equations (8.9), (8.10).

The full wavefunction (8.23) thus depends on three quantum numbers n,l,m. Here, n represents the principal quantum number. In Dirac notation, we will write it as $|n, l, m\rangle$. From (8.30a) and (8.43a) we have:

$$n = 1, 2, 3, ...$$
$$l = 0, 1, 2, ..., n - 1$$
$$m = 0, \pm1, \pm2, ..., \pm l \qquad (8.45)$$

Therefore, each of the energy levels defined by (8.9) or (8.44) is degenerated for n^2 times (the earlier mentioned spin degeneration is not taken into account here).

Remark The degeneration factor $g_n = n^2$ is easily calculated from (8.45), setting
$$g_n = \sum_{l=0}^{n-1} (2l + 1),$$
Then, for an even value of n, we have:

$$g_n = \sum_{l=0}^{n-1} (2l + 1) = 1 + 3 + 5 + ... + (2n - 3) + (2n - 1)$$

$$= \underbrace{\underbrace{1 + (2n - 1)}_{=2n} + \underbrace{3 + (2n - 3)}_{=2n} + ...}_{\frac{n}{2} \ times} = \frac{n}{2} 2n = n^2$$

For an odd n, we separate the last term and further proceed analogously:

$$g_n = \sum_{l=0}^{n-1} (2l + 1) = \underbrace{\underbrace{1 + (2n - 3)}_{=2n-2} + \underbrace{3 + (2n - 5)}_{=2n-2} + ... + (2n - 1)}_{\frac{(n-1)}{2} \ times}$$

$$= \underbrace{\left(\frac{n - 1}{2}\right)(2n - 2)}_{=n^2 - 2n + 1} + (2n - 1) = n^2$$

Including the electron spin (see later Sect. 8.6.1), we come to the result that each energy level corresponding to the quantum number n is degenerated for $2n^2$ times. Note that this result coincides with what we obtained earlier from our illustrative treatment in Sect. 8.3.2, when assuming a ground state degeneracy $Z_1 = 2$.

From (8.44) it is further obvious that:

$$\kappa = \frac{\sqrt{-2\mu E}}{\hbar} = \frac{\pi \mu e^2}{\varepsilon_0 h^2} \frac{Z}{n} = \frac{Z}{a_0 n}$$

The final radial wavefunctions R are given by Laguerre polynomials according to the Table 8.3 (after Demtröder).

8.4   Consistency Considerations

8.4.1   Characteristic Spatial Dimensions of the Hydrogen Ground State

So far we have presented a simple version of the Bohr theory, followed by the development of a quantum mechanical treatment of an electron in the Coulomb potential. We have already mentioned that the Bohr theory provides predictions of emission frequencies that are in excellent agreement to the measurements. Once the quantum mechanical treatment as provided so far results in the same predictions for the hydrogen energy levels, everything seems to be consistent.

On the other hand, in our quantum mechanical treatment that operates in terms of the wavefunction, the electrons distance from the nucleus is no more a fixed parameter, but obeys a certain probability density. As an example of a consistency check, let us consider the ground state of the hydrogen atom, according to (Table 8.3):

$$|n, l, m\rangle = |1, 0, 0\rangle = \frac{1}{\sqrt{\pi}} \left(\frac{1}{a_0}\right)^{\frac{3}{2}} e^{-\frac{r}{a_0}} \tag{8.46}$$

The probability dw to observe the electron in a volume element dV is

$$dw = \frac{1}{\pi a_0^3} e^{-\frac{2r}{a_0}} dV = \frac{1}{\pi a_0^3} e^{-\frac{2r}{a_0}} r^2 \sin\theta dr d\phi d\theta$$

After integrating over the angles, we obtain the probability to observe the electron in the spherical shell between r and $r + dr$ as:

$$dw = \frac{4}{a_0^3} e^{-\frac{2r}{a_0}} r^2 dr$$

Hence, the corresponding probability density is given by: $\frac{4}{a_0^3} e^{-\frac{2r}{a_0}} r^2$. It has a maximum at

$$\frac{d}{dr}\left[\frac{4}{a_0^3} e^{-\frac{2r}{a_0}} r^2\right] = 0 \Rightarrow 2r e^{-\frac{2r}{a_0}} = r^2\left(\frac{2}{a_0}\right) e^{-\frac{2r}{a_0}} \Rightarrow r = a_0$$

Thus, although the electron in the ground state may be found at different distances from the nucleus, the corresponding probability density is largest at the Bohr radius, which is a quite reasonable result (compare also Table 1.1).

This distance, which corresponds to a maximum of the probability density, must not be confused with the expectation value of r. Indeed, in the ground state, from (5.13) and Table 8.3 we have

$$\langle r \rangle = \frac{1}{\pi a_0^3} \int r e^{-\frac{2r}{a_0}} dV = \frac{1}{\pi a_0^3} \int r e^{-\frac{2r}{a_0}} r^2 \sin\theta \, dr \, d\varphi \, d\theta$$

$$= \frac{4}{a_0^3} \int e^{-\frac{2r}{a_0}} r^3 dr$$

Making use of : $\int_0^{\infty} x^n e^{-px} dx = \frac{n!}{p^{n+1}}$ (see Integral (1.1))

Results in

$$\Rightarrow \langle r \rangle = \frac{4}{a_0^3} \int_0^{\infty} e^{-\frac{2r}{a_0}} r^3 dr = \frac{24}{a_0^3} \left(\frac{a_0}{2}\right)^4 = \frac{3}{2} a_0$$

which differs from the Bohr's radius, although it has the same order of magnitude. Indeed, approximately we can write $(3/2)*a_0 = 0.08$ nm, which is the value for the atomic radius used in Sect. 2.4 for our first estimations.

8.4.2 Important Expectation Values

For completeness, Table 8.4 summarizes some more important expectation values obtained for the eigenstates of the hydrogen-like atom according to Table 8.3.

Table 8.4 Some important expectation values (compare [2]):

$\langle r \rangle$	$\frac{1}{2Z}[3n^2 - l(l+1)]a_0$	
$\langle r^2 \rangle$	$\frac{n^2}{2Z^2}[5n^2 + 1 - 3l(l+1)]a_0^2$	
$\langle r^{-1} \rangle$	$\frac{Z}{n^2}\frac{1}{a_0}$	
$\langle U(r) \rangle = -\frac{Ze^2}{4\pi\varepsilon_0}\langle r^{-1} \rangle$	$-\frac{e^2}{4\pi\varepsilon_0 a_0}\frac{Z^2}{n^2} =$	(compare (5.20e))
	$-2\frac{Z^2}{n^2}Ry = 2E_n$	
$\langle T_{kin} \rangle = E_n - \langle U(r) \rangle$	$-E_n$	
$\langle r^{-2} \rangle =$	$\frac{Z^2}{n^3\left(l+\frac{1}{2}\right)}\frac{1}{a_0^2}$	
$\langle r^{-3} \rangle$	$\frac{Z^3}{n^3 l(l+1)\left(l+\frac{1}{2}\right)}\frac{1}{a_0^3}$	

Note that the result $2\langle T_{kin} \rangle = -\langle U \rangle$ represents the quantum mechanical analogon to the classical expression (8.4): $2T_{kin} = -U$. It is at the same time a special case of (5.20e), i.e., is consistent with the quantum virial theorem

Remark From here a shortcut for re-deriving the formulas for the Rydberg energy as well as Bohr's radius may be obtained. Indeed, in the hydrogen atom ground state, we have $n = 1$ and $E_1 = -Ry$. Let us now start from the idea that $< T_{kin,1} > = Ry$ and $< U_1 > = -2\,Ry$, which is in full analogy to the classical considerations presented in Sect. 8.3.1, formula (8.4). Making use of the plausible formula:

$$\langle U_1 \rangle = -\frac{e^2}{4\pi\,\varepsilon_0 a_0}$$

we find:

$$\langle U_1 \rangle = -\frac{e^2}{4\pi\,\varepsilon_0 a_0} = -2Ry \;\Rightarrow\; Ry = \frac{e^2}{8\pi\,\varepsilon_0 a_0}.$$

On the other hand, from Bohr's quantization rule we have another plausible expression:

$$\langle T_{kin,1} \rangle = Ry = \frac{L^2}{2I} = \frac{\hbar^2}{2\mu a_0^2}.$$

Hence,

$$Ry = \frac{e^2}{8\pi\,\varepsilon_0 a_0} = \frac{\hbar^2}{2\mu a_0^2} \;\Rightarrow\; a_0 = \frac{4\pi\,\varepsilon_0 \hbar^2}{\mu e^2}$$

Remark Note by the way that (8.9) is also valid for anti-hydrogen atoms that are built from a positron and an antiproton. Indeed, it should be only the product of the charges of the two constituents that matters. As an experimental fact, anti-hydrogen has the same emission spectrum as normal hydrogen atoms have. This is even true for the so-called fine structure (see later Sect. 8.6.1) of the hydrogen energy levels [3].

8.4.3 Highly Excited Hydrogen-Like Atoms

Let us finally have a closer look at Bohr's quantization rule (8.6). Together with (8.7) and (8.9) it results in the couple of equations:

$$L = n\hbar$$

$$r_n = \frac{n^2}{Z} a_0$$

$$E_n = -Ry \frac{Z^2}{n^2}$$

Obviously, in the Bohr's theory, n must be different from zero, and therefore, $\mathbf{L}$ must be different from a zero-vector as well. There is consequently no s-orbital in this simple (and any classical) approach. So what is the relation between Bohr's quantization rule and the quantum mechanical result (8.28):

$$\hat{L}^2\Phi = \hbar^2 l(l+1)\Phi??$$

Let us regard the case of the largest possible l-values, namely $l = l_{max} = n\text{-}1$.
In this case, from (8.39), (8.40), and (8.43) we have the radial part of the wavefunction according to:

$$R \propto r^{l_{max}} e^{-\kappa r} = r^{n-1} e^{-\frac{r}{na_0}}$$

This function has only one maximum, and therefore, the corresponding orbit can be regarded as the analogon to a classical circular orbit.

Let us have a short look at the function $r^2|R|^2$. It is again proportional to the probability density to observe the electron at the distance r from the nucleus. Differentiating this expression quickly shows that this probability has its maximum at $r = n^2 a_0$. In order to visualize $r^2|R|^2$ for different n, it makes sense to normalize it to its maximum value, which may be set to be equal to one. Hence we look at the function $f(x)$ defined by

$$r^2|R|^2 \underset{x \equiv \frac{r}{a_0}}{\longrightarrow} f(x) = \frac{x^{2n} e^{-\frac{2x}{n}}}{n^{4n} e^{-2n}},$$

which has a maximum at $x = n^2$ with $f(x = n^2) = 1$. Figure 8.7 shows $f(x)$ for different values of the principal quantum number n. Obviously, at large principal quantum numbers, the probability to observe the electron close to the nucleus is rather vanishing.

At the same time, for $l = l_{max} = n\text{-}1$, (8.28) yields:

$$\hat{L}^2\Phi = \hbar^2 n(n-1)\Phi$$

Fig. 8.7 Visualization of the probability density to observe the electron at distance $r = xa_0$ from the nucleus in the case of highly excited circular orbits

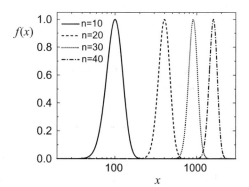

For large values of n, we obtain

$$\hat{\mathbf{L}}^2 \Phi = \hbar^2 n(n-1)\Phi\big|_{n\to\infty} \to \hbar^2 n^2 \Phi$$

Which is obviously equivalent to Bohr's quantization rule. Therefore, quantum states with a large principal quantum number n and a maximum possible angular quantum number l come closest to the circular orbits we have regarded in the Bohr's theory.

In case of a hydrogen atom, assuming $l = n - 1$ and finally investigating the case of highly excited electronic states ($n \to \infty$), from Table 8.4 we quickly find:

$$\langle (\Delta r)^2 \rangle = \langle r^2 \rangle - \langle r \rangle^2 = \frac{1}{4}(2n^3 + n^2)a_0^2 \Rightarrow \frac{\sqrt{\langle r^2 \rangle - \langle r \rangle^2}}{\langle r \rangle}\Bigg|_{n\to\infty} \to 0$$

Even in higher order atoms, the wavefunction of a highly excited single electron has practically no spatial overlap with the orbits of the remaining (not excited) electrons (Fig. 8.7). Therefore, its energy levels are in well correspondence to (8.9). Atomic single electron excitation states corresponding to $n \gg 10$ are therefore called Rydberg states.

Such that in a Rydberg state with maximum l, the electrons orbit comes closest to the classical idea of a circular orbit with a well-defined radius. This convergence may be regarded as a special realization of the already mentioned correspondence principle.

8.5   Application to Practical Problems

8.5.1   Rigid Spherical Rotor

Let us use the possibility to discuss—quasi by the way—another simple but important model system, namely the rigid rotor. A classical rigid rotor is a particle that moves in a way that it always has the same distance from a fixed center. In our two-body system, that center coincides with the mass center, while the mass of this particle coincides with the reduced mass of the two bodies. We therefore again start from (2.9) and demand that $r = \text{const}$ (compare (2.11)).

Let us now write down the eigenenergies of a rigid spherical rotor. From the Schrödinger equation for the radial part of the wavefunction of a particle in a central field, from (8.24) we have:

$$2\mu r^2 \left[E - U(r) - \frac{1}{R}\frac{\hat{p}_r^2}{2\mu}R \right] = \frac{1}{\Phi}\hat{\mathbf{L}}^2 \Phi = \text{const.} = \hbar^2 l(l+1) \Rightarrow$$

$$\left[\frac{\hat{p}_r^2}{2\mu} + \frac{\hbar^2 l(l+1)}{2\mu r^2} + U(r) \right]R = ER = \left[-\frac{\hbar^2}{2\mu}\Delta_r + \frac{\hbar^2 l(l+1)}{2\mu r^2} + U(r) \right]R$$

$$(8.47)$$

For a rigid rotor, we require: $r = \text{const} \rightarrow U(r) = \text{const}$; $\hat{p}_r^2 = 0$. We then have the rotational energy levels:

$$E = E_l = \underbrace{U(r)}_{=const} + \frac{\hbar^2 l(l+1)}{2\mu r^2} = \underbrace{U(r)}_{=const} + \frac{\hbar^2 l(l+1)}{2I} = \underbrace{U(r)}_{=const} + E_{rot}(l) \quad (8.48)$$

According to (8.26) or (8.30a), the degeneration of each of these energy levels is $2l + 1$.

The term $\frac{\hbar^2 l(l+1)}{2I}$ in (8.48) describes the rotational energy levels $E_{rot}(l)$ of a rigid rotor.

Remark It might seem that the conditions $r = \text{const}$ and $\hat{p}_r^2 = 0$ could cause problems with respect to the uncertainty relation. But there should be no problem. First of all, the position and momentum of the "fixed center" (here the mass center) itself must suffice the uncertainty relation, such that coordinate and momentum of the rigid rotor are also never fixed at the same time. Concerning the observables r and p_r^2, let us for simplicity discuss the 1D case and return to task 5.7.2 from Chap. 5. A commutator like $\left[\hat{x}, \hat{p}_x^2\right]$ is indeed different from zero. We have:

$$\left[\hat{x}, \hat{p}_x^2\right] = 2i\hbar\hat{p}_x$$

Such that from (5.40) it follows that:

$$\left\langle(\Delta x)^2\right\rangle\left\langle\left(\Delta p_x^2\right)^2\right\rangle \geq \hbar^2\langle p_x\rangle^2$$

Concerning $\langle p_x\rangle$, we make use of Ehrenfest's theorem (5.19):

$$\frac{d}{dt}\langle x\rangle = \frac{\langle p_x\rangle}{m} \Rightarrow \langle p_x\rangle = m\frac{d}{dt}\langle x\rangle$$

Once we speak on a rigid rotor, the minimum requirement is that $\langle x\rangle = const$, such that

$$\left\langle(\Delta x)^2\right\rangle\left\langle\left(\Delta p_x^2\right)^2\right\rangle \geq \hbar^2\langle p_x\rangle^2 = \hbar^2\left(m\frac{d}{dt}\langle x\rangle\right)^2 = 0$$

Our requirements of $r = \text{const}$ and $\hat{p}_r^2 = 0$ do not contradict that relation.

But what about our central topic, the interaction of a rigid rotor with light? We have found the energy levels, but have no idea on transition matrix elements or selection rules. Clearly, the matrix elements might be calculated from (6.18) making use of the spherical functions (8.26) or Table 8.2. This procedure, dear reader, may be found in other textbooks on quantum mechanics. We will here rather guess the selection rules, by making use of the short cut provided by the correspondence principle. We will require, that for highly excited rotational states, quantum mechanical (QM) and classical (class) transition frequencies and rotational energies shall coincide.

Let us this way answer the question, in how far emission or absorption of light may change the quantum number l by a value of Δl. We then have for highly excited rotational quantum states:

$$l \to \infty \Rightarrow \omega_{QM} \to \omega_{class};$$

$$\omega_{QM} = \frac{E_{rot,QM}(l + \Delta l) - E_{rot,QM}(l)}{\hbar}$$

$$= \frac{\hbar}{2I}[(l + \Delta l)(l + 1 + \Delta l) - l(l + 1)] \Rightarrow \omega_{QM}\big|_{l \to \infty} \to \frac{\hbar}{I}l\Delta l$$

$$E_{rot,QM} = \frac{\hbar^2 l(l + 1)}{2I}; \quad E_{rot,QM}\big|_{l \to \infty} \to \frac{\hbar^2 l^2}{2I}$$

From the correspondence principle, for large l we may demand:

$$\frac{\hbar^2 l^2}{2I} \to E_{rot,class} = \frac{I\omega_{class}^2}{2}$$

Consequently,

$$\omega_{class}^2 = \frac{\hbar^2 l^2}{I^2} \Rightarrow \omega_{class} = \pm\frac{\hbar l}{I}$$

When comparing with:

$$\omega_{QM}\big|_{l \to \infty} \to \frac{\hbar}{I}l\Delta l \to \omega_{class} = \pm\frac{\hbar l}{I}$$

We find the selection rule:

$$\Delta l = \pm 1 \tag{8.49}$$

Hence, absorption or emission of light changes the rotational quantum number by an absolute value of one.

8.5.2   Particle in a Spherical Box Potential

Let us now discuss another model system, namely a particle confined in "spherical potential box." In contrast to the simple model task discussed in Sect. 4.3.3, we now have a 3D situation with the potential:

$$U = 0; \quad 0 \leq r \leq a$$
$$U \to \infty; \quad r > a$$

The task is nevertheless simple when we restrict on s-states. In an s-state, $l = 0$, and the angular part of the wavefunction is simply a constant, so that $\psi = \psi(r)$. Therefore, the application of the angular part of the Laplacian to the wavefunction gives no contribution, and we only have to consider its radial part. From (8.33) and $l = 0$ we therefore have the Schrödinger equation:

$$r \leq a: \; E\psi = \left[-\frac{\hbar^2}{2m} \left(\frac{d^2}{dr^2} + \frac{2}{r} \frac{d}{dr} \right) \right] \psi$$

Introducing again $\chi = r\psi$ (8.34) we find:

$$r \left[-\frac{\hbar^2}{2m} \left(\frac{d^2}{dr^2} + \frac{2}{r} \frac{d}{dr} \right) \right] \psi = E\chi$$

or

$$-\frac{\hbar^2}{2m} \frac{d^2}{dr^2} \chi = E\chi$$

(compare (8.36)). Then.
$\chi \infty \sin kr$ with $\sin ka = 0$. Consequently, $k = (n\pi)/a$ with $n = 1,2,3...$ and:

$$\frac{\hbar^2}{2m} \left(\frac{n\pi}{a} \right)^2 = E_n = \frac{h^2 n^2}{8ma^2}$$

(compare (4.7)). The wavefunctions are: $\psi = \frac{\chi}{r} = \text{const.} \frac{\sin\left(\frac{n\pi}{a}\right)r}{r}$.
From the normalization condition

$$1 = \int_0^a \int_0^{2\pi} \int_0^\pi \psi^2 r^2 \sin\theta \, dr \, d\phi \, d\theta = 4\pi \int_0^a \psi^2 r^2 dr = 4\pi \text{const.}^2 \int_0^a \sin^2\left(\frac{n\pi}{a}\right) r \, dr =$$

$$= 4\pi \text{const.}^2 \frac{a}{n\pi} \underbrace{\int_0^{n\pi} \sin^2 \xi \, d\xi}_{= \frac{n\pi}{2}} = 2\pi a \text{const.}^2 = 1 \Rightarrow \text{const.} = \frac{1}{\sqrt{2\pi a}} \Rightarrow$$

$$\psi = \frac{1}{\sqrt{2\pi a}} \frac{\sin\left(\frac{n\pi}{a}\right)r}{r}$$

$$(8.50)$$

8.5.3 The Hydrogen Atom in a Magnetic Field: Bohr Theory and Gyromagnetic Factor

Let us return to the hydrogen atom and assume an electron on a circular orbit. In a classical picture, its rotation is equivalent to an electrical current given by:

$$I = -\frac{e}{t_{period}} = -ef = -e\frac{\omega}{2\pi} = -e\frac{v}{2\pi r}.$$

The magnetic dipole moment is

$$d_{magn} = IA = -e\frac{v}{2\pi r}\pi r^2 = -e\frac{v}{2}r = -e\frac{vm_e}{2m_e}r = -\frac{eL}{2m_e} \text{ or } d_{magn} = -\frac{eL}{2m_e}$$
$$(8.51)$$

The potential energy of the magnetic dipole in an external magnetic field is:

$$U = -\mathbf{d}_{magn}\mathbf{B}$$

If

$$\mathbf{B} = \begin{pmatrix} 0 \\ 0 \\ B \end{pmatrix} \Rightarrow U = \frac{eL_zB}{2m_e} = \frac{e\hbar}{2m_e}mB = \mu_B mB \qquad (8.52)$$

With μ_B—Bohr's magneton, and eigenvalues $l_z = \hbar m$; $-l \leq m \leq l$. Correspondingly an external (time-independent) magnetic field leads to a splitting of the initially degenerated quantum levels with different quantum number m into a series of $2l+1$ energy levels. This is the essence of the normal Zeeman effect.

The gyromagnetic factor γ_l is defined as the ratio of the absolute value of the magnetic moment and the angular momentum of the electron on its orbit. From (8.51) and (8.52) we have:

$$\gamma_l = \frac{|d_{magn}|}{|L|} = \frac{e}{2m_e} = \frac{\mu_B}{\hbar} \qquad (8.53)$$

8.6 Advanced Material

8.6.1 Electron Spin and Spin–orbit Interaction

Let us start from two experimental results:

Experimental Fact (1st Peculiarity): A beam of silver atoms in their ground state ($l = 0$) propagating through a gradient magnetic field splits off into two parts (Stern–Gerlach experiment). Note that the degeneracy of states with a given l is $2l+1$

(number of projections of the angular momentum on the z-axis); i.e., it is an *odd* number whenever l is integer.

The splitting obviously indicates that the angular momentum of the silver valence electron is different from zero—although it is in an s-state. In order to interpret the experimental findings one had to assume that the electron has—in addition to its angular momentum arising from its orbital movement around the nucleus—some internal momentum that was called its spin. An even number of projections would only be possible, if the corresponding angular momentum (i.e., the electron spin **S**) would be characterized by a half-integer quantum number s = 1/2 with projections on the z-axis according to:

$$m_s = \begin{cases} -\frac{1}{2} \\ +\frac{1}{2} \end{cases} \Rightarrow 2s + 1 = 2 \Rightarrow s = \frac{1}{2} \Rightarrow |S| = \hbar\sqrt{s(s+1)} = \frac{\sqrt{3}}{2}\hbar \quad (8.54)$$

Experimental Fact (2nd Peculiarity): The gyromagnetic factor corresponding to the electron spin (Einstein–de Haas experiment) is given by (compare (8.53))

$$\gamma_s = \frac{|d_{magn,s}|}{|S|} \approx \frac{e}{m_e} = \frac{2\mu_B}{\hbar} = 2\gamma_l \quad (8.55)$$

Consequences

Fine Structure of Hydrogen Energy Levels

In a strong sense, the fine structure of hydrogen energy levels is a relativistic effect and would need the application of relativistic quantum mechanics. This is clearly beyond the frames of this course. The following argumentation therefore has to be understood rather as an illustration of the nature of the effect, but nevertheless we can learn a lot from that.

The fine structure of the hydrogen atoms energy levels arises from the interaction of the electrons spin magnetic and orbital magnetic momenta (spin–orbit interaction). Indeed, the interaction operator $\hat{V}_{ls}$ describing the potential energy U_{ls} of the electrons spin magnetic moment $d_{magn,s}$ in the magnetic field B_l caused by its movement on the orbit is given by:

$$U_{ls} = -\mathbf{d}_{magn,s}\mathbf{B}_l \propto \frac{Ze^2}{r^3}\mathbf{SL} \quad (8.56)$$

$$\Rightarrow \hat{V}_{ls} \propto \hat{S}\hat{L} \quad (8.56a)$$

Here, $\hat{S}$ is the operator of the electron spin.

Remark Let us have a short look at the origin of the r^{-3} dependence. If we change to a coordinate system where the electron is at rest, we will observe that it is circled by the nucleus with a charge Ze at a distance r. Then, the movement of the nucleus generates a magnetic field acting on the electron that is given by [4]:

$$\mathbf{B}_l = -\frac{\mu_0 Ze}{4\pi}\frac{\mathbf{r}\times\mathbf{v}}{r^3} = -\frac{\mu_0 Ze}{4\pi m_e}\frac{\mathbf{L}}{r^3}.$$

Then, by using

$$U_{ls} = -\mathbf{d}_{\text{magn},s}\mathbf{B}_l$$

and (8.55), we obtain (8.56).

Equation (8.56) describes a coupling of the orbital and spin magnetic momenta. We therefore introduce the operator of the total (orbit and spin) angular momentum of a single electron $\hat{J}$ as:

$$\hat{\mathbf{J}} = \hat{\mathbf{L}} + \hat{\mathbf{S}} \tag{8.57}$$

We will require that $\hat{\mathbf{J}}$ obeys the typical quantization rules of an angular momentum. Thus we have eigenvalues of $\hat{\mathbf{J}}^2$ equal to $\hbar^2 j(j+1)$. Because of the two possible projections of the single electron spin, in a hydrogen atom we can find the following possible quantum numbers j:

$$l > 0 : j = l - \frac{1}{2}; \, l + \frac{1}{2}$$
$$l = 0 : j = \frac{1}{2}. \tag{8.57a}$$

This is a special case of angular momentum summation in quantum mechanics. We will shortly introduce the general case later in Sect. 9.2.

In order to investigate the structure of the scalar product **SL**, we now make use of

$$\hat{\mathbf{J}}^2 = \left(\hat{\mathbf{L}} + \hat{\mathbf{S}}\right)^2 \Rightarrow \hat{\mathbf{L}}\hat{\mathbf{S}} = \frac{1}{2}\left(\hat{\mathbf{J}}^2 - \hat{\mathbf{L}}^2 - \hat{\mathbf{S}}^2\right)$$

Let us at the moment postulate the existence of common eigenfunctions of the operators $\hat{\mathbf{S}}^2, \hat{\mathbf{J}}^2, \hat{\mathbf{L}}^2$ (see task 8.7.10 of this chapter) and the Hamiltonian of the relativistic hydrogen atom. Then, when operating on these joint eigenfunctions, we may apply the following substitution:

$$\hat{\mathbf{L}}\hat{\mathbf{S}} = \frac{1}{2}\left(\hat{\mathbf{J}}^2 - \hat{\mathbf{L}}^2 - \hat{\mathbf{S}}^2\right) \to \frac{1}{2}\hbar^2[j(j+1) - l(l+1) - s(s+1)] \tag{8.58}$$

For $l > 0$, according to (8.56), (8.57a), and (8.58) any energy level given by (8.9) splits off into two components, defining the fine structure of energy levels caused by the spin–orbit interaction. The corrections to the energy levels may be calculated by means of the perturbation theory, which includes the calculation of $\left\langle \frac{1}{r^3} \right\rangle$. The latter is provided in Table 8.4 and contributes a further Z^3-term. Thus, together with (8.56), we have a Z^4-dependence of the spin–orbit interaction. The full proportionality factor (let us call it a) in (8.56a) is proportional to the square of the fine-structure constant and the Rydberg energy of the corresponding state according to (note the structural similarity to $\left\langle \frac{1}{r^3} \right\rangle$ as provided in Table 8.4):

$$a = -E_n \frac{Z^2 \alpha^2}{nl\left(l + \frac{1}{2}\right)(l + 1)} \propto Z^4 \tag{8.59}$$

The appearance of the fine-structure constant α (compare (8.13)) in (8.59) is a manifestation of the relativistic nature of the effect.

Remark In a weak *external* time-independent magnetic field, the coupling between spin and orbital magnetic momenta remains preserved. Then, each energy level defined by j splits off into $2j + 1$ sublevels with a gyromagnetic factor defined by a specific quantity called the Lande factor. In an emission spectrum, this splitting of the energy terms leads to a multiplicity of spectral lines, giving rise to what is called the anomalous Zeeman effect.

In a strong external magnetic field, however, the spin–orbit coupling is destroyed (Paschen Back effect), and the introduction of $\mathbf{J}$ makes no more sense. Then, orbital and spin magnetic momenta align to the external field separately, and no anomalous Zeeman effect is observed anymore.

Nomenclature: Let us mention in this context the usual nomenclature for characterizing atomar electronic states. In order to indicate the quantum numbers n,j,l, and s, the following writing is accepted:

$$n^{2s+1} X_j; \ X = S, P, D, F \text{ according to } l = 0, 1, 2, 3, ..$$

The term $2s + 1$ indicates the number of possible spin orientations. In a hydrogen atom, $s = 1/2$ and $2s + 1 = 2$, indicating two possible spin projections. In higher order atoms, the full spin may be larger.

8.6.2 Spinors and Pauli Matrices

The experimental facts mentioned at the beginning of the previous section indicate that an electron has—besides its spatial degrees of freedom—an additional degree of freedom that allows accepting only two values: spin "up" and spin "down" (Fig. 8.8). The corresponding wavefunctions are formally written as $|\uparrow\rangle$ and $|\downarrow\rangle$.

Spin up $|\uparrow\rangle$ Spin down $|\downarrow\rangle$

Fig. 8.8 Daily experience illustration of spin-up (left) and spin-down (right) quantum states. Cartoon by Dr. Alexander Stendal. Printed with permission

We require that the corresponding spin operator shall commute with all operators describing spatial degrees of freedom. Hence,

$$\left[\hat{S}, \hat{r}\right] = \left[\hat{S}, \hat{p}\right] = \left[\hat{S}, \hat{L}\right] = 0 \tag{8.60}$$

And so on. In accordance with (8.54), we set for a single electron:

$$\hat{S}^2|\uparrow\rangle = \frac{3}{4}\hbar^2|\uparrow\rangle; \quad \hat{S}^2|\downarrow\rangle = \frac{3}{4}\hbar^2|\downarrow\rangle$$
$$\hat{S}_z|\uparrow\rangle = \frac{\hbar}{2}|\uparrow\rangle; \quad \hat{S}_z|\downarrow\rangle = -\frac{\hbar}{2}|\downarrow\rangle \tag{8.61}$$

An illustrative mathematical description of the electrons spin states is possible in termini of spinors [5, p. 272–277].

The electron spin operator may be written in the following manner:

$$\hat{\mathbf{S}} = \begin{pmatrix} \hat{S}_x \\ \hat{S}_y \\ \hat{S}_z \end{pmatrix} = \frac{\hbar}{2}\sigma = \frac{\hbar}{2}\begin{pmatrix} \sigma_1 \\ \sigma_2 \\ \sigma_3 \end{pmatrix} \tag{8.62}$$

Here, the σ_n is the Pauli matrices:

$$\sigma_1 = \begin{pmatrix} 0 & 1 \\ 1 & 0 \end{pmatrix}; \quad \sigma_2 = \begin{pmatrix} 0 & -i \\ i & 0 \end{pmatrix}; \quad \sigma_3 = \begin{pmatrix} 1 & 0 \\ 0 & -1 \end{pmatrix} \tag{8.62a}$$

Obviously,

$$\sigma_1\sigma_1 = \sigma_2\sigma_2 = \sigma_3\sigma_3 = \begin{pmatrix} 1 & 0 \\ 0 & 1 \end{pmatrix} \tag{8.63}$$

For the z-component of the electron spin, we therefore have the operator:

$$\hat{S}_z = \frac{\hbar}{2}\sigma_3 = \frac{\hbar}{2}\begin{pmatrix} 1 & 0 \\ 0 & -1 \end{pmatrix} \tag{8.64}$$

Let us express the spin eigenfunction as a 2×1-matrix, i.e., a so-called spinor:

$$|\psi_{spin}\rangle = \begin{pmatrix} c_1 \\ c_2 \end{pmatrix}$$

The normalized eigenfunctions and eigenvalues are found from the following equations (compare (8.61)-(8.64)):

$$\hat{S}^2\begin{pmatrix} c_1 \\ c_2 \end{pmatrix} = \hbar^2 s(s+1)\begin{pmatrix} c_1 \\ c_2 \end{pmatrix} = \frac{\hbar^2}{4}(\sigma_1\sigma_1 + \sigma_2\sigma_2 + \sigma_3\sigma_3)\begin{pmatrix} c_1 \\ c_2 \end{pmatrix}$$

$$= \frac{\hbar^2}{4}\begin{pmatrix} 3 & 0 \\ 0 & 3 \end{pmatrix}\begin{pmatrix} c_1 \\ c_2 \end{pmatrix} = \frac{3\hbar^2}{4}\begin{pmatrix} c_1 \\ c_2 \end{pmatrix} \Rightarrow s(s+1) = \frac{3}{4} \Rightarrow s = \frac{1}{2}$$

$$\hat{S}_z\begin{pmatrix} c_1 \\ c_2 \end{pmatrix} = \hbar m_s\begin{pmatrix} c_1 \\ c_2 \end{pmatrix} = \frac{\hbar}{2}\sigma_3\begin{pmatrix} c_1 \\ c_2 \end{pmatrix} = \frac{\hbar}{2}\begin{pmatrix} 1 & 0 \\ 0 & -1 \end{pmatrix}\begin{pmatrix} c_1 \\ c_2 \end{pmatrix} = \frac{\hbar}{2}\begin{pmatrix} c_1 \\ -c_2 \end{pmatrix}$$

$$\Rightarrow m_s = \pm\frac{1}{2} \Rightarrow |\psi_{spin}\rangle = \begin{pmatrix} c_1 \\ c_2 \end{pmatrix} = \begin{cases} \begin{pmatrix} 1 \\ 0 \end{pmatrix} \equiv |\uparrow\rangle; \ m_s = +\frac{1}{2} \\ \begin{pmatrix} 0 \\ 1 \end{pmatrix} \equiv |\downarrow\rangle; \ m_s = -\frac{1}{2} \end{cases}$$

Hence, we find two eigenfunctions, corresponding to the directions "spin upwards" $\left(|\uparrow\rangle = \begin{pmatrix} 1 \\ 0 \end{pmatrix}\right)$ and "spin downwards" $\left(|\downarrow\rangle = \begin{pmatrix} 0 \\ 1 \end{pmatrix}\right)$. Note that these spinors should be understood as specific sets of two number, and *not* as two-dimensional spatial vectors!

As an example, let us apply this knowledge to the case of a system built from *two* electrons. For simplicity, we will restrict to the particular case of a state, where both spins are directed upwards (i.e., along the positive z-axis). We write the spin wavefunction as:

$$|\psi_{spin}\rangle = |\uparrow\rangle_1|\uparrow\rangle_2$$

Here, the index indicates the number of the electron.

We have now to introduce two sets of Pauli matrices, each acting on one of the spins. Hence, $\sigma(1)$ acts on the spin coordinates of the first electron and $\sigma(2)$ on the second. We now have (compare (8.57)):

$$\hat{S}_z|\uparrow\rangle_1|\uparrow\rangle_2 = \hbar m_s|\uparrow\rangle_1|\uparrow\rangle_2 = \frac{\hbar}{2}\{\sigma_3(1) + \sigma_3(2)\}|\uparrow\rangle_1|\uparrow\rangle_2 \qquad (8.65)$$

Note that

$$\sigma_3|\uparrow\rangle = \sigma_3\begin{pmatrix} 1 \\ 0 \end{pmatrix} = \begin{pmatrix} 1 & 0 \\ 0 & -1 \end{pmatrix}\begin{pmatrix} 1 \\ 0 \end{pmatrix} = \begin{pmatrix} 1 \\ 0 \end{pmatrix} = |\uparrow\rangle$$

So we find

$$\hat{S}_z|\uparrow\rangle_1|\uparrow\rangle_2 = \hbar m_s|\uparrow\rangle_1|\uparrow\rangle_2 = \frac{\hbar}{2}\{[\sigma_3(1)|\uparrow\rangle_1]\,|\uparrow\rangle_2 + |\uparrow\rangle_1[\sigma_3(2)|\uparrow\rangle_2]\}$$

$$= \frac{\hbar}{2}\{|\uparrow\rangle_1|\uparrow\rangle_2 + |\uparrow\rangle_1|\uparrow\rangle_2\} = \hbar|\uparrow\rangle_1|\uparrow\rangle_2 \Rightarrow m_S = +1$$

We come to the result that our assumed spin function $|\psi_{spin}\rangle = |\uparrow\rangle_1|\uparrow\rangle_2$ is an eigenfunction of $\hat{S}_z$. The corresponding eigenvalue, and thus the projection of the full spin on the z-axis, is now $+\hbar$.

Let us now turn to the square of the spin. We have to write (compare (8.62)):

$$\hat{\mathbf{S}} = \begin{pmatrix} \hat{S}_x \\ \hat{S}_y \\ \hat{S}_z \end{pmatrix} = \frac{\hbar}{2}\begin{pmatrix} \sigma_1(1) + \sigma_1(2) \\ \sigma_2(1) + \sigma_2(2) \\ \sigma_3(1) + \sigma_3(2) \end{pmatrix}$$

$$\Rightarrow S^2 = \frac{\hbar^2}{4}\left\{6\begin{pmatrix} 1 & 0 \\ 0 & 1 \end{pmatrix} + 2\sigma_1(1)\sigma_1(2) + 2\sigma_2(1)\sigma_2(2) + 2\sigma_3(1)\sigma_3(2)\right\} \qquad (8.66)$$

We note that

$$\sigma_1|\uparrow\rangle = \sigma_1\begin{pmatrix} 1 \\ 0 \end{pmatrix} = \begin{pmatrix} 0 & 1 \\ 1 & 0 \end{pmatrix}\begin{pmatrix} 1 \\ 0 \end{pmatrix} = \begin{pmatrix} 0 \\ 1 \end{pmatrix} = |\downarrow\rangle$$

$$\sigma_2|\uparrow\rangle = \sigma_2\begin{pmatrix} 1 \\ 0 \end{pmatrix} = \begin{pmatrix} 0 & -i \\ i & 0 \end{pmatrix}\begin{pmatrix} 1 \\ 0 \end{pmatrix} = i\begin{pmatrix} 0 \\ 1 \end{pmatrix} = i|\downarrow\rangle$$

Then,

$$\hat{S}^2|\uparrow\rangle_1|\uparrow\rangle_2$$

$$= \frac{\hbar^2}{4}\left\{6\begin{pmatrix} 1 & 0 \\ 0 & 1 \end{pmatrix}|\uparrow\rangle_1|\uparrow\rangle_2 + 2\sigma_1(1)\sigma_1(2)\,|\uparrow\rangle_1|\uparrow\rangle_2 + 2\sigma_2(1)\sigma_2(2)\,|\uparrow\rangle_1|\uparrow\rangle_2 + 2\sigma_3(1)\sigma_3(2)\,|\uparrow\rangle_1|\uparrow\rangle_2\right\}$$

$$= \frac{\hbar^2}{4}\{6|\uparrow\rangle_1|\uparrow\rangle_2 + 2\sigma_1(1)|\uparrow\rangle_1\sigma_1(2)|\uparrow\rangle_2 + 2\sigma_2(1)|\uparrow\rangle_1\sigma_2(2)|\uparrow\rangle_2 + 2\sigma_3(1)|\uparrow\rangle_1\sigma_3(2)|\uparrow\rangle_2\}$$

$$= \frac{\hbar^2}{4} \{6|\uparrow\rangle_1|\uparrow\rangle_2 + 2|\downarrow\rangle_1|\downarrow\rangle_2 + 2i|\downarrow\rangle_1 i|\downarrow\rangle_2 + 2|\uparrow\rangle_1|\uparrow\rangle_2 \} = \frac{\hbar^2}{4} \{6 + 2\}|\uparrow\rangle_1|\uparrow\rangle_2$$

$$= 2\hbar^2|\uparrow\rangle_1|\uparrow\rangle_2 = \hbar^2 S(S+1)|\uparrow\rangle_1|\uparrow\rangle_2 \Rightarrow S = 1$$

We find that our assumed spin function $|\psi_{spin}\rangle = |\uparrow\rangle_1|\uparrow\rangle_2$ is also an eigenfunction of $\hat{\mathbf{S}}^2$. Concerning the corresponding eigenvalue, we come to the reasonable result that the spin quantum number of two electrons with parallel spin is equal to 1.

8.6.3 Remarks on Selection Rules for Optical Transitions (Electric Dipole Interaction)

When neglecting *any* magnetic interactions, the basic selection rules for an electron in a central field may be written as:

$$\Delta n - \text{arbitrary}$$
$$\Delta l = \pm 1$$
$$\Delta m_s = 0 \qquad\qquad (8.67)$$
$$\Delta m = \begin{cases} 0; & \text{linear light polarization} \\ \pm 1; & \text{circular light polarization} \end{cases}$$

Hereby, the selection rules concerning n, l, and m may be obtained from a direct calculation in terms of (6.18) and (6.19) using electric dipole interaction as perturbation and unperturbed wavefunctions like those given in Table 8.3. Moreover, we recognize the selection rule (8.49) $\Delta l = \pm 1$, which we have derived from the correspondence principle for the special case of a rigid rotor. Contrarily to the electron, a photon has an integer spin $s_{phot} = 1$ and, therefore, (8.49) can be interpreted as a consequence of angular momentum conservation.

The condition on m_s arises from the fact that the electron spin does not couple to the electric field of the light wave. The mentioned selection rules find practical application in the Paschen–Back effect (see task 8.7.11). Here, because of the strong external magnetic field, the electronic spin–orbit coupling is destroyed, and the applied external static (i.e., time-independent) magnetic field does not cause quantum transitions.

The selection rules are modified when spin–orbit interaction comes into play [5, pp. 303–309]. From

$$\left[\hat{L}_x, \hat{L}_y\right] = i\hbar\hat{L}_z; \quad \left[\hat{L}_y, \hat{L}_z\right] = i\hbar\hat{L}_x; \quad \left[\hat{L}_z, \hat{L}_x\right] = i\hbar\hat{L}_y$$

and the analogous relations for the spin components, we find:

$$\hat{\mathbf{J}}^2 = \left(\hat{\mathbf{L}} + \hat{\mathbf{S}}\right)^2 = \hat{\mathbf{L}}^2 + \hat{\mathbf{S}}^2 + 2\hat{\mathbf{L}}\hat{\mathbf{S}} \Rightarrow \begin{cases} \left[\hat{S}_z, \hat{\mathbf{J}}^2\right] \neq 0 \\ \left[\hat{L}_z, \hat{\mathbf{J}}^2\right] \neq 0 \end{cases}$$

Indeed,

$$\left[\hat{L}_z, \mathbf{\hat{L}\hat{S}}\right] = \left[\hat{L}_z, \hat{L}_x\hat{S}_x + \hat{L}_y\hat{S}_y + \hat{L}_z\hat{S}_z\right] = i\hbar\left(\hat{L}_y\hat{S}_x - \hat{L}_x\hat{S}_y\right)$$

$$\left[\hat{S}_z, \mathbf{\hat{L}\hat{S}}\right] = \left[\hat{S}_z, \hat{L}_x\hat{S}_x + \hat{L}_y\hat{S}_y + \hat{L}_z\hat{S}_z\right] = -i\hbar\left(\hat{L}_y\hat{S}_x - \hat{L}_x\hat{S}_y\right)$$

Therefore,

$$\left[\hat{L}_z + \hat{S}_z, \mathbf{\hat{L}\hat{S}}\right] = \left[\hat{J}_z, \mathbf{\hat{L}\hat{S}}\right] = 0 \Rightarrow \left[\hat{J}_z, \mathbf{\hat{J}}^2\right] = 0$$

We come to the result that the square of the total momentum does not commute with the z-components of **L** and **S**, but with its own z-component. This is a particular case of the more general situation we will discuss later in Sect. 9.2 (Fig. 9.3 on right). At the moment it is sufficient for us to note that any quantum state with a well-defined $\mathbf{\hat{J}}^2$ may also have a well-defined value of its z-component.

Therefore, in the presence of spin–orbit interaction, the selection rules are formulated in terms of j and the corresponding magnetic quantum number m_j, which characterizes the eigenvalues of $\hat{J}_z$. Additionally, from angular momentum conservation, we still have:

$$\Delta l = \pm 1$$

The corresponding selection rules for j may be written as:

$$\begin{aligned}
\Delta j &= 0, \pm 1 \text{ (except } j = 0 \to j = 0) \\
\Delta m_j &= 0, \pm 1 \text{ (except } \Delta m_j = 0 \; if \; \Delta j = 0)
\end{aligned} \tag{8.68}$$

These selection rules may again be derived in terms of (6.18), using so-called spherical spinors as unperturbed wavefunctions. In short, their appearance is related to the following combinations of quantum numbers:

$$\Delta j = \pm 1 : \quad j = l \pm \frac{1}{2} \overset{\Delta l = \pm 1}{\curvearrowright} j + \Delta j = l + \Delta l \pm \frac{1}{2}$$

$$\Delta j = 0 \quad : \quad j = l \pm \frac{1}{2} \overset{\Delta l = \pm 1}{\curvearrowright} j + \Delta j = l + \Delta l \mp \frac{1}{2}$$

It is nevertheless useful to provide a heuristic illustration of (8.68):

- When accepting that the spin is rather inert with respect to optical transitions, the selection rule $\Delta j = \pm 1$ is a direct conclusion from (8.49). Then, the rule $\Delta m_j = 0, \pm 1$ also appears as a direct generalization of the corresponding expression for $\Delta m = 0, \pm 1$ in (8.67).
- What is new compared to (8.67) is the transition with $\Delta j = 0$. Once (8.49) holds, it must be accompanied with a change of the projection of

the electron spin with regard to the direction of the orbital momentum l (see later Fig. 9.4 for a purely geometrical illustration). One can illustrate their origin when remembering that the orbital movement of the electron gives rise to a magnetic field B_l that couples to the electron spin according to (8.56). If, as a result of an optical transition, l is changed in accordance with (8.49), the magnetic field B_l will also change its value. This gives rise to a time-dependent magnetic perturbation potential, which may cause the corresponding change in the spin projection relative to $\mathbf{L}$. In this picture, a magnetic dipole interaction (see (8.56)) is involved, and in fact, the corresponding spectral lines are weaker than those corresponding to $\Delta j = \pm 1$.

- Clearly, if $\Delta j = 0$, the angular momentum inherent to the absorbed (or emitted) photon requires that at least the direction of the total momentum must change. Therefore, $\Delta j = 0$ cannot be accompanied by $\Delta m_j = 0$. As a particular case, $j = 0 \rightarrow j = 0$ must be forbidden at all.

The selection rules (8.68) are valid for the anomalous Zeeman effect.

8.7 Tasks for Self-Check

8.7.1 Please assign the assertion given in the left column of the following table to one or several of the mentioned basic model systems by ticking the corresponding box:

Assertion	1D harmonic oscillator	Particle in 1D box potential with impermeable walls	H atom
Example: $V(r) = -\frac{1}{4\pi\varepsilon_0}\frac{e^2}{r}$			X
$E_n = \hbar\omega_0\left(n + \frac{1}{2}\right)$			
$\psi_1 = \frac{1}{\sqrt{\pi}}\left(\frac{1}{a_0}\right)^{\frac{3}{2}}e^{-r/a_0}$			
$E_n \propto n^2$			
$x_{n,n+2} = 0\,\forall n$			
$E_n \propto -n^{-2}$			
$\psi_n(x) = \sqrt{\frac{2}{L}}\sin\frac{n\pi}{L}x$			
$\left(x_{n,n-1}\right)^2 = \frac{n\hbar}{2m\omega_0}$			
$\nu_{nm} = R_\infty\left(\frac{1}{m^2} - \frac{1}{n^2}\right)$			
$\hat{H} = \left[-\frac{\hbar^2}{2m}\frac{d^2}{dx^2} + \frac{m\omega_0^2}{2}x^2\right]$			

Assertion	1D harmonic oscillator	Particle in 1D box potential with impermeable walls	H atom
$\psi_{311} =$ $\frac{\sqrt{2}}{81\sqrt{\pi}}\left(\frac{1}{a_0}\right)^{\frac{3}{2}}\left(6-\frac{r}{a_0}\right)\frac{r}{a_0}e^{-r/3a_0}\sin\vartheta e^{i\phi}$			

8.7.2 Let the attractive force between a neutron (mass m_n) and an electron (mass m_e) be given by the law of gravitation. Consider the smallest orbit which the electron can have—according to Bohr's theory—when moving around the neutron ([6, task 8.6])!
- Write a formula for the centrifugal force acting on the electron!
- Find expressions for kinetic, potential and total energies!
- Set up an equation which corresponds to the Bohr postulate for quantization of circular orbits!
- How large is the radius r of the "ground state" orbit with $n = 1$?

8.7.3 A positronium is a bound electron–positron pair. The positron is the antiparticle of the electron, so it has the same rest mass and the opposite electric charge of the electron. On the assumption that positron and electron circle their common center of gravity, calculate their distance as well as their energy levels ([after [6, task 8.8])

8.7.4 Estimate the difference between the emission wavelength of the transition $n = 2 \rightarrow n = 1$ in an ordinary hydrogen atom and a deuterium atom.

8.7.5 Prove the commutation relations:

$$\left[\hat{L}_x, \hat{L}_y\right] = i\hbar\hat{L}_z$$

$$\left[\hat{L}_y, \hat{L}_z\right] = i\hbar\hat{L}_x$$

$$\left[\hat{L}_z, \hat{L}_x\right] = i\hbar\hat{L}_y$$

8.7.6 Prove the commutation relation: $\left[\hat{L}_z, \hat{L}^2\right] = 0$.

8.7.7 The electronic ground state of the hydrogen atom is described by the wavefunction

$$\psi(r) = \frac{1}{\sqrt{\pi}}a_0^{-\frac{3}{2}}e^{-\frac{r}{a_0}}$$

where a_0 is Bohr's radius. Calculate the probability w that the electron is observed within a sphere of radius R centered at the position of the proton. Estimate the radius for which $w = 1/2$ is observed (after [7, task 11.2])!

8.7.8 Consider the hydrogen $2p_z$ state with $n = 2$, $l = 1$, $m = 0$. Calculate the expectation value of the potential energy in this state! (after [7, task 13.4] - Hint: The wavefunction is given in Table 8.3).

8.7.9   For the hydrogen 3 s state, derive the expression for the radial part of the wavefunction

$$R \propto \left(1 - \frac{2r}{3a_0} + \frac{2r^2}{27a_0^2}\right)e^{-\frac{r}{3a_0}} \propto \left(27 - \frac{18r}{a_0} + \frac{2r^2}{a_0^2}\right)e^{-\frac{r}{3a_0}}$$

directly from (8.42)!

8.7.10 10 Check the commutation relations:

$$\left[\hat{L}_x, x\right] = 0$$

$$\left[\hat{L}_z, V(r)\right] = 0$$

$$\left[\hat{\mathbf{J}}^2, \hat{\mathbf{L}}^2\right] = 0$$

8.7.11 Consider a hydrogen atom state with a given quantum number l. A very strong magnetic field is applied such that you observe splitting of the energy levels according to the Paschen Back effect. How many different energy levels do you observe as a result of the splitting of each level with that l when assuming (8.55) $\gamma_s = 2\gamma_l$?

References

Specific References

1. R. Beliveau, D. Gingras, *Der Tod* (Kösel-Verlag München, Das letzte Geheimnis des Lebens, 2012), p. 226
2. А.С. Давыдов, Квантовая Механика, Москва Физматгиз 1963, p.159
3. Feinstruktur der Antimaterie, Spektrum der Wissenschaften 2020 (5); p. 10
4. P.A. Tipler, G. Mosca, *Physik, 7* (Springer, Auflage, 2015), p. 874
5. А.А. Соколов, И.М. Тернов, В.Ч. Жуковский: Квантовая механика, Москва "Наука" 1979
6. H.Haken, H.C. Wolf, *The Physics of Atoms and Quanta* (Springer, 2005)
7. U. Fano, L. Fano, *Physics of Atoms and Molecules* (University of Chicago Press, An introduction to the structure of matter, 1973)

General Literature

8. L.D. Landau, E.M. Lifshitz, *Quantum Mechanics (Volume 3 of A Course of Theoretical Physics)* (Pergamon Press, 1965)
9. W. Demtröder, *Atoms, Molecules, and Photons* (Springer, 2010)
10. S. Flügge, *Practical Quantum Mechanics* (Springer, 1999)

The Helium Atom

9

Abstract

We provide an approximate treatment of the helium atom as a simple example of a many-body system. The configuration space is introduced, and conclusions from the principal indistinguishability of quantum particles are derived. Several approaches are presented for estimating the ground-state energy of the helium atom.

9.1   Starting Point

The helium atom consists of one nucleus ($Z = 2$) and two electrons (Fig. 9.1).

It is an example of a three-body system, where attractive forces are acting between the nucleus and each of the electrons, while between the two electrons, we have a repulsive force acting. There exists no general analytical solution to this problem in classical mechanics, and we cannot expect to find an exact solution of the Schrödinger equation for this system. Hence, we will have to make use of model simplifications here. One of them is that we regard the nucleus as fixed and place the origin of the coordinate system into that nucleus. Then, as it shown in Fig. 9.1, the position of each of the electrons is characterized by a vector $\mathbf{r}_i$, while in the case of the helium atom, $i = 1, 2$. In a quantum mechanical language, that means that each electron in the system contributes three coordinates defining $\mathbf{r}_i$ into the set of arguments of the wavefunction Ψ.

Note that i is simply a counting index here; because of the later discussed indistinguishability of the electrons, we cannot "mark" the electrons and distinguish them by their number. Instead, a physical state described by "electron 1" with coordinates $\mathbf{r}_1$ and "electron 2" with coordinates $\mathbf{r}_2$ is *identical* to a state with "electron 1" with coordinates $\mathbf{r}_2$ and "electron 2" with coordinates $\mathbf{r}_1$.

In terms of a *very* simplified treatment, let us for a moment moreover forget the electron spin and further assume that the repulsive interaction between the two

© The Author(s), under exclusive license to Springer Nature Switzerland AG 2022   209
O. Stenzel, *Light–Matter Interaction*, UNITEXT for Physics,
https://doi.org/10.1007/978-3-030-87144-4_9

Fig. 9.1 Helium atom

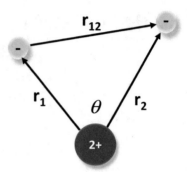

electrons could be neglected. In this case the Hamiltonian of the system would look like:

$$\hat{H} \equiv \hat{H}^{(0)} = \hat{T}_{\text{kin}} + U = \hat{T}_1 + \hat{T}_2 + U_1 + U_2 \tag{9.1}$$

Here, to each of the electrons, we ascribe a kinetic energy T and a potential energy U, both depending only on the coordinates of the corresponding electron. Then, (9.1) may be written as:

$$\hat{H}^{(0)} = \underbrace{\hat{T}_1 + U_1}_{\equiv \hat{H}_1} + \underbrace{\hat{T}_2 + U_2}_{\equiv \hat{H}_2} \equiv \hat{H}_1 + \hat{H}_2 \tag{9.2}$$

where each of the $\hat{H}_i$ depends only on the coordinates $\mathbf{r}_i$ of the ith electron. In this situation, when considering that the wavefunction of the two electrons depends on the coordinates of both electrons, it is useful to factorize the wavefunction according to:

$$\psi(\mathbf{r}_1, \mathbf{r}_2)^{(0)} = \psi_1(\mathbf{r}_1)\psi_2(\mathbf{r}_2) \tag{9.3}$$

$\psi_1(\mathbf{r}_1)$ and $\psi_2(\mathbf{r}_2)$ will be further called single-particle wavefunctions. Substituting into the simplified Schrödinger equation:

$$\hat{H}^{(0)}\psi^{(0)} = E^{(0)}\psi^{(0)}; \quad \psi^{(0)} = \psi(\mathbf{r}_1, \mathbf{r}_2)^{(0)} \tag{9.3a}$$

yields:

$$\hat{H}^{(0)}\psi^{(0)} = \left(\hat{H}_1 + \hat{H}_2\right)\psi_1(\mathbf{r}_1)\psi_2(\mathbf{r}_2) = E^{(0)}\psi_1(\mathbf{r}_1)\psi_2(\mathbf{r}_2)$$
$$= \psi_2(\mathbf{r}_2)\hat{H}_1\psi_1(\mathbf{r}_1) + \psi_1(\mathbf{r}_1)\hat{H}_2\psi_2(\mathbf{r}_2)$$

However, $\hat{H}_1$ and $\hat{H}_2$ are nothing else than the Hamiltonians of a single electron in the central field of the nucleus with atomic order $Z = 2$. Hence, when associating the single electron wavefunctions $\psi_1(\mathbf{r}_1)$ and $\psi_2(\mathbf{r}_2)$ with the eigenfunctions of the hydrogen-like atom with the eigenvalues E_1 and E_2, we immediately find:

$$\hat{H}^{(0)}\psi^{(0)} = E^{(0)}\psi_1(\mathbf{r}_1)\psi_2(\mathbf{r}_2) = \psi_2(\mathbf{r}_2)\hat{H}_1\psi_1(\mathbf{r}_1) + \psi_1(\mathbf{r}_1)\hat{H}_2\psi_2(\mathbf{r}_2)$$
$$= (E_1 + E_2)\psi_1(\mathbf{r}_1)\psi_2(\mathbf{r}_2)$$
$$E^{(0)} = E_1 + E_2 \tag{9.4}$$

Hence the Ansatz (9.3) solves the simplified Schrödinger Eq. (9.3a) provided that

$\psi_1(\mathbf{r}_1)$ and $\psi_2(\mathbf{r}_2)$ are eigenfunctions of $\hat{H}_1$ and $\hat{H}_2$, respectively. Then, the energy levels are given by $E^{(0)} = E_1 + E_2$.

Let us apply this knowledge to the ground state of the helium atom $E_{\mathrm{He},0}^{(0)}$. Taking (8.9) into account, we then have:

$$E^{(0)} = E_1 + E_2 \Rightarrow E_{\mathrm{He},0}^{(0)} = -2Z^2 Ry = -8Ry = -108.8\,\mathrm{eV} \tag{9.5}$$

In this approximation, the helium atoms ground-state energy is identical to twice the ground-state energy of a singly ionized He ion, the latter being equal to $-4Ry = -54.4\,\mathrm{eV}$. The result can be verified experimentally by measuring the ionization energy of a helium atom, i.e., the energy which must be supplied to a helium atom such that it gets rid of both electrons when starting from its ground state. This energy is equal to 78.9 eV as verified experimentally. From here we may draw two conclusions:

- The estimated value $E_{\mathrm{He},0}^{(0)}$ is of the correct order of magnitude, but nevertheless very inaccurate. This is obviously caused by neglecting the Coulomb interaction between the electrons.
- Let us illustrate the process of ionization of the helium atom as a two-step-process. In the first step, one electron (we call it the "first" one) leaves the atom, while the other one (the "second") still remains in the singly ionized helium atom. The energy necessary for the second step of ionization, namely releasing the "second" electron from helium, must be equal to $Z^2 Ry = 4\,Ry = 54.4$ eV. This is clear from the theory of hydrogen-like atoms, compare (8.9). Then, when comparing with the experimental total ionization energy, the energy necessary to release the "first" of the helium electrons turns out to be around 24.5 eV only. Obviously, the presence of the "second" electron leads to an effective reduction of the positive charge of the nucleus, such that the "first" electron may escape with a rather small individual ionization energy. When setting $Z_{\mathrm{eff}}^2 Ry = 24.5\,\mathrm{eV}$, we estimate $Z_{\mathrm{eff}} \approx 1.342$, a value in between the atomic orders of hydrogen ($Z = 1$) and helium ($Z = 2$). This result brings us to the physical idea that the "second" electron has a screening effect on the nuclear charge, thus giving rise to the smaller amount of energy necessary for

performing the first ionization step. In the second step, the remaining ("second") electron, of course, feels the full charge of the helium nucleus, and needs the full $4Ry$ to escape from the nucleus.

In fact the electrons in the helium atom belong to the class of indistinguishable quantum particles. Therefore, our formal distinction into a "first" and "second" is only a bookkeeping artifact; in reality we have to expect that in a helium atom, each of the electrons has a screening effect on the other. Within this assumption, in Sect. 9.3.2, a variation approach will be presented that allows estimating the helium atom ground-state energy much more accurate than the very simplified treatment presented so far.

9.2   Physical Ideas

In the previous chapters, the theoretical treatment was effectively restricted to the discussion of single-particle systems; i.e., we discussed a single quantum particle (usually an electron) moving in some potential. Our new topic—the helium atom— is our first example of a many-particle system. Although in our particular case, "many" means "two," it is a huge qualitative difference to the single-particle case. Therefore, *several* new concepts will have to be taken into account. Consequently, Sect. 9.4 will be rather extended again.

Configuration space: The single-particle wavefunction depends on three coordinates only. Therefore, in the single-particle case, we had:

$$\Psi = \Psi(\mathbf{r}, t)$$

In a many-particle case (let it be N particles), the wavefunction will naturally depend on the coordinates of all particles [1]. Hence, we now have:

$$\Psi = \Psi(\mathbf{r}_1, \mathbf{r}_2, ..., \mathbf{r}_N, t) \tag{9.6}$$

The coordinates $\{\mathbf{r}_1, \mathbf{r}_2, ..., \mathbf{r}_N\}$ span what we call the configuration space, which has a dimensionality of $3N$. Thus, the N-particle wavefunction is defined in the corresponding $3N$-dimensional configuration space. Instead of (4.6), the probability density w to find the N-particle system in a certain configuration will then be given by:

$$dw(\mathbf{r}_1 \in dV_1, \mathbf{r}_2 \in dV_2, ..., \mathbf{r}_N \in dV_N) = |\Psi(\mathbf{r}_1, \mathbf{r}_2, ..., \mathbf{r}_N, t)|^2 dV_1 dV_2...dV_N \tag{9.7}$$

Correspondingly, for the calculation of expectation values, integration in (5.13) has to be performed over $dV_1 dV_2...dV_N$.

Identical (or indistinguishable) particles: In contrast to classical particles, quantum particles like electrons are indistinguishable. The indices $1 ... N$ in (9.6) are introduced for bookkeeping purposes only. Once the two electrons in the helium atom are indistinguishable, we have to expect that a wavefunction like (9.6) must suffice certain symmetry relations with respect to interchanging the electrons positions. This is a qualitatively new circle of problems and will be topic of Sect. 9.3.1. Figure 9.2 provides an illustration of the concepts of classical distinguishable and quantum mechanical indistinguishable particles.

Three-body problem: In contrast to the hydrogen atom, the helium atom can no more be regarded a two-body problem. The simplifications we had introduced in Sect. 2.3.1, and particularly the very helpful concept of the reduced mass, will no more find application now. Although, because of the repulsive interaction between the two electrons, the potential is no more of central symmetry, but becomes angle-dependent.

Fig. 9.2 Distinguishable (on top) and indistinguishable (on bottom) objects. Cartoon by Dr. Alexander Stendal. Printed with permission

Angular Momentum addition: [2, pp. 165–169].

Once the total angular momentum provided by the electrons in a complicated atom appears as a superposition of the contributions of the single electrons, the question of angular momentum addition in quantum mechanics becomes essential for understanding the internal dynamics of atoms. This angular momentum addition is conceptually simple in classical physics—one simply has to perform a geometrical vector addition. This is no more possible in quantum mechanics, because all angular moments have to obey quantization rules like (8.27) and (8.30). This clearly restricts the number of possible angular momentum addition results. We have already been confronted with that circumstance, namely when performing the addition of orbital and spin angular momenta of an electron in Sect. 8.6.1 (compare (8.57a)).

The strong theory of momentum addition in quantum mechanics is clearly outside the topics of this book, nevertheless we need to apply certain results from this theory here. We will restrict to the case of the addition of two angular momenta only. For our further treatment, what we need to know is the following recipe:

Assume two different angular momenta $\mathbf{L}_1$ and $\mathbf{L}_2$. Let $\mathbf{L}_1$ be characterized by the quantum number L_1, and $\mathbf{L}_2$ by L_2. Both these momenta shall sum up to the total angular momentum $\mathbf{L} = \mathbf{L}_1 + \mathbf{L}_2$. Then, the principally possible quantum numbers L that characterize the total momentum are given by:

$$L = L_{\min}, L_{\min} + 1, \ldots, L_{\max} - 1, L_{\max}$$
$$L_{\max} = L_1 + L_2; \quad L_{\min} = |L_1 - L_2|$$

For the particular case of $L_1 = l$ and $L_2 = 1/2$, from here we immediately obtain (8.57a) as a special case.

Note that at the same time, the magnetic quantum numbers m_1 and m_2 sum up to the projection of the resulting momentum m according to:

$$m = m_1 + m_2 = -L, \ldots, +L$$

This allows constructing a simple geometrical illustration of the corresponding angular momentum addition process, when associating the angular momentum vectors as arrows with a length proportional to $\sqrt{L(L+1)}$ and a projection on the z-axis proportional to m—see Fig. 9.3 on left.

Figure 9.3 clearly indicates that generally, a different length—and corresponding different values of L—of the resulting angular momentum vector may be obtained without violating the above-mentioned rules. Thereby, the set of states $|L_1 m_1 L_2 m_2\rangle$ with given values of L_1, m_1, L_2 and m_2 forms a basis for representing the final states.

Note that in a many-body system, it is the total angular momentum that is conserved and not necessarily any of the individual momenta of its individual constituents. It therefore makes sense to turn to another basis $|L_1 L_2 L m\rangle$, when the quantum numbers L_1, L_2, L and m are fixed. Note that in this situation, which is sketched in Fig. 9.3 on right, the individual projections m_1 and m_2 may no more

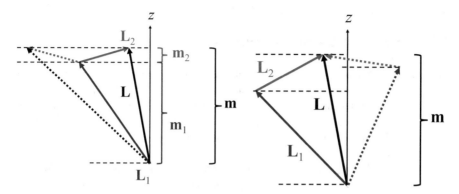

Fig. 9.3 On left: angular momentum addition with well-defined values of L_1, m_1 (in navy), L_2 and m_2 (in red). In general, L (in black) may accept different values. On right:: angular momentum addition with well-defined values of L_1, L_2, L, and m. In general, m_1 and m_2 may be uncertain

be well defined, although their sum is. Thereby, the quantum state of a system corresponding to a given L, a given projection of the total momentum m, as well as fixed L_1 and L_2 may be described by a wavefunction written as:

$$|L_1 L_2 Lm\rangle = \sum_{m_1, m_2} \underbrace{(L_1 L_2 m_1 m_2 | Lm)}_{\text{Clebsch-Gordan coefficients}} |L_1 m_1 L_2 m_2\rangle$$

$$(L_1 L_2 m_1 m_2 | Lm) \neq 0 \Rightarrow m = m_1 + m_2$$

The quantities $(L_1 L_2 m_1 m_2 | Lm)$ are called Clebsch–Gordan coefficients.

Solution of the time-independent Schrödinger equation: As already mentioned, we will not be able to find an exact analytic solution to the Schrödinger equation for the electron dynamics in the helium atom. Therefore, the application of approximate methods will be essential. We will restrict here to the calculation of the helium ground-state energy, neglecting spin effects in the first step. In Sect. 9.3.2, we will present two different approaches for that, namely

- A perturbation approach according to Sect. 5.3.8. Here, the unperturbed wavefunctions will correspond to (9.1) and (9.3a), and we will regard the Coulomb interaction between the electrons as the perturbation.
- A variation approach. This way we will essentially model the screening mechanism explained at the end of Sect. 9.1.

Then, we will separately discuss effects caused by the electron spins. In contrast to the hydrogen atom, we will find out that spin effects in many-electron atoms do no more provide small corrections to the energy levels, but have an essential impact on the energy levels (Sect. 9.4.3).

9.3   Theoretical Considerations

9.3.1   The Pauli Exclusion Principle

In Sect. 4.5.3, we have already supposed that a quantum state (for the model case of a particle in a box) as defined by the value of the quantum number n may be populated with a maximum of two electrons (degeneration factor $g = 2$). We have assigned this degeneration to the two possible orientations of the electron spin. Once in a helium atom, we have two electrons, we will have to return to this question now.

Let us regard two identical particles 1 and 2, and let us assume that each of them is characterized by a set of quantum numbers n, l, and m (for simplicity we will write $\{n\}$ for this triple of quantum numbers), the spin orientation m_s, and its coordinates $\mathbf{r}$. The full wavefunction of the system is thus:

$$\Psi = \Psi(\{n_1\}, m_{s1}, \mathbf{r}_1; \{n_2\}, m_{s2}, \mathbf{r}_2) \tag{9.8}$$

Once the particles are indistinguishable, the physical state should not be influenced by interchanging the positions of the two particles. Then, the Hamiltonian of the system should commute with the transposition operator $\hat{P}$ [2, pp. 373–378, 3, pp. 381–384], which is defined by:

$$\hat{P}\Psi(\{n_1\}, m_{s1}, \mathbf{r}_1; \{n_2\}, m_{s2}, \mathbf{r}_2) = \Psi(\{n_2\}, m_{s2}, \mathbf{r}_1; \{n_1\}, m_{s1}, \mathbf{r}_2) \tag{9.9}$$

Let us remark that twice the action of $\hat{P}$ results in the initially assumed wavefunction (9.8). Hence

$$\hat{P}\hat{P}\Psi = \Psi$$

The eigenvalue problem for $\hat{P}$ can be written as:

$$\hat{P}\Psi = P\Psi \Rightarrow \hat{P}\hat{P}\Psi = P^2\Psi = \Psi \Rightarrow P^2 = 1 \Rightarrow P = \pm 1 \tag{9.10}$$

We find, that the eigenvalues of $\hat{P}$ are given by $P = \pm 1$. Therefore, the eigenfunctions of $\hat{P}$ are either symmetric or antisymmetric with respect to interchanging the positions of the particles. Particles with symmetric wavefunctions according to:

$$\Psi^{(\text{sym})}(\{n_1\}, m_{s1}, \mathbf{r}_1; \{n_2\}, m_{s2}, \mathbf{r}_2) = \Psi^{(\text{sym})}(\{n_2\}, m_{s2}, \mathbf{r}_1; \{n_1\}, m_{s1}, \mathbf{r}_2) \tag{9.11}$$

are called bosons. For example, photons and α-particles belong to the class of bosons. They are characterized by an integer value of their spin.

Particles with an antisymmetric wavefunctions according to:

$$\Psi^{(\text{asym})}(\{n_1\}, m_{s1}, \mathbf{r}_1; \{n_2\}, m_{s2}, \mathbf{r}_2) = -\Psi^{(\text{asym})}(\{n_2\}, m_{s2}, \mathbf{r}_1; \{n_1\}, m_{s1}, \mathbf{r}_2) \tag{9.12}$$

are called fermions. For example, electrons, protons or neutrons belong to the class of fermions. The latter are characterized by a half-integer value of their spin.

For fermions, from (9.12) we immediately obtain the Pauli exclusion principle. Indeed, assuming identical quantum numbers for both particles, i.e., when setting: $\{n_2\} = \{n_1\} = \{n\}$ and $m_{s2} = m_{s1} = m_s$, from (9.12) we obtain:

$$\Psi^{(asym)}(\{n\}, m_s, \mathbf{r}_1; \{n\}, m_s, \mathbf{r}_2) = -\Psi^{(asym)}(\{n\}, m_s, \mathbf{r}_1; \{n\}, m_s, \mathbf{r}_2)$$

$$\Rightarrow \Psi^{(asym)}(\{n\}, m_s, \mathbf{r}_1; \{n\}, m_s, \mathbf{r}_2) = 0$$

(9.12a)

The wavefunction of two identical fermions becomes zero when both fermions are in identical quantum states. Hence, in any quantum system, the states of the fermions must differ in at least one quantum number. For example, in the 1D particle-in-a-box system from Sect. 4.3.3, in a quantum state characterized by the quantum number n, a maximum of only two electrons (with different spin orientation) may be found.

For bosons, according to (9.11), no corresponding exclusion principle may be formulated. For the particular case of photons, this is directly reflected by the set of eigenvalues of the photon number operator (in a given photon mode) according to (5.51). Therefore, when deriving the Planck distribution in Sect. 7.3.1, we could assume an arbitrary number of photons N_p populating any given mode. This leads to another phenomenon specific to bosons, namely the Bose–Einstein condensation, which is observed when at lowest temperature; all bosons tend to occupy the ground state of the system. For example, superconductivity or superfluidity is related to the Bose–Einstein condensation.

On the other hand, the occupation of a fermionic quantum state can only be either zero or one. An occupation number operator like (5.51), defined on the basis of (5.48) and (5.43), must not be applied to fermions. In fact, it is only relation (5.43) that needs to be modified in order to construct a fermionic occupation number operator (see later task 9.7.3).

Let us now assume that each of the particles may be characterized by a single-particle wavefunction $\Psi = \Psi(\{n_i\}, m_{si}, \mathbf{r}_i)$. In analogy to what has been presented in Sect. 9.1, the N-particle wavefunction may be written as a product of single-particle wavefunctions provided that there is no interaction between the particles. However, a product like (9.3) will not suffice symmetry relations like (9.11) and (9.12). Therefore, when considering the symmetry requirements, but neglecting any other interaction between the particles forming the system, the symmetric or antisymmetric wavefunctions of the system of the two particles of type (9.8) can be written as:

Fermions:

$$\Psi^{(asym)} = \frac{1}{\sqrt{2}}[\Psi(\{n_1\}, m_{s1}, \mathbf{r}_1)\Psi(\{n_2\}, m_{s2}, \mathbf{r}_2)$$

$$-\Psi(\{n_1\}, m_{s1}, \mathbf{r}_2)\Psi(\{n_2\}, m_{s2}, \mathbf{r}_1)]$$

(9.13)

Bosons:

$$\Psi^{(\text{sym})} = \frac{1}{\sqrt{2}}[\Psi(\{n_1\}, m_{s1}, \mathbf{r}_1)\Psi(\{n_2\}, m_{s2}, \mathbf{r}_2)$$
$$+\Psi(\{n_1\}, m_{s1}, \mathbf{r}_2)\Psi(\{n_2\}, m_{s2}, \mathbf{r}_1)] \tag{9.14}$$

Obviously, (9.13) suffices the Pauli exclusion principle.

Note that the antisymmetric wavefunction as characteristic for non-interacting fermions may be written in the more compact manner:

$$\Psi^{(\text{asym})} = \frac{1}{\sqrt{2}}\begin{vmatrix} \Psi(\{n_1\}, m_{s1}, \mathbf{r}_1) & \Psi(\{n_1\}, m_{s1}, \mathbf{r}_2) \\ \Psi(\{n_2\}, m_{s2}, \mathbf{r}_1) & \Psi(\{n_2\}, m_{s2}, \mathbf{r}_2) \end{vmatrix} \tag{9.15}$$

This equation may be easily generalized to the case of N non-interacting fermions. That results in:

$$\Psi^{(\text{asym})} = \frac{1}{\sqrt{N!}}\begin{vmatrix} \Psi(\{n_1\}, m_{s1}, \mathbf{r}_1) & \Psi(\{n_1\}, m_{s1}, \mathbf{r}_2) & \dots & \Psi(\{n_1\}, m_{s1}, \mathbf{r}_N) \\ \Psi(\{n_2\}, m_{s2}, \mathbf{r}_1) & \Psi(\{n_2\}, m_{s2}, \mathbf{r}_2) & \dots & \Psi(\{n_2\}, m_{s2}, \mathbf{r}_N) \\ \cdot & \cdot & & \cdot \\ \cdot & \cdot & \dots & \cdot \\ \cdot & \cdot & & \cdot \\ \Psi(\{n_N\}, m_{sN}, \mathbf{r}_1) & \Psi(\{n_N\}, m_{sN}, \mathbf{r}_2) & \dots & \Psi(\{n_N\}, m_{sN}, \mathbf{r}_N) \end{vmatrix}$$
$$\tag{9.16}$$

The determinants in (9.15) and (9.16) are called *Slater* determinants. The Slater determinant becomes zero whenever two fermions have identical quantum numbers, such that it is consistent with the Pauli principle.

9.3.2 The Helium Atom Ground State Without Spin Contributions

Hamiltonian

According to (9.1), we have the kinetic energy corresponding to the electron movement:

$$\hat{T}_{\text{kin}} \cong -\frac{\hbar^2}{2m_e}[\Delta_1 + \Delta_2] = \hat{T}_1 + \hat{T}_2 \tag{9.17}$$

In contrast to (9.1), we find for the potential energy

$$U = -\frac{e^2}{4\pi\varepsilon_0}\left[\frac{Z}{r_1} + \frac{Z}{r_2} - \frac{1}{r_{12}}\right] = U_1 + U_2 + U_{12} \tag{9.18}$$

Let θ be the angle formed between $\mathbf{r}_1$ and $\mathbf{r}_2$ (Fig. 9.1). Then

$$r_{12}^2 = r_1^2 + r_2^2 - 2r_1r_2\cos\theta \tag{9.19}$$

Obviously, the potential does no more correspond to a central field, because it has angular dependence.

The helium atom ionization energy E_{ion}, i.e., the energy necessary to ionize the atom starting from its ground state, may be measured and has been found to be equal to 78.9 eV. Therefore, the ground state energy of helium $E_{He,0}$ should be -78.9 eV.

"Zeroth" Approximation

In a zeroth approximation, we neglect the Coulomb interaction between the two electrons. In fact, this is what we have already done in Sect. 9.1. From (9.4), (9.17), and (9.18) we write:

$$\psi(\mathbf{r}_1, \mathbf{r}_2)^{(0)} = \psi_1(\mathbf{r}_1)\psi_2(\mathbf{r}_2)$$

$$\Rightarrow -\frac{\hbar^2}{2m_e}\Delta_1\psi_1(\mathbf{r}_1) - \frac{Ze^2}{4\pi\varepsilon_0 r_1}\psi_1(\mathbf{r}_1) = E_1\psi_1(\mathbf{r}_1)$$

$$-\frac{\hbar^2}{2m_e}\Delta_2\psi_2(\mathbf{r}_2) - \frac{Ze^2}{4\pi\varepsilon_0 r_2}\psi_2(\mathbf{r}_2) = E_2\psi_2(\mathbf{r}_2)$$

$$E^{(0)} = E_1 + E_2 \Rightarrow E_{He,0}^{(0)} = -2Z^2 Ry = -8Ry = -108.8\,\text{eV}$$

$$(9.20)$$

This zero-order approximation value $E_{He,0}^{(0)}$ is of the correct order, but nevertheless very inaccurate.

First-Order Perturbation Theory

Let us tackle $E_{He,0}^{(0)}$ according to (9.20) as a "zeroth" approximation, while regarding the time-independent interaction potential

$$\hat{V} = \frac{e^2}{4\pi\varepsilon_0 r_{12}} \tag{9.21}$$

as a perturbation. The first-order correction term to the ground-state energy might then be estimated (compare (5.27)) as the expectation value of (9.21) calculated in terms of the unperturbed wavefunction (9.20). Combining (9.20) with the ground-state wavefunction from Table 8.3, we have:

$$\psi(\mathbf{r}_1, \mathbf{r}_2)^{(0)} = \psi_1(\mathbf{r}_1)\psi_2(\mathbf{r}_2) = \frac{1}{\sqrt{\pi}}\left(\frac{Z}{a_0}\right)^{3/2}e^{-Zr_1/a_0}\frac{1}{\sqrt{\pi}}\left(\frac{Z}{a_0}\right)^{3/2}e^{-Zr_2/a_0}$$

$$= \frac{1}{\pi}\left(\frac{Z}{a_0}\right)^3 e^{-Z(r_1+r_2)/a_0} \tag{9.21a}$$

According to (9.7), the first-order correction term (5.27) is now obtained after integration over the configuration space (for a derivation see Sect. 9.5):

$$E^{(1)} = \int\limits_{V_1} \int\limits_{V_2} dV_1 dV_2 \left(\frac{e^2}{4\pi \varepsilon_0 r_{12}} \right) \left(\frac{1}{\pi} \left(\frac{Z}{a_0} \right)^3 e^{-\frac{Z}{a_0}(r_1+r_2)} \right)^2 = \frac{5}{4} Z Ry = 2.5 Ry$$

(9.22)

Hence, from (9.20) we find the ground-state energy:

$$E = E^{(0)} + E^{(1)} = -8Ry + 2.5Ry = -5.5Ry = -74.8\,\text{eV} \qquad (9.23)$$

This is already much closer to the experimental value than the zero-order approximation. Nevertheless, the perturbation potential (9.21) cannot be regarded a small correction term in (9.18), which explains the remaining differences.

Variation Approach
From (9.17) and (9.18), we find the full Hamiltonian according to:

$$\hat{H} = \hat{T}_1 + \hat{T}_2 + U_1 + U_2 + U_{12}$$

From the definition of the ground state, we have:

$$E_{\text{He},0} = \min\{E_{\text{He},n}\} \le \int \psi^* \hat{H} \psi \, dV = \left\langle \hat{H} \right\rangle$$

where Ψ describes any physically possible state of the system. In particular, we have

$$E_{\text{He},0} \le \min\left\{ \int \psi^* \hat{H} \psi \, dV \right\} \equiv E_{\min} \qquad (9.24)$$

Physical idea for estimating $E_{\min}$: The basic idea is that the electrons screen each other from the field of the nucleus. We introduce the hydrogen-like single-particle test wavefunction:

$$\psi_1(\mathbf{r}) = \frac{1}{\sqrt{\pi}} \left(\frac{Z'}{a_0} \right)^{3/2} e^{-Z'r/a_0} \qquad (9.25)$$

Here, we have introduced a modified atomic number Z'. The idea is now to find a Z' such that the expectation value of the corresponding system energy is minimized. As explained in Sect. 9.1, we would expect that the thus found Z' has a value in-between the values 1 and 2.

From (9.17) and (9.18), we have:

$$\left\langle \hat{H} \right\rangle = \langle T_1 \rangle + \langle T_2 \rangle + \langle U_1 \rangle + \langle U_2 \rangle + \langle U_{12} \rangle$$

Once the electrons in Fig. 9.1 are in completely identical physical conditions, we can also write:

$$\left\langle \hat{H} \right\rangle = 2\langle T_1 \rangle + 2\langle U_1 \rangle + \langle U_{12} \rangle$$

From (8.4) we set:

$$T_{kin,H} = |E_H| \Rightarrow 2\langle T_1 \rangle = 2Ry\left(Z'\right)^2$$

Further:

$$2\langle U_1 \rangle = 2\left(-\frac{1}{4\pi\varepsilon_0}\right) \int \psi_1^2(\mathbf{r}_1)\frac{Ze^2}{r_1}\psi_2^2(\mathbf{r}_2)\mathrm{d}V_1\mathrm{d}V_2$$

$$= 2\left(-\frac{1}{4\pi\varepsilon_0}\right) \int \psi_1^2(\mathbf{r}_1)\frac{Ze^2}{r_1}\mathrm{d}V_1$$

Note that ψ_1 depends on Z', and not on Z. Therefore, from (8.4) we have for the hydrogen-like atom:

$$U_H = -2T_{kin,H} \Rightarrow \left(-\frac{1}{4\pi\varepsilon_0}\right) \int \psi_1^2(\mathbf{r}_1)\frac{Z'e^2}{r_1}\mathrm{d}V_1 = -2\langle T_1 \rangle = -2Ry\left(Z'\right)^2$$

$$\Rightarrow 2\langle U_1 \rangle = 2\left(-\frac{1}{4\pi\varepsilon_0}\right) \int \psi_1^2(\mathbf{r}_1)\frac{Ze^2}{r_1}\mathrm{d}V_1$$

$$= 2\frac{Z}{Z'}\underbrace{\left(-\frac{1}{4\pi\varepsilon_0}\right) \int \psi_1^2(\mathbf{r}_1)\frac{Z'e^2}{r_1}\mathrm{d}V_1}_{=-2\langle T_1 \rangle} = -4\frac{Z}{Z'}\langle T_1 \rangle = -4RyZZ'$$

And finally, according to (9.22)

$$\langle U_{12} \rangle = \frac{5}{4}Z'Ry$$

Hence

$$E(Z') = \left\langle \hat{H} \right\rangle = 2\langle T_1 \rangle + 2\langle U_1 \rangle + \langle U_{12} \rangle = 2Ry\left[\left(Z'\right)^2 - 2ZZ' + \frac{5}{8}Z'\right]$$

$$= 2Ry\left[\left(Z'\right)^2 - 2Z'\left(Z - \frac{5}{16}\right)\right]$$

For any wavefunction given by (9.25), we can thus calculate the expectation value of the system energy. According to (9.24), this expression should be minimized for estimating the helium atoms ground-state energy. That yields:

$$\frac{\partial}{\partial Z'}E(Z') = 0 \Rightarrow 2Z' - 2Z + \frac{5}{8} = 0 \Rightarrow Z' = Z - \frac{5}{16} = 1.6875 \Rightarrow$$

Table 9.1 Summary on helium ionization energy estimation

Approach	$E_{\text{ionization}}/\text{eV}$
Experiment	78.9
Zero-order approximation	108.8
First-order perturbation	74.8
Variation approach	>77.45

$$E(Z')_{\min} = 2Ry\left[\left(Z - \frac{5}{16}\right)^2 - 2\left(Z - \frac{5}{16}\right)^2\right] = -2\left(Z - \frac{5}{16}\right)^2 Ry$$

$$Z = 2 \Rightarrow E(Z')_{\min} = -2\frac{729}{256}Ry \approx -5.7Ry \approx -77.45\,\text{eV}$$

$$E_0 < -77.45\,\text{eV} \Rightarrow E_{ion} > 77.45\,\text{eV} \tag{9.26}$$

9.4   Consistency Considerations

9.4.1   Screening

According to (9.26), the obtained "effective" atomic order Z' in the test wavefunction (9.25) is equal to 1.6875 and thus smaller than 2. This result is consistent with the physical mechanism assumed behind the calculation, namely that the charge distribution of the electrons wavefunctions leads to an efficient screening of the potential of the nucleus.

9.4.2   Comparison of the Presented Approaches

In summary, we have found the following estimations for the ionization energy Table 9.1:

Obviously, the first-order perturbation approach already comes closer to the true value than the zero-order approach from Sect. 9.1, but the differences are still significant. Clearly, the term U_{12} does not need to be substantially smaller than the other terms in the Hamiltonian of the He atom, such that a first-order perturbation approach is surely not the method of choice. The variation approach, on the other hand, is based on a physically transparent mechanism and leads to the best reproduction of the experimental values among the methods described here.

9.4.3   Symmetry Requirements

The above discussion did not take into account requirements arising from the symmetry of the wavefunction as formulated in (9.13). In fact, we know that the two electron wavefunction must be antisymmetric with respect to interchanging the electrons. When taking the spin into account, the wavefunction of the two electrons can—in a symbolic manner - be written as [4, p. 1013]:

$$\psi = f(\mathbf{r}_1, \mathbf{r}_2) f_s(s_1, s_2) \tag{9.27}$$

where f stands for the spatial part, and f_s for the spin part of the wavefunction. For more details see Sect. 9.6.1. Once the full wavefunction must be antisymmetric with respect to interchanging the two electrons, we have two possibilities:

- f is a symmetric function, and f_s an antisymmetric function. This situation corresponds to an antiparallel arrangement of the two electron spins, such that the full spin S is zero ($S = 0$). Helium atoms with such a wavefunction form what is called para-helium.
- f is an antisymmetric function, and f_s a symmetric function. This situation corresponds to a parallel arrangement of the two electron spins, such that the full spin is one ($S = 1$). Helium atoms with such a wavefunction form what is called ortho-helium.

It turns out that para- and ortho-helium have extremely different spectroscopic properties, such that formerly it was believed, that these two helium modifications are two completely different materials.

So what gives rise to the mentioned spectroscopic differences? When looking at para-helium, we have the following picture:

Para-helium: $S = 0$; $\mathbf{J} = \mathbf{L} + \mathbf{S} = \mathbf{L}$; electron spins are antiparallel, $2S + 1 = 1$. In order to distinguish between quantum numbers relevant for the single electrons (s, l, j), we use capital letters (S, L, J) for the full spin, full orbital angular momentum and full total angular momentum, as formed by summarizing the corresponding individual contributions of the electrons.

In the case $2S + 1 = 1$, we speak on a singlet state. In this context, $2S + 1$ is also called the multiplicity of the corresponding state. The lowest by energy states are (according to the nomenclature $^{2S+1}X_J$; S, P, D, F according to $l = 0$, 1, 2, 3,..., compare Sect. 8.6.1):

(1s, 1s): $^1S_0$: This is the ground state. Both electrons may occupy the $n = 1$ state, because their spins are antiparallel.

(1s, 2s): $^1S_0$: One electron with $n = 1$, the other with $n = 2$ and $l = 0$.

(1s, 2p): $^1P_1$: One electron with $n = 1$, the other with $n = 2$ and $l = 1$.

Ortho-helium: $S = 1$; $\mathbf{J} = \mathbf{L} + \mathbf{S} \Rightarrow J = \{L\text{-}1, L, L + 1\}$. The electron spins are parallel, such that $2S + 1 = 3$, which is characteristic for a triplet state (except $L = 0$). Note that in this case, the state (1s, 1s) is forbidden, because the spins are parallel. Then, it is the Pauli exclusion principle that forbids the formation of such a state; otherwise we would have two electrons in exactly the same quantum state.

Therefore, the lowest by energy states are:

(1s, 2s): $^3S_1$.

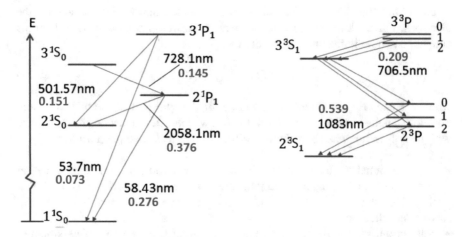

Fig. 9.4 Lowest energy levels (terms) of the helium atom. The strongest emission lines are visualized by arrows. Red numbers indicate the absolute value of selected oscillator strength data according to [6]

$(1s, 2p)$: $^3P_2$; $^3P_1$; $^3P_0$.

The corresponding energy levels for Para- and Ortho-helium (compare [5, p. 184]) are visualized in the Fig. 9.4. It is obvious that para- and ortho-helium have quite different spectroscopic properties, caused by the lack of a $(1s, 1s)$ ground state of ortho-helium, as well as the triplet character of its energy levels. Therefore, in helium, spin effects can by no means be interpreted as small corrections to energy terms, as soon as we get involved into many-electron atoms.

Remark When writing down the lowest energy electron states of para- and ortho-helium, we have implicitly postulated an angular momentum coupling mechanism that is called **LS-coupling** (or Russel–Sounders coupling): Its essence is in the assumption that first the single electrons orbital and spin momenta sum up to the full orbital and spin momenta according to $\mathbf{L} = \mathbf{l}_1 + \mathbf{l}_2$; $\mathbf{S} = \mathbf{s}_1 + \mathbf{s}_2$. Then, the total angular momentum is formed according to: $\mathbf{J} = \mathbf{L} + \mathbf{S}$. This coupling mechanism is a model assumption which best works in elements with a small atomic order Z.

9.4.4 Selection Rules

The electric dipole-allowed transitions comply with the general selection rules (LS-coupling case):

$\Delta J = 0, \pm 1$ Without $J = 0 \rightarrow J = 0$

$\Delta m_J = 0, \pm 1$ Without $m_J = 0 \rightarrow m_J = 0$ @ $\Delta J = 0$

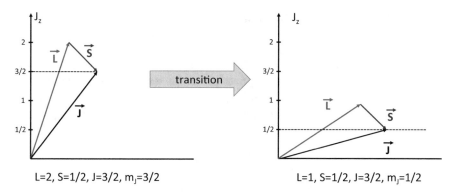

Fig. 9.5 Illustration of a transition according to $\Delta S = 0$, $\Delta L = -1$, $\Delta J = 0$, $\Delta m_J = -1$ (compare [5, p. 723]). J_z is given in units of $\hbar$

$$\Delta S = 0$$

$$\Delta L = 0, \pm 1$$

$\Delta l = \pm 1$ for the electron which changes its configuration.

Note the similarity to (8.67) and (8.68). $\Delta S = 0$ is again an expression of the fact that the spin does not couple to the electric field of the light wave, such that a singlet state cannot be transferred to a triplet state by electric dipole interaction and vice versa. $\Delta L = 0$ corresponds to the rather seldom situation that the absorption of one photon is accompanied by an orbital momentum change of two electrons such that $\Delta l_1 = +1$ and $\Delta l_2 = -1$ result in $\Delta L = 0$ (compare [7, pp. 315, 349]).

In full consistency with (8.68), we again have the allowed transition $\Delta J = 0$. The heuristic illustration is the same as provided in Sect. 8.6.3. Figure 9.5 provides an illustration of such a transition in terms of a vector diagram.

Remark In heavy atoms, Z becomes large, and therefore the spin orbit interaction $(\propto Z^4)$ according to (8.59) may dominate over the Coulomb interaction ($\propto Z$, compare (9.22)), so that momentum coupling is rather accomplished as *jj*-coupling according to:

$\mathbf{j}_i = \mathbf{l}_i + \mathbf{s}_i$; $\mathbf{J} = \Sigma \mathbf{j}_i$. Then, we have the modified selection rules:

$\Delta J = 0, \pm 1$ Without $J = 0 \rightarrow J = 0$

$\Delta m_J = 0, \pm 1$ Without $m_J = 0 \rightarrow m_J = 0$ @ $\Delta J = 0$

$\Delta j = 0, \pm 1$ for the single electron.

9.5   Application to Practical Problems

The following calculation rather presents a practice-related application to the material developed in Sect. 5.5.8, but once it is applied to the helium atom, this nevertheless seems to be the appropriate place. So let us use this paragraph to calculate the integral (9.22). This way we tackle $E_{He,0}^{(0)}$ according to (9.5) as a "zeroth" approximation, while regarding the time-independent interaction potential

$$\hat{V} = \frac{e^2}{4\pi\varepsilon_0 r_{12}}$$

as a perturbation. The first-order correction term to the ground-state energy might then be estimated (compare (5.27)) as the expectation value of (9.21) calculated in terms of the unperturbed wavefunction (9.21a):

$$\psi(\mathbf{r}_1, \mathbf{r}_2)^{(0)} = \psi_1(\mathbf{r}_1)\psi_2(\mathbf{r}_2) = \frac{1}{\sqrt{\pi}}\left(\frac{Z}{a_0}\right)^{3/2}e^{-Zr_1/a_0}\frac{1}{\sqrt{\pi}}\left(\frac{Z}{a_0}\right)^{3/2}e^{-Zr_2/a_0}$$

$$= \frac{1}{\pi}\left(\frac{Z}{a_0}\right)^3 e^{-Z(r_1+r_2)/a_0}$$

The first-order correction term is given by:

$$E^{(1)} = \int\limits_{V_1}\int\limits_{V_2} dV_1 dV_2 \left(\frac{e^2}{4\pi\varepsilon_0 r_{12}}\right)\left(\frac{1}{\pi}\left(\frac{Z}{a_0}\right)^3 e^{-\frac{Z}{a_0}(r_1+r_2)}\right)^2$$

$$= \frac{e^2}{4\pi^3\varepsilon_0}\left(\frac{Z}{a_0}\right)^6 \int\limits_{V_1}\int\limits_{V_2} \frac{e^{-\frac{2Z}{a_0}(r_1+r_2)}}{r_{12}} dV_1 dV_2 = ?$$

From Fig. 9.1, we recognize that

$$r_{12}^2 = r_1^2 + r_2^2 - 2r_1 r_2 \cos\theta$$

Furthermore, $dV_1 = r_1^2 \sin\theta_1 dr_1 d\theta_1 d\varphi_1$
For integrating over V_2, let us direct the z_2-axis along $\mathbf{r}_1$. Then,

$$dV_2 = r_2^2 \sin\theta_2 dr_2 d\theta_2 d\varphi_2 = r_2^2 \sin\theta dr_2 d\theta d\varphi_2$$

The integrand in the expression for $E^{(1)}$ does not depend on φ_1, φ_2, and θ_1. Obviously, we can therefore write:

$$\underbrace{\int_0^{2\pi} d\varphi_1}_{=2\pi} \underbrace{\int_0^{2\pi} d\varphi_2}_{=2\pi} \underbrace{\int_0^\pi \sin\theta_1 d\theta_1}_{=2} = 8\pi^2$$

Remark The change from the angular arguments θ_1 and θ_2 to θ_1 and θ looks a bit like magic, but in fact integrating over θ_1 and θ_2 will sum up over the same geometrical configurations like integrating over θ_1 and θ.

We find:

$$E^{(1)} = \frac{2e^2}{\pi\varepsilon_0}\left(\frac{Z}{a_0}\right)^6 \int_0^\infty r_1^2 e^{-\frac{2Z}{a_0}r_1}\,dr_1 \int_0^\infty r_2^2 e^{-\frac{2Z}{a_0}r_2}\,dr_2 \int_0^\pi \frac{\sin\theta\,d\theta}{\sqrt{r_1^2 + r_2^2 - 2r_1r_2\cos\theta}}$$

$$= \frac{2e^2}{\pi\varepsilon_0}\left(\frac{Z}{a_0}\right)^6 \int_0^\infty r_1^2 e^{-\frac{2Z}{a_0}r_1}\,dr_1 \int_0^\infty r_2^2 e^{-\frac{2Z}{a_0}r_2}\,dr_2 \int_{-1}^1 \frac{dx}{\sqrt{r_1^2 + r_2^2 - 2r_1r_2x}}$$

With $x = \cos\theta$. Substituting further: $\xi = r_1^2 + r_2^2 - 2r_1r_2x$
We find:

$$I = \int_{-1}^1 \frac{dx}{\sqrt{r_1^2 + r_2^2 - 2r_1r_2x}} = \frac{1}{2r_1r_2}\int_{(r_1-r_2)^2}^{(r_1+r_2)^2} \frac{d\xi}{\sqrt{\xi}} = \frac{1}{r_1r_2}\sqrt{\xi}\,\Big|_{(r_1-r_2)^2}^{(r_1+r_2)^2}$$

$$= \begin{cases} \frac{2}{r_1}; & r_1 > r_2 \\ \frac{2}{r_2}; & r_2 > r_1 \end{cases}$$

(Note here that in real algebra, $y = \sqrt{x}$ is the positive one of the two solutions of $y^2 = x$).
Thus,

$$E^{(1)} = \frac{2e^2}{\pi\varepsilon_0}\left(\frac{Z}{a_0}\right)^6 \int_0^\infty r_1^2 e^{-\frac{2Z}{a_0}r_1}\,dr_1 \int_0^\infty r_2^2 e^{-\frac{2Z}{a_0}r_2}\,dr_2\, I$$

$$= \frac{4e^2}{\pi\varepsilon_0}\left(\frac{Z}{a_0}\right)^6 \int_0^\infty r_1 e^{-\frac{2Z}{a_0}r_1}\,dr_1 \underbrace{\int_0^{r_1} r_2^2 e^{-\frac{2Z}{a_0}r_2}\,dr_2}_{I_1}$$

$$+ \frac{4e^2}{\pi\varepsilon_0}\left(\frac{Z}{a_0}\right)^6 \int_0^\infty r_1^2 e^{-\frac{2Z}{a_0}r_1}\,dr_1 \underbrace{\int_{r_1}^\infty r_2 e^{-\frac{2Z}{a_0}r_2}\,dr_2}_{I_2}$$

So we have two contributions to $E^{(1)}$; let us explicitly calculate the first one:

$$E_1^{(1)} = \frac{4e^2}{\pi \varepsilon_0} \left(\frac{Z}{a_0}\right)^6 \int_0^\infty r_1 e^{-\frac{2Z}{a_0}r_1} dr_1 \underbrace{\int_0^{r_1} r_2^2 e^{-\frac{2Z}{a_0}r_2} dr_2}_{I_1}$$

Obviously,

$$I_1 = \int_0^{r_1} r_2^2 e^{-\frac{2Z}{a_0}r_2} dr_2 = \int_0^{r_1} x^2 e^{-\frac{2Zx}{a_0}} dx$$

$$= \frac{a_0^3}{8Z^3} \int_0^{\frac{2Zr_1}{a_0}} \xi^2 e^{-\xi} d\xi \quad (\xi = 2Zx/a_0)$$

From partial integrating:

$$\int \xi^2 e^{-\xi} d\xi = -\left(2 + 2\xi + \xi^2\right) e^{-\xi}$$

Such that

$$I_1 = \frac{a_0^3}{8Z^3} (2 + 2\xi + \xi^2) e^{-\xi} \Big|_{\frac{2Zr_1}{a_0}}^{0}$$

$$= \frac{a_0^3}{4Z^3} \left[1 - \left(1 + \frac{2Zr_1}{a_0} + \frac{2Z^2r_1^2}{a_0^2} \right) e^{-\frac{2Zr_1}{a_0}} \right]$$

This results in:

$$E_1^{(1)} = \frac{4e^2}{\pi \varepsilon_0} \left(\frac{Z}{a_0}\right)^6 \int_0^\infty r_1 e^{-\frac{2Z}{a_0}r_1} dr_1 I_1 = \frac{e^2}{\pi \varepsilon_0} \left(\frac{Z}{a_0}\right)^3$$

$$\int_0^\infty r_1 e^{-\frac{2Z}{a_0}r_1} dr_1 \left[1 - \left(1 + \frac{2Zr_1}{a_0} + \frac{2Z^2r_1^2}{a_0^2} \right) e^{-\frac{2Zr_1}{a_0}} \right]$$

$$= \frac{e^2}{\pi \varepsilon_0} \left(\frac{Z}{a_0}\right)^3 \int_0^\infty r_1 e^{-\frac{2Z}{a_0}r_1} dr_1$$

$$- \frac{e^2}{\pi \varepsilon_0} \left(\frac{Z}{a_0}\right)^3 \int_0^\infty e^{-\frac{4Z}{a_0}r_1} dr_1 \left(r_1 + \frac{2Zr_1^2}{a_0} + \frac{2Z^2r_1^3}{a_0^2} \right)$$

Making use of (1.1):

$$\int_0^\infty x^n e^{-px} dx = \frac{n!}{p^{n+1}}$$

We find:

$$\frac{e^2}{\pi \varepsilon_0} \left(\frac{Z}{a_0}\right)^3 \int_0^\infty r_1 e^{-\frac{2Z}{a_0}r_1} dr_1 = \frac{e^2}{\pi \varepsilon_0} \left(\frac{Z}{a_0}\right)^3 \frac{1}{\left(\frac{2Z}{a_0}\right)^2}$$

$$= \frac{e^2}{4\pi \varepsilon_0} \frac{Z}{a_0} = 2ZRy$$

$$\frac{e^2}{\pi \varepsilon_0} \left(\frac{Z}{a_0}\right)^3 \int_0^\infty r_1 e^{-\frac{4Z}{a_0}r_1} dr_1 = \frac{e^2}{\pi \varepsilon_0} \left(\frac{Z}{a_0}\right)^3 \frac{1}{\left(\frac{4Z}{a_0}\right)^2} = \frac{e^2}{16\pi \varepsilon_0} \frac{Z}{a_0}$$

$$\frac{e^2}{\pi \varepsilon_0} \left(\frac{Z}{a_0}\right)^3 \int_0^\infty e^{-\frac{4Z}{a_0}r_1} dr_1 \frac{2Zr_1^2}{a_0} = \frac{2e^2}{\pi \varepsilon_0} \left(\frac{Z}{a_0}\right)^4 \frac{2}{\left(\frac{4Z}{a_0}\right)^3} = \frac{e^2}{16\pi \varepsilon_0} \frac{Z}{a_0}$$

$$\frac{e^2}{\pi \varepsilon_0} \left(\frac{Z}{a_0}\right)^3 \int_0^\infty e^{-\frac{4Z}{a_0}r_1} dr_1 \frac{2Z^2 r_1^3}{a_0^2} = \frac{2e^2}{\pi \varepsilon_0} \left(\frac{Z}{a_0}\right)^5 \frac{6}{\left(\frac{4Z}{a_0}\right)^4} = \frac{3}{4} \frac{e^2}{16\pi \varepsilon_0} \frac{Z}{a_0}$$

Therefore,

$$E_1^{(1)} = \frac{e^2}{\pi \varepsilon_0} \left(\frac{Z}{a_0}\right)^3 \int_0^\infty r_1 e^{-\frac{2Z}{a_0}r_1} dr_1$$

$$- \frac{e^2}{\pi \varepsilon_0} \left(\frac{Z}{a_0}\right)^3 \int_0^\infty e^{-\frac{4Z}{a_0}r_1} dr_1 \left(r_1 + \frac{2Zr_1^2}{a_0} + \frac{2Z^2 r_1^3}{a_0^2}\right)$$

$$= \frac{e^2}{\pi \varepsilon_0} \frac{Z}{a_0} \left(\frac{1}{4} - \frac{1}{16} - \frac{1}{16} - \frac{3}{4}\frac{1}{16}\right) = \frac{e^2}{\pi \varepsilon_0} \frac{Z}{a_0} \frac{16 - 8 - 3}{4 * 16}$$

$$= \frac{e^2}{\pi \varepsilon_0} \frac{Z}{a_0} \frac{5}{4 * 16} = \frac{e^2}{8\pi \varepsilon_0} \frac{Z}{a_0} \frac{5}{8} = \frac{5}{8} ZRy$$

The second term (E_2, related to I_2) gives the same result. This is already evident from the symmetry of the task, because the first and the second electrons are in identical physical conditions. Hence, we finally have:

$$E^{(1)} = \int_{V_1} \int_{V_2} dV_1 dV_2 \left(\frac{e^2}{4\pi \varepsilon_0 r_{12}}\right) \left(\frac{1}{\pi} \left(\frac{Z}{a_0}\right)^3 e^{-\frac{Z}{a_0}(r_1+r_2)}\right)^2$$

$$= \frac{4e^2}{\pi \varepsilon_0} \left(\frac{Z}{a_0}\right)^6 \underbrace{\int_0^\infty r_1 e^{-\frac{2Z}{a_0}r_1} dr_1 \int_0^{r_1} r_2^2 e^{-\frac{2Z}{a_0}r_2} dr_2}_{I_1}$$

$$+ \frac{4e^2}{\pi \varepsilon_0} \left(\frac{Z}{a_0}\right)^6 \underbrace{\int_0^\infty r_1^2 e^{-\frac{2Z}{a_0}r_1} dr_1 \int_{r_1}^\infty r_2 e^{-\frac{2Z}{a_0}r_2} dr_2}_{I_2}$$

$$= \frac{5}{4} Z Ry = 2.5 Ry$$

This result coincides with (9.22).

9.6 Advanced Material

9.6.1 Structure of the Wavefunction of Two Electrons in a Helium Atom (LS-Coupling)

From (9.27) we have the symbolic writing:

$$\psi = f(\mathbf{r}_1, \mathbf{r}_2) f_s(s_1, s_2)$$

According to Sect. 9.3.1, this wavefunction must be antisymmetric with respect to interchanging the two electrons [3, p. 385–388]. Hence,

$$\begin{aligned}
\psi &= f_{n_1,n_2}(\mathbf{r}_1, \mathbf{r}_2) f_s(s_1, s_2) \\
&= -f_{n_2,n_1}(\mathbf{r}_1, \mathbf{r}_2) f_s(s_2, s_1) \\
&= -f_{n_1,n_2}(\mathbf{r}_2, \mathbf{r}_1) f_s(s_2, s_1)
\end{aligned} \tag{9.28}$$

where n_1 and n_2 stand for quantum numbers not related to the spin. s_1 and s_2 symbolize the spin states of the corresponding electrons.

According to (9.27), we search the solutions as:

$$\begin{aligned}
\psi^{(+)} &= f_{n_1,n_2}^{(\text{sym})}(\mathbf{r}_1, \mathbf{r}_2) f_s^{(\text{asym})}(s_1, s_2) \\
\psi^{(-)} &= f_{n_1,n_2}^{(\text{asym})}(\mathbf{r}_1, \mathbf{r}_2) f_s^{(\text{sym})}(s_1, s_2)
\end{aligned} \tag{9.29}$$

With (compare Sect. 9.3.1)

$$f_{n_1,n_2}^{(\text{sym})}(\mathbf{r}_1, \mathbf{r}_2) = \frac{1}{\sqrt{2}} [\Psi(\{n_1\}, \mathbf{r}_1)\Psi(\{n_2\}, \mathbf{r}_2) + \Psi(\{n_1\}, \mathbf{r}_2)\Psi(\{n_2\}, \mathbf{r}_1)]$$

$$f_{n_1,n_2}^{(\text{asym})}(\mathbf{r}_1, \mathbf{r}_2) = \frac{1}{\sqrt{2}} [\Psi(\{n_1\}, \mathbf{r}_1)\Psi(\{n_2\}, \mathbf{r}_2) - \Psi(\{n_1\}, \mathbf{r}_2)\Psi(\{n_2\}, \mathbf{r}_1)]$$

$$\tag{9.30}$$

For the spin part, we introduce the following symbolic writing:

1st electron, state spin up: $|\uparrow\rangle_1$

1st electron, state spin down: $|\downarrow\rangle_1$

2nd electron, state spin up: $|\uparrow\rangle_2$

2nd electron, state spin down: $|\downarrow\rangle_2$.

Let us remember the writing of these wavefunctions as spinors. We have $|\uparrow\rangle = \begin{pmatrix} 1 \\ 0 \end{pmatrix}$ and $|\downarrow\rangle = \begin{pmatrix} 0 \\ 1 \end{pmatrix}$.

From the Pauli matrices (8.62a), we have

$$\sigma_1|\uparrow\rangle = |\downarrow\rangle$$
$$\sigma_2|\uparrow\rangle = i|\downarrow\rangle$$
$$\sigma_3|\uparrow\rangle = |\uparrow\rangle$$
$$\sigma_1|\downarrow\rangle = |\uparrow\rangle$$
$$\sigma_2|\downarrow\rangle = -i|\uparrow\rangle$$
$$\sigma_3|\downarrow\rangle = -|\downarrow\rangle \tag{9.31}$$

Then, in this symbolic writing, we find the following expressions for the wavefunctions (9.29):

Antisymmetric spin wavefunction (spins are antiparallel):

$$\psi^{(+)} = f_{n_1,n_2}^{(sym)}(\mathbf{r}_1, \mathbf{r}_2) f_s^{(asym)}(s_1, s_2)$$
$$= \frac{1}{\sqrt{2}} f_{n_1,n_2}^{(sym)}(\mathbf{r}_1, \mathbf{r}_2)\left(|\uparrow\rangle_1|\downarrow\rangle_2 - |\downarrow\rangle_1|\uparrow\rangle_2\right) \tag{9.32}$$

According to calculations in terms of (8.62)–(8.66), (9.31), this asymmetric with respect to the spin function solution corresponds to $S = 0$, and $m_S = 0$.

Symmetric spin wavefunction (spins are parallel):

$$\psi_1^{(-)} = f_{n_1,n_2}^{(asym)}(\mathbf{r}_1, \mathbf{r}_2) f_s^{(asym)}(s_1, s_2)$$
$$= \begin{cases} f_{n_1,n_2}^{(asym)}(\mathbf{r}_1, \mathbf{r}_2)|\uparrow\rangle_1|\uparrow\rangle_2 : & m_S = 1 \\ \frac{1}{\sqrt{2}} f_{n_1,n_2}^{(asym)}(\mathbf{r}_1, \mathbf{r}_2)\left(|\uparrow\rangle_1|\downarrow\rangle_2 + |\downarrow\rangle_1|\uparrow\rangle_2\right) : & m_S = 0 \\ f_{n_1,n_2}^{(asym)}(\mathbf{r}_1, \mathbf{r}_2)|\downarrow\rangle_1|\downarrow\rangle_2 : & m_S = -1 \end{cases} \tag{9.33}$$

According to (8.62)–(8.66), (9.31), these symmetric with respect to the spin function solutions correspond to $S = 1$, and $m_S = 0, \pm 1$.

In all these expressions, the spatial part of the wavefunction is given in terms of (9.30).

We see that for the antisymmetric spin configuration (spins are antiparallel, $S = 0$), we find one single solution, corresponding to a singlet state (para-helium). For the symmetric spin configurations (spins are parallel, $S = 1$), we have three solutions with different projections of the total spin on the z-axis, thus corresponding to a triplet of states (ortho-helium).

9.6.2 Hund's Rules

This short section is to summarize selected information about atoms with more than two electrons. We will restrict our attention to the electronic ground-state configurations.

First of all, let us remember the nomenclature as introduced in Sect. 8.6.1. In application to many-electron atoms, we now write the terms as:

$$^{2S+1}X_J; \quad X = S, P, D, F \text{ according to } L = 0, 1, 2, 3, ... \qquad (9.34)$$

The following Table 9.2 summarizes ground-state electronic configurations of the elements up to $Z = 54$ (Xenon). Note that completely filled electronic "shells" do not add any nonzero S- or L-contributions. Nonzero contributions arise only from partially filled (sub-)shells. Therefore, the ground state of the noble gas atoms is always $^1S_0$.

The problem is that the individual electronic spins and orbital momenta sum up in a vector-like manner, i.e., with possibly different individual directions. Different total momenta may result in different energies, and it is not a priori clear which of them corresponds to the lowest energy, i.e., the ground state. Fortunately, for light atoms, some simple rules exist for identifying the ground-state configuration.

Indeed, the ground states of atoms with predominant LS-coupling obey the so-called Hund's rules, which indicate the $(2S + 1)$, L, and J-values in (9.34) as relevant in the ground state:

1. $^{2S+1}X_J$: For a given electron configuration, the term with maximum multiplicity $2S + 1$ has the lowest energy.
2. $^{2S+1}X_J$: For a given multiplicity, the term with the largest L has the lowest energy.
3. $^{2S+1}X_J$: If the outermost shell is half-filled or less, then the level with the lowest J corresponds to the lowest energy. If the outermost shell is more than half-filled, the highest J corresponds to the lowest energy.

It is nevertheless tricky to construct the ground-state configuration for a complicated atom from these three rules, because first of all, any discussed electronic state must additionally comply with the symmetry requirements introduced (9.12) and (9.27) and, in particular, with Pauli's exclusion principle. But if you scan through the Table 9.2, you will notice that the ground states are—often enough—characterized by large multiplicities, as well as large values of L and J.

As an illustration, consider the silicon atom ground state. According to Table 9.2, except the 3p orbital, all other subshells are filled. It is hence the 3p orbital with two electrons (subscripts 1 and 2) that defines S, L, and J.

Table 9.2 Electronic ground-state configurations (LS-coupling) of atoms with atomic numbers $Z \in [1, 54]$ (after [7, p. 342])

Z	Element	Shells K $n=1$ s	L $n=2$ s	L $n=2$ p	M $n=3$ s	M $n=3$ P	M $n=3$ d	N $n=4$ s	N $n=4$ p	N $n=4$ d	O $n=5$ s	O $n=5$ p	L-S ground-state configuration
1	H	1											$^2S_{1/2}$
2	He	2											$^1S_0$
3	Li	2	1										$^2S_{1/2}$
4	Be	2	2										$^1S_0$
5	B	2	2	1									$^2P_{1/2}$
6	C	2	2	2									$^3P_0$
7	N	2	2	3									$^4S_{3/2}$
8	O	2	2	4									$^3P_2$
9	F	2	2	5									$^2P_{3/2}$
10	Ne	2	2	6									$^1S_0$
11	Na	2	2	6	1								$^2S_{1/2}$
12	Mg	2	2	6	2								$^1S_0$
13	Al	2	2	6	2	1							$^2P_{1/2}$
14	Si	2	2	6	2	2							$^3P_0$
15	P	2	2	6	2	3							$^4S_{3/2}$
16	S	2	2	6	2	4							$^3P_2$
17	Cl	2	2	6	2	5							$^2P_{3/2}$
18	Ar	2	2	6	2	6							$^1S_0$
19	K	2	2	6	2	6		1					$^2S_{1/2}$
20	Ca	2	2	6	2	6		2					$^1S_0$
21	Sc	2	2	6	2	6	1	2					$^2D_{3/2}$
22	Ti	2	2	6	2	6	2	2					$^3F_2$
23	V	2	2	6	2	6	3	2					$^4F_{3/2}$
24	Cr	2	2	6	2	6	4	1					$^7S_3$
25	Mn	2	2	6	2	6	5	2					$^6S_{5/2}$
26	Fe	2	2	6	2	6	6	2					$^5D_4$
27	Co	2	2	6	2	6	7	2					$^4F_{9/2}$
28	Ni	2	2	6	2	6	8	2					$^3F_4$
29	Cu	2	2	6	2	6	10	1					$^2S_{1/2}$
30	Zn	2	2	6	2	6	10	2					$^1S_0$
31	Ga	2	2	6	2	6	10	2	1				$^2P_{1/2}$

(continued)

Table 9.2 (continued)

Z	Element	Shells											L-S ground-state configuration
		K $n=1$	L $n=2$		M $n=3$			N $n=4$			O $n=5$		
		s	s	p	s	P	d	s	p	d	s	p	
32	Ge	2	2	6	2	6	10	2	2				$^3P_0$
33	As	2	2	6	2	6	10	2	3				$^4S_{3/2}$
34	Se	2	2	6	2	6	10	2	4				$^3P_2$
35	Br	2	2	6	2	6	10	2	5				$^2P_{3/2}$
36	Kr	2	2	6	2	6	10	2	6				$^1S_0$
37	Rb	2	2	6	2	6	10	2	6		1		$^2S_{1/2}$
38	Sr	2	2	6	2	6	10	2	6		2		$^1S_0$
39	Y	2	2	6	2	6	10	2	6	1	2		$^2D_{3/2}$
40	Zr	2	2	6	2	6	10	2	6	2	2		$^3F_2$
41	Nb	2	2	6	2	6	10	2	6	4	1		$^6D_{1/2}$
42	Mo	2	2	6	2	6	10	2	6	5	1		$^7S_3$
43	Tc	2	2	6	2	6	10	2	6	6	1		$^6D_{9/2}$
44	Ru	2	2	6	2	6	10	2	6	7	1		$^5F_5$
45	Rh	2	2	6	2	6	10	2	6	8	1		$^4F_{9/2}$
46	Pd	2	2	6	2	6	10	2	6	10			$^1S_0$
47	Ag	2	2	6	2	6	10	2	6	10	1		$^2S_{1/2}$
48	Cd	2	2	6	2	6	10	2	6	10	2		$^1S_0$
49	In	2	2	6	2	6	10	2	6	10	2	1	$^2P_{1/2}$
50	S	2	2	6	2	6	10	2	6	10	2	2	$^3P_0$
51	Sb	2	2	6	2	6	10	2	6	10	2	3	$^4S_{3/2}$
52	Te	2	2	6	2	6	10	2	6	10	2	4	$^3P_2$
53	J	2	2	6	2	6	10	2	6	10	2	5	$^2P_{3/2}$
54	Xe	2	2	6	2	6	10	2	6	10	2	6	$^1S_0$

From $l_1 = l_2 = 1$; $s_1 = s_2 = 1/2$ we get the following nominal possibilities for combining the individual momenta into S and L:

$S \rightarrow$ L $\downarrow$	0	1
0	^1S	^3S
1	^1P	^3P
2	^1D	^3D

Then, the first of Hund's rules would cancel out the singlet states with $S = 1$, while the second rule would favor the ^3D state because of the largest L-value.

But it is easily seen that the ^{3}D state must be forbidden by the Pauli principle, because both electrons have identical n and l-values, as well as identical projection m and m_s. Then, the ground state should correspond to the ^3P configuration, which obviously does not contradict Pauli's exclusion principle.

Remark In fact this is not a sufficient conditions, because we have to require, that the wavefunction attributed to the corresponding quantum state must be different from zero and antisymmetric with respect to interchanging the electrons position. A deeper analysis of the symmetry of the wavefunctions leads to the result that the ^3S and ^1P states in the silicon atom are symmetry-forbidden, at least for identical principal quantum numbers, while ^3P is indeed allowed [5, p. 201]. Hence it corresponds to the ground state of the silicon atom.

We thus have $S = 1$ and $L = 1$, which allows for $J = 0, 1, 2$.

In order to find J, we make use of Hund's third rule. The $3p$ shell is filled with only two electrons, while it can be filled with up to $2(2l + 1) = 6$ electrons. It is thus less than half-filled, and therefore, according to Hund's third rule, the lowest J (i.e., $J = 0$) corresponds to the state with the lowest energy. The ground state of the silicon atom is therefore $^3P_0$.

9.7 Tasks for Self-check

9.7.1 Multiple-choice test: Mark all answers which seem you correct!

The emission lines of the Lyman spectral series are observed in the	Infrared
	Visible
	Ultraviolet
	γ-range
The emission lines of the Paschen spectral series are observed in the	Visible
	Ultraviolet
	γ-range
In a hydrogen atom, Bohr's radius is approximately equal to	0.05 nm
	$5 * 10^{-11}$ m
	10 nm
In spherical coordinates, the volume element dV is given by the expression (θ is the angle between $\mathbf{r}$ and the z-axis):	$dV = dr\, d\phi\, d\theta$
	$dV = r^2 \sin\theta\, dr\, d\phi\, d\theta$
	$dV = 2\hbar^3 \pi r\, dr$
	$dV = \sin\theta\, d\phi\, d\theta$
In a hydrogen atom, a state $|n = 2; l = 4; m = -7\rangle$ is	Possible
	Impossible
The s-orbital	Has spherical symmetry
	Corresponds to $l = 0$

	Corresponds to $l = 2$
The Pauli exclusion principle is valid for	Bosons
	Fermions
Electrons belong to the class of	Fermions
	Bosons
	Neither of them
Slater determinants describe the wavefunctions of systems of non-interacting	Fermions
	Bosons
	Neither of them
The electronic configuration $^1S_7$ is	Possible
	Impossible

9.7.2 True or wrong? Make your decision!

Assertion	True	Wrong
$\left[\hat{x}, \hat{p}_z^2\right] = 0$?		
$\left[\hat{z}, \hat{L}_z\right] = 0$?		
$\left[\hat{L}_x, \hat{L}_y\right] = 0$?		
$\left[\hat{L}_z, \hat{L}^2\right] = i\hbar\left(\hat{L}_x + \hat{L}_y\right)$?		
$\left[\hat{L}_z, U(r)\right] = 0$?		
In any circular Bohr orbit, the electrons kinetic energy is equal to its potential one		
From the hydrogen emission spectrum, only certain lines of the Balmer series fall into the visible spectral range		
In its ground state, the electron in a hydrogen atom can be observed in both s- and p-orbitals		
Dipole-allowed quantum transitions in a hydrogen atom are only observed when the principal quantum number changes for a value of ± 1		
The wavefunction of a system of two fermions is always symmetric with respect to interchanging the particles		
The wavefunction of a system of non-interacting Bosons may be constructed in terms of a Slater determinant		

9.7.3 In Sect. 5.6.3, we introduced the photon number operator (5.51) by using the relations (5.43) and (5.48). Obviously, this operator has arbitrary integer eigenvalues ≥ 0 and thus describes the occupancy of bosonic quantum states. On the contrary, its fermionic counterpart should only allow for occupancies of a quantum state equal to zero or one. Show that a corresponding fermionic occupation number operator may be constructed in an absolutely equivalent manner, by only replacing the commutation rule (5.43) by a corresponding anti-commutation rule according to $\hat{b}\hat{b}^+ + \hat{b}^+\hat{b} = 1$ (compare also [4, p. 1265])

9.7.4 Calculate the minimum angle formed between the angular momentum vector **L** and the z-axis in a D-state [8]

9.7.5 Calculate the angle between **J** and **L** in a $^4D_{3/2}$ state [7].

9.7.6 For a given pair of quantum numbers L and S, calculate $\sum_{\{J\}} (2J + 1)$, where $\{J\}$ is the set of all possible J-values [7].

References

Specific References

1. D. Dürr, D. Lazarovici, *Verständliche Quantenmechanik* (Springer, 2018), pp. 12–16
2. А.С. Давыдов, Квантовая Механика, Москва Физматгиз (1963)
3. А.А. Соколов, И.М. Тернов, В.Ч. Жуковский, Квантовая механика, Москва "Наука" (1979)
4. M. Bartelmann, B. Feuerbacher,T. Krüger, D. Lüst, A. Rebhan, A. Wipf, *Theoretische Physik* (Springer, 2015)
5. W. Demtröder, *Experimentalphysik 3, Atome, Moleküle und Festkörper* (Springer, 2016)
6. G.W.F. Drake (ed.) *Springer Handbook of Atomic, Molecular, and Optical Physics* (Springer, 2006), p. 216
7. H. Haken, H.C. Wolf, *The Physics of Atoms and Quanta* (Springer, 2005)
8. P.A. Tipler, G. Mosca, *Physik, 7* (Springer, Auflage, 2015), pp. 1218–1219

General Literature

9. L.D. Landau, E.M. Lifshitz: *Quantum Mechanics (Volume 3 of A Course of Theoretical Physics)* (Pergamon Press, 1965)
10. S. Flügge, *Practical Quantum Mechanics* (Springer, 1999)
11. W. Demtröder, *Atoms, Molecules, and Pphotons* (Springer, 2010)

Part IV
Introduction to Molecular Physics and Spectroscopy

"Narr und Eulenspiegel" (Fool and Eulenspiegel)
Sculptures and Photograph by Astrid Leiterer, Jena, Germany (www.astrid-art.de). Photograph reproduced with permission

The process of chemical bonding between different atoms results in the formation of molecules. The number of atoms, their masses, the bonding strength, and symmetry considerations are essential for understanding the optical behavior of molecules in different spectral regions.

The Hydrogen Molecule

10

Abstract

Basic knowledge on molecular bonding mechanisms is presented. The adiabatic or Born–Oppenheimer approximation is introduced, and the theory of covalent bonding in the hydrogen molecule is derived in detail. Also, empiric interaction potentials like the Lennard–Jones or Morse potentials are shortly discussed.

10.1 Starting Point

In Chap. 8, we have started discussing the physics of atoms from the hydrogen atom—the simplest atom, which consists of 2 particles only—namely an electron and a proton (the nucleus). Similarly, the simplest neutral molecule is the hydrogen molecule, as consisting of two hydrogen atoms. Figure 10.1 visualizes this situation.

In Fig. 10.1, arrows symbolize the individual interaction terms that contribute to the full potential energy of the system. The situation will become even more complicated when other molecules built from (a possibly larger) number of higher order atoms are taken into consideration. In fact, the four-body system as presented in Fig. 10.1 corresponds to the simplest situation of a neutral molecule, but even this simplest case does not allow for an analytic solution of the corresponding Schrödingers equation. Hence, similarly to the helium atom, we will have to make use of approximate solutions of the Schrödinger equation.

In the general case, the molecule may be considered as an agglomerate of electrons and atomic nuclei. The Hamiltonian of the molecule may be written in the following general manner [1, 2]:

$$\hat{H} = \hat{T}_e + \hat{T}_c + U\left(\mathbf{r}_{e,i}, \mathbf{r}_{c,j}\right) \tag{10.1}$$

Here, $\hat{T}_e$ represents the operator of the electrons kinetic energy of all electrons and $\hat{T}_c$ that of the nuclei. The term U contains the potential energy of all electrons

© The Author(s), under exclusive license to Springer Nature Switzerland AG 2022   241
O. Stenzel, *Light–Matter Interaction*, UNITEXT for Physics,
https://doi.org/10.1007/978-3-030-87144-4_10

Fig. 10.1 Classical
illustration of the
H_2—molecule

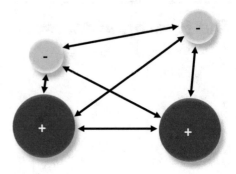

and nuclei and is thus depending on the electron coordinates $(\mathbf{r}_{e,i})$ as well as
on those of the nuclei $(\mathbf{r}_{c,j})$. The index i counts the electrons, j the nuclei. The
stationary Schrödingers equation may be written as:

$$\left(E - \hat{H}\right)\psi\left(\mathbf{r}_{e,i}, \mathbf{r}_{c,j}\right) = 0 \qquad (10.2)$$

It will be our aim in this chapter to get an idea of the approximate solution of
(10.2) for the H_2 molecule as illustrated in Fig. 10.1. In order to achieve this goal,
we will have to draw together our knowledge on classical estimations concerning
molecule dynamics (Chap. 2), as well as the hydrogen atom and the consequences
of the indistinguishability of identical particles that form a many-body system
(Chaps. 8 and 9).

10.2   Physical Idea

As it has already been discussed in Chap. 2, electrons and nuclei strongly differ
in their masses for several orders of magnitude. Therefore, the movement of electrons will be much faster than that of the nuclei. That will allow us to facilitate
the approximate solution of (10.2): We will assume that during a characteristic
time relevant for the electron movement, the positions of the nuclei will practically remain unchanged. On the other hand, in the time span characteristic for
the movement of the nuclei, any electron (please accept for a moment the rather
classical illustration provided in Fig. 10.1) will have enough time for performing
a large number of "loops" around their average positions, such that the nuclei feel
some effective impact from the electron system, which does not depend on the
"actual coordinates" of the electrons. This will allow us subdividing (10.2) into
two equations: one for the electrons dynamics, and another for the movement of
the nuclei.

10.3 Theoretical Considerations

10.3.1 The Adiabatic Approximation

Let us now make use of the following ansatz [1, 2]:

$$\psi\left(\mathbf{r}_{e,i}, \mathbf{r}_{c,j}\right) = \psi_e\left(\mathbf{r}_{e,i}, \mathbf{r}_{c,j}\right)\psi_c\left(\mathbf{r}_{c,j}\right) \tag{10.3}$$

In (10.3), the wavefunction of the whole system $\psi\left(\mathbf{r}_{e,i}, \mathbf{r}_{c,j}\right)$ is represented as a product of that of the electronic (e) and that of the nuclei (c) subsystems. Once during the fast movement of the electrons, the nuclei positions will practically remain fixed, the electronic part of the wavefunction $\psi_e\left(\mathbf{r}_{e,i}, \mathbf{r}_{c,j}\right)$ may be calculated assuming the spatial configuration of the nuclei *as a fixed set of parameters* ($\mathbf{r}_{c,j} = \mathbf{const} \,\forall j$). This is the essence of the so-called adiabatic or Born–Oppenheimer approximation. Therefore, in (10.3), the electron wavefunction appears principally dependent on the actual nucleus coordinates, while the latter are tackled as fixed parameters on the time scale relevant for the electron movement. Combining (10.1)–(10.3), we obtain:

$$\frac{1}{\psi_e}\left[E - \hat{T}_e - U\left(\mathbf{r}_{e,i}, \mathbf{r}_{c,j}\right)\right]\psi_e = \frac{1}{\psi_c}\hat{T}_c\psi_c \tag{10.4}$$

It is obvious that the term in (10.4) on right depends only on the nuclei coordinates. But if so, the same must be valid for the term on left as well. We can therefore write:

$$\frac{1}{\psi_e}\left[E - \hat{T}_e - U\left(\mathbf{r}_{e,i}, \mathbf{r}_{c,j}\right)\right]\psi_e = \frac{1}{\psi_c}\hat{T}_c\psi_c \equiv W\left(\mathbf{r}_{c,j}\right) \tag{10.5}$$

Any of the thus defined functions W (there may be more than one) can principally be represented as the superposition of a constant first term and a second one, which depends on the nuclei coordinates:

$$W\left(\mathbf{r}_{c,j}\right) \equiv E_c - u\left(\mathbf{r}_{c,j}\right) \tag{10.6}$$

This results in the following equation for the electronic wavefunction:

$$\left[E - E_c + u\left(\mathbf{r}_{c,j}\right) - \hat{T}_e - U\left(\mathbf{r}_{e,i}, \mathbf{r}_{c,j}\right)\right]\psi_e$$
$$\equiv \left[E_e\left(\mathbf{r}_{c,j}\right) - \hat{T}_e - U\left(\mathbf{r}_{e,i}, \mathbf{r}_{c,j}\right)\right]\psi_e = 0; \quad \mathbf{r}_{c,j} = \mathbf{const} \,\forall j \tag{10.7}$$

In (10.7), the expression for the energy eigenvalues of the electronic subsystem $E_e\left(\mathbf{r}_{c,j}\right)$ is given by:

$$E_e\left(\mathbf{r}_{c,j}\right) \equiv E - E_c + u\left(\mathbf{r}_{c,j}\right) \tag{10.8}$$

On the other hand, the wavefunction for the movement of the nuclei cannot depend on the actual position of the electrons, because the latter may change many times during the time span characteristic for the movement of a nucleus. Therefore, the nuclei rather feel some averaged response from the electron movement, which defines the potential energy relevant for their (oscillatory) movement. From (10.5) and (10.6), we have:

$$\left[\hat{T}_c + u(\mathbf{r}_{c,j})\right]\psi_c(\mathbf{r}_{c,j}) = E_c\psi_c(\mathbf{r}_{c,j}) \tag{10.9}$$

Obviously, the term $u(\mathbf{r}_{c,j})$ has the meaning of the potential energy of the interaction between the nuclei. It is related to the solution of the electronic part of the Schrödinger equation via (10.8). Then, E_c has to be interpreted as the total (=kinetic + potential) energy of the nuclei.

This way, the adiabatic approximation allowed us to subdivide Schrödingers (10.2) of the full molecule into two separate (10.7) and (10.9), which describe the electronic and nuclei subsystems, respectively. Solving (10.7) with fixed nuclei coordinates (i.e., as a point-by-point calculation) gives principal access to the coordinate dependence of $u(\mathbf{r}_{c,j})$, and knowledge of the latter allows solving (10.9). In different electronic eigenstates (different solutions of (10.7)), $u(\mathbf{r}_{c,j})$ may be different as well, so that the solution of the nuclei (10.9) depends on the concrete electronic eigenstate. This may give rise to different equilibrium positions of the nuclei in different electronic states and, therefore, to differences in molecular size and shape.

On the other hand, (10.8) defines the full energy in a given molecular eigenstate as some superposition of electron and nucleus contributions. It is the movement of the nuclei that defines what we have earlier associated with vibrational and rotational degrees of freedom (Sect. 2.5).

10.3.2 Simplest Example of Covalent Bonding: The H_2^+- Molecule Ion

As a first example, we will turn to a quantum system that is even simpler than the H_2 molecule, namely to the H_2^+—molecule ion. It is obtained by removing one of the electrons from the system shown in Fig. 10.1. As the result, we obtain a simpler three-body system as visualized in Fig. 10.2.

This is a single-electron system, and we will concentrate on solving the Schrödingers equation for the single electron in the field supplied by the two protons. In accordance to the adiabatic approximation, we will assume that the configuration of the nuclei is fixed. Hence, we set:

$$r_{ab} = \text{const } \forall j \tag{10.10}$$

A particular consequence of (10.10) is that no kinetic energy resulting from the relative movement of the nuclei with respect to each other will be taken into account in this section.

Fig. 10.2 H_2^+—Molecule
ion

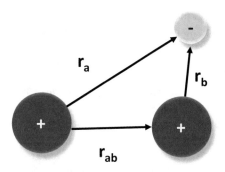

The Schrödinger equation for the electron (i.e., ψ_e, although we will skip the subscript in the following for simplicity) may be written as:

$$-\left[\frac{\hbar^2}{2m_e}\Delta + \frac{e^2}{4\pi\varepsilon_0 r_a} + \frac{e^2}{4\pi\varepsilon_0 r_b}\right]\psi = E\psi \qquad (10.11)$$

Because of (10.10), the Coulomb interaction of the two protons only gives a constant contribution, which we will not take into account explicitly at the moment. Because, in our present case, we deal with a single-electron system; we do not need to care about consequences from the Pauli principle and therefore neglect any spin effects.

Let us remark that we know the wavefunctions ψ_a and ψ_b of the hydrogen-type Schrödinger equations:

$$-\frac{\hbar^2}{2m_e}\Delta\psi_a - \frac{e^2}{4\pi\varepsilon_0 r_a}\psi_a = \hat{H}_a\psi_a = E_a\psi_a$$
$$-\frac{\hbar^2}{2m_e}\Delta\psi_b - \frac{e^2}{4\pi\varepsilon_0 r_b}\psi_b = \hat{H}_b\psi_b = E_b\psi_b \qquad (10.12)$$

In our treatment, we will restrict on the electronic ground state of the molecule ion. In particular, the eigenvalues E_a and E_b should be identical and correspond to the hydrogen ground state according to (8.46). Then, the solutions of (10.12) may be regarded as special cases of the solution of (10.11): When r_{ab} is large enough, the electron will practically be bound to either of the nuclei (a) or (b), so that the system shown in Fig. 10.2 becomes nothing else than a hydrogen atom plus a proton located at a large distance from the hydrogen atom. In this case, the radii r_a and r_b are of quite different orders of magnitude, such that one of the terms describing the potential energy in (10.11) may be neglected.

If, however, the protons come close enough to each other, such that their distance becomes comparable to the spatial extensions of the relevant atomic orbitals, then the situation changes. None of the two contributions to the potential energy of the electron may anymore be neglected, such that we have to assume that in a tricky way, the electron "belongs" to both nuclei. The idea of our further treatment is then the following: We will solve the Schrödinger equation for the hydrogen molecule ion (10.11) with the constraint (10.10) in terms of the ansatz:

$$\psi = c_1\psi_a + c_2\psi_b \tag{10.13}$$

We thus search the solution of (10.11) as a superposition of the solutions of (10.12). This ansatz is called LCAO (LCAO = Linear combination of atomic orbitals). It is an exact solution if $\mathbf{r}_{ab}$ is very large, and we will take it as an approximate ansatz for smaller distances as well. Equation (10.13) is the mathematical formulation of our idea that the electron "belongs" to both nuclei.

When substituting (10.13) into (10.11), we obtain:

$$-\left[\frac{\hbar^2}{2m_e}\Delta + \frac{e^2}{4\pi\varepsilon_0 r_a} + \frac{e^2}{4\pi\varepsilon_0 r_b}\right](c_1\psi_a + c_2\psi_b)$$
$$= c_1\left[\hat{H}_a - \frac{e^2}{4\pi\varepsilon_0 r_b}\right]\psi_a + c_2\left[\hat{H}_b - \frac{e^2}{4\pi\varepsilon_0 r_a}\right]\psi_b = E(c_1\psi_a + c_2\psi_b) \tag{10.14}$$

Let us regard the eigenvalues E_a and E_b as some initial approximation of the electron energy in the ground state of the molecule. Hence, we assume:

$$\hat{H}_a\psi_a = E^{(0)}\psi_a$$
$$\hat{H}_b\psi_b = E^{(0)}\psi_b \tag{10.15}$$

Then, when combining (10.14) and (10.15), we find:

$$c_1\left[E^{(0)} - E - \frac{e^2}{4\pi\varepsilon_0 r_b}\right]\psi_a + c_2\left[E^{(0)} - E - \frac{e^2}{4\pi\varepsilon_0 r_a}\right]\psi_b = 0 \tag{10.16}$$

Introducing:

$$\Delta E = E - E^{(0)} \tag{10.17}$$

We find:

$$c_1\left[\Delta E + \frac{e^2}{4\pi\varepsilon_0 r_b}\right]\psi_a + c_2\left[\Delta E + \frac{e^2}{4\pi\varepsilon_0 r_a}\right]\psi_b = 0 \tag{10.18}$$

We will further restrict ourselves to real wavefunctions ψ_a and ψ_b, as it is valid in the hydrogen atom ground state (compare (8.46)). Then,

$$\int_V \psi_a^2 dV = \int_V \psi_b^2 dV = 1$$

We define:

$$\int_V \psi_a\psi_b dV \equiv S \tag{10.19}$$

S quantifies the spatial overlap of the wavefunctions ψ_a and ψ_b (overlap integral). For wavefunctions like (8.46), we obviously have:

$$S \geq 0 \qquad (10.19a)$$

We further define:

$$\int_V \psi_a \psi_a \left(-\frac{e^2}{4\pi\varepsilon_0 r_b}\right) dV = \int_V \psi_b \psi_b \left(-\frac{e^2}{4\pi\varepsilon_0 r_a}\right) dV \equiv C$$

$$\int_V \psi_a \psi_b \left(-\frac{e^2}{4\pi\varepsilon_0 r_b}\right) dV = \int_V \psi_a \psi_b \left(-\frac{e^2}{4\pi\varepsilon_0 r_a}\right) dV \equiv D \qquad (10.20)$$

While C quantifies the electrostatic energy of the electron with respect to the "other" proton, the term D quantifies a specific quantum mechanical phenomenon, called exchange interaction. D is usually called the exchange integral. For wavefunctions like (8.46), we find:

$$C \leq 0; \quad D \leq 0 \qquad (10.20a)$$

Now, after having multiplied (10.18) with ψ_a or ψ_b and integrated over the volume, we obtain:

$$c_1(\Delta E - C) + c_2(S\Delta E - D) = 0$$
$$c_1(S\Delta E - D) + c_2(\Delta E - C) = 0 \qquad (10.21)$$

A nontrivial solution of (10.21) requires that

$$\begin{vmatrix} (\Delta E - C) & (S\Delta E - D) \\ (S\Delta E - D) & (\Delta E - C) \end{vmatrix} = 0 \Rightarrow (\Delta E - C)^2 - (S\Delta E - D)^2 = 0$$

From here, we find the correction to the electron energy ΔE:

$$\Delta E = \frac{C - SD \pm (CS - D)}{1 - S^2} = \begin{cases} \frac{C-D}{1-S} \\ \frac{C+D}{1+S} \end{cases} \qquad (10.22)$$

Then, from (10.17), we have:

$$E = E^{(0)} + \Delta E = E^{(0)} + \frac{C \pm D}{1 \pm S} \qquad (10.23)$$

Let us now look on the wavefunctions according to (10.13). When substituting (10.22) into the first equation from (10.21), we obtain:

$$\Delta E = \frac{C - D}{1 - S} \Rightarrow c_1 \left(\frac{C - D}{1 - S} - C \right) + c_2 \left(S \frac{C - D}{1 - S} - D \right) = 0$$

$$\Rightarrow c_1(C - D - C + SC) + c_2(CS - DS - D + DS) = 0$$

$$\Rightarrow c_1(SC - D) + c_2(CS - D) = 0 \Rightarrow c_1 = -c_2 \Rightarrow \psi \propto (\psi_a - \psi_b)$$

$$\Delta E = \frac{C + D}{1 + S} \Rightarrow c_1 \left(\frac{C + D}{1 + S} - C \right) + c_2 \left(S \frac{C + D}{1 + S} - D \right) = 0$$

$$\Rightarrow c_1(C + D - C - SC) + c_2(CS + DS - D - DS) = 0$$

$$\Rightarrow -c_1(SC - D) + c_2(CS - D) = 0 \Rightarrow c_1 = +c_2 \Rightarrow \psi \propto (\psi_a + \psi_b)$$

According to Fig. 10.2, we have to expect that S, C, and D depend on the distance r_{ab} between the nuclei. Therefore, our correction term ΔE according to (10.22) will also depend on r_{ab}. At sufficiently large distances, it will approach zero because C and D are vanishing at distances larger than the characteristic decay length of the atomic wavefunctions.

The full energy of the molecule will of course also contain the potential energy of the two nuclei.

$$E_{total} = E^{(0)} + \frac{C \pm D}{1 \pm S} + \frac{e^2}{4\pi \varepsilon_0 r_{ab}} \equiv E(r_{ab}) \tag{10.24}$$

When the nuclei are placed at an infinitely large distance, from (10.24), it is obvious that $E_{total} = E^{(0)}$. When the distance between the nuclei becomes smaller, the electrostatic energy corresponding to the repulsive interaction between the nuclei will increase. Moreover, the energy correction term ΔE will split off into the two possible solutions (10.22). Thereby, in case of the antisymmetric solution $\psi = \psi^{(asym)} \propto (\psi_a - \psi_b)$, we have the correction term $\Delta E = \frac{C-D}{1-S}$, which is small by absolute value because the integrals C and D tend to compensate each other (compare (10.19a) and (10.20a). It is therefore likely to assume that in (10.15), the repulsive Coulomb interaction between the nuclei gives the dominant contribution, such that the potential energy monotonically increases when the nuclei approach each other. In such a situation, no potential minimum is formed, and a stable bonding configuration cannot be achieved. On the contrary, in case of the symmetric solution $\psi = \psi^{(sym)} \propto (\psi_a + \psi_b)$, we have the correction term $\Delta E = \frac{C+D}{1+S} \leq 0$, which is certainly negative at finite distances (compare (10.19a) and (10.20a)) and therefore tends to (at least partially) compensate the repulsive Coulomb interaction between the nuclei. In this case, we have a chance to observe a potential minimum in the potential energy, which would correspond to molecular bonding.

This result has a transparent physical interpretation. In the asymmetric state, the electron has only a minimum probability to be found in-between the two nuclei (Fig. 10.3, black curve). Therefore, the repulsive potential between the nuclei will

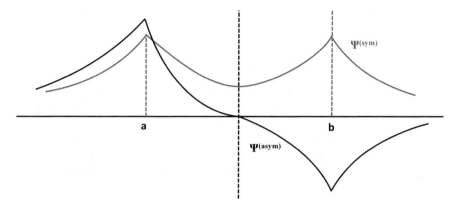

Fig. 10.3 Symmetric and antisymmetric superpositions of the hydrogen ground-state wavefunctions (8.46)

dominate, and no molecule ion may be formed. For this reason, the antisymmetric state $\psi^{(asym)} \propto (\psi_a - \psi_b)$ is called an antibonding electronic state.

In case of the symmetric solution $\psi^{(sym)} \propto (\psi_a + \psi_b)$, there is however a good chance to observe the electron in-between the nuclei, which leads to a reduction of the repulsive force between the nuclei. Therefore, a potential minimum may exist that marks the equilibrium distance $r_{ab,0}$ between the nuclei in the molecule ion. Such a state is called a bonding state. At lowest distances r_{ab}, as it may be guessed from (10.24), the repulsive potential between the nuclei will dominate (Fig. 10.4).

Fig. 10.4 Ground-state energy $E(r_{ab})$ of the hydrogen molecule ion in bonding and antibonding states according to (10.24)

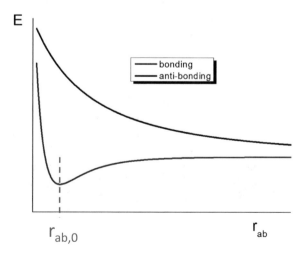

10.4   Consistency Considerations

Obviously, the model assumptions formulated in the previous sections as well as the obtained results need some discussion:

- Our treatment of the hydrogen molecule ion has essentially based on the adiabatic approximation, particularly on condition (10.10). This led us to expressions like (10.24) and—as a result of a qualitative discussion—to dependencies $E = E(r_{ab})$ like those shown in Fig. 10.4. Thereby, the distance between the nuclei r_{ab} was tackled as a fixed parameter when solving the Schrödinger equation for the electron in the potential provided by the two nuclei. Consequently, both E and ψ depend on that parameter r_{ab}. Note that (10.24) is a particular case of (10.8).

  On the other hand, condition (10.10) does not allow for well-known phenomena like thermal expansion of macroscopic many-atomic systems, or vibrations of nuclei with respect to each other, which give rise to specific MIR absorption spectra (compare Sect. 2.5.1). A treatment of these phenomena needs the application of the Schrödinger equation of the nuclei subsystem (10.9). At the moment, let us recognize the encouraging principal similarity of the red curve in Fig. 10.4 and the black one in Fig. 2.7. Thus, when applying a classical picture, an elongation of the nuclei from the equilibrium value $r_{ab,0}$ should result in an increase of their potential energy. Then, a restoring force should be generated that pulls the nuclei back to their equilibrium distance, thus giving rise to vibrations of the nuclei with respect to each other. Practically, through the application of (10.8), $E = E(r_{ab})$ gives direct access to the potential $u(r_{c,j})$ that is responsible for the movements of the nuclei.
- The different effects of the symmetric and antisymmetric constellations in the electrons wavefunctions may be visualized in a simple classical picture (Fig. 10.5):

Thus, in the asymmetric situation on top of Fig. 10.5, nucleus b feels strong repulsion from nucleus a and strong attraction from the electron. It will obviously be pushed to the right. Nucleus a feels strong repulsion from nucleus b but only weak attraction from the electron. It will obviously be pushed to the left. Nothing like a molecule may be formed.

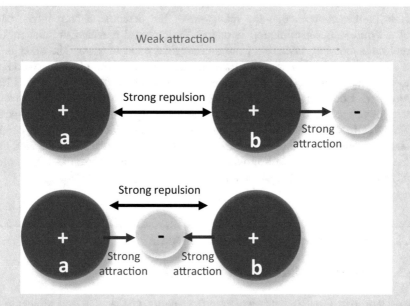

Fig. 10.5 Classical illustration of antisymmetric (on top) and symmetric (on bottom) electronic constellations in the H_2^+ molecule ion

In the symmetric constellation on bottom, however, the strong internuclear repulsion is somehow compensated by the strong attraction by the electron (again some kind of screening). It appears likely that this symmetric constellation should be favored in molecular bonding.

- However, the classical illustration from Fig. 10.5 alone cannot explain the existence of stable molecules. This already follows from the impossibility to obtain stable classical atoms (Sect. 2.4) necessary as the starting point for the formation of molecules. Moreover, it is obvious that:
  - A situation like that shown in Fig. 10.5 may be tackled as a snaphot of the configuration of a classical system, but of course, it is quite unstable.
  - Fig. 10.5 does not give access to understanding typical molecular spatial extensions.
- In terms of the quantum mechanical treatment, the efficiency of molecular bonding is predominatly defined by the values of the integrals C and D. It is the effective spatial extension of the atomic wavefunctions that acts as the crucial spatial parameter here. Therefore, in the quantum mechanical picture, the distance between the nuclei in a molecule is of the same order as typical atomic sizes are (compare again Table 1.1). This is in full agreement to corresponding experimental findings.
- According to the LCAO ansatz (10.13), the electron in the hydrogen molecule ion somehow "belongs" to both nuclei. In this situation, one speaks on covalent bonding.

- Note in this context, that the reduction of the potential energy in the case of a symmetric wavefunction is one of the physical mechanisms that makes molecular bonding favorable from an energetic point of view. The other is in a decrease in kinetic energy of the electron. Indeed, if the electron "belongs" to both nuclei, it is allowed to move in a larger spatial region and is thus less localized. According to our discussion in Sect. 4.3.3, that is expected to result in a lower ground-state kinetic energy (compare also Task 5.7.9 in this regard).
- Note that in the LCAO wavefunction (10.13), the absolute values of the coefficients c_1 and c_2 turned out to be identical. This is an expression of the fact, that the situations "the electron is with proton a" and "the electron is with proton b" reflect the same physical situation.

10.5   Application to Practical Problems

10.5.1 Interatomic Interaction Potentials

Let us return to Fig. 10.4. It visualizes the energy of the diatomic molecule in dependence of the internuclear distance r_{ab}. In the bonding electronic state, it has a local minimum at the equilibrium distance $r_{ab,0}$. Any elongation of the nuclei from their equilibrium positions will increase the energy and therefore give rise to a restoring force, thus resulting in oscillations of the nuclei. Therefore, the particular shape of the $E(r_{ab})$—dependence as given by (10.24) is of immense significance for understanding the optical spectra of real molecules. In fact, it should define the potential $u(r_{ab})$ that defines the interaction forces between the nuclei and thus their oscillations with respect to each other in a diatomic molecule (and, of course, also many-atomic molecules). Thereby, in the case of covalent bonding, $E(r_{ab})$ may principally be calculated by solving (10.14) in a point-by-point manner, assuming different fixed values of r_{ab}. Then, the potential $u(r_{ab})$ relevant for the vibrations of the nuclei should have the same shape, but for practical purposes, it may differ from $E(r_{ab})$ by a suitably chosen constant (compare (10.6)).

In practice, for many relevant tasks, it is however sufficient to replace the true $u(r_{ab})$ dependence by a suitably parametrized model dependence that reflects the main features of the true $u(r_{ab})$ curve. It is the purpose of this short section to introduce a couple of such practically relevant models.

Let us summarize some basic requirements on such $u(r_{ab})$ curves. Besides usual mathematical requirements such as continuity and differentiability of the curves, we would it like to have the following conditions fulfilled:

1. For $r_{ab} \to \infty$, $E(r_{ab})$ should approach a constant value that corresponds to the sum of the energies of the single isolated atoms that form the molecule. In many applications, it is useful to define the interaction potential $u(r_{ab})$ such that $u(r_{ab} \to \infty) \to 0$ (no interaction at infinite distance).
2. The $u(r_{ab})$ curve should have one local minimum at $r_{ab} = r_{ab,0}$.

3. For $r_{ab} \to 0$, $u(r_{ab})$ should be positive and become "very large."

Note that these requirements are consistent with the shape of the $E(r_{ab})$ dependence in a bonding configuration as shown in Fig. 10.4. Thereby, (10.24) was derived assuming covalent bonding. But clearly, other bonding types (ionic bonding, van der Waals bonding) must result in $E(r_{ab})$-dependencies which are qualitatively close to the shape of $E(r_{ab})$ as shown in Fig. 10.4. Therefore, the requirements (1)–(3) are more general and do not only concern covalent bonding of diatomic molecules.

Let us have a closer look at some examples, applicable for diatomic molecules:

10.5.2 A Simple Model Potential

The requirements (1)–(3) are obviously fulfilled for any potential given by:

$$u(r_{ab}) = -\frac{\sigma_1}{(r_{ab})^{n_1}} + \frac{\sigma_2}{(r_{ab})^{n_2}}; \quad n_2 > n_1 > 0; \sigma_1, \sigma_2 > 0; n_1, n_2 \text{ integer} \quad (10.25)$$

The constant σ_1 defines the strength of the attractive force between the nuclei a and b, while the constant σ_2 is responsible for the repulsive force. In accordance to our classical considerations from Sect. 2.5.1 (Fig. 2.7), we further have:

$$u(r_{ab} \to \infty) = 0$$
$$u(r_{ab} \to 0) \to \infty$$

The minimum of the potential (10.15) defines the equilibrium distance $r_{ab,0}$. Differentiating (10.25) and setting the first derivative equal to zero results in:

$$(r_{ab,0})^{n_2 - n_1} = \frac{n_2}{n_1} \frac{\sigma_2}{\sigma_1}$$

Accordingly, $u(r_{ab})$ must have a local minimum at $r_{ab} = r_{ab,0}$. We will call that minimum value of the potential U_0, i.e.,

$$U_0 \equiv u(r_{ab} = r_{ab,0}) < 0$$

Then, the model parameters σ_1 and σ_2 may be expressed through $r_{ab,0}$ and U_0. This results in another writing of (10.15), namely:

$$u(r_{ab}) = \frac{U_0}{n_1 - n_2} \left[n_1 \left(\frac{r_{ab,0}}{r_{ab}} \right)^{n_2} - n_2 \left(\frac{r_{ab,0}}{r_{ab}} \right)^{n_1} \right]; \quad n_2 > n_1 > 0; n_1, n_2 \text{integer}$$

$$(10.26)$$

Fig. 10.6 Illustration of
ionic bonding between a
chlorium and a potassium ion

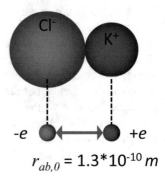

$r_{ab,0} = 1.3 * 10^{-10}\,m$

10.5.3 Model Potentials for Ionic Bonding

In ionic bonding, the attractive force between the ions is primarily caused by the electrostatic Coulomb attraction; hence, $n_1 = 1$ (Fig. 10.6).

A simple potential for modeling this type of bonding is therefore:

$$u(r_{ab}) = \frac{U_0}{1 - n_2}\left[\left(\frac{r_{ab,0}}{r_{ab}}\right)^{n_2} - n_2\left(\frac{r_{ab,0}}{r_{ab}}\right)\right]; \; n_2 > 1; n_2 \text{integer} \qquad (10.27)$$

For different ions, the value of n_2 changes varies between 7 and 12 [3].

In addition to the Coulomb attraction between the ions, the Born–Heisenberg theory of ionic bonding takes the effects of dipoles induced in the ions into account. That results in the appearance of additional attractive terms in the potential. In this model, n_2 is set equal to 9, such that the following potential is proposed [3]:

$$u(r_{ab}) = -\frac{\sigma_{11}}{r_{ab}} - \frac{\sigma_{12}}{(r_{ab})^4} - \frac{\sigma_{13}}{(r_{ab})^7} + \frac{\sigma_2}{(r_{ab})^9} \qquad (10.28)$$

10.5.4 The Lennard-Jones Potential

It is my hope, dear reader, that you have already become familiar with that potential by having solved task 2.7.3 in Chap. 2. In terms of the more general writing provided by (10.26), setting $n_1 = 6$ and $n_2 = 12$, we may write

$$u(r_{ab}) = \frac{U_0}{6}\left[12\left(\frac{r_{ab,0}}{r_{ab}}\right)^6 - 6\left(\frac{r_{ab,0}}{r_{ab}}\right)^{12}\right] \qquad (10.29)$$

(10.29) is called a Lennard–Jones potential. Although (10.29) represents an empirical formula, the r^{-6} dependence may be analytically derived for the particular case of van der Waals bound molecules. We will not perform that calculation, but essentially, the r^{-6} dependence is obtained from a simple induced dipole inter-action when taking the r^{-3} decay of the electric field around a point dipole into account [4].

10.5.5 The Morse Potential

The Morse potential is another widely used model potential. We will return to it in the next chapter, when modeling nuclei vibrations in terms of an anharmonic oscillator model. For the moment, we indicate the corresponding formula defining the Morse potential [5]:

$$u(r_{ab}) = U_0 \left[2e^{-\beta(r_{ab} - r_{ab,0})} - e^{-2\beta(r_{ab} - r_{ab,0})} \right] \tag{10.30}$$

β is a real positive parameter in reciprocal length units (i.e., the reciprocal value of some characteristic damping length). In contrast to the potentials given by (10.25)–(10.29), the Morse potential does not diverge at $r_{ab} \to 0$.

Its advantage is that the Schrödinger equation for a particle with mass m moving in the Morse potential may be exactly solved. The energy levels of bound states depend on one quantum number $n \geq 0$ with the constraint $\left(n + \frac{1}{2} < \frac{\sqrt{-2mU_0}}{\beta\hbar} \right)$ and are given by:

$$E(n) = U_0 + \hbar\omega_e \left(n + \frac{1}{2} \right) - \hbar\omega_e x_e \left(n + \frac{1}{2} \right)^2 \tag{10.31}$$

where ω_e is a characteristic angular frequency defined by:

$$\omega_e = \beta \sqrt{-2\frac{U_0}{m}} \tag{10.32}$$

Except the third term in (10.31), the energy values coincide with those obtained for a harmonic oscillator with a vertex of the potential curve located at U_0 and a resonance frequency ω_e (compare (5.34)). Correspondingly, the term

$$x_e = -\frac{\hbar\omega_e}{4U_0} = \frac{\hbar\omega_e}{4D} \ll 1 \tag{10.33}$$

is called the anharmonicity of the Morse potential. For $x_e = 0$, (10.31) turns into the corresponding formula for the harmonic oscillator. The restriction on the quantum number n for forming a bound state may now be rewritten as:

$$n + \frac{1}{2} < \frac{\sqrt{-2mU_0}}{\beta\hbar} = -\frac{2U_0}{\hbar\omega_e} = \frac{2D}{\hbar\omega_e} = \frac{1}{2x_e}$$

Figure 10.7 shows examples of the shapes of different of the mentioned potentials, with identical assumed position and depth of the potential minimum. Note that ionic bonding corresponds to a rather far-reaching interaction potential, because of the slow convergence at large distances.

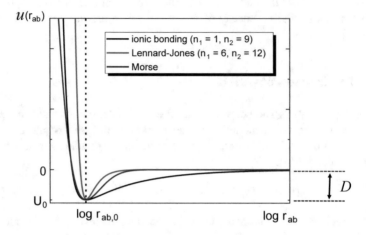

Fig. 10.7 Examples of often used model potentials. The vertical dashed straight line indicates the position of the common potential minimum. Again, (compare Fig. 2.7), $D = -U_0$ (do not confuse with the exchange integral)

10.6   Advanced Material: The Hydrogen Molecule

The treatment of the hydrogen molecule basically follows the LCAO treatment earlier applied to the hydrogen molecule ion. We will only mention the basic steps and results of such an approach. The basic difference is in the presence of two indistinguishable electrons. Therefore, there is a need to modify the approximate ansatz (10.13). Keeping in mind the structure of the spatial part of the wavefunction like (9.30), instead of (10.13), we now write:

$$\psi^{(\text{sym})} = \frac{1}{\sqrt{2(1+S)}}[\psi_a(1)\psi_b(2) + \psi_a(2)\psi_b(1)]$$

$$\psi^{(\text{asym})} = \frac{1}{\sqrt{2(1-S)}}[\psi_a(1)\psi_b(2) - \psi_a(2)\psi_b(1)] \tag{10.34}$$

Here,

$$S \equiv \int_{V_1, V_2} \psi_a(1)\psi_a(2)\psi_b(1)\psi_b(2)\mathrm{d}V_1\mathrm{d}V_2 \tag{10.35}$$

(10.35) is the generalization of the overlap integral (10.19). As in the case of the helium atom, integration has now to be performed over the configuration space of the coordinates of two electrons.

Here, (1) and (2) denote the two electrons, while a and b the two protons (compare Fig. 10.8). The symbols in (10.34) denote situations like "electron 1 is with nucleus b" $\Rightarrow \psi_b(1)$ and so on.

A detailed LCAO calculation shows that the structure of (10.24) remains valid with somewhat generalized definitions of C and D (not to be confused with the dissociation energy). Thus, instead of (10.20), we now have (symbols as explained in Fig. 10.8):

$$\frac{e^2}{4\pi\varepsilon_0}\int_{V_1,V_2}[\psi_a(1)\psi_b(2)]^2\left(-\frac{1}{r_{b1}}-\frac{1}{r_{a2}}+\frac{1}{r_{12}}\right)dV_1dV_2$$
$$=\frac{e^2}{4\pi\varepsilon_0}\int_{V_1,V_2}[\psi_a(2)\psi_b(1)]^2\left(-\frac{1}{r_{a1}}-\frac{1}{r_{b2}}+\frac{1}{r_{12}}\right)dV_1dV_2\equiv C \quad (10.36)$$

$$\frac{e^2}{4\pi\varepsilon_0}\int_{V_1,V_2}\psi_a(1)\psi_a(2)\psi_b(1)\psi_b(2)\left(-\frac{1}{r_{b1}}-\frac{1}{r_{a2}}+\frac{1}{r_{12}}\right)dV_1dV_2$$
$$=\frac{e^2}{4\pi\varepsilon_0}\int_{V_1,V_2}\psi_a(1)\psi_a(2)\psi_b(1)\psi_b(2)\left(-\frac{1}{r_{a1}}-\frac{1}{r_{b2}}+\frac{1}{r_{12}}\right)dV_1dV_2\equiv D$$
$$(10.37)$$

Fig. 10.8 Illustration of the used symbols

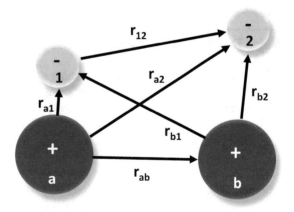

At infinite distance between the nuclei, of course, we now have the ground-state energies of two hydrogen atoms. Therefore, instead of (10.24), our final result is:

$$E_{\text{total}} = 2E^{(0)} + \frac{C \pm D}{1 \pm S} + \frac{e^2}{4\pi\varepsilon_0 r_{ab}} \equiv E(r_{ab}) \qquad (10.38)$$

where S, C, and D are given by (10.35)–(10.37). And again, it is the symmetric spatial part of the wavefunction that corresponds to the bonding configuration. The dependence of the total energy of the molecule on the fixed distance of the nuclei is qualitatively the same as presented earlier in Fig. 10.4.

The rest of the discussion is similar to what has been established for the helium atom. Once we have now two electrons (i.e., two fermions), their full wavefunction must be antisymmetric with respect to interchanging the positions of the electrons. The general structure of such wavefunctions is indicated in (9.30), (9.32) and (9.33). Thus, similarly to the discussion performed in Sect. 9.4.3, in the case of the symmetric (bonding) spatial wavefunction, the spin function must be antisymmetric, i.e., the spins must be antiparallel in a bonding state. Contrarily, the antibonding antisymmetric spatial part of the wavefunction corresponds to a symmetric spin wavefunction, i.e., parallel spins.

Concerning numerical values, in the hydrogen molecule, we have $r_{ab,0} \approx 0.074$ nm; $U_0 \approx -4.52$ eV [6] (compare Table 1.1). Note the consistency of $r_{ab,0} \approx 0.074$ nm with the expectation value of r in the hydrogen atom ground state: $<r> \approx 0.08$ nm as calculated in Sect. 8.4.1.

The two electrons with antiparallel spins are called to form an electron pair.

10.7   Tasks for Self-check

10.7.1 Recall the approaches for estimating atomic and molecular characteristic spatial dimensions we have mentioned so far!

10.7.2 Check the normalization of the wavefunctions (10.34)!

10.7.3 Consider the motion of a particle confined in the following symmetric (double-well) potential:

$$U = \begin{cases} 0; & 0 \le x \le L;\ L+b \le x \le 2L+b \\ U_0; & L < x < L+b \\ \infty; & x < 0 \text{ and } x > 2L+b \end{cases}$$

Find the allowed energy levels assuming $U_0 > 0$ and $E < U_0$!

10.7.4 Consider an alternative model of the hydrogen molecule ion by replacing the Coulomb potentials by delta potential according to task 4.4. For calculations, consider a one-dimensional potential $U(x) = -A\delta(x - x_0) - A\delta(x + x_0);\ A > 0$

Calculate the electronic energy in dependence of the distance $2x_0$! (after [7])

10.7.5 In Sect. 10.3.2: Instead of (10.13), you might also use an ansatz $\psi = c_1\psi_a - c_2\psi_b$. What would you obtain then?

References

Specific References

1. А.А. Соколов, И.М. Тернов, В.Ч. Жуковский, Квантовая механика, Москва "Наука" (1979), pp. 426–427
2. O. Stenzel, *Optical Coatings. Material Aspects in Theory and Practice* (Springer 2014), pp. 227–238
3. V. Kondratyev, *The Structure of Atoms and Molecules* (2nd printing, Mir Publishers 1967), Chap. 7
4. W. Demtröder, *Atoms, Molecules, and Photons* (Springer 2010), pp. 327–330
5. Л.Д.Ландау, Е.М.Лифшиц, Квантовая Механика, Нерелятивистская Теория, Москва "Наука" (1974), pp. 96–97
6. P.A. Tipler, G. Mosca, *Physik, 7* (Springer, Auflage, 2015), p. 1256
7. H. Haken, H.C. Wolf, *The Physics of Atoms and Quanta* (Springer 2005), Task 24.1

General Literature

8. L.D. Landau, E.M. Lifshitz, *Quantum Mechanics, vol. 3 of A Course of Theoretical Physics* (Pergamon Press, 1965)
9. H. Haken, H.C. Wolf, *The Physics of Atoms and Quanta* (Springer 2005)

Optical Spectra of Molecules

11

Abstract

Important spectral features of molecular spectra are introduced, including rotation of the whole molecule, vibrations of the atomic nuclei, and the response of valence electrons. The treatment is focused on diatomic molecules as a well-accessible model system, but complemented by information about selected specifics of rotational-vibrational spectra of polyatomic molecules.

11.1   Starting Point

In order to get a general idea on the specifics of molecular absorption spectra, let us start with some purely classical considerations. A molecule is built from several atoms; i.e., it consists of a certain number of electrons and a certain number of atomic nuclei. Let the number of atomic nuclei in the molecule be N_a.

In a classical picture, the molecule can be understood as a couple of point masses connected by elastic springs. The spring constant is an analogue to the bond strength and is controlled—in covalent bounding—by the spatial overlap of the valence electrons wavefunctions, i.e., those "outermost" electrons that are responsible for the formation of covalent bonding (compare Chap. 10). Once the molecule should be stable to small perturbations with respect to size and shape as a whole, it is reasonable assuming that the spring constants responsible for the movements of nuclei and valence electrons are of the same order of magnitude. Then, as a consequence of the different masses of electrons and nuclei, the vibrational eigenfrequencies of the motion of nuclei should be much smaller than that of the valence electrons. A rough estimate is (compare Sects. 2.5.1 and 2.5.2):

$$\frac{\omega_{\text{valence electron}}}{\omega_{\text{nucl}}} = \frac{\lambda_{\text{nucl}}}{\lambda_{\text{valence electron}}} \cong \sqrt{\frac{m_{\text{nucl}}}{m_e}} \approx 100 \tag{11.1}$$

© The Author(s), under exclusive license to Springer Nature Switzerland AG 2022
O. Stenzel, *Light–Matter Interaction*, UNITEXT for Physics,
https://doi.org/10.1007/978-3-030-87144-4_11

The vibrations of nuclei correspond to frequencies in the middle infrared (further *vibronic excitations*, the corresponding spectra are often called *vibrational* spectra), and those of valence electrons to the visible or ultraviolet spectral regions:

$$\omega_{nucl} \equiv \omega_{vibr} \in MIR$$

$$\omega_{valence\ electron} \equiv \omega_{electr} \in VIS/UV \qquad (11.2)$$

Therefore, purely vibrational spectra are traditionally recorded by means of MIR spectrophotometers, while the set of MIR eigenfrequency characteristic for a certain molecule (the so-called *fingerprint* spectrum) may be used for identification purposes. This technique is, of course, also used in solid-state spectroscopy in order to identify substances, contaminations, bonding configurations, crystalline symmetries, and the like. It is characteristic for all these infrared spectra that no optical excitation of electronic eigenmodes occurs, while only vibrational degrees of freedom are excited.

In a purely electronic spectrum, on the contrary, only electronic vibrations are excited by absorption of light, while the motion of nuclei will not be affected. Such excitations should occur in the VIS/UV, but they are rarely observed. The point is that the excitation of valence electrons is in most cases accompanied by an additional excitation of vibrations of the nuclei, so that in VIS/UV absorption, we usually observe a superposition of electronic and vibronic excitations. The reason for this behavior may be understood from the quantum mechanical description: It turns out that the equilibrium positions of the nuclei (which define size and shape of the molecule) depend on the respective quantum state of the electrons (compare Sect. 10.3.1). When the electronic state is changed (which happens rather rapidly), the (initially maybe resting) nuclei come into motion to move toward their new equilibrium positions. This way they gain kinetic energy, which does not permit them to stop at the new equilibrium position; instead, they will oscillate around these positions. In quantum mechanics, this behavior is predicted as a consequence of the relevance of such powerful concepts like the Born–Oppenheimer approximation and the Franck–Condon principle. The important point at the moments is that this way a vibration of nuclei may be established, and the corresponding energy must been taken from the electromagnetic field. The absorbed energy is thus used for the excitation of a superposition of electronic *and* nuclei oscillations, and hence, the corresponding absorption frequency may be symbolically written as:

$$\omega_{absorption} \equiv \omega_{electr} + \sum \omega_{vibr} \in VIS/UV \qquad (11.3)$$

Equation (11.3) does not simply describe a single absorption line, but a rather complicated absorption feature, which is blue-shifted with respect to the "purely" electronic excitations, and is formed as the superposition of a multiplicity of spectrally overlapping excitations of different vibronic eigenmodes. If the molecule has the possibility to freely rotate in space, the picture becomes even more involved.

Fig. 11.1 Coronene
molecule. Please do not
confuse with the mad corona
virus; the latter is larger in
size (approximately 125 nm
in diameter) and nearly
spherical

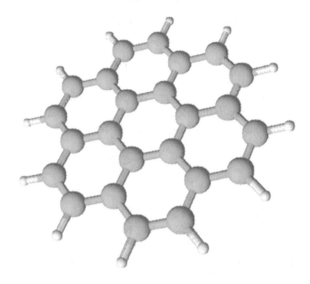

But why a multiplicity of vibrational modes? Let us illustrate this by means
of an example [1]. In Fig. 11.1, a *coronene* molecule which has the sum formula
$C_{24}H_{12}$ is shown. A coronene molecule is planar and formed from the total of N_a
$= 36$ atomic nuclei.

For estimating the number of vibrational degrees of freedom, the nuclei may be
regarded as point masses. Then, once every point mass has 3 independent degrees
of freedom in space, the system of 36 point masses is fully described by 108
coordinates, i.e., has 108 degrees of freedom. Three of them define the position
of the molecule as a whole in space (they form the three *translational* degrees
of freedom), and three of them are necessary to describe the orientation of the
molecule (three *rotational* degrees of freedom). The rest as given by $3N_a$-6 defines
the number of internal vibrational degrees of freedom. This leads us to 102 vibrational degrees of freedom for the coronene molecule. Correspondingly, we expect
a set of 102 vibrational normal or eigenfrequencies in relation (11.3), and even
when many of them are degenerated as a consequence of the high symmetry of
the coronene molecule, we will still find a multiplicity of different eigenfrequencies which is responsible for the complicated structure of the UV spectrum of
the coronene molecule. For completeness, we mention here that in the case of
linear molecules, the discussion is slightly different: In this case, only two rotational degrees of freedom are apparent, and therefore, the number of vibrational
eigenmodes is calculated by a slightly different term $3N_a$-5.

11.2 Physical Idea

The basic physical ideas concerning the spectral properties of molecules have
been formulated earlier when discussing quantum transitions. We start from (5.23),

which defines the set of transition angular frequencies ω_{nm}:

$$\omega_{nm} = \frac{E_n - E_m}{\hbar}$$

In order to get access to the possible transition frequencies, we need to have information about the energy levels that correspond to optically allowed transitions. From (10.8), we obtain the energy levels:

$$E = E_e(\mathbf{r}_{c,j}) + \underbrace{E_c - u(\mathbf{r}_{c,j})}_{=T_{kin,c}} \tag{11.4}$$

The calculations in Sects. 10.3.2 and 10.6 have been performed assuming fixed positions of the nuclei; hence, what we calculated there is the term $E_e(\mathbf{r}_{c,j})$, i.e., the total energy of the rigid molecule. In other words, no kinetic energy of the nuclei has been taken into account so far. But once E_c has been interpreted as the total mechanical energy of the nuclei, then $E_c - u(\mathbf{r}_{c,j})$ has the physical meaning of the kinetic energy of the nuclei. Thus, (11.4) asserts that the total energy of the molecule may be understood as the energy of the rigid molecule plus the kinetic energy of the nuclei. E_c is practically accessible from solving (10.9), while the corresponding kinetic energy formally contains contributions from translational, vibrational, and rotational degrees of freedom. In order to get rid from the contributions of translational movements of the whole molecule, we again choose a coordinate system such that the mass center of the molecule is at rest. Then, because of the large difference between characteristic vibrational and rotational frequencies, (Chap. 2), the kinetic energy of the nuclei appears to be approximately composed from contributions of the (faster) relative movement of the nuclei with respect to each other (*vibrations*), as well as from the (much slower) rotation of the molecule as a whole.

Thus, according to (11.4), we obtain the full energy of the molecular state when adding the kinetic energy of the nuclei to $E_e(\mathbf{r}_{c,j})$. For our further purposes, we now rewrite (11.4) in the following manner:

$$E = E_{\text{electronic}} + E_c \tag{11.4a}$$

Once E_c contains all mechanical (i.e., kinetic and potential) energy accumulated in the nuclei, the rest of terms on the right side of (11.4) should have the meaning of an electronic energy. Subdividing E_c into vibrational and rotational contributions, we finally have:

$$E \approx E_{\text{electronic}} + E_{\text{vibr}} + E_{\text{rot}} \tag{11.5}$$

Combined with (5.23), that yields:

$$\omega_{nm} = \frac{E_n - E_m}{\hbar}$$

$$\approx \frac{E_{n,\,\text{electronic}} - E_{m,\text{electronic}} + E_{n,\text{vibr}} - E_{m,\text{vibr}} + E_{n,\text{rot}} - E_{m,\text{rot}}}{\hbar}$$

$$\equiv \underbrace{\omega_{\text{electronic}}}_{\in\ \text{UV/VIS}} + \underbrace{\omega_{\text{vibr}}}_{\in\ \text{(NIR)/MIR}} + \underbrace{\omega_{\text{rot}}}_{\in\ \text{FIR}} \tag{11.6}$$

(11.6) is a generalized version of (11.3).

The idea of the further treatment is to quantify the individual contributions to (11.6) in terms of different models, such that the structure of molecule spectra in different characteristic spectral regions becomes obvious.

11.3 Theoretical Material

11.3.1 Vibration and Rotation of the Diatomic Molecule

Let us now apply this knowledge to the discussion of molecular spectra. We have (10.9):

$$\left(\hat{T}_c + u\big(\mathbf{r}_{c,j}\big)\right)\psi_c\big(\mathbf{r}_{c,j}\big) = E_c\psi_c\big(\mathbf{r}_{c,j}\big)$$

Note that $u\big(\mathbf{r}_{c,j}\big)$ has been introduced in terms of (10.6) only. Of course, this relation does not provide an unambiguous definition of $u\big(\mathbf{r}_{c,j}\big)$, but once we interpret it as a potential that describes the relative movement of the nuclei, it makes sense to calibrate it such that

$$u\big(\mathbf{r}_{c,j} \to \infty\big) = 0 \tag{11.7}$$

Note that in this case, in bound states we usually have $E_c < 0$.

In the special case of a diatomic molecule, $j = 1, 2$, and we set (compare (2.7)):

$$\mathbf{r} = \mathbf{r}_{c,2} - \mathbf{r}_{c,1}; \quad u = u(|\mathbf{r}|) = u(r)$$

Then, the movement of the two nuclei in the central potential u may be regarded as a two-body problem. In full analogy to the treatment of the hydrogen atom ((8.33) or (8.35)), we write [2]:

$$\Delta = \Delta_r + \frac{1}{r^2}\Delta_{\theta,\phi}$$

$$\psi_c = \psi_c(r, \theta, \phi) = R(r)Y_l^m(\theta, \phi)$$

$$\Rightarrow E_C R = \left[-\frac{\hbar^2}{2\mu}\Delta_r + \frac{\hbar^2 l(l+1)}{2\mu r^2} + u(r)\right]R$$

Practically, l now acts as quantum number quantifying the rotation state of the molecule. It is however common in molecular spectroscopy to use the letter J for the rotational quantum number. Hence, we write instead:

$$E_C R = \left[-\frac{\hbar^2}{2\mu} \Delta_r + \frac{\hbar^2 J(J+1)}{2\mu r^2} + u(r) \right] R; \quad J = 0, 1, 2, \ldots$$

and have:

$$\Delta_r R + \frac{2\mu}{\hbar^2} \left[E_C - u(r) - \frac{\hbar^2 J(J+1)}{2\mu r^2} \right] R = 0$$

According to (8.34), we set again $\chi = rR$ and obtain (compare (8.36)):

$$\left. \begin{aligned}
\frac{d^2 \chi}{dr^2} &= r \left[\frac{2}{r} \frac{dR}{dr} + \frac{d^2 R}{dr^2} \right] = r \Delta_r R \\
r \Delta_r R &+ \frac{2\mu}{\hbar^2} \left\{ [E_C - u(r)] - \frac{J(J+1)}{r^2} \right\} r R = 0 \\
\Rightarrow \frac{d^2 \chi}{dr^2} &+ \frac{2\mu}{\hbar^2} \left[E_C - u(r) - \frac{\hbar^2 J(J+1)}{2\mu r^2} \right] \chi = 0
\end{aligned} \right\}$$

Let us now remember Fig. 2.7 or Fig. 10.7, which show a typical shape of $u(r)$ which has a minimum at $r = r_0$. We further set: $r = r_0 + x$ and assume $|x| \ll r_0$, such that:

$$\frac{\hbar^2 J(J+1)}{2\mu r_0^2} = E_{\text{rot}} \approx \frac{\hbar^2 J(J+1)}{2\mu r^2} \tag{11.8}$$

Remark This is clearly an approximation, because with respect to rotation, the molecule is now regarded as a rigid rotor. But this model assumption makes sense because during one rotation of the molecule, a large number of oscillations are performed, so that the true distance between the atoms may be replaced by some average performed during one oscillation: $\frac{1}{r^2} \rightarrow \left\langle \frac{1}{r^2} \right\rangle_{\text{oscillation}}$. This way, at the moment we clearly neglect the action of centrifugal forces. One should also keep in mind that the term $\left\langle \frac{1}{r^2} \right\rangle_{\text{oscillation}}$ will depend on the vibrational state of the molecule. Nevertheless in our approximation, we assume that $\frac{1}{r_0^2} \approx \left\langle \frac{1}{r^2} \right\rangle_{\text{oscillation}}$

We then obtain:

$$\frac{d^2 \chi}{dx^2} + \frac{2\mu}{\hbar^2} [E_C - E_{\text{rot}} - u(x)] \chi \approx 0$$

Introducing now the energy of the nuclei vibration:

$$E_{\text{vibr}} = E_C - E_{\text{rot}}$$

We obtain the equation for the molecule vibration:

$$-\frac{\hbar^2}{2\mu}\frac{d^2\chi}{dx^2} + u(x)\chi = E_{\text{vibr}}\chi \qquad (11.9)$$

where $u(r)$ represents a potential like shown in Fig. 2.7 or 10.7. As already mentioned in Sect. 10.5.1, it could be obtained from solving a Schrödinger equation like (10.14) in a point-by-point manner, but in many cases it is convenient to make use of available model potentials like the Morse potential (10.30) or even a harmonic oscillator approach.

11.3.2 Rotation Spectra of a Gas of Diatomic Molecules

Assuming a rigid rotor, from (11.8) we have:

$$E_{\text{rot}} = \frac{\hbar^2}{2I}J(J+1); \quad J = 0, 1, 2, \ldots$$
$$\psi_{\text{rot}} = \psi_{J,m}(\theta, \phi); \quad m = 0, \pm 1, \pm 2, \ldots, \pm J$$

Clearly, each rotational energy level of a diatomic molecule given by J has a degeneration of $(2J + 1)$.

The rotational spectral term (compare Sect. 8.3.2) is given by:

$$G(J) \equiv \frac{E_{\text{rot}}}{hc} = \frac{\hbar}{4\pi c I}J(J+1) \equiv BJ(J+1) \qquad (11.10)$$

with B—rotational constant.

For example, the rotational constant of a the CO molecule is

$$B_{\text{CO}} \approx 1.9\,\text{cm}^{-1}$$

In a purely rotational spectrum, neither the vibronic nor the electronic energy of the molecule will change. Therefore from (11.6), we have:

$$\omega_{nm} = \frac{E_n - E_m}{\hbar} = \frac{E_{n,\text{rot}} - E_{m,\text{rot}}}{\hbar} \equiv \underbrace{\omega_{\text{rot}}}_{\in \text{FIR}} \qquad (11.11)$$

Then, the transition wavenumber between the rotational energy levels $J = 0$ and $J = 1$ is:

$$\nu = G(1) - G(0) = 2B \approx 3.8\,\text{cm}^{-1}$$

It clearly falls into the far infrared (FIR) spectral range.

Let us have a look on the selection rules. Assuming electric dipole interaction, we have the perturbation operator (6.20) (keep in mind that in a rigid rotor, $r =$ const, $\mathbf{e}$—unit vector parallel to $\mathbf{d}$):

$$\hat{V} = -\mathbf{dE} = -qr\mathbf{eE} = -|\mathbf{d_{perm}}|\mathbf{eE} \tag{11.12}$$

From here:

$$\hat{V} \neq 0 \Rightarrow |\mathbf{d_{perm}}| \neq 0 \tag{11.13}$$

Thus, in order to record a rotation spectrum, the molecule must have a permanent dipole moment $\mathbf{d_{perm}}$. Therefore, molecules like CO and HCl show a pure rotation spectrum, while N_2, H_2, or O_2 do not.

Note that in our approximation, a molecule like HCl is modeled as a two-body system, where two different point masses are connected to each other with a fixed distance. Such a system is also called the dumbbell model of a diatomic molecule (Fig. 11.2, on left).

Concerning ΔJ, so from Sect. 8.5.1, (8.49), we know the selection rule:

$$\Delta J = \pm 1 \tag{11.14}$$

In a pure rotation spectrum, absorption must be accompanied by an increase in the rotation energy, and therefore, we have the selection rule $\Delta J = +1$.

Therefore, when combining (11.10) and (11.11), allowed transition wavenumbers may be calculated according to the recipe:

$$\nu_{J+1,J} = G(J + 1) - G(J)$$

Fig. 11.2 Rotation spectra of diatomic molecules: on left—illustration of the dumbbell model, on right—spectral terms and allowed transitions. Cartoon by Dr. Alexander Stendal. Printed with permission

$$= B(J+1)(J+2) - BJ(J+1) = 2B(J+1);$$
$$J = 0, 1, 2, ... \tag{11.15}$$

(11.15) obviously describes a set of equidistant absorption lines located in the far infrared spectral region. They are visualized by colored arrows in Fig. 11.2 on right.

What about the absorption line intensity? For a single molecule, one would have to calculate the corresponding Einstein coefficients (7.22) in terms of the dumbbell model. This is outside the scope of this course. But for a gas of freely rotating molecules held at temperature T, a crude estimation may be provided on the basis of (6.31). There we assumed for the absorption coefficient of an ensemble of two-level systems:

$$\alpha \propto |\mathbf{d}_{21}|^2 (N_1 - N_2)$$

The simplest model would be to set the matrix element constant and calculate the population difference from the product of energy-level degeneration multiplied with a Boltzmann factor, such that

$$N(J) \propto (2J+1)e^{-\frac{E_{\text{rot}}(J)}{k_B T}} = (2J+1)e^{-\frac{G(J)}{\vartheta}} \tag{11.16}$$

Here,

$$\vartheta \equiv \frac{k_B}{hc} T; \quad \frac{k_B}{hc} = 0.695 \ (\text{cm K})^{-1} \tag{11.16a}$$

Remark A more refined treatment allows it to calculate the absorption coefficient (compare (6.31)) of a gas of rigid rotating diatomic molecules including the transition matrix elements as well as the population difference between the participating quantum levels explicitly. Then, the absorption spectrum may be quantified in terms of the absorption coefficient as follows (11.17):

$$\alpha(v) = N \frac{8v^2 B^2 \Gamma d_{\text{perm}}^2}{3\varepsilon_0 c\hbar\theta} \sum_J \frac{(J+1)^2 \left[e^{-\frac{BJ(J+1)}{\theta}} - e^{-\frac{B(J+1)(J+2)}{\theta}} \right]}{\left[(2B(J+1))^2 - v^2 \right]^2 + 4v^2\Gamma^2} \tag{11.17}$$

Here, the single absorption lines are assumed as Lorentzians with a (homogeneous) linewidth 2Γ, and N is the concentration of molecules. Figure 11.3 shows the absorption coefficient calculated for a gas of fictive molecules with $B = 2 \ \text{cm}^{-1}$, as a function of the wavenumber in cm^{-1}, assuming the temperature 300 K (a), on the left, and 500 K (b), on the right. $\Gamma = 0.2 \ \text{cm}^{-1}$. The absorption coefficient is given in relative units, and the value 1.0 corresponds to the maximum in the 300 K spectrum.

An increase in temperature leads to the appearance of more lines, but the maximum intensity of the lines corresponding to lower J-values decreases because of saturation effects (compare Sect. 6.5.1).

A derivation of (11.17) will be provided in Sect. 14.6.2.

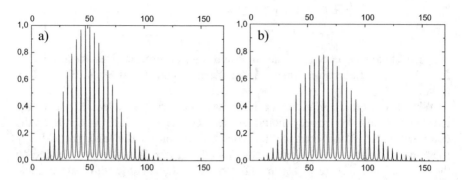

Fig. 11.3 Examples of rotation spectra calculated by means of (11.17)

11.3.3 Molecule Vibration (Diatomic Molecules)

Let us return to (11.9):

$$-\frac{\hbar^2}{2\mu}\frac{d^2\chi}{dx^2} + u(x)\chi = E_{\text{vibr}}\chi$$

In a purely vibronic spectrum without any rotation of the molecule and any change in the electronic state, from (11.6) we have:

$$\omega_{nm} = \frac{E_{n,\text{vibr}} - E_{m,\text{vibr}}}{\hbar} \equiv \underbrace{\omega_{nm,\text{vibr}}}_{\in(\text{NIR}),\,\text{MIR}} = hc\nu_{nm,\text{vibr}} \qquad (11.18)$$

In order to calculate the vibrational transition frequencies, we need knowledge on the vibronic energy levels E_{vibr}. Their position depends on the concrete shape of $u(x)$. Let us discuss two special cases.

<u>Harmonic oscillator:</u>
In the harmonic oscillator approximation, we set:

$$u(x) = \frac{\mu\omega_0^2}{2}x^2 - D; \quad \max\left\{\begin{array}{c} -r_0 \\ -\sqrt{\frac{2D}{\mu\omega_0^2}} \end{array}\right\} \le x \le \sqrt{\frac{2D}{\mu\omega_0^2}}$$

According to Fig. 10.7, the dissociation energy $D = -U_0$. Then, we have the vibrational eigenenergies (compare Sect. 5.5.1)):

$$E_{\text{vibr}} = E_{\text{vibr}}(v) = -D + \hbar\omega_0\left(v + \frac{1}{2}\right); \quad v = 0, 1, 2, ..., \text{int}\left(\frac{D}{\hbar\omega_0} - \frac{1}{2}\right)$$

Here, v is the vibrational quantum number. In molecular spectroscopy, it is common to characterize the vibration state by a quantum number v instead of n, as we did it in the case of the harmonic oscillator.

According to Sect. 5.5.1, for electric dipole interaction, the corresponding selection rule for light absorption is: $\Delta v = +1$

Anharmonic oscillator in terms of the Morse potential:
The anharmonic potentials as shown in Figs. 2.7 or 10.7 may be approximated by the harmonic one in the vicinity of the potential minimum only. Otherwise, we have required:

$$u(r \to \infty) = 0$$
$$u(r \to 0) \to \text{"large"}$$
$$u(r = r_0) = -D \equiv -hc D_e$$

The Morse potential is a satisfacting model potential for an anharmonic oscillator, which has an exact analytical solution of the corresponding Schrödinger equation (compare Sect. 10.5.5). It may be written as (see also Fig. 10.7 and set $D = -U_0$):

$$u(x) = D\left(e^{-2\beta x} - 2e^{-\beta x}\right) \; ; \; x = r - r_0$$

The vibrational terms obtained for the Morse potential are given by (10.31):

$$G(v) = -D_e + v_e\left(v + \frac{1}{2}\right) - v_e x_e\left(v + \frac{1}{2}\right)^2$$

with
$x_e = \frac{v_e}{4D_e} \ll 1$—degree of anharmonicity. If $x_e \neq 0$, the energy levels are no more equidistant.

The appearance of anharmonicity has several consequences to the infrared optical properties of molecules. Namely,

- The vibration wavenumbers corresponding to transitions starting from different states $|v\rangle$ to the corresponding $|v + 1\rangle$ are no more identical.
- The strong selection rule $\Delta v = \pm 1$ breaks down. Instead, vibrational overtone spectra ($|\Delta v| > 1$ are allowed, too. These overtone absorptions may also fall into the NIR spectral region, while the fundamental vibration wavenumber ($v = 0 \to v = 1$) is in the MIR. Therefore, in equations like (11.6), it has explicitly been indicated that vibrational absorption lines may occur in the NIR.
- Possible vibrational transition wavenumbers are defined by:

$$v_{nm} = G(v = n) - G(v = m)$$
$$= v_e(n - m) - v_e x_e\left[\left(n + \frac{1}{2}\right)^2 - \left(m + \frac{1}{2}\right)^2\right]$$
$$= v_e(1 - x_e)(n - m) - v_e x_e\left[n^2 - m^2\right] \qquad (11.19)$$

- Because of the asymmetry of the Morse potential, $<r>$ will now depend on the vibrational state: $<r> = f(v)$. The same applies for $\left\langle \frac{1}{r^2} \right\rangle_{\text{oscillation}}$.

11.3.4 Rotational-Vibrational Spectra (Diatomic Molecules)

In a rotational-vibrational spectrum without any change in the electronic state, from (11.6) we have:

$$\omega_{nm} = \frac{E_{n,\text{vibr}} - E_{m,\text{vibr}} + E_{n,\text{rot}} - E_{m,\text{rot}}}{\hbar} \equiv \omega_{nm,\text{vibr}} + \omega_{nm,\text{rot}}$$

$$= \underbrace{\omega_{nm}}_{\in(\text{NIR}), \text{MIR}} = hc\nu_{nm} \tag{11.20}$$

The transition wavenumber is, in our approximation, simply a sum of the vibrational transition wavenumber according to (11.18) and the rotational transition wavenumber according to (11.11). Let us again discuss two special cases.

Harmonic oscillator:
The full energy of the motion of the nuclei (except translational motion) is:

$$E_C = E_C(v, J) = E_{\text{vibr}}(v) + E_{\text{rot}}(J)$$

$$= -D + \hbar\omega_0\left(v + \frac{1}{2}\right) + \frac{\hbar^2 J(J+1)}{2I}$$

$$- D + hc\left[\nu_0\left(v + \frac{1}{2}\right) + BJ(J+1)\right] \tag{11.21}$$

Here, $\nu_0 = \frac{\omega_0}{2\pi c}$

In a rotational-vibrational spectrum, the absorption of a photon leads to changes in both the vibrational and rotational energies of the molecule.

The selection rules for the rotational-vibrational absorption spectrum of a diatomic molecule obviously are:

Absorption: $\Delta v = 1$; $\Delta J = \pm 1$

This results in absorption wavenumbers:

$$\nu_{v+1, J'=J\pm 1; v, J} = \underbrace{G_{\text{vibr}}(v+1) - G_{\text{vibr}}(v)}_{\equiv \nu_0} + G_{\text{rot}}(J \pm 1) - G_{\text{rot}}(J)$$

$$= \nu_0 + \begin{cases} 2B(J+1); & \Delta J = J' - J = 1; \quad R\text{-branch} \\ -2BJ; & \Delta J = J' - J = -1; \quad P\text{-branch} \end{cases} \tag{11.22}$$

Here, we have assumed that the rotational constant does not depend on the vibrational excitation level. Because of $B \ll \nu_0$, the wavenumber region where the rotational-vibrational spectrum of a gas of diatomic molecules is observed is dominated by ν_0 and thus located in the MIR. The rotational transitions lead to a fine substructure of that vibrational transition. The P-branch, which is formed by rotational transitions with $\Delta J = -1$, is red-shifted with respect to the pure vibrational transition frequency ν_0. The R-branch, which is formed by rotational transitions with $\Delta J = +1$, is blue-shifted with respect to the pure vibrational transition frequency ν_0. Once at room temperature, in a rotational-vibrational spectrum, the thermal occupation of the higher energy level may be neglected, the relative intensity I of the spectral lines may be approximated (compare (6.31)) according to:

$$\alpha \propto |d_{21}|^2 (N_1 - N_2) \propto N_1 \Rightarrow I(J) \propto (2J+1)e^{-\frac{E_{rot}(J)}{k_B T}} = (2J+1)e^{-\frac{G(J)}{\vartheta}}$$

$$(11.23)$$

Figure 11.4 shows a calculated rotational-vibrational spectrum assuming $\nu_0 = 2200\,\text{cm}^{-1}$; $B = 2\,\text{cm}^{-1}$; $T = 300\,\text{K}$; $\Gamma = 0.5\,\text{cm}^{-1}$. The rotational substructure is well resolved.

Figure 11.5 shows the same with $\Gamma = 5\,\text{cm}^{-1}$. Here, the width of the individual absorption lines constituting the spectrum is larger than their wavenumber separation; hence, the rotational substructure of the spectrum is no more resolved. The resulting Doublette structure obtained this way in Fig. 11.5 is called a Bjerrum–Doublette [3]. As it may be estimated from (11.23), the wavenumber difference between the two peaks of a Bjerrum–Doublette is $\Delta\nu \approx \sqrt{8B\vartheta}$. Hence, the spectrum gives access to the temperature of the gas.

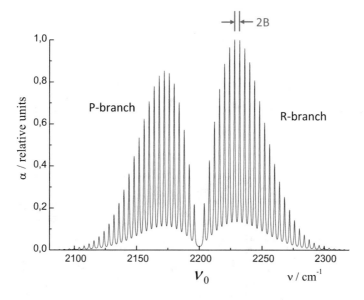

Fig. 11.4 Rotational-vibrational spectrum of a gas of diatomic molecules with $\Gamma < B$

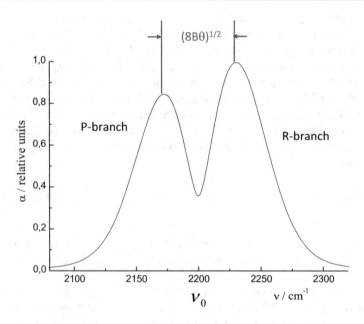

Fig. 11.5 Rotational-vibrational spectrum of a gas of diatomic molecules with $\Gamma > 2B$

If the temperature is known, the analysis of the Bjerrum–Doublette gives access to the mass moment of inertia. Indeed, we find:

$$I = \frac{h\vartheta}{\pi^2 c \Delta v^2} = \frac{4k_B T}{\Delta \omega^2} \text{ (Bjerrum formula)}$$

How to avoid confusing the P- and R-branches? The mnemonic trick is simple: In a purely **r**otational ("R") absorption spectrum, the rotational quantum number J can only *increase* (for a value of 1); hence, $"R" \Leftrightarrow \Delta J = +1$. All other branches are then named in alphabetical order:

Branch	$\Delta J=$	Remark
O	−2	For example, in Raman spectroscopy
P	−1	This paragraph
Q	0	Rotational-vibrational absorption spectrum of polyatomic molecules
R	+1	This paragraph
S	+2	For example, in Raman spectroscopy

Fig. 11.6 Corresponding theoretical infrared spectra. The quadratic dependence of the transition wavenumber on the rotational quantum number is sometimes visualized in terms of a so-called Fortrat parabola [4], pp. 343–353]

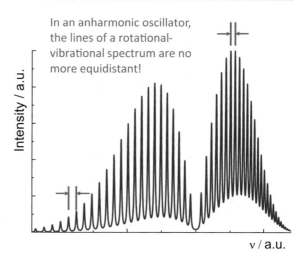

Anharmonic oscillator in terms of the Morse potential:

Because of the asymmetry of the Morse potential, <r> will now depend on the vibrational state: $<r> = f(v)$. The same applies for $\left\langle \frac{1}{r^2} \right\rangle_{\text{oscillation}}$ even when the vibrational excitation level is small. Consequently, $I = I(v)$ and $B = B(v)$. Usually, a larger vibrational quantum number v will correspond to a smaller B. Instead of (11.22), for the rotational-vibrational spectrum, we therefore have:

$$
\begin{aligned}
\nu_{v=n,J'=J\pm1;v=m,J} &= G(v = n) - G(v = m) \\
&\quad + B(v = n)J'(J' + 1) - B(v = m)J(J + 1) \\
&= \nu_{nm} + \begin{cases} J^2[B(v = n) - B(v = m)] \\ \quad + J[3B(v = n) - B(v = m)] \\ \quad + 2B(v = n); \ J' = J + 1 \ (R) \\ J^2[B(v = n) - B(v = m)] \\ \quad - J[B(v = n) + B(v = m)]; \\ J' = J - 1 \ (P) \end{cases}
\end{aligned} \tag{11.24}
$$

The lines within the rotational substructure are no more equidistant, too. Equation (11.24) describes effects of the rotational-vibrational interaction in molecular infrared spectroscopy (Figs. 11.6 and 11.7).

11.3.5 Electronic Transitions

In the general case of an electronic transition, the electronic state changes combined with changes in the vibronic as well as rotational quantum states. Hence, it is now (11.6) that is to be applied in its full beauty.

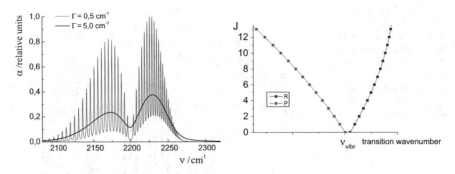

Fig. 11.7 Left: rotational-vibrational spectrum of a gas of diatomic molecules, considering rotational-vibrational interaction with $B(v = n) < B(v = m)$; $n > m$. Right: corresponding Fortrat parabolas. The squares mark the positions of the individual absorption lines defined by (11.24)

For simplicity, in this section we will refrain from explicitly mentioning the rotational movement. What will be called "vibrational" motion here, in fact, has to be understood as the complete complex of rotational and vibrational excitations of the nuclei subsystem, provided, of course, that the molecule is allowed to rotate freely.

Let us start from (11.4a). Correspondingly, any electronic eigenstate m is split into vibrational states of the nuclei motion, numbered by v. Thereby, the interaction potential u depends on the actual electronic configuration.

Let us now consider an electronic quantum transition between two different electronic molecular eigenstates, say $m = 1$ and $n = 2$. Our general assumption was that during the time span relevant for an electronic transition, the coordinates of the nuclei do not change. We have:

$$E_1 = E_{e,1} + E_{c,1}$$
$$E_2 = E_{e,2} + E_{c,2} \qquad\qquad (11.25)$$

Let us regard the case of light absorption and assume that $E_2 > E_1$. So, the process starts from state 1. At low temperatures, it is reasonable to assume that prior to the absorption process, the molecule is in the vibrational ground state ($v_1 = 0$). Then, the expectation value of the nucleus coordinate will be close to the equilibrium value, so that we further assume $r_c = r_{c,0,m=1}$. Once the electron movement is much faster than the nuclei movement, it is further reasonable to assume that during the quantum transition, the nucleus coordinate does not change, but remains fixed. Such a transition is visualized in Fig. 11.8.

From (11.25), it appears now straightforward to calculate the expected transition (absorption) frequencies. One has to expect a series of absorption lines, because the absorption process may end at different vibronic excitation levels in the excited electronic state. But it is important to have a look on the transition matrix elements

Fig. 11.8 Absorption of light in a molecule: Electronic transition. Dashed horizontal lines visualize the vibrational energy levels of the nuclei vibrations

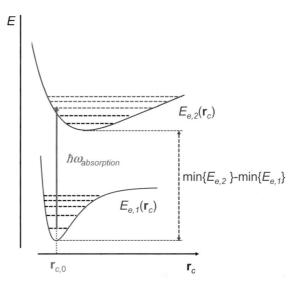

before, because the latter define the selection rules. Some transitions may be characterized by a very small matrix element and are thus not expected to contribute to the absorption spectrum in a significant manner. So, let us write the expression for the dipole operator:

$$\mathbf{d} = \sum_j q_{c,j}\mathbf{r}_{c,j} + \sum_i q_{e,i}\mathbf{r}_{e,i} \tag{11.26}$$

By using the ansatz (10.3) for the wavefunction, the following expression for the transition matrix element is obtained (electric dipole interaction):

$$\int \psi_{e2}^*\psi_{c2}^* \sum q_c\mathbf{r}_c\psi_{e1}\psi_{c1}d\mathbf{r}_c d\mathbf{r}_e + \int \psi_{e2}^*\psi_{c2}^* \sum q_e\mathbf{r}_e\psi_{e1}\psi_{c1}d\mathbf{r}_c d\mathbf{r}_e \neq 0 \tag{11.27}$$

Here, we have skipped the subscripts j, as well as i. In order to have an allowed quantum transition, the mentioned matrix element must be different from zero, as indicated in (11.27). Let us discuss the behavior of the first and second terms in (11.27) separately.

Concerning the first term, the adiabatic approximation requires setting $\mathbf{r}_c$ as constants with respect to the electronic transitions. Hence, the first term in (11.27) may be written as:

$$\int \psi_{c2}^* \sum q_c\mathbf{r}_c\psi_{c1} \left\{ \int \psi_{e2}^*\psi_{e1}d\mathbf{r}_e \right\} d\mathbf{r}_c = 0 \tag{11.27a}$$

This term must be zero because of the orthogonality of the electronic eigenfunctions obtained at the same nucleus coordinates. Therefore, the transition matrix element is controlled by the second term in (11.27) only. It can be written as:

$$\int \psi_{c2}^* \psi_{c1} \left\{ \int \psi_{e2}^* \sum q_e \mathbf{r}_e \psi_{e1} d\mathbf{r}_e \right\} d\mathbf{r}_c \neq 0 \qquad (11.27b)$$

Once the first term in (11.27) is zero, the second must be different from zero in order to make the transition allowed. This requires the purely electronic transition element $\int \psi_{e2}^* \sum q_e \mathbf{r}_e \psi_{e1} d\mathbf{r}_e$ to be different from zero, which defines an electronic selection rule (you may have noticed that we do not explicitly consider spin-related effects in our simplified treatment—these effects are highly essential for calculating correct numerical values of the matrix elements, but not for understanding the general ideas introduced here). And it also requires that there is a spatial overlap of the nuclei wavefunctions ψ_c in the first and second quantum states. This is the essence of the famous Franck–Condon principle. It may suppress the efficiency of certain transitions significantly and gives rise to spectral features in molecular spectra that are sometimes called the Franck–Condon envelope. This necessary spatial overlap of vibrational wavefunctions in the first and second states allows us building an illustrative picture on how electronic transitions in molecules are composed.

And this is what is schematically shown in Fig. 11.8. It shows assumed $E_e(\mathbf{r}_c)$ dependencies (as congruent to $u(\mathbf{r}_c)$) for the first and second electronic states, respectively (in navy). More precisely, the assumed in the figure $E_e(\mathbf{r}_c)$ dependencies rather represent a particular situation characteristic for diatomic molecules, but this makes the approach more illustrative and embodies the same physical principles. Note that the minima of the $E_e(\mathbf{r}_c)$ curves do not necessarily coincide. The nuclei vibrational energy levels in both electronic states are visualized by the horizontal dashed lines. Note that each of those dotted lines corresponds to an energy level as defined by (11.25). Possible electronic quantum transitions may therefore be illustrated by arrows, which start and end at different dashed lines within the participating electronic eigenstates. Once the nucleus coordinate should not change during the transition, those arrows should be *vertical*.

Let us now return to our assumption that the absorption process starts from the lowest vibrational energy level in the lower electronic state. This is visualized in the figure by the red arrow, which starts exactly from that level. As caused by the mismatch of the minima positions of the $E_{e,1}(\mathbf{r}_c)$ and $E_{e,2}(\mathbf{r}_c)$ curves, the arrow will definitely not hit the minimum of the $E_{e,2}(\mathbf{r}_c)$ curve. In the illustration, the absorption process will therefore fail to end at the lower nuclei vibrational energy levels in the second state, because the corresponding nuclei wavefunctions have only a very small spatial overlap. Only higher vibrational levels, as highlighted by the red-dashed horizontal lines, have a chance to contribute to the absorption process in a significant manner. Each of these transitions contributes to the full absorption spectrum at a certain frequency and an intensity, which is controlled by the spatial overlap of the nucleus wavefunctions in (11.27b), thus giving rise to the specific shape of the Franck–Condon envelope and consequently the absorption spectrum. But most of these transitions will correspond to a somewhat higher absorption frequency as it could be expected from the energy difference between the minima of the E_e ($\mathbf{r}_c$) curves only. We therefore have:

$$\min\left(\hbar\omega_{\text{absorption}}\right) \approx \min E_{e,2}(\mathbf{r}_c) - \min E_{e,1}(\mathbf{r}_c) \qquad (11.28)$$

This visual illustration is of course very simplified, and in fact, transitions between black-dashed lines also give their contribution to the absorption. But this contribution is small, because of the rather vanishing spatial overlap of the "lower" vibronic wavefunctions. Thus, when assuming a transition from the lowest vibronic level of the electronic ground state to the lowest vibronic level of the excited electronic state, the overlap integral in (11.27b) will be controlled by the convolution of the exponentially decaying tails of the vibronic wavefunctions, which gives a rather vanishing contribution to the full spectrum.

The moral is that the absorption spectrum of a molecule should show a rather complicated spectral shape, which is caused by the possible excitation of different vibronic levels as the result of light absorption. The relative efficiency of these individual contributions is controlled by expressions like (11.27b). As the consequence, broad and complicated spectra may be observed, such as exemplified in Fig. 11.8. The set of principally possible absorption frequencies follows from (11.6).

Let us finish this section with a short treatment of fluorescence. Imagine that an absorption process like that shown in Fig. 11.8 has taken place. The molecule must now somehow "manage" the excess in energy it has gained as the result of light absorption. Once it must come back to equilibrium with its environment with time, energy relaxation processes will come into play. This is illustrated in Fig. 11.9.

Usually, the first relaxation step is nonradiative: The molecule keeps being electronically excited, but the nucleus vibration relaxes down to lower vibronic energy levels. Note that this process (shown by the dashed arrow in Fig. 11.9) is accompanied by changes in the expectation values for the nucleus coordinates: The nuclei find their new equilibrium positions. When this process is finished, the molecule may (if this is allowed by matrix elements like (11.27), and alternative relaxation processes are significantly slower) return into the electronic ground state by *spontaneous emission of a photon*. This is then observed as fluorescence from the molecule. Again, the Franck–Condon principle holds, so that this transition is again selective with respect to the accessible vibrational energy levels in the electronic ground state: The possibly well-accessible levels are again marked in red, while those definitely corresponding to unlike transitions are black. And it is immediately seen from the geometry of the figure that the red arrow, corresponding to fluorescence, is shorter than the navy arrow, which symbolizes absorption. Moreover, instead of (11.28) we now have:

$$\max(\hbar\omega_{\text{fluorescence}}) \approx \min E_{e,2}(\mathbf{r}_c) - \min E_{e,1}(\mathbf{r}_c) \qquad (11.29)$$

Fig. 11.9 Absorption and
emission of light in a
molecule in accordance with
the Franck–Condon principle

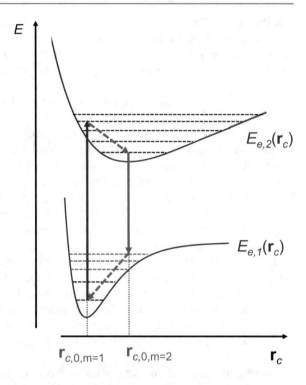

Fig. 11.9 Absorption and emission of light in a molecule in accordance with the Franck–Condon principle

The result is that fluorescence spectra are usually red-shifted with respect to the corresponding absorption spectra.

All in all, absorption and fluorescence spectra deliver a different information. Direct absorption measurements detect the migration of the previously absorbed energy through different relaxation channels. Thus, calorimetric methods measure the sample heating caused by light absorption and thus detect the fraction of the absorbed energy that participates in nonradiative relaxation processes. Fluorescence measurements are exactly complimentary: They detect what is relaxing through radiation.

Fluorescence spectroscopy has its own terminology. In fluorescence spectroscopy, it is common to call the absorption frequency (navy arrow in Fig. 11.9) an excitation frequency. Note that fluorescence spectroscopy may be performed in two different ways: One can fix the absorption (excitation) frequency and measure the corresponding spectrum of emitted light. This way one records an *emission spectrum*. On the other hand, one can measure the fluorescence intensity at a fixed wavelength, while scanning the absorption (excitation) frequency. The then observed spectrum resembles certain features of an absorption spectrum and is called *excitation spectrum*.

11.4 Consistency Considerations

Let us now return to Fig. 11.8, which illustrates an electronic transition in a molecule according to the Franck–Condon principle. From (11.6), we know:

$$\omega_{nm} = \frac{E_n - E_m}{\hbar} \approx \frac{E_{n,\text{electronic}} - E_{m,\text{electronic}} + E_{n,\text{vibr}} - E_{m,\text{vibr}} + E_{n,\text{rot}} - E_{m,\text{rot}}}{\hbar}$$

$$\equiv \underbrace{\omega_{\text{electronic}}}_{\in\ \text{UV/VIS}} + \underbrace{\omega_{\text{vibr}}}_{\in\ \text{(NIR)/MIR}} + \underbrace{\omega_{\text{rot}}}_{\in\ \text{FIR}}$$

How to visualize the purely rotational spectrum and the vibrational–rotational spectrum without changes in the electronic state in such a graph?

An attempt of such an illustration is provided in Fig. 11.10. We again restrict on two electronic states, each of them with a corresponding vibrational–rotational substructure. Let us now explicitly distinguish between vibrational and rotational energy levels of the nuclei motion. This is shown on right of Fig. 11.10. The black horizontal dashed lines visualize the vibrational energy levels, and the green lines the rotational substructure.

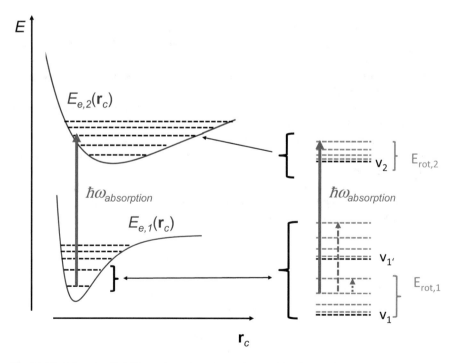

Fig. 11.10 More detailed illustration of an electronic transition in a molecule

Table 11.1 Overview on optical transitions in a molecule

Type	Transition angular frequency	Spectral range
Rotational spectra	$\omega_{nm} \approx \frac{E_{n,\text{rot}} - E_{m,\text{rot}}}{\hbar}$	FIR
Rotational-vibrational spectra	$\omega_{nm} \approx \frac{E_{n,\text{vibr}} - E_{m,\text{vibr}} + E_{n,\text{rot}} - E_{m,\text{rot}}}{\hbar}$	Fundamental vibration: MIR
		Overtone spectra: MIR/NIR
Electronic transitions	$\omega_{nm} \approx \frac{E_{n,\text{electronic}} - E_{m,\text{electronic}} + E_{n,\text{vibr}} - E_{m,\text{vibr}} + E_{n,\text{rot}} - E_{m,\text{rot}}}{\hbar}$	VIS/UV

Obviously, a transition according to (11.6) corresponds to the solid red vertical arrow. In this case, electronic as well as vibrational and rotational states are allowed to change during the transition.

The rotational-vibrational transition according to (11.20) takes place without changes of the electronic state. In this case, only the vibrational and rotational states are allowed to change. This is visualized by the dashed red arrow of medium length.

The purely rotational transition according to (11.11) takes place without changing neither the electronic nor the vibrational state. It is visualized by the shortest vertical arrow on the right of Fig. 11.10. From this image, it might also become clear that these three types of transition give rise to spectral features in quite different spectral regions.

This result has been obtained earlier in terms of our classical estimation of characteristic frequencies in molecule dynamics (Table 2.2). We may now complete this table as demonstrated in Table 11.1. All in all, we recognize that the rather simple illustration presented in Fig. 11.10 contains all discussed types of absorption spectra.

What about our classical estimations from Chap. 2? Concerning vibrational and electronic spectra, everything seems to be consistent with the quantum mechanical results. Concerning the rotational spectra, the situation is a bit more complicated. The classical estimation was:

$$\nu_{\text{rot, classics}} = \frac{\Omega}{2\pi c} = \frac{1}{2\pi c}\sqrt{\frac{k_B T}{I}}$$

In terms of the rotational constant B, this can be written as:

$$\nu_{\text{rot, classics}} = \sqrt{2B\vartheta} \qquad (11.30)$$

where ϑ is given by (11.16a).

Generally, in classics, there is no restriction on the value of the angular momentum, so that the classical estimation obviously corresponds to some average rotation frequency of molecules in a gas held at temperature T. On the contrary, the quantum mechanical expression is (11.15):

$$\nu_{J+1,J} = B(J+1)(J+2) - BJ(J+1) = 2B(J+1); \quad J = 0, 1, 2, \ldots$$

which describes a series of equidistant absorption lines of a different intensity. This is of course something different from the classical case. But note that the spacing between the individual absorption lines is the narrower, the larger the mass moment of inertia of the molecule is. Thus, in the case of sufficiently large (surely not diatomic) molecules we may hope to observe some convergence between the classical and quantum mechanical predictions.

Now, in order to establish a relation between the classical and quantum mechanical expressions, we somehow need to bring the temperature into the play. Let us therefore again regard a gas of diatomic molecules in thermal equilibrium.

In this case, the relative number of molecules with rotational quantum number J will be given by (11.16):

$$N(J) \propto (2J+1)e^{-\frac{E_{rot}(J)}{k_B T}} = (2J+1)e^{-\frac{G(J)}{\vartheta}}$$

Let us find the quantum number J_{max} for which (11.16) reaches its maximum. Formally regarding J as a continuous variable, we have:

$$\frac{\partial}{\partial J} N(J) = 0 \Rightarrow J_{max} = \sqrt{\frac{\vartheta}{2B}} - \frac{1}{2}$$

J_{max} corresponds to the rotational quantum number where $N(J)$ has its maximum. The rotational transition wavenumbers corresponding to the transition $J_{max} \rightarrow J_{max} + 1$ are, according to (11.15), given by: $\nu_{J_{max}+1,J_{max}} = 2B(J_{max}+1) = B + \sqrt{2B\vartheta} = B + \nu_{rot,\,classics}$

Obviously, at room temperature we have

$$\sqrt{2B\vartheta} \gg B \Leftrightarrow I \gg \frac{\hbar^2}{4k_B T}$$

Therefore, $\nu_{J_{max}+1,J_{max}}\big|_{I \gg \frac{\hbar^2}{4k_B T}} \rightarrow \nu_{rot,\,classics}$

We come to the conclusion that for sufficiently large (and heavy) molecules, the rotational transition wavenumber $\nu_{J_{max}+1,J_{max}}$ converges to the classical estimation. This has already been visualized earlier in Fig. 11.5. There, it was illustrated that the spacing between the maxima of the envelope of the rotational-vibrational spectrum of a gas of diatomic molecules held at temperature T is approximately equal to $\sqrt{8B\vartheta} = 2\sqrt{2B\vartheta} = 2\nu_{rot,\,classics}$ (Fig. 11.11). Hence, the classical rotation wavenumber indicates the position of each of the maxima in the Bjerrum–Doublette relative to the purely vibrational transition wavenumber. Again, a result is consistent with the earlier mentioned correspondence principle.

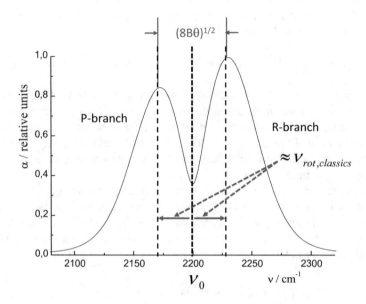

Fig. 11.11 Bjerrum–Doublette and the classical rotation wavenumber

Let us come to a last point. We have discussed (11.6), (11.20), and (11.11) mainly from the viewpoint of the corresponding absorption processes. In Fig. 11.9, we have also introduced spontaneous emission processes, but only from higher excited to lower excited *electronic* states. What about purely vibronic or rotational spontaneous emission spectra?

In fact, such processes are of minor practical relevance. The reason has been discussed in Chap. 7 (compare (7.11)). It is the ω^3 dependence in the expression for the Einstein coefficient A_{21} that makes spontaneous emission processes efficient at VIS/UV frequencies, which are relevant for transitions between different electronic states. Vibrational or rotational transitions, instead, give rise to spectral features in the MIR or FIR. Here, the frequency is much smaller, such that the ω dependence rather leads to a suppression of spontaneous emission processes, because other (nonemissive) relaxation processes are now faster.

11.5   Application to Practical Problems

11.5.1 Example 1: Estimation of Interatomic Distance

We will use Sect. 11.5 to indicate some analytic applications of rotational-vibrational molecular spectroscopy. In the first example, we will make use of the features of the rotation spectra to get information about structural parameters of the molecule.

Let us return to figures like Fig. 11.3 or 11.4. No matter whether we have a purely rotational (Fig. 11.3) or a rotational-vibrational (Fig. 11.4) spectrum recorded, the wavenumber spacing between two resolved adjacent absorption lines is always equal to $2B$. This is even true for the nonrigid rotor at low rotational quantum numbers (compare (11.24) and later Task 11.7.1). Therefore, the measured wavenumber spacing between the rotational lines provides information about the rotational constant and thus, because of (11.10), on the mass moment of inertia of the molecule. In the case of a diatomic molecule, provided that the reduced mass is known, the interatomic distance becomes directly accessible.

11.5.2 Example 2: Estimation of the Temperature

If a gas of molecules is held a temperature T, free rotation is allowed, and the vibrational–rotational spectrum looks like what is presented in Figs. 11.4 or 11.5. Thereby, the wavenumber spacing between the two maxima of the envelope of the spectrum gives access to the product $B\vartheta$. If B is known, the temperature may be estimated or vice versa.

11.5.3 Example 3: Estimation of the Dissociation Energy

This is an example on the use of the Morse potential for interpreting vibrational spectra, without considering the rotational substructure. Thus, from (11.19) we have the vibrational transition wavenumbers for the Morse potential according to:

$$\nu_{nm} = G(v = n) - G(v = m) = \nu_e(1 - x_e)(n - m) - \nu_e x_e\left[n^2 - m^2\right] \quad (11.31)$$

Hence, each transition wavenumber depends on two parameters, namely the characteristic wavenumber ν_e and the anharmonicity x_e. In turn, two measured transition wavenumbers will give access to both ν_e and x_e, and consequently, all other transition frequencies may be predicted by means of (11.19). And, in particular, the dissociation energy may be predicted by means of (10.33). This will be the essence of this section.

Nevertheless, the present task is a bit more tricky. Imagine that you have measured the absorption spectrum of a gas of $(CF_3)_3CH$ molecules. This is no more a diatomic molecule, but the stretching vibration of the C–H group is much higher in frequency than the other normal vibrations of this molecule, such that the C–H stretching vibration may be considered as a vibration of the H atom relative to the rest of the molecule. This way we nevertheless have some similarity with the situation in a diatomic molecule. Assume now that you observe the fundamental transition wavenumber of the stretch vibration of the CH group as $\nu_{1,0}$ = 2992 cm^{-1}. You also register the transition wavenumber corresponding to the first overtone as $\nu_{2,0}$ = 5882 cm^{-1}. From these data, assuming a Morse potential and neglecting any rotations, estimate the energy (in eV) necessary to dissociate a single CH group if the gas is held at room temperature [5].

Provided that the gas is held at room temperature, we may assume that the thermal energy is much smaller than the energy necessary to excite a vibration of the CH group. Hence, prior to the absorption process, the Morse oscillator is certainly in its ground state. The energy necessary to dissociate the CH group starting from the vibrational ground state is different from the classical dissociation energy D (compare Fig. 10.7). Provided that $E_{\text{vibr}} = 0$ corresponds to dissociation (Fig. 10.7), then the energy necessary for dissociation from the vibrational ground state is simply equal to $E_{\text{diss}} = -E_{\text{vibr, ground state}} = -hcG(v = 0)$. From

$$G(v) = -D_e + v_e\left(v + \frac{1}{2}\right) - v_e x_e\left(v + \frac{1}{2}\right)^2$$

we have the ground-state energy as

$$hcG(v = 0) = \left[-D_e + \frac{v_e}{2} - \frac{v_e x_e}{4}\right]hc < 0$$

Once $G = 0$ corresponds to dissociation, the former equation obviously defines the energy $E_{\text{diss}} = -hcG(v = 0) = \left[D_e - \frac{v_e}{2} + \frac{v_e x_e}{4}\right]hc$ necessary for dissociation. From (10.33) $x_e = \frac{v_e}{4D_e}$ we have then:

$$G(v = 0) = -D_e(1 - x_e)^2$$

or $E_{diss} = hcD_e(1 - x_e)^2$
So, one has to find x_e and D_e in order to solve the task.
Let us return to (11.31). When setting $m = 0$, we have:
$v_{n,0} = v_e\left[(1 - x_e)n - x_e n^2\right]$. For two given transition wavenumbers, if $v_{n_1,0}$
and $v_{n_2,0}$, it follows that: $a \equiv \frac{v_{n_2,0}}{v_{n_1,0}} = \frac{[(1-x_e)n_2 - x_e n_2^2]}{[(1-x_e)n_1 - x_e n_1^2]}$ or $x_e = \frac{an_1 - n_2}{a(n_1 + n_1^2) - (n_2 + n_2^2)}$
When setting $n_1 = 1$ (fundamental vibration) and $n_2 = 2$ (first overtone) as well as $a = 5882/2992 \approx 1.966$, we find $x_e \approx 0.0164$.
Then, from $v_{n_1,0} = v_e\left[(1 - x_e)n_1 - x_e n_1^2\right] \Rightarrow v_e \approx 3093.5 \text{ cm}^{-1}$.
We further obtain:

$$x_e = \frac{v_e}{4D_e} \Rightarrow D_e \approx 47156.5 \text{ cm}^{-1}$$

and finally $E_{\text{diss}} = hcD_e(1 - x_e)^2 \approx 5.66\,\text{eV}$
This is by the way a value close to what we have assumed in our classical estimation of molecular vibrational frequencies in Sect. 2.5.1.
I hope these examples are sufficient to provide an idea on the value of infrared spectroscopy for analytic purposes.

11.6 Advanced Material: Polyatomic Molecules

11.6.1 Rotation

In the general case of a polyatomic molecule, instead of the scalar mass moment of inertia, we have the tensor of the mass moment of inertia. The diagonalized tensor may be written as:

$$\mathbf{I} = \begin{pmatrix} I_A & 0 & 0 \\ 0 & I_B & 0 \\ 0 & 0 & I_C \end{pmatrix}$$

Let (ξ, η, ζ) be the main axes of rotation of the molecule. Note that the coordinate system (ξ, η, ζ) is fixed with the molecule. Then, the Hamiltonian is

$$\hat{H}_{rot} = \frac{\hat{L}_\xi^2}{2I_A} + \frac{\hat{L}_\eta^2}{2I_B} + \frac{\hat{L}_\zeta^2}{2I_C}$$

Asymmetric top molecules: In case of an asymmetric top molecule ($I_A \neq I_B \neq I_C \neq I_A$), there is no analytic solution to the corresponding Schrödinger equation. The water molecule H_2O provides an example of an asymmetric top.

Symmetric top molecules: In case of a symmetric top molecule $I_A = I_B \neq I_C$, we have:

$$\begin{aligned}
\hat{H}_{rot} &= \frac{\hat{L}_\xi^2}{2I_A} + \frac{\hat{L}_\eta^2}{2I_A} + \frac{\hat{L}_\zeta^2}{2I_C} - \frac{\hat{L}_\xi^2}{2I_A} + \frac{\hat{L}_\eta^2}{2I_A} + \frac{\hat{L}_\zeta^2}{2I_A} \\
&\quad - \frac{\hat{L}_\zeta^2}{2I_A} + \frac{\hat{L}_\zeta^2}{2I_C} = \frac{\hat{\mathbf{L}}^2}{2I_A} + \left(\frac{1}{I_C} - \frac{1}{I_A} \right) \frac{\hat{L}_\zeta^2}{2}
\end{aligned} \tag{11.32}$$

For example, NH_3 is a symmetric top molecule. And, by the way, the $(CF_3)_3CH$ molecule from Sect. 11.5.3 also represents an example of a symmetric top molecule. From (11.32), we find the energy of rotation of a symmetric top according to [6]:

$$\begin{aligned}
E_{rot} &= \frac{\hbar^2}{2I_A} J(J+1) + \frac{\hbar^2}{2} \left(\frac{1}{I_C} - \frac{1}{I_A} \right) K^2; \quad K = -J, -(J-1), \ldots, J \Rightarrow \\
G_{rot} &= \frac{E_{rot}}{hc} = BJ(J+1) + (A-B)K^2 \\
B &= \frac{\hbar}{4\pi c I_A}; \quad A = \frac{\hbar}{4\pi c I_C}
\end{aligned} \tag{11.33}$$

Symmetric top molecules are subdivided into prolate ($I_A = I_B > I_C$) and oblate ($I_A = I_B < I_C$) top molecules. Thus, $ClCH_3$ is a prolate top, and C_6H_6 an oblate top. Also, all linear molecules are at the same time prolate tops. The degeneration of their rotational energy levels is $2(2J+1)$ @ $K \neq 0$ and $(2J+1)$ @ $K = 0$.

In order to not confuse the criteria for prolate and oblate tops, the simple mnemonic trick is just to remember that linear molecules represent extreme cases of prolate top. Their mass moment of inertia is $0 = I_C < I_A = I_B$.

Remark Note that the wavefunctions of the symmetric top are obtained as common wavefunctions of the operators for $\mathbf{L}^2, L_\xi$, and L_z, where z corresponds to a Cartesian coordinate in the laboratory coordinate system. This topic is beyond the scope of this course, but let us mention that the wavefunctions corresponding to the rotation of the symmetric top $|J, K, M\rangle$ depend on three quantum numbers J, K, and M [7]. Once the rotation energy is independent on M (which denotes the projection of $\mathbf{L}$ on the z-axis), the degeneration of each rotational energy level with respect to M is $2J + 1$. If $K \neq 0$, the rotation energy does not depend on the sign of K; therefore, we have an additional twofold degeneration with respect to K. Hence, the full degeneration of an energy level with $K \neq 0$ is $2(2J + 1)$.

Note that, in contrast to the diatomic molecule, we now have two rotational constants A and B!

Spherical top molecules: In the case of a spherical top molecule, we have:

$$(I_A = I_B = I_C) \Rightarrow E_{rot} = \frac{\hbar^2}{2I_A}J(J + 1);$$

$$G_{rot} = BJ(J + 1); B = \frac{\hbar}{4\pi c I_A}$$

In this case, each energy level is degenerated for $(2J + 1)^2$ times. Examples are provided by SF_6 or CH_4.

For observing a purely rotational absorption spectrum, the permanent electric dipole moment of the molecule must be different from zero. Therefore, exactly spherical top molecules do not show a purely rotational absorption spectrum. Symmetric top molecules may show rotational absorption spectra, and the selection rules are $\Delta J = +1$ and $\Delta K = 0$.

11.6.2 Molecule Vibration

According to Sect. 11.1, the number of vibrational degrees of freedom of a molecule built from a number of N_a atoms is $3N_a$-6 (or $3N_a$-5, if the molecule is linear). A diatomic molecule ($N_a = 2$) is surely linear such that $3N_a$-5 $= 1$. The well-known conclusion is that in a diatomic molecule, we have only one vibrational degree of freedom. If $N_a > 2$, instead of a single vibrational wavenumber ν_e, we have a set of vibrational wavenumbers $\{\nu_{e,j}\}$.

Thus, for example, the CO_2 molecule is linear, and thus, it has $3N_A$-5 $= 4$ vibrational degrees of freedom:

- An asymmetric stretching vibration (IR-active)
- A symmetric stretching vibration (IR-inactive)
- A twofold degenerated bending vibration (IR-active).

Here, the terminus "IR-active" denotes that the intensity of the according absorption line in the IR spectrum is different from zero (i.e., the transition dipole matrix element and the oscillator strength are different from zero). It is rather illustrative that an asymmetric stretching vibration of a molecule like CO_2 forms an oscillating dipole moment. The IR-inactive symmetric stretching vibration cannot be excited by absorption of a photon, because no vibrating dipole moment is induced this way. Instead, the vibration rather leads to a modulation of the *volume* occupied by the molecule, which makes the corresponding transition observable in the so-called Raman spectrum of the molecule (see Sect. 12.5.3). Therefore, this mode is called to be Raman-active. Note that in any centrosymmetric system (such as CO_2), no vibration mode may be both IR-active and Raman-active (rule of mutual exclusion).

In the general case of an anharmonic oscillator (not necessarily Morse), the vibrational term is usually approximated in terms of a series like (11.34):

$$G\left(v_1, v_2, ..., v_{3N_a-6}\right) = \sum_{j=1}^{3N_a-6} v_{e,j}\left(v_j + \frac{1}{2}\right)$$

$$+ \sum_{j=1}^{3N_a-6}\sum_{k=1}^{3N_a-6} C_{j,k}\left(v_j + \frac{1}{2}\right)\left(v_k + \frac{1}{2}\right) + \cdots \quad (11.34)$$

Here, the first term denotes the harmonic contribution to the spectral term, while the second and possible further terms correspond to anharmonic contributions. In addition to pure overtones, where the quantum number of a fixed jth vibration mode changes for a value $\Delta v_j > 1$, (11.34) also allows for combination tone spectra, where more than one vibrational modes become excited as a result of a single photon absorption process.

11.6.3 Rotational-Vibrational Spectra

As a characteristic of rotational-vibrational spectra of polyatomic molecules, so-called Q-branches, corresponding to $\Delta J = 0$, may be allowed. Thus, in a symmetric top, rotational-vibrational transitions are subdivided into those forming the parallel bands (the vibrational dipole moment change is parallel to the molecule symmetry axis) and perpendicular bands (the vibrational dipole moment change is perpendicular to the molecule symmetry axis). The selection rules are:

Parallel band: $\Delta K = 0$; $\Delta J = -1, 0, +1$ (except $K = 0$, then $\Delta J = -1, +1$).
Perpendicular band: $\Delta J = -1, 0, +1$; $\Delta K = -1, +1$.

Hence, every vibrational transition is accompanied by a series of rotational transitions, as defined by the mentioned selection rules. The *relative* intensities $I_{J,K}$ of the individual transitions $J,K \rightarrow J', K'$ may be approximated by:

$$I_{J,K} \approx (2J + 1)g(K)A_{KJ}e^{-\frac{G_{rot}(J,K)}{\vartheta}} \qquad (11.35)$$

Here, $g(K)$ is the degeneracy with respect to K, i.e., $g(K = 0) = 1$ and $g(K \neq 0)$ = 2. G_{rot} is given by (11.33). The A_{KJ} are Hönl–London factors as given in Table 11.2 (compare [8]).

Figures 11.12 and 11.13 show simulated rotational-vibrational spectra of the parallel band of a fictive symmetrical top molecule without and with rotational-vibrational interaction.

These mentioned selection rules seem confusing, and it is therefore prospective to develop an illustration of their origin. Again, these illustrations cannot be regarded as a derivation, and they are neither exact nor complete. Let us start from the simplest case, the diatomic molecule. Let ζ coincide with the molecular axis. Then, $I_c = 0$. In this case,

$$\mathbf{L} = \mathbf{I}\Omega \Rightarrow \mathbf{L} = \begin{pmatrix} L_\xi \\ L_\eta \\ 0 \end{pmatrix}$$

Thus, the vector of the angular momentum is always perpendicular to the symmetry axis ζ. On the other hand, the dipole moment induced by the vibration of the two nuclei $\mathbf{d}_{vibr}$ is parallel to ζ. We thus have: $\mathbf{L} \perp \mathbf{d}_{vibr}$. Let us now come to our illustration (Fig. 11.14). We will try to reproduce the mentioned selection rules by purely geometrical considerations, without any calculation of matrix elements.

Table 11.2 Hönl–London factors

ΔJ	Parallel band	Perpendicular band	
	$\Delta K = 0$	$\Delta K = +1$	$\Delta K = -1$
+1	$A_{KJ} = \frac{(J+1)^2-K^2}{(J+1)(2J+1)}$	$A_{KJ} = \frac{(J+2+K)(J+1+K)}{(J+1)(2J+1)}$	$A_{KJ} = \frac{(J+2-K)(J+1-K)}{(J+1)(2J+1)}$
0	$A_{KJ} = \frac{K^2}{J(J+1)}$	$A_{KJ} = \frac{(J+1+K)(J-K)}{J(J+1)}$	$A_{KJ} = \frac{(J+1-K)(J+K)}{J(J+1)}$
-1	$A_{KJ} = \frac{J^2-K^2}{J(2J+1)}$	$A_{KJ} = \frac{(J-1-K)(J-K)}{J(2J+1)}$	$A_{KJ} = \frac{(J-1+K)(J+K)}{J(2J+1)}$

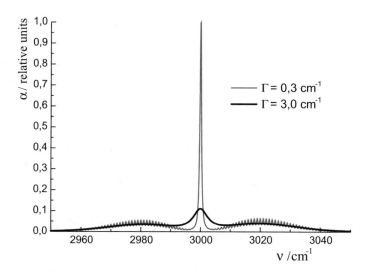

Fig. 11.12 *P*-, *R*-, and *Q*-branch in a simulated rotational-vibrational spectrum of a symmetric top, assuming fixed rotational constants. The rotational lines appear to be equidistant. In this approximation, the position of the Q-band marks the position of the purely vibronic transition

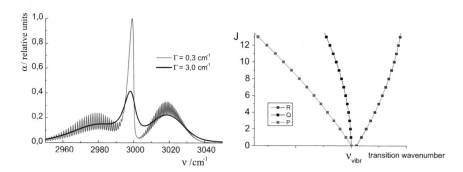

Fig. 11.13 Left: *P*-, *R*-, and *Q*-branches in a simulated rotational-vibrational spectrum of a symmetric top, assuming different rotational constants in the vibrational ground and excited states. The rotational lines are no more equidistant. Therefore, the transitions constituting the Q-band do no more exactly coincide, which leads to a shift as well as to a decrease in intensity of the maximum of the *Q*-band. Right: corresponding Fortrat parabolas.

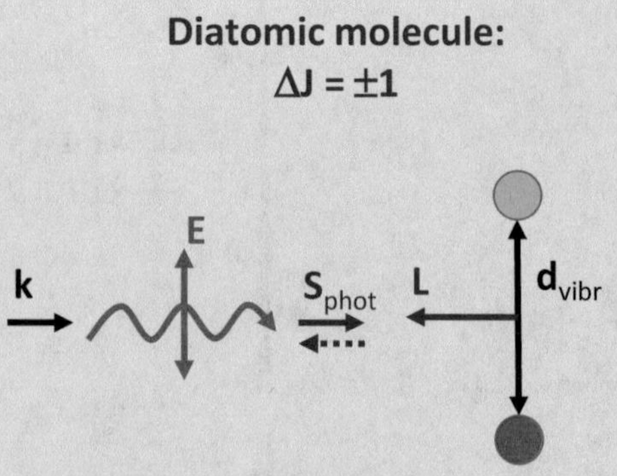

Diatomic molecule:

$$\Delta J = \pm 1$$

Fig. 11.14 Illustration of the physical origin of the selection rule $E \| d_{vibr}$ in a diatomic molecule

It seems reasonable that a linearly polarized light wave is most efficiently interacting with the molecular vibration if $E \| D_{vibr}$. The linear polarization of the light can be represented as the superposition of two circularly polarized waves, corresponding to two possible directions of the photon spin S_{phot}, namely parallel or antiparallel to the wavevector k. Thus, in the circularly polarized wave, E rotates in a plane perpendicular to k, while the molecule rotates in the plane perpendicular to L. Then, interaction between the wave and the molecule rotation should be most effective when these two planes coincide, i.e., when L is parallel or antiparallel to k. In this case, the photon spin is either parallel or antiparallel to L. After absorption of the photon, angular momentum conservation requires that $L' = L \pm S_{phot}$. Once the photon has an integer spin, this results in $J' = J \pm 1$.

Let us now adapt this geometrical picture to the parallel band in a symmetric top molecule (Fig. 11.15, prolate top).

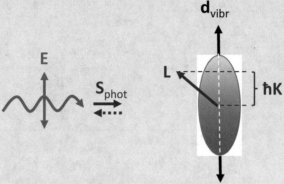

Fig. 11.15 Illustration of the physical origin of the selection rule $\Delta K = 0$ in the parallel band of a symmetric top molecule. The white dashed line indicated the symmetry axis of the top

Once the vibrational dipole moment is here parallel to the molecule axis again, the same argumentation as before leads us to the conclusion that the photon spin is perpendicular to the symmetry axis of the top. Then, it is reasonable assuming that the absorption of the photon cannot change the projection of $\mathbf{L}$ on the molecule axis, such that $\Delta K = 0$ must be expected. On the other hand, $\mathbf{L}$ is no more necessarily perpendicular to the molecule axis. Therefore, photon spin and $\mathbf{L}$ may form different angles, such that the photon spin may have projections parallel as well as perpendicular to $\mathbf{L}$. As a result, $\mathbf{L}$ may change its length ($\Delta J = \pm 1$) and/or its direction ($\Delta J = 0$). Compared to the diatomic molecule, the appearance of a Q-branch appears as the result of the modified angles between the molecule symmetry axis and the angular momentum of the molecule. Only if $K = 0$, we have the same geometrical conditions as in the diatomic molecule, and correspondingly $\Delta J = 0$ becomes forbidden. Consequently, the corresponding Hönl–London factor (Table 11.2) becomes equal to zero.

When accepting this illustration, the discussion of the perpendicular band becomes straightforward (Fig. 11.16):

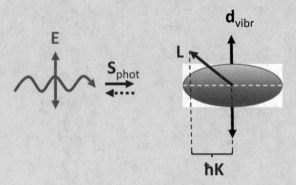

Fig. 11.16 Illustration of the physical origin of the selection rule $\Delta K \neq 0$ in the perpendicular band of a symmetric top molecule. The white dashed line indicated the symmetry axis of the top

Now, in the general case, the photon spin may have a component along the molecule axis, and thus, $\Delta K = \pm 1$ becomes allowed. For J, we have $\Delta J = 0, \pm 1$ again, even if $K = 0$.

11.7   Tasks for Self-check

11.7.1   Multiple-choice test: Mark all answers which seem to you correct!

In the bonding configuration of H_2, the electron spins are	Parallel
	Antiparallel
The mass moment of inertia of a diatomic molecule is typically of the order	10^{-27} kgm^2
	10^{-47} kgm^2
	10^{-67} kgm^2
In an anharmonic oscillator, vibrational overtone absorption	Is strongly forbidden
	May be observed
Purely rotational spectra of diatomic molecules are typically observed in the	Far infrared
	Ultraviolet
In a C_{60} molecule, the number of vibrational degrees of freedom is equal to	180
	186
	174
The rotational-vibrational absorption spectrum of a gas of diatomic molecules may	Show a P-branch
	Be temperature-dependent
	Be observed in the infrared spectral region

11.7.2 True or wrong? Make your decision!

Assertion	True	Wrong
Molecular fingerprint spectra are typically recorded in the middle infrared spectral range		
The oxygen molecule O_2 shows an intense purely rotational absorption spectrum in the FIR		
While fundamental molecular vibration modes lead to absorptions in the MIR, overtones may also contribute to absorption phenomena in the NIR		

11.7.3 Estimate the correction to the rotational terms of a nonrigid rotor compared to the rigid rotor! Assume that the action of the centrifugal force results is a small relative change of the mass moment of inertia.

11.7.4 Consider the electronic ground state of the HCl molecule. In the vicinity of the potential minimum (compare Fig. 2.2), the potential shall be approximated by the harmonic potential:

$$U(x) = \frac{\kappa}{2}x^2; \quad x \equiv r - r_0$$

Assume further that the vibration frequency f is $9 * 10^{13}$ s^{-1}. Estimate the classical vibration amplitude x_0 for the state $v = 1$, and show that it is small compared to the equilibrium distance between the nuclei $r_0 \approx 1.28$ A°! (compare [4], task 9.8).

11.7.5 What is the wavelength of a photon which induces a transition between two adjacent energy levels of a harmonic oscillator with the following properties: $\mu = m_p$; $\kappa = 532$ N/m?

11.7.6 As a result of the anharmonicity of the interatomic potentials, thermal expansion of solids is observed [9]. Assuming an anharmonic model potential near the potential minimum given by (compare Sect. 2.2):

$$U(x) = \frac{\kappa}{2}x^2 - \gamma x^3; \quad \gamma x^3 << \frac{\kappa}{2}x^2; \quad x \equiv r - r_0$$

perform a classical estimation of the average increase in <r> as a function of temperature.

11.7.7. Consider a fullerene C_{60} molecule. Estimate its mass moment of inertia (for rotation around a central axis), assuming the molecule as a hollow sphere with very thin walls and a diameter of 0.71 nm.

11.7.8 Imagine that you have two gas cuvettes, one of them filled with gaseous HBr and the other one with HCl. Unfortunately, the cuvettes have got confused, and you have to find out which of the cuvettes contain which of the gases. You therefore record infrared absorption spectra of both systems at room temperature and observe in both cases a well-resolved rotational substructure of the absorption feature.

You recognize that in cuvette No. 1, the spacing $\Delta \nu$ between two adjacent absorption lines in the rotational spectrum is smaller than in cuvette No. 2. Which of the cuvettes contains HCl and which HBr?

Mass of Br atom	Mass of Cl atom	HBr interatomic distance	HCl interatomic distance
$\approx 80 \; m_p$	$\approx 35 \; m_p$	0.141 nm	0.127 nm

11.7.9 Imagine that you have two gas cuvettes, one of them filled with gaseous HCl and the other one with DCl (D = deuterium). Unfortunately, the cuvettes have got confused, and you have to find out which of the cuvettes contain which of the gases. You therefore record infrared absorption spectra of both systems at room temperature and observe in both cases a typical rotational-vibrational absorption feature.

You recognize that in cuvette No. 1, the absorption feature is centered at a wavenumber around $\nu_1 \approx 3000 \; cm^{-1}$, while in cuvette No 2, the corresponding absorption feature is centered around $\nu_2 \approx 2000 \; cm^{-1}$, which of the cuvettes contains HCl and why? Comment on the ratio between ν_1 and ν_2!

11.7.10 Return to the data provided in Sect. 11.5.3. At which wavenumber you would expect to observe the second overtone of the CH stretching vibration? [5]

References

Specific References

1. O. Stenzel, *Optical Coatings. Material Aspects in Theory and Practice* (Springer, 2014), pp. 227–230
2. А.А. Соколов, И.М. Тернов, В.Ч. Жуковский, Квантовая механика, Москва "Наука" (1979), pp. 427–431
3. V. Kondratyev, *The Structure of Atoms and Molecules*, 2nd printing (Mir Publishers, 1967), Chap. 8
4. W. Demtröder, *Experimentalphysik 3, Atome, Moleküle und Festkörper* (Springer, 2016)
5. О. Штенцель, Н.И.Коротеев, Спектроскопия высоких обертонов квазилокальных колебаний в многоатомных молекулах, Учет вклада колебательно-врачательных переходов, Вестн. Моск. Ун-та. Сер. 3. Физика. Астрономия. Т26, No. 2 (1985), pp. 66–69
6. Л.Д.Ландау, Е.М.Лифшиц Квантовая Механика: Нерелятивистская Теория, Москва "Наука" (1974), pp. 470–473
7. А.С. Давыдов, Квантовая Механика, Москва Физматгиз (1963), pp. 179–186
8. G. Herzberg, *Infrared and Raman Spectra of Polyatomic Molecules* (G. van Nostrand Company 1945), 421–426
9. Ch. Weißmantel, C. Hamann, *Grundlagen der Festkörperphysik* (VEB Deutscher Verlag der Wissenschaften, Berlin, 1979) pp. 317–319

General Literaure

10. G. Herzberg, *Infrared and Raman Spectra of Polyatomic Molecules* (G. van Nostrand Company, 1945)
11. W. Demtröder, (*Atoms, Molecules, and Photons*) (Springer, 2010)
12. L.D. Landau, E.M. Lifshitz, *Quantum Mechanics (Vol. 3 of A Course of Theoretical Physics)* (Pergamon Press, 1965)

Intermezzo: Polarizabilities

"Mühlenstraße in Jena" (Mill street in Jena)

Painting and Photo by Astrid Leiterer, Jena, Germany (www.astrid-art.de). Photo reproduced with permission.

The polarizability concept is extremely useful for quantifying the optical response of microscopic objects. In the continuum theory, it provides a link to macroscopic optical constants by using the concept of local electric fields. The latter may be calculated as the field inside a fictive microscopic cavity created in a macroscopic piece of material. These fictive cavities are assumed to host the microscopic dipoles, and in many manageable models, they are of spherical or ellipsoidal shapes.

From Atoms and Molecules to Continuous Media

12

Abstract

The concept of the microscopic polarizability is introduced to provide a link between microscopic and macroscopic dielectric properties of matter. The relation between static electric polarizability and static dielectric constant is derived in terms of the Clausius–Mossotti equation. The concept is generalized to typical optical frequencies. A semiclassical expression of the microscopic polarizability is derived basing on a perturbative approach.

12.1 Starting Point

This chapter marks some turning point in the internal logical structure of this course. While the previous chapters exclusively dealt with basic features of the optical response of microscopic objects such as atoms or molecules, we will now turn to the optical properties of spatially extended media. We have good motivations for that:

- Exact solutions of the Schrödinger equation are only available for selected quantum systems with a restricted number of degrees of freedom.
- Many technological applications, however, base on functional units that are large enough for being inaccessible to a stringent quantum mechanical treatment.
- Therefore, for accurately predicting the optical properties of a macroscopic system (forward search), we need to have a recipe that links the results of the quantum mechanical treatment of its microscopic constituents to its macroscopic optical properties.

© The Author(s), under exclusive license to Springer Nature Switzerland AG 2022
O. Stenzel, *Light–Matter Interaction*, UNITEXT for Physics,
https://doi.org/10.1007/978-3-030-87144-4_12

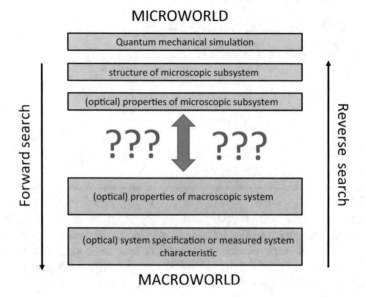

Fig. 12.1   Motivation to Chap. 12

- Also, many optical characterization experiments are performed on rather macro-scopic samples. If you wish to conclude from maybe a measured spectrum on the properties of the microscopic units that build the sample (reverse search), you will again need some recipe that links the macroscopic optical parameters to the microworld.

All in all, what we need is to complete the following scheme (Fig. 12.1).

What we have done so far was to develop a theory about the optical behavior of objects that belong to the microworld. What remains to be done is to create a corresponding picture of light–matter interactions in the macroworld and to connect it to the microscopic description. This way we proceed moving along our logical path defined by the bottom-up approach.

This chapter will therefore introduce and/or recall several concepts useful for interlinking the microworld with the macroworld. We will start by introducing the microscopic polarizability and demonstrate its use for predicting macroscopic dielectric properties of matter in the special case of electrostatics.

But let us firstly distinguish the major states of matter (Table 12.1) in termini of their bulk modulus and their shear modulus:

All of these states of matter are built, on a microscopic level, from atomic nuclei and corresponding electrons. Concerning the electrons, in the following classical picture we will strongly distinguish between free and bound electrons. Bound electrons "belong" to a nucleus: If they are displaced from their equilibrium position, they will suffer a restoring force that pulls them back to the equilibrium

Table 12.1 Basic states of matter

State	Volume	Shape	Bulk modulus	Shear modulus
Solid	Defined	Defined	Large	Large
Liquid	Defined	Undefined	Large	–
Gas	Undefined	Undefined	Moderate	–

Fig. 12.2 Illustration of the concept of bound and free electrons. Cartoon by Dr. Alexander Stendal. Printed with permission

position (Fig. 12.2). Free electrons feel no restoring force, they may freely move through a macroscopic medium and give thus rise to a specific phenomenon called DC electric conduction.

Therefore, with respect to its static electrical conductivity, classic matter may further be classified into electrical insulators (only bound charge carriers) and electrical conductors (with free charge carriers). This is a very rough classification for the moment, we will later see that it makes sense to introduce the class of semiconductors as some separate group with electric properties somewhere in-between insulators and conductors.

With respect to its density, we will distinguish between condensed matter (solids, liquids) and gaseous macroscopic media.

12.2 Physical Idea

For a description of macroscopic optical phenomena, we will entirely use a classical continuum approach. The basic idea is visualized in Fig. 12.3. Imagine a

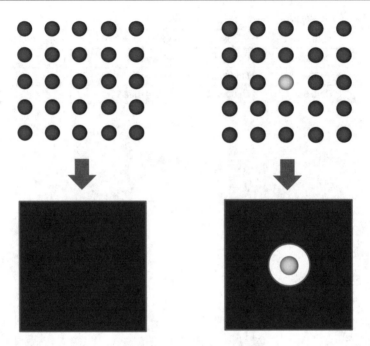

Fig. 12.3   Illustration of the continuum approach. Details see text

macroscopic system built up from a tremendous number of small microscopic objects. This system is shown in Fig. 12.3 left on top. The microscopic building units are shown as blue spheres—for example, this can be molecules in a liquid. Let us now imagine that this system is interacting with an electromagnetic wave such that the characteristic sizes of and distances between the microscopic units are much smaller than the wavelength of the incident wave. Then, the electromagnetic wave will be unable to resolve the spatial structure of the system. In this case, instead of describing the optical properties of the real system, we will replace it by the model of a continuous medium, i.e., a fictive optically homogeneous material (left on bottom) which interacts with the mentioned wave in a nearly identical manner as the real system (left on top) does.

The replacement of the real system by that homogeneous "pap" is the first step in our further description. In the next chapter, we will introduce so-called optical constants, which formally describe the optical properties of such a continuous material.

Of course, we still need to "transport" our knowledge on the optical behavior of microscopic (classical or quantum) objects into that macroscopic picture. This is visualized in the right half of Fig. 12.3. Let us select one of the microscopic units (the bright blue one) from the original system (right on top). And let us further assume, that we are able to calculate the interaction of that microscopic object with an incident electromagnetic field.

The problem is that the selected microscopic unit may "feel" an electric field that is different from that of the external light wave incident to the full system. Indeed, the incident light wave interacts with *all* microscopic units, and not only with the selected one. As a consequence, what the selected unit "feels" is the superposition of the incident field with the fields created by the other microscopic objects. As a result, the selected object feels a specific *local field* E_{loc} that needs to be calculated.

This idea is easily transferred into the continuum language. Let us imagine that the selected unit is placed into a fictive microscopic cavity that has been created within the continuous medium (right on bottom). Then, the field inside this cavity is associated with the mentioned local field. Provided we are able to calculate the field in that cavity, we find the field locally incident to the microscopic unit. Once we know the reaction of the microscopic unit to any incident field, we can calculate the reaction of the microscopic object even when it is part of a macroscopic system. This is the basic idea of interlinking microscopic and macroscopic optical descriptions, and it will find its final expression later in terms of the Clausius–Mossotti and Lorentz–Lorenz equations.

12.3 Theoretical Material

12.3.1 Microscopic Dipoles and Microscopic Polarizability

Let us now imagine an isolated microscopic object, being illuminated by a monochromatic light wave according to $\mathbf{E} = \mathbf{E_0}e^{-i\omega t}$. Its characteristic dimensions should be much smaller than the wavelength of the light, and therefore the field is considered as spatially homogeneous over the extensions of the object. We have to expect that the electric field $\mathbf{E}$ of the incident wave results in the formation of a dipole moment $\mathbf{d}$ in the object. We generally write:

$$\mathbf{d} = \mathbf{d(E)} \tag{12.1}$$

When formally expanding (12.1) into a Tailors series, we obtain:

$$\mathbf{d} = \mathbf{d(E)} = \sum_{j=0}^{\infty} a_j \mathbf{E}^j \equiv \underbrace{\mathbf{d}_{perm}}_{\substack{\text{permanent electric} \\ \text{dipole moment}}} + \underbrace{\varepsilon_0\left(\beta\mathbf{E} + \beta^{(2)}\mathbf{EE} + \beta^{(3)}\mathbf{EEE} + ...\right)}_{\text{induced electric dipole moment}}$$

$$\tag{12.2}$$

Here the field-independent term forms a permanent dipole moment, while the field-dependent terms form the induced dipole moment of the microscopic object. Here, β is the linear (complex) microscopic polarizability of the object. It appears responsible for describing basic linear optical properties of microscopic objects. The $\beta^{(j)}$—terms are called hyperpolarizabilities and are essential for the effects of nonlinear optics (see later Chap. 18). Note that in order to be accurate, in the

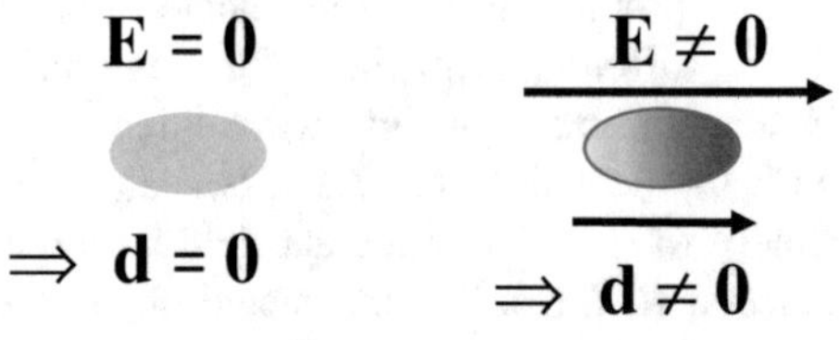

Fig. 12.4 Formation of an
induced dipole

$$E = 0 \qquad\qquad E \neq 0$$

$$\Rightarrow \mathbf{d} = 0 \qquad\qquad \Rightarrow \mathbf{d} \neq 0$$

nonlinear terms, we would have to use an electric field as given by (6.20a), i.e., as $\mathbf{E} = \mathbf{E}_{0,\mathrm{real}} \cos \omega t$ instead of $\mathbf{E} = \mathbf{E}_0 e^{-i\omega t}$. This detail will be topic of Chap. 18, but in essence, according to Euler's formula, we will have to add the conjugate complex term to $\mathbf{E} = \mathbf{E}_0 e^{-i\omega t}$ in (12.2).

Moreover, in the general case, the β-values in (12.2) will be tensors of different order. In order to not overload the formulas, we will prefer a scalar writing of the polarizabilities, i.e., presume certain optical isotropy.

Let us now assume that the object has no permanent electric dipole moment, but forms an induced dipole moment $\mathbf{d}$ as the result of the interaction with the monochromatic electric field $\mathbf{E}$ (Fig. 12.4). For sufficiently weak field strength $\mathbf{E}$ (again see later Chap. 18), the linear term in (12.2) will be the dominating one, and from (12.2) we find (12.3):

$$\mathbf{d} = \varepsilon_0 \beta \mathbf{E} \qquad\qquad (12.3)$$

Generally, β may depend on the angular frequency and may even show resonant behavior (compare Sect. 2.5). A simple and physically transparent classical model for describing the frequency dependence of the linear polarizability is provided by the so-called oscillator model.

12.3.2 The Oscillator Model

The oscillator model derived in the following is very general. It may be applied to the intramolecular motion of nuclei (in infrared spectroscopy) as well as to bound electrons. So that we will simply speak in the following on (microscopic) induced dipole moments, and do not care about their physical origin.

So let us regard the motion of a single charge carrier, which is bound to its equilibrium position ($x = 0$) by an elastic restoring force. An oscillating field may lead to small ($x \ll \lambda$) movements of the charge carriers, thus inducing dipoles that interact with the field. The equation of motion of a single charge carrier with mass m is given by (2.13):

$$q E = q E_0 e^{-i\omega t} = m\ddot{x} + 2\gamma m\dot{x} + m\omega_0^2 x$$

This is the equation for forced oscillations of a damped harmonic oscillator with the eigenfrequency ω_0. From (2.14), we obtain the microscopic dipole moment d

$= qx$:

$$d = \frac{q^2 E}{m} \frac{1}{\omega_0^2 - \omega^2 - 2i\omega\gamma} \tag{12.4}$$

Note that within a macroscopic medium, the acting electric field has to be associated with what we have called the local field in Sect. 12.2. When comparing (12.3) and (12.4), we quickly find the polarizability according to:

$$\beta = \frac{q^2}{\varepsilon_0 m} \frac{1}{\omega_0^2 - \omega^2 - 2i\omega\gamma} \tag{12.5}$$

Equation (12.5) describes a resonant behavior of the microscopic dipole, when the angular frequency ω of the field approaches the angular eigenfrequency of the dipole. In this resonance condition, the interaction between radiation and matter is expected to be most effective.

Note that the linear polarizability has the dimension of the volume. The model that was described here is usually called the Lorentzian oscillator model. More specifically, because of the presence of one single resonance frequency, it is sometimes called the single oscillator model.

Let us therefore generalize the oscillator model to the so-called multioscillator model. Instead of one resonance frequency, we have now a set of M discrete resonance frequencies $\{\omega_{0j}\}$ for each oscillator. The natural generalization of (12.5) becomes:

$$\beta = \frac{q^2}{\varepsilon_0 m} \sum_{j=1}^{M} \frac{f_j}{\omega_{0j}^2 - \omega^2 - 2i\omega\gamma_j} \tag{12.6}$$

The real factor f_j describes the relative strength of the absorption lines according to the different degrees of freedom. If we deal with a molecule, for example, different normal vibrations of the nuclei or various electronic oscillations may thus be taken into account.

12.3.3 Electrostatics of Dielectric Media: Macroscopic Polarization of the Medium

Let us now turn to electrostatics, i.e., the reaction of the medium to a constant electric field. In insulators, free electrons are absent, and an applied static electric field does cause electric current, but rather the formation of microscopic dipole moments d. Note that in terms of the oscillator model, the static polarizability

$$\beta_{stat} = \beta(\omega = 0)$$

according to (12.5) or (12.6) is always real and larger than zero. Therefore, according to (12.3), a static microscopic dipole moment will be induced whenever an electric field is applied.

The dipole moment per unit volume forms the macroscopic polarization vector **P**:

$$\mathbf{P} = \frac{\sum_j \mathbf{d}_j}{V} = N \langle \mathbf{d} \rangle \tag{12.7}$$

N is here the concentration. The displacement vector **D** is defined as:

$$\mathbf{D} = \varepsilon_0 \mathbf{E} + \mathbf{P} \tag{12.8}$$

As it is known from courses on electricity, at a surface or interface between two media, the normal component of the displacement vector is continuous.

Let us now consider the interface between vacuum and a macroscopic dielectric (Fig. 12.5). In vacuum, we apply an external electric field $\mathbf{E}_0$, while the polarization of vacuum is naturally equal to zero. At the interface to the dielectric, the induction of static dipoles in the dielectric ($\mathbf{P} \neq \mathbf{0}$) results in the formation of surface charges. Therefore, the electric field **E** in the dielectric is different from the external electric field $\mathbf{E}_0$. This is illustrated in Fig. 12.5.

The continuity of the induction vector at the interface results in:

$$D_\perp = \underbrace{\varepsilon_0 E_{0,\perp}}_{\substack{\text{external medium} \\ \text{(here vacuum)}}} = \underbrace{\varepsilon_0 E_\perp + P_\perp}_{\text{in the dielectric}} \tag{12.9}$$

Let us now consider that, in analogy to (12.3), the induced polarization **P** is proportional to the electric field in the dielectric. Once **P** is, according to (12.7), some ensemble average, it makes sense to associate **E** with the average field in the medium, which we now consider as a continuum. We therefore write:

$$P = \varepsilon_0 \chi_{\text{stat}} E \Rightarrow \mathbf{D} = \varepsilon_0 \mathbf{E} + \mathbf{P} = \varepsilon_0 \varepsilon_{\text{stat}} \mathbf{E}; \quad \varepsilon_{\text{stat}} = 1 + \chi_{\text{stat}} > 1 \tag{12.10}$$

Fig. 12.5 Formation of uncompensated surface charges at the surface of a polarized dielectric. Note that in the bulk of the dielectric, positive and negative charges compensate each other

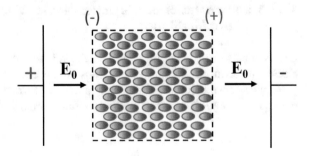

This way we have introduced the (sometime called *relative*) static dielectric constant ε_{stat}. The parameter χ_{stat} is called the (linear) static dielectric susceptibility of the dielectric. Consequently, when taking (12.9) into account, we find

$$\Rightarrow E_{0,\perp} > E_\perp$$

In the medium, the electric field is (in average) smaller than outside. The physical reason is transparent and obvious from Fig. 12.5: It is due to the formation of an uncompensated surface charge formed as a result of the polarization of the medium. The static dielectric constant ε_{stat} turns out to be a function of the polarizability of the microscopic units forming the condensed system and their concentration N, as it will be shown in Sect. 12.3.4.

12.3.4 Clausius–Mossotti Equation

Let us assume a system of identical induced microscopic dipoles. According to (12.7), the macroscopic polarization may be written as:

$$\mathbf{P} = N\mathbf{d} \tag{12.11}$$

The induced microscopic dipole moment itself may be expressed through the microscopic polarizability β:

$$d = \varepsilon_0 \beta_{stat} E_{loc} \Rightarrow P = N\varepsilon_0 \beta_{stat} E_{loc} \tag{12.12}$$

On the other hand, in our continuum approach, we have (12.10):

$$P = \varepsilon_0 \chi_{stat} E = \varepsilon_0 (\varepsilon_{stat} - 1) E \tag{12.13}$$

Here we have assumed isotropy of the continuous medium, such that we refrain from the vector character of field and polarization.

The problem is as follows: (12.13) describes the macroscopic response of the medium, and the electric field fixed in (12.5) is the average field in that medium. It is formed from the external field and the field of all the dipoles in the medium. On the contrary, (12.12) describes a microscopic dipole moment, and the field is the microscopic (or local) field acting on the selected dipole. The question is, whether or not these fields are identical.

In the general case, these fields are different, and the aim of this section is to derive an equation that allows us to calculate the microscopic (local) field for the special case of optically isotropic materials.

Let us regard a single induced dipole in the medium. The field acting on the dipole is built from two constituents: the external field and the field caused by all other dipoles, except the considered one (Fig. 12.6 on top). Of course,

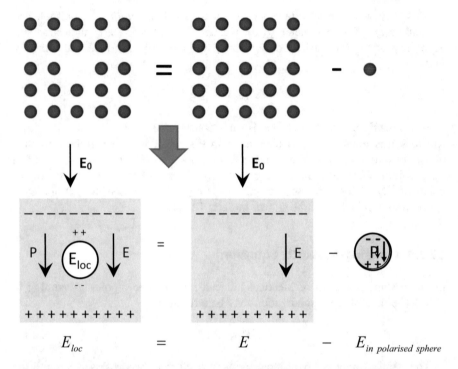

Fig. 12.6 Idea of calculating the local field by means of the superposition principle

nobody would start the calculation from the external field, subsequently adding the response of 10^{23} dipoles. Instead, we will make use of the superposition principle. We will subtract the field of our regarded dipole from the average field in the medium, and this way we can find the field that is acting on the selected dipole itself.

In continuum electrostatics, this calculation is easy to perform, replacing the dipole by a small polarized sphere (in accordance with the assumed isotropy).

In Fig. 12.6 on bottom, our approach is visualized. On the left, we have the continuous medium with a small microscopic spherical cavity inside. The field inside this cavity corresponds to the mentioned microscopic (or local) field, because it may be considered as the field in the compact medium less the field in a homogeneously polarized sphere.

This allows calculating the local field in an explicit manner. When postulating a homogeneous polarization in the continuous medium, the field E must also be homogeneous, and we have:

$$E_{\mathrm{loc}} = E - E_{\mathrm{sphere}} \tag{12.14}$$

As the field in a homogeneously polarized sphere is equal to $-P/3\varepsilon_0$ (compare task 12.7.1), from (12.14) we obtain:

$$E_{loc} = E + \frac{P}{3\varepsilon_0} \tag{12.14a}$$

When taking (12.13) into account, that results in:

$$E_{loc} = \frac{\varepsilon_{stat} + 2}{3} E \tag{12.15}$$

Combining (12.12), (12.13) and (12.15), we finally obtain:

$$P = \varepsilon_0(\varepsilon_{stat} - 1)E = N\varepsilon_0\beta_{stat}E_{loc} = N\varepsilon_0\beta_{stat}\frac{\varepsilon_e + 2}{3}E$$

$$\frac{N\beta_{stat}}{3} = \frac{\varepsilon_{stat} - 1}{\varepsilon_{stat} + 2} \quad or \quad \varepsilon_{stat} = 1 + \frac{N\beta_{stat}}{1 - \frac{N\beta_{stat}}{3}} \tag{12.16}$$

For small concentrations ($N \to 0$), which is valid for diluted gases, from (12.16) we have:

$$\varepsilon_{stat}|_{N \to 0} \to 1 + N\beta_{stat} \tag{12.16a}$$

(12.16) is called the Clausius–Mossotti Equation. It establishes the necessary relation between macro-and microworlds by relating the static dielectric constant to the static microscopic polarizability and vice versa. This way measurements on a macroscopic sample may give information on microscopic parameters, such as the microscopic polarizability.

Some important relations in electrostatics of dielectric media are summarized in Table 12.2.

Table 12.2 Survey of important relations

Physical quantity (isotropic dielectrics in electrostatics only)	In vacuum	In condensed medium
Dielectric susceptibility	$\chi_{stat} = 0$	$\chi_{stat} > 0$
Dielectric constant $\varepsilon_{stat} = 1 + \chi_{stat}$	$\varepsilon_{stat} = 1$	$\varepsilon_{stat} > 1$
Polarization vector	$\mathbf{P} = 0$	$\mathbf{P} = \varepsilon_0\chi_{stat}\mathbf{E} = \varepsilon_0(\varepsilon_{stat} - 1)\mathbf{E}$
Electric displacement (induction)	$\mathbf{D} = \varepsilon_0\mathbf{E}$	$\mathbf{D} = \varepsilon_0\varepsilon_{stat}\mathbf{E}$
Local electric field (spherical symmetry)	$\mathbf{E}_{loc} = \mathbf{E}$	$\mathbf{E}_{loc} = \frac{\varepsilon_{stat}+2}{3}\mathbf{E}$
Boundary conditions at an interface between media 1 and 2	$D_{\perp 1} = D_{\perp 2}; E_{\text{II}1} = E_{\text{II}2}$	

12.3.5 Electrical Conductors

Let us now turn to the effects caused by free electrons. The physical assumption in the Drude theory of electric conduction is that the movement of the electrons is dominated by two contributions: an isotropic chaotic thermal movement that does not contribute to electric charge transport, and a drift movement that gives rise to macroscopic electric charge transport, i.e., to electric conduction. In a direct current, which results from a time-independent electric field, the drift velocity is constant, such that all forces that might influence the drift velocity sum up to zero. Our assumption was that there is no restoring force acting on free charge carriers (here electrons), such that in analogy to (2.13) we now write:

$$\sum \mathbf{F} = q\mathbf{E} - \frac{m_e}{\tau}\mathbf{v}_D = -e\mathbf{E} - \frac{m_e}{\tau}\mathbf{v}_D = \mathbf{0} \qquad (12.17)$$

In analogy to Sect. 2.3.3, τ corresponds to some characteristic decay time. In the present physical situation, it is associated with some average time between two collisions of conduction electrons with atomic cores. From (12.17) we find:

$$-\frac{e}{m_e}\tau\,\mathbf{E} = \mathbf{v}_D$$

The term $\frac{e}{m_e}\tau$ defines the mobility of the electron in a classical language. When introducing the current density $\mathbf{j}$ via:

$$\mathbf{j} = q N_e \mathbf{v}_D$$

With N_e—carrier concentration; we find:

$$\mathbf{j} = N_e \frac{e^2}{m_e}\tau\,\mathbf{E} \equiv \sigma_{stat}\mathbf{E} \qquad (12.18)$$

where we have introduced the static electrical conductivity σ_{stat}:

$$\sigma_{stat} = N_e \frac{e^2}{m_e}\tau \qquad (12.19)$$

Note that the static conductivity depends on the square of the charge of the charge carriers. Therefore, measurements of the conductivity do not provide information on whether the charges are positive or negative.

What about a temperature dependence of the conductivity? It is reasonable expecting that an increase in temperature leads to a reduction of the collision time τ, and thus, according to (12.19), to a decrease in electric conductivity. This is the situation typically observed in metals. In so-called semiconductors, however, the concentration of free charge carriers N_e tends to strongly increase when the temperature is increased. This effect usually overcompensates the decrease in τ, such

Fig. 12.7 Faster than a conduction electron: The Roman snail

that in a semiconductor, the conductivity according to (12.19) tends to increase with increasing temperature. Finally, from (12.18) we obtain:

$$j = \sigma_{\text{stat}} E = \frac{I}{A} = \sigma_{\text{stat}} \frac{U}{l} \Rightarrow \frac{U}{I} = \frac{l}{A\sigma_{\text{stat}}} = R = \frac{l}{A}\rho \Rightarrow \frac{1}{\sigma_{\text{stat}}} = \rho$$

We thus find Ohm's law with R—resistance, I—current, U—voltage, ρ—specific resistance.

Remark Note that in many practically relevant situations, the drift velocity of electrons in an electric conductor is much smaller than their thermal velocity (compare task 12.7.2). It is comparable to the propagation velocity of a snail (see Fig. 12.7). The Roman snail shown in this figure moves with a rather high velocity of three meters per hour, which even exceeds the typical electron drift velocities.

12.4 Consistency Considerations

<u>Microscopic objects</u>: Let us start with some considerations on what we are calling a microscopic polarizability. The term "microscopic" indicates that the polarizing object is rather "small". But what is meant with "small" in this connection? In other words: What is the reference length, that might be considered as "large" compared to the spatial extension of the "small" object?

In many standard textbooks, the polarizability is introduced as some atomic or molecular polarizability, thus describing the specific response of single atoms or molecules to an electric field. Atoms and molecules are (sub-)nanometer-sized objects (Table 1.1) and are clearly associated with the microworld. In other branches, such as cluster physics, the polarizability concept is successfully applied to much larger units, namely clusters with sizes up to at least 10–20 nm. It would clearly be misleading to speak on atomic or molecular polarizabilities in that context. For that reason, dear reader, I prefer the terminus of a "microscopic polarizability".

Of course, this terminus may be misinterpreted as well. When associating "microscopic" with "micrometer-sized", we will clearly cause confusion. In fact, in our treatment, we use the terminus "microscopic polarizability" as a parameter characterizing the properties of objects with *spatial extensions much smaller than the wavelength* of the light which interacts with the object. Thus, when speaking on photonics in practical applications, the wavelength of the light is usually larger than 100 nm, and object sizes of around 20 nm seem to represent the upper limit of the applicability of that concept. On the other hand, the application of the polarizability in our understanding may become problematic even when being applied to atoms, namely when they interact with X-rays.

Thus, it is the wavelength of the light that provides the characteristic length, which should be large compared to the spatial extensions of the polarizing object. We have to pay a price for that: in electrostatics, the "wavelength" is infinitely large, and there is no upper limit for the applicability of our polarizability concept. Indeed, when calculating the local field in a dielectric (Fig. 12.6), no assumption was made on the diameter on the fictive cavity that hosts the dipole. In fact, in electrostatics, such cavities may be rather large. Consequently, the static polarizability of objects with arbitrary size can be calculated this way, and we will demonstrate this in Sect. 12.5.1. In such cases, we will simply speak on a polarizability, and skip the terminus "microscopic".

In optics, the wavelength has a finite value, and in fact the electric field is no more spatially homogeneous. Therefore, in order to make use of equations like (12.15) or (12.16) at optical frequencies, we must require that the diameter of the "fictive sphere" is much smaller that the wavelength. Once the fictive sphere hosts the microscopic dipole, the latter must also be much smaller than the wavelength, which is consistent with our understanding of the "microscopic" polarizability.

Thus, although the field in a light wave is rapidly oscillating, in application to sufficiently small objects, we make use of the model case of a spatially homogeneous field—exactly like in electrostatics. This approach is called the quasistatic approximation.

Shape of the fictive cavity: In Sect. 12.3.4, we have calculated the local field by creating a fictive cavity in the medium, while the cavity was assumed to be of spherical shape. This assumption results in expression (12.15). But why a spherical shape? And what about other shapes?

A sphere is something isotropic, and it makes sense to apply this approach to isotropic materials. In anisotropic materials, other cavity shapes may be reasonable. The point is that other cavity shapes will result in other expressions for the local field. Let us shortly have a look at Fig. 12.8 to understand the physical reason.

In the picture on the left of Fig. 12.8, we find the familiar situation of a spherical cavity. The difference between $\mathbf{E}$ and $\mathbf{E}_{\text{loc}}$ is explained by the presence of surface charges at the border of the cavity. We have:

Spherical cavity: $E_{\text{loc}} = \frac{\varepsilon_{\text{stat}} + 2}{3} E$.

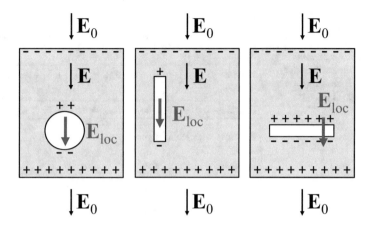

Fig. 12.8 Illustration of the physical origin for the dependence of the local field on the cavity shape

Let us now have a look at the situation shown in the picture on right. It shows a pancake-shaped cavity with an axis parallel to the applied field. The effect of the surface charges should be even stronger that in the case of a spherical cavity. In the extreme case, the charges at the border of the cavity completely compensate those at the outer borders of the whole macroscopic object. We therefore have:

Pancake: $E_{\text{loc}} = E_0 = \varepsilon_{\text{stat}} E$.

Let us now turn to the picture in the middle. It shows a needle-like cavity with an axis parallel to the field. The effect of the surface charges is rather negligible here. Instead, because of the continuity of the tangential components of the electric field (Table 12.2 the row on bottom), we have:

Needle: $E_{\text{loc}} = E$.

Thus, because of the differences in the relevance of surface charges, the local field in a cavity will be strongly dependent on the assumed cavity shape.

The different expressions for the local field may be condensed into one expression, namely:

$$E_{\text{loc}} = [1 + (\varepsilon_{\text{stat}} - 1)L]E$$

where L is a so-called depolarization factor. The previously discussed particular cases correspond to L-values of $L = 1/3$ (sphere), $L = 1$ (pancake), and $L = 0$ (needle).

Table 12.3 summarizes important values for depolarization factors in different situations.

But what about conduction electrons? In Sect. 12.3.5 (electrostatics of electric conductors), we did not care about local fields at all. The reason is simple to be understood at least in a classical picture. Indeed, a free electron is allowed to travel over large distances in the material and therefore rather feels the average

Table 12.3 Depolarization factors L. For completeness, the general expression for calculating L for an ellipsoid along the three main axes l_a, l_b, and l_c is included (for a derivation see [1])

Type of cavity	E parallel to the cavity axis	E perpendicular to the cavity axis
Sphere	$L = \frac{1}{3}$; $E_{loc} = \frac{\varepsilon+2}{3}E$	
Needle	$L = 0$; $E_{loc} = E$	$L = \frac{1}{2}$; $E_{loc} = \frac{\varepsilon+1}{2}E$
Pancake	$L = 1$; $E_{loc} = \varepsilon E$	$L = 0$; $E_{loc} = E$
Ellipsoid with main axes l_a, l_b, l_c	$L_\xi = $ $\frac{l_a l_b l_c}{2} \int_0^\infty \dfrac{ds}{\left(s+l_\xi^2\right)\sqrt{(s+l_a^2)(s+l_b^2)(s+l_c^2)}}$; $\xi = a, b, c$ $E_{loc} = \left[1 + (\varepsilon - 1)L_\xi\right]E$; $L_a + L_b + L_c = 1$ $\Rightarrow N\beta = \frac{\varepsilon-1}{1+(\varepsilon-1)L}$ or $\varepsilon = 1 + \frac{N\beta}{1-N\beta L}$	

macroscopic field **E** than a local field. This idea is nevertheless also consistent with the local field picture: Just imagine that the electron drifts along the direction of the field. Then, according to Table 12.3, its movement appears to be confined in a needle-like cavity, hence $L = 0$ and $E_{loc} = E$.

Note in this connection, that (12.16a) may be obtained in different ways. It immediately follows from (12.16) in the limiting case $N \to 0$. But is also obtained for an arbitrary concentration, just assuming $L = 0$. Indeed, assuming $L = 0$; $E_{loc} = E$, we have:

$$P = \varepsilon_0(\varepsilon_{\text{stat}} - 1)E = N\varepsilon_0\beta_{\text{stat}}E_{\text{loc}} = N\varepsilon_0\beta_{\text{stat}}E \Rightarrow \varepsilon_{\text{stat}} = 1 + N\beta_{\text{stat}}$$

Oscillating fields: Of course, our primary goal is to apply this stuff not to electrostatics, but to optics. The good news is, that in application to dielectrics, the structure of our expressions will not change as long as we are able to make use of the quasistatic approximation. Therefore, in Table 12.3, the subscript "stat" has already been skipped in all expressions. Of course, so far we have not introduced an ε valid at optical frequencies, but that will be topic of Chap. 13.

And concerning bulk metals? We will simplify the situation even more. Once in an oscillating field, the free electrons are expected to perform an oscillating movement as well, we will tackle them as oscillating dipoles and finally include their response into the polarization vector and the not yet introduced ε valid at optical frequencies. This will be done in Chap. 14. Note that the thus introduced induced dipole moment of a free electron formally becomes infinitely large at $\omega \to 0$.

Semiclassical considerations: So far we have performed a purely classical discussion of the polarizability concept. A particular feature of this treatment is, that in (12.12), we have implicitly assumed that the polarizability of an isolated single dipole coincides with the polarizability of a dipole which is part of a macroscopic system. In general, this assumption is obviously misleading. When taking the polarizability of a single silicon atom, you will never be able to predict the properties of a silicon crystal with a given density simply by applying the Clausius–Mossotti Eq. (12.16). The reason is, that in the process of covalent bonding of the silicon atoms when forming a crystal, new electronic wavefunctions are formed (compare Chap. 10 and later Chap. 16), and they will certainly result in polarizing properties that differ from those of the isolated atom. Therefore, our simple Clausius-Mossotti approach is applicable for any processes where dipoles are physically mixed, without changing the electronic properties of the individual dipoles. As soon as chemical bonding comes into play, the approach presented so far is insufficient.

This rises the question on how the polarizability of a quantum system may be calculated. A corresponding sample calculation will be presented in Sect. 12.6. The basic idea is that the calculation of $\mathbf{d}$ by solving the classical Newton's equation of motion (2.13) will be replaced by calculating the quantum mechanical expectation value $\langle \mathbf{d} \rangle$. As a result, we will see that the properties of the electronic wavefunctions have direct impact on the polarizability of the quantum system, as we have it expected.

12.5 Application to Practical Problems

12.5.1 The Static Polarizability of a Dielectric Sphere in Vacuum

Let us for a moment continue with the classical description of polarizabilities and assume the model task of a single dielectric sphere polarizing in an external homogenous static electric field. What is the polarizability of that sphere?

The problem is easily solved in terms of Eq. (12.16). We have:

$$\frac{N\beta_{\text{stat}}}{3} = \frac{\varepsilon_{\text{stat}} - 1}{\varepsilon_{\text{stat}} + 2}$$

Once we deal with a single sphere, the number of dipoles is 1, and the occupied volume $\frac{4\pi}{3}r^3$, while r is the radius of the sphere. Therefore,

$$N = \frac{1}{V} = \frac{1}{\frac{4\pi}{3}r^3} \Rightarrow \frac{N\beta_{\text{stat}}}{3} = \frac{\beta_{\text{stat}}}{4\pi r^3} = \frac{\varepsilon_{\text{stat}} - 1}{\varepsilon_{\text{stat}} + 2} \Rightarrow \beta_{\text{stat}} = 4\pi r^3 \frac{\varepsilon_{\text{stat}} - 1}{\varepsilon_{\text{stat}} + 2}$$

$$(12.20)$$

We thus come to the conclusion that the static polarizability of a dielectric sphere is of the order of its volume.

The case of a metal sphere is also easy to handle. The static polarizability of a single free electron obviously becomes imaginary and approaches infinity by absolute value, as easily seen from (12.5) when setting $\omega_0 = 0$. Then, according to (12.16a), for the static "dielectric constant" of a metal we find: $\varepsilon_{\text{stat}} \to i\infty$. Consequently, for a metal sphere, $\beta_{\text{stat}} = 4\pi r^3$. Thus, even for a metal sphere, the static polarizability turns out be of the order of the volume of the sphere.

> As a rule of thumb, we therefore may state: The larger the volume of the assumed microscopic object is, the larger its static polarizability will be.

Note that throughout our derivation, no restriction on the radius of the sphere was introduced. In application to atomic dimensions, we obtain a classical estimate of the static polarizability of an atom according to:

$$\beta_{\text{stat, atom}} = 4\pi r_{\text{atom}}^3 = 3V_{\text{atom}}$$

This expression naturally coincides with the polarizability obtained in the early Thomson's model of an atom [2].

12.5.2 The Static Dielectric Constant of an Ensemble of Permanent Dipole

In gases or liquids, permanent microscopic dipoles may cause an induced macroscopic polarization because of the formation of a preferred dipole orientation.

Let us assume a material that is built from permanent microscopic electric dipoles. The permanent dipoles are allowed to rotate with some damping. This is the typical situation in a gas or liquid built from polar molecules (e.g., water). When no external electric field is applied, the stochastic thermally activated movement of the dipoles will not be able to create a macroscopic polarization. However, in an external electric field, the dipoles will more or less align with the field, creating a resulting induced macroscopic polarization.

The energy of a permanent microscopic dipole $\mathbf{d}_{\text{perm}}$ in a local field $\mathbf{E}_{\text{loc}}$ may be given as:

$$U = -\mathbf{d}_{\text{perm}}\mathbf{E}_{\text{loc}} = -d_{\text{perm}}E_{\text{loc}}\cos\theta = U(\theta)$$

where θ is the angle formed between the dipole moment and the field. In the case of thermal equilibrium and weak electric fields, the probability density $w(\theta)$ is:

$$w(\theta) \propto e^{-\frac{U(\theta)}{k_B T}} \approx 1 - \frac{U(\theta)}{k_B T} = 1 + \frac{d_{\text{perm}}E_{\text{loc}}\cos\theta}{k_B T}$$

This way we postulate here that thermal effects are dominating over the electric dipole interaction. Our derivation is thus only valid for sufficiently large temperatures such that the proposed linear approximation of the Boltzmann factor makes sense.

Let us now remark that the classical average of the dipole moment may be written as $\langle d \rangle = d_{perm} < cos\theta >$. We further find:

$$\langle \cos\theta \rangle = \frac{\int_0^\pi \cos\theta \, w(\theta) \sin\theta d\theta}{\int_0^\pi w(\theta) \sin\theta d\theta} = \frac{\int_0^\pi \cos\theta \left(1 + \frac{d_{perm} E_{loc} \cos\theta}{k_B T}\right) \sin\theta d\theta}{\int_0^\pi \left(1 + \frac{d_{perm} E_{loc} \cos\theta}{k_B T}\right) \sin\theta d\theta} = \frac{1}{3} \frac{d_{perm} E_{loc}}{k_B T}$$

Hence,

$$\langle d \rangle = \frac{1}{3} \frac{d_{perm}^2 E_{loc}}{k_B T} \tag{12.21}$$

The $1/T$-dependence in (12.21) is an expression of the Curie's law.

The combined effect of linear induced electronic and orientational polarization contributions of a dipole is sometimes merged together in terms of the Debye Eq. (12.22). This way, different physical polarization mechanisms are combined. From (12.12), (12.16), and (12.21), we obtain this Debye equation as:

$$\frac{\varepsilon_{stat} - 1}{\varepsilon_{stat} + 2} = \frac{N\beta_{stat}}{3} \Rightarrow \frac{\varepsilon_{stat} - 1}{\varepsilon_{stat} + 2} = \frac{N}{3}\left(\beta_{stat} + \frac{1}{3}\frac{d_{perm}^2}{k_B T \varepsilon_0}\right) \tag{12.22}$$

Note that if $\beta_{stat} << \frac{1}{3}\frac{d_{perm}^2}{k_B T \varepsilon_0}$ is fulfilled, from (12.22) we find:

$$\frac{\varepsilon_{stat} - 1}{\varepsilon_{stat} + 2} \approx \frac{1}{9}\frac{N d_{perm}^2}{k_B T \varepsilon_0} \Rightarrow \varepsilon_{stat} = 1 + \chi_{stat} \approx 1 + \frac{\frac{N d_{perm}^2}{3k_B T \varepsilon_0}}{1 - \frac{N d_{perm}^2}{9k_B T \varepsilon_0}} \Rightarrow \chi_{stat} \approx \frac{\frac{N d_{perm}^2}{3k_B \varepsilon_0}}{T - \frac{N d_{perm}^2}{9k_B \varepsilon_0}} \tag{12.22a}$$

Obviously, this expression for the static susceptibility approaches infinity when $T \to \frac{N d_{perm}^2}{9k_B \varepsilon_0}$. This way the classical expression (12.22a) predicts the formation of a spontaneous polarization of the medium as a result of the interaction of the permanent dipoles even in the absence of an external field, provided of course that the temperature becomes small enough. The term $\frac{N d_{perm}^2}{9k_B \varepsilon_0}$ is in fact negligibly small for diluted gases, but may account for more than 10^3 K when the concentration achieves $N \approx 10^{23}$ cm^{-3} corresponding to values typical for condensed matter. Note in this connection that (12.22a) coincides by structure with the Curie–Weiss law valid in ferroelectrics and ferromagnetics ($\chi_{stat} \approx \frac{const}{T - T_C}$; T_C—Curie temperature). The classical term $\frac{N d_{perm}^2}{9k_B \varepsilon_0}$, which predicts a classical transition temperature to ferroelectricity, is thus even of the correct order when comparing with realistic values for Curie temperatures—see Table 12.4.

Table 12.4 Curie temperatures of selected ferroelectrics ([3])	Ferroelectric material	Curie temperature T_C/K
	HCl	98
	$NaNO_2$	437
	$Sr_2Nb_2O_7$	1615

12.5.3 Idea on Light Scattering

Elastic light scattering: Let us now return to Eq. (12.3) and consider a polarizing microscopic object that is illuminated by light. Let us further assume that the microscopic object is much smaller than the wavelength of the incident light. Then, according to (12.3), the induced dipole moment oscillates with the angular frequency of the light wave. In a classical picture, this will result in the emission of light according to (2.16). The frequencies of the incident and emitted (scattered) light are—in this case—identical.

This is a mechanism which results in elastic scattering of the incident light. When combining (2.17) and (12.3), we get for the intensity I_S of the light scattered by a single microscopic scatterer:

$$I_S \propto \omega^4 \beta^2(\omega) I_e \tag{12.23}$$

where I_e marks the incident light intensity. The obtained ω^4-dependence of the intensity is characteristic for what is called Rayleigh scattering of light. The shorter the wavelength, the more efficient the scattering mechanism is. This is the reason for the blue color of the sky.

Rayleigh scattering is not the only important elastic light scattering mechanism. Water drops in a cloud are also scattering light, but these water drops are usually larger than the light wavelength. Therefore, (12.23) is not applicable to them, and in fact the frequency dependence of the scattering efficiency becomes much weaker. As a result, when being illuminated by sunlight, clouds usually appear as white objects on the blue background of the sky (compare Fig. 7.3, on top).

The presence of scatterers in an optical path will impair the visibility of a remote object, as visualized in Fig. 12.9. There are two basic effects responsible:

- In the presence of scattering, a part of the light traveling from the object to the observer is lost as a result of scattering. The corresponding intensity decrease is again described by some type of Beer's law [compare (6.30)], $I = I(z) = I_0 e^{-\gamma z}$, where γ is called the scattering coefficient, which in turn is directly proportional to the concentration of scatterers and their individual scattering efficiencies.
- Due to scattering, light portions from background illumination may be coupled into the optical path and arrive at the observer, thus superimposing with the light directly arriving from the object.

Fig. 12.9 Mittelberg in the Alps (Kleinwalsertal, Austria). On left: clear view; on right: misty day. The top of the church tower is better reproduced in foggy conditions; the tower appears to be longer in the picture on right

Both effects in combination are responsible for the general loss of information evident on the right of Fig. 12.9 compared to the situation on left. But note that the dark top of the church tower is better recognized in the fog compared to the clear view. The reason is simple: on left, the dark church tower top is practically invisible on the dark background of the mountain behind (the "Bärenkopf" (engl. Bear head)). On right, the fog between the church and the mountain destroys the information about that dark background, instead, the church tower now appears on the white background of the fog. This background correction provided by the fog overcompensates its disturbing effects.

<u>Inelastic light scattering:</u> Let us now assume that the regarded microscopic object is simply a molecule. As it has been established experimentally, in addition to elastic light scattering, inelastic light scattering may be observed, resulting in frequency changes of the light after having interacted with the molecule. In particular, as the result of the spontaneous Raman effect, new lines (Raman lines) may occur in the spectrum, which are shifted in frequency (so-called Raman shift) for frequency portions that coincide with the frequency of vibrational and/or rotational excitations. For simplicity, in the following text passages we will again refer to vibrational excitations only.

How can we explain this effect?

In analogy to the adiabatic approximation, where the electronic wavefunctions depend on coordinates of the electrons and, in a parametric sense, on those of the nuclei, we now assume a classical picture where the electronic part of the molecular microscopic polarizability (which defines the response to the quickly oscillating fields relevant for an incident VIS or UV light beam) also depends on the coordinates of the nuclei (which move much slower). When using the letter Q for generalized nuclei coordinates, we may write for the electronic polarizability:

$$\beta = \beta(Q) = \beta_0 + \left.\frac{\partial \beta}{\partial Q}\right|_0 (Q - Q_0) + \dots$$

Let us further assume that the nuclei perform oscillations around the equilibrium values of the generalized nuclei coordinates Q_0 according to:

$$(Q - Q_0) = a \cos(\omega_{vibr}t); \quad \omega_{vibr} << \omega$$

Assuming an oscillating real incident field according to $E = E_0 \cos(\omega t)$, from (12.3) we have:

$$d = \varepsilon_0 \beta E = \varepsilon_0 \beta_0 E_0 \cos(\omega t) + \varepsilon_0 \left.\frac{\partial \beta}{\partial Q}\right|_0 a E_0 \cos(\omega t) \cos(\omega_{vibr}t)$$

$$+ ... \Rightarrow d = \underbrace{\varepsilon_0 \beta_0 E_0 \cos(\omega t)}_{\substack{\text{absorption,} \\ \text{Rayleigh scattering}}} + \underbrace{\varepsilon_0 \left.\frac{\partial \beta}{\partial Q}\right|_0 a E_0 \frac{\cos(\omega - \omega_{vibr})t}{2}}_{\text{Raman, Stokes lines}}$$

$$+ \underbrace{\varepsilon_0 \left.\frac{\partial \beta}{\partial Q}\right|_0 a E_0 \frac{\cos(\omega + \omega_{vibr})t}{2}}_{\text{Raman, Antistokes lines}} + ... \tag{12.24}$$

The first term in (12.24) gives rise to absorption as well as Rayleigh scattering processes. The second term generates lines in the spectrum that are red-shifted with respect to ω. The third term generates lines in the spectrum that are blue-shifted with respect to ω.

Here, the following terminology has been accepted:

- The incident light with angular frequency ω corresponds to what is called the Excitation line.
- Stokes lines are red-shifted with respect to ω: $\{\omega_s\} = \omega - \{\omega_{vibr}\}$
- Anti-Stokes lines are blue-shifted with respect to ω: $\{\omega_a\} = \omega + \{\omega_{vibr}\}$

This situation is visualized in Fig. 12.10.

Anti-Stokes lines are usually substantially weaker in intensity than Stokes lines.

Obviously, the Stokes process results in the excitation of molecular vibrations, and the necessary energy corresponds to a loss in frequency in the scattered photon compared to the exciting photon. Anti-Stokes processes start from vibrationally excited molecules, and a vibration quantum becomes added to the photon energy

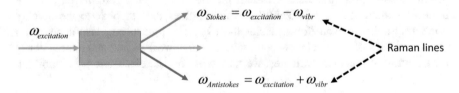

Fig. 12.10 Illustration of a Raman process

of the incident light. In thermodynamic equilibrium, in a spectrum of a molecular gas, Anti-Stokes lines are therefore much weaker. In analogy to (12.23), we have:

$$I_{\text{Stokes}} \propto \omega_s^4 \sigma_R(\omega, \omega_S) I_e$$

$$I_{\text{Anti-Stokes}} \propto \omega_a^4 \sigma_R(\omega, \omega_a) e^{-\frac{\nu_{\text{vibr}}}{\vartheta}} I_e$$

Here we have introduced the Raman cross section σ_R. As the polarizability, it is frequency dependent and shows resonant behavior when one of the participating frequencies becomes resonant with an allowed (electronic) transition in the molecule. Hence, the Raman spectrum may be resonantly enhanced by selecting a suitable excitation frequency. In such a situation, we speak on a resonant Raman effect.

We come to the conclusion, that vibrational (and rotational) spectra of molecules may be investigated by different types of spectroscopy. In infrared (absorption) spectroscopy, the light frequency must be tuned to the investigated transition frequency. IR activity of the investigated transition $m \rightarrow n$ requires the matrix element d_{nm} to be different from zero (Fig. 12.11 on left).

In the Raman process, selection rules are different from those relevant for IR absorption. In accordance with its two-photon character (Fig. 12.11 on right), the condition for Raman activity of the transition $m \rightarrow n$ is:

$$\sum_l d_{nl} d_{lm} \neq 0 \tag{12.25}$$

Therefore, Raman spectroscopy is complementary to IR absorption spectroscopy, because it has other selection rules. For example, in a first-order Raman spectrum, the rotational quantum number may change to a value up to 2. Indeed, in (12.25), when state m corresponds to the rotational quantum number J, in state l we may achieve $J + 1$, so that $J + 2$ becomes accessible in state n. Correspondingly, in rotational–vibrational Raman spectra, S-branches ($\Delta J = +2$) may be observed as well as O-branches ($\Delta J = -2$).

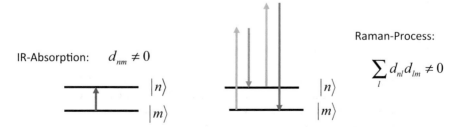

IR-Absorption: $d_{nm} \neq 0$

Raman-Process:
$$\sum_l d_{nl} d_{lm} \neq 0$$

Fig. 12.11 Dipole interaction selection rules for infrared absorption and spontaneous Raman scattering

In a Raman transition, according to (12.25), we therefore have to require:

$$\mathbf{d}_{nl}\mathbf{d}_{lm} \propto \mathbf{r}_{nl}\mathbf{r}_{lm} = \int \psi_n^*(\mathbf{r})\, \mathbf{r}\psi_l(\mathbf{r})dV \int \psi_l^*(\mathbf{r})\, \mathbf{r}\psi_m(\mathbf{r})dV \neq 0$$

In a centrosymmetric system (compare Sect. 6.6.1), the state l must have a parity that is different from the parities of both states m and n. Consequently, m and n must have the same parity. But in this case, IR activity is forbidden. On the other hand, IR activity requires different parities of n and m, which makes the transition Raman-inactive. We come to the conclusion that in a centrosymmetric system (molecule), when assuming electric dipole interaction, no transition may be both IR- and Raman-active. This is the essence of the so-called rule of mutual exclusion.

While IR-active transitions are characterized by an oscillating induced dipole moment (such as in the case of the asymmetric stretching vibration of CO_2), Raman-active vibrations are accompanied—as a rule of thumb—by a modulation of the volume occupied by the molecule, such as in the case of the symmetric stretching vibration of CO_2. This gives rise to a modulation of the electronic polarizability, i.e. $\left.\frac{\partial \beta}{\partial \varrho}\right|_0 \neq 0$ and therefore, according to (12.24), to Raman activity.

12.6   Advanced Material: A Quantum Mechanical Expression for the Polarizability

Let us finally generalize (12.6) to the semiclassical case. According to (12.6), the following classical expression for the microscopic polarizability was obtained:

$$\beta = \frac{q^2}{\varepsilon_0 m} \sum_{j=1}^{M} \frac{f_j}{\omega_{0j}^2 - \omega^2 - 2i\omega\gamma_j}$$

In the course of our treatment of the response of a two-level system with electromagnetic radiation according to Chap. 6, we found that the efficiency of that interaction is determined by at least three factors:

- The square of the absolute value of the matrix element of the perturbation (in our case the electric dipole) operator (which is proportional to the *oscillator strength* $f_{nl} = \frac{2m}{\hbar}\omega_{nl}|x_{nl}|^2$).
- The relation between the frequency of the incident electromagnetic wave and the energy spacing between the energy levels of the unperturbed material system which define the resonance frequencies (*resonance condition*).
- When regarding the average polarizability of an assembly of many two-level systems, then we also have to consider the *population difference* of the energy levels involved into the quantum transition.

This may be used to provide a guess on the structure of the semiclassic expression for the polarizability. For this purpose, in Eq. (12.6), we will make the following substitutions:

$$\omega_{oj} \rightarrow \omega_{nl} = \frac{E_n - E_l}{\hbar}$$

$$\gamma_j \rightarrow \gamma_{nl}$$

$$q^2 f_j \rightarrow q^2 f_{nl}[W(l) - W(n)] = \frac{2m|d_{nl}|^2 \omega_{nl}}{\hbar}[W(l) - W(n)]$$

$$\sum_j \rightarrow \sum_l \sum_{n>l}$$

Here, it is assumed for simplicity that the quantum states are counted in a way that $E_n \geq E_l$ is fulfilled. The values $W(l)$ are the statistical probabilities that the l-th energy level is populated.

Then, the semiclassical expression for the polarizability, dielectric function, or optical constants is provided by setting:

$$\beta = \frac{2}{\varepsilon_0 \hbar} \sum_l \sum_{n>l} |d_{nl}|^2 \omega_{nl} \frac{[W(l) - W(n)]}{\omega_{nl}^2 - \omega^2 - 2i\omega\gamma_{nl}} \tag{12.26}$$

This expression already coincides with the correct result, although we can by no means assert that we have derived it. It was a guess, but concerning an accurate derivation, let us start with two remarks.

Firstly, we must recognize that so far we have not introduced any relaxation mechanism into our quantum mechanical picture. On the other hand, as it directly follows from the Fourier analysis (compare Sect. 2.3.3), the Lorentzian lineshapes in (12.26) correspond to an exponential damping of the polarization. In a quantum mechanical language, this may be derived in the frames of the density matrix concept, which is however beyond the material presented in this course. A derivation of (12.26) in the framework of the density matrix is provided in [4].

Secondly, the derivation of expression (12.26) *without* the imaginary terms responsible for relaxation may be performed in the usual language of wavefunctions, when calculating the quantum mechanical expectation value of the dipole moment ⟨**d**⟩ of a quantum system perturbed by an oscillating field according to (6.20a). This is, for example, presented in the highly recommended rather classical textbook on quantum mechanics [5].

We are lazy enough not to repeat that whole derivation in this book. Let us however make use of a shortcut in order to demonstrate how Eq. (12.26) may be derived (however without the relaxation terms γ). For that purpose, we return to the stuff from Sect. 6.3.2 and, in particular, to a wavefunction of the perturbed quantum system as defined by (6.12) and (6.21).

The idea is to calculate the expectation value of the dipole moment and to set it equal to (12.3) in order to find an expression for the polarizability. Hence, we set:

$$\langle \mathbf{d} \rangle = \int_V \Psi^* \hat{\mathbf{d}} \Psi dV = \varepsilon_0 \beta \mathbf{E} \tag{12.27}$$

There is a small inconsistency in that shortcut. Namely, in (12.3), the assumed field is strongly monochromatic as described by the cosine function. In (6.21), however, we have in fact assumed a cosine field multiplied with a step function that guarantees that the light is switched on exactly at $t = 0$. Then, (6.21) has been assumed to be an approximate solution valid for a certain time span after the oscillating field has been switched on such that the system still "remembers" its initial conditions. Therefore, when honestly substituting (6.12) and (6.21) into (12.27), we will obtain that (12.27) is not fulfilled.

We will escape from that inconsistency in the following manner:

- In (12.3) the field is assumed to be monochromatic according to a cosine-law that theoretically extends from "$t = -\infty$" until now. In any real system, because of relaxation mechanisms, the information on the initial conditions of the quantum system would surely have been lost during this time span.
- In (6.21) the initial conditions of the quantum system (here at $t = 0$) have explicitly been taken into account. Indeed, we assumed a wavefunction of a superposition state in terms of (6.12) according to:

$$\Psi(\mathbf{r}, t) = \sum_n a_n(t) \Psi_n(\mathbf{r}, t)$$

While requiring $t = 0 :\Rightarrow |a_l| = 1; a_{n \neq 1} = 0$. The solution (6.21):

$$a_m(t) \approx a_l \frac{\mathbf{d}_{ml} \mathbf{E_0}}{\hbar} \left\{ \frac{e^{i(\omega_{ml} - \omega)t} - 1}{\omega_{ml} - \omega} + \frac{e^{i(\omega_{ml} + \omega)t} - 1}{\omega_{ml} + \omega} \right\}; \quad m \neq l$$

is clearly consistent with those initial conditions. It is therefore not astonishing that (12.27), (6.12), and (6.21) cannot be fulfilled together.

- Our physical assumption is now that although (6.21) is dominated by the initial conditions of the quantum system, it also contains—as a seed—the information about the response of the system to a well-established stationary monochromatic perturbation. We expect that this seed will result in a term in $\langle \mathbf{d} \rangle = \int_V \Psi^* \hat{\mathbf{d}} \Psi dV$ that is proportional to $\mathbf{E} = \mathbf{E_0} e^{-i\omega t}$. The idea is thus to consider only that term in (12.27) in order to calculate the polarizability. Our starting point is therefore:

$$\varepsilon_0 \beta \mathbf{E_0} e^{-i\omega t} = \int_V \Psi^* \hat{\mathbf{d}} \Psi dV \Bigg|_{\propto \mathbf{E_0} e^{-i\omega t}} = \langle \mathbf{d} \rangle |_{\propto \mathbf{E_0} e^{-i\omega t}}$$

On the other hand, from (6.12) it follows:

$$\Psi(\mathbf{r}, t) \approx \Psi_l(\mathbf{r}, t) + \sum_{n \neq l} a_n(t) \Psi_n(\mathbf{r}, t)$$

Consequently:

$$\Psi^* \Psi \approx \underbrace{\Psi_l^* \Psi_l}_{\Rightarrow d_{\text{perm}}} + \underbrace{\Psi_l^* \sum_{n \neq l} a_n(t) \Psi_n + \Psi_l \sum_{n \neq l} a_n^*(t) \Psi_n^*}_{\text{contains terms proportional to } \mathbf{E}_0 e^{-i\omega t}} + \underbrace{\cdots}_{\propto E_0^2} \tag{12.27a}$$

Obviously, when calculating the induced dipole moment $\langle \mathbf{d} \rangle$, we have to concentrate on the central term in the above expression. The rest is mathematics. When setting for simplicity $a_l = 1$, (6.21) may be rewritten as:

$$a_n(t) \approx a_l \frac{\mathbf{d}_{nl} \mathbf{E_0}}{\hbar} \left\{ \frac{e^{i(\omega_{nl} - \omega)t} - 1}{\omega_{nl} - \omega} + \frac{e^{i(\omega_{nl} + \omega)t} - 1}{\omega_{nl} + \omega} \right\}$$

$$= 2 \frac{\mathbf{d}_{nl} \mathbf{E_0}}{\hbar} \frac{1}{\omega_{nl}^2 - \omega^2} \left[e^{i\omega_{nl}t} (\omega_{nl} \cos \omega t - i\omega \sin \omega t) - \omega_{nl} \right]$$

Once we are only interested in the oscillating according to $\mathbf{E} = \mathbf{E_0} e^{-i\omega t}$ term, we ignore the constant term and write:

$$\langle \mathbf{d} \rangle = \int_V \Psi^* \hat{\mathbf{d}} \Psi dV \rightarrow \sum_{n \neq l} \left(a_n \mathbf{d}_{\ln} e^{i\omega_{\ln}t} + c.c. \right)$$

$$= 2 \sum_{n \neq l} \frac{|\mathbf{d}_{nl}|^2 \mathbf{E_0}}{\hbar} \frac{1}{\omega_{nl}^2 - \omega^2} 2\text{Re}[(\omega_{nl} \cos \omega t - i\omega \sin \omega t)] \Bigg|_{\propto E_0 e^{-i\omega t}} =$$

$$= 4 \frac{\mathbf{E_0}}{\hbar} \sum_{n \neq l} \frac{|\mathbf{d}_{nl}|^2 \omega_{nl}}{\omega_{nl}^2 - \omega^2} \cos \omega t \Bigg|_{\propto E_0 e^{-i\omega t}}$$

And therefore:

$$\langle \mathbf{d} \rangle = \varepsilon_0 \underbrace{\frac{2}{\varepsilon_0 \hbar} \sum_{n \neq l} \frac{|\mathbf{d}_{nl}|^2 \omega_{nl}}{\omega_{nl}^2 - \omega^2}}_{= \beta} \underbrace{2\mathbf{E_0}}_{= \mathbf{E}_{0,\text{real}}} \cos \omega t = \varepsilon_0 \underbrace{\frac{2}{\varepsilon_0 \hbar} \sum_{n \neq l} \frac{|\mathbf{d}_{nl}|^2 \omega_{nl}}{\omega_{nl}^2 - \omega^2}}_{= \beta} \mathbf{E_0} \left(e^{-i\omega t} + e^{-i\omega t} \right)$$

$$\Rightarrow \beta = \frac{2}{\varepsilon_0 \hbar} \sum_{n \neq l} \frac{|\mathbf{d}_{nl}|^2 \omega_{nl}}{\omega_{nl}^2 - \omega^2}$$

$$\tag{12.28}$$

This expression for the polarizability was derived assuming that $a_l = 1$, i.e., the system is in state l. In reality, the quantum system may be found in different states. We therefore perform a rather classical averaging procedure, assigning to each possible state l a population probability $W(l)$. Clearly, $\sum_l W(l) = 1$. When remembering that $\omega_{nl} = -\omega_{ln}$, we obtain:

$$\beta = \frac{2}{\varepsilon_0\hbar}\sum_{n\neq l}\frac{|\mathbf{d}_{nl}|^2\omega_{nl}}{\omega_{nl}^2-\omega^2} \to \frac{2}{\varepsilon_0\hbar}\sum_l W(l)\sum_{n\neq l}\frac{|\mathbf{d}_{nl}|^2\omega_{nl}}{\omega_{nl}^2-\omega^2}$$

$$= \frac{2}{\varepsilon_0\hbar}\sum_l\sum_{n>l}\frac{|\mathbf{d}_{nl}|^2\omega_{nl}}{\omega_{nl}^2-\omega^2}[W(l)-W(n)]$$

This is of course a real expression for the polarizability. It differs from expressions like (12.6) and (12.26) by the term $2i\omega\gamma_{nl}$ in the denominator. In order to make our expression consistent with the reasonable assumption on an exponential damping of the freely oscillating dipole, the only thing we can do is to phenomenologically add that term and thus to complete the Lorentzian-like frequency dependence of the polarizability as it has already been obtained in Sect. 2.3.2, Eq. (2.14). We thus come to the final expression:

$$\beta = \frac{2}{\varepsilon_0\hbar}\sum_l\sum_{n>l}\frac{|\mathbf{d}_{nl}|^2\omega_{nl}}{\omega_{nl}^2-\omega^2}[W(l)-W(n)]$$

$$\to \frac{2}{\varepsilon_0\hbar}\sum_l\sum_{n>l}\frac{|\mathbf{d}_{nl}|^2\omega_{nl}}{\omega_{nl}^2-\omega^2-2i\omega\gamma_{nl}}[W(l)-W(n)]$$

This expression exactly coincides with our guess (12.26).

Once more, a more stringent derivation of (12.26) is possible in terms of the density matrix approach.

So far we have regarded a single dipole exposed to an external field according to (12.27). If the dipole is part of a macroscopic system, the field in (12.27) has to be interpreted as the corresponding local field.

12.7   Tasks for Self-check

12.7.1

(a) Find an expression for the electric field inside a homogeneously polarized dielectric sphere! Note: The task is easily solved when regarding the single polarized sphere as a superposition of two homogeneously charged spheres with slightly shifted central points.

(b) Find an expression for the electric field inside a homogeneously polarized dielectric cylinder! Assume the cylinder as infinitely long, while the polarization vector is directed perpendicular to the axis of the cylinder!

Note: The task is easily solved when regarding the single polarized cylinder as a superposition of two homogeneously charged cylinders with parallel but not coinciding axes.

12.7.2 Assume a current $I = 1$ A flowing through a copper wire with a diameter d of 1.63 mm. Estimate the drift velocity of the electrons, assuming that there is approximately 1 free electron per copper atom, a mass density of $\rho = 8.93$ g cm^{-3}, and a mass number of copper of 63.5. (after [6]).

12.7.3 Imagine that you have recorded the Raman spectrum of a C_{60} fullerite sample by means of an Argon ion laser, operating at an excitation wavelength of 514 nm. You observe a strong Stokes line at a wavelength of 556 nm. At which wavelength you will observe that Stokes line, when the excitation wavelength is changed to 488 nm?

12.7.4 Consider an electron confined in the potential of a harmonic oscillator $U(x) = \kappa x^2/2$. Starting from Eq. (12.28), derive an expression for the static polarizability of that confined electron in the quantum state $|k\rangle$ when the electric field is parallel to the x-axis (after [7])!

References

Specific References

1. Л.Д. Ландау, Е.М.Лифшиц, Электродинамика Сплошных Сред Москва "Наука" 1982, p. 42
2. P. Grosse *Freie Elektronen in Festkörpern* (Springer, Berlin, 1979), p. 17–18
3. H. Warlimont, W. Martienssen, (eds.) *Springer Handbook of Materials Data* (Springer, 2018), p. 897–930
4. O. Stenzel, *The Physics of Thin Film Optical Spectra. An introduction* (Springer, Berlin, 2016), pp. 255–269
5. А.С. Давыдов, *Квантовая Механика* (Москва Физматгиз 1963), pp. 353–360
6. P.A. Tipler, G. Mosca, *Physik*, vol. 7 (Springer, Auflage, 2015), pp. 801–802
7. U. Fano, L. Fano, *Physics of Atoms and Molecules. An Introduction to the Structure of Matter* (University of Chicago Press 1973, task 17.1)

General Literature to this Chapter

8. P.A. Tipler, G. Mosca, *Physik*, vol. 7 (Springer, Auflage, 2015)
9. R.P. Feynman, R.B. Leighton, M. *Sands: The Feynman Lectures of Physics*, vol. 2 (Addison—Wesley Publishing Company, Inc. 1964)
10. M. Born, E. Wolf, *Principles of Optics* (Pergamon Press, Oxford, London, Edinburgh, New York, Paris, Frankfurt, 1968)

Part V
Optical Properties of Continuous Media

"Friedrich-Schiller-Universität Jena". Colored etching and Photograph by Astrid Leiterer, Jena, Germany (www.astrid-art.de). Photograph reproduced with permission

Continuous media contain a thermodynamically relevant number of atoms. In a continuum description, the summarized impact of the individual optical responses of that tremendous number of atoms may be replaced by continuous and smooth effective optical functions, such as the well-known refractive index. The optical effects described in terms of geometrical optics are based on the introduction of suchlike functions and include well-known phenomena such as the image inversion of an object created by a water drop.

Linear Optical Constants I

13

Abstract

Dielectric function and optical constants of continuous media are introduced. Dispersion is obtained as a natural conclusion from memory effects. Applications include the discussion of orientation polarization, the treatment of normal incidence interface reflections, as well as a simple treatment of the optical properties of material mixtures.

13.1 Starting Point

The light–matter interaction may be theoretically treated at different levels of difficulty.

A purely classical description makes use of Maxwell's equations for the description of the electrical and magnetic fields and classical models (e.g., Newton's equations of motion) for the dynamics of the charge carriers present in any terrestrial matter (Chap. 2). On the contrary, a quantum mechanical treatment is possible within the framework of the quantization of the electromagnetic field (so-called second quantization—Sect. 5.6.3) and a quantum theoretical treatment of matter (Chaps. 8–11). This description is necessary, when spontaneous optical effects have to be described (spontaneous emission according to Sect. 7.6.2, spontaneous Raman scattering, or spontaneous paramagnetic interactions in nonlinear optics) on a theoretically strong level. In applied spectroscopy, the accurate quantum mechanical description is often replaced by the so-called semiclassical treatment. Here, the properties of matter are described in terms of quantum mechanical models, while the fields are treated within the framework of Maxwell's theory (compare Sect. 6.3.2). Maxwell's equations are therefore essential in both classical and semiclassical approaches, and for that reason we start our discussion

© The Author(s), under exclusive license to Springer Nature Switzerland AG 2022
O. Stenzel, *Light–Matter Interaction*, UNITEXT for Physics,
https://doi.org/10.1007/978-3-030-87144-4_13

from these equations, which are given below:

$$\text{div}\mathbf{B} = 0 \quad \text{div}\mathbf{D} = 0$$

$$\text{curl}\mathbf{E} = -\frac{\partial \mathbf{B}}{\partial t} \quad \text{curl}\mathbf{H} = \frac{\partial \mathbf{D}}{\partial t}$$

$$\mathbf{B} = \mu_0(\mathbf{H} + \mathbf{M}) \quad \mathbf{D} = \varepsilon_0\mathbf{E} + \mathbf{P} \tag{13.1}$$

Here, $\mathbf{E}$ and $\mathbf{H}$ represent the vectors of the electric and magnetic fields, while $\mathbf{D}$ and $\mathbf{B}$ stand for the electric displacement and the magnetic induction, respectively. $\mathbf{P}$ is the polarization, and $\mathbf{M}$ the magnetization of the medium. In (13.1), neither the free charge carrier density nor their current density is explicitly present. Keeping in mind, that optics deal with rapidly oscillating electric and magnetic fields, due to the short periods, "free" charges will oscillate around their equilibrium position quite similar to bound charges (compare the explanations in Sect. 12.4). So in our description as used here, the displacement vector contains information on both free and bound charges. The very few cases, where the static response of matter with free electrons becomes important in the frames of this course, cannot be treated within (13.1) and will need separate discussion.

13.2   Physical Idea

According to the general philosophy in this book, we concentrate on electric dipole interactions again. In the following, we will therefore assume that the media are generally non-magnetic ($\mathbf{M}$ is a zero-vector) and isotropic. When neglecting magnetization effects, from (13.1) one obtains straightforwardly:

$$\text{curlcurl}\mathbf{E} = \text{graddiv}\mathbf{E} - \Delta\mathbf{E} = -\mu_0\frac{\partial^2 \mathbf{D}}{\partial t^2} \tag{13.2}$$

This is an equation with two unknown vectors. In order to proceed further, at this point, we need to establish a second equation, namely a relationship between the vectors $\mathbf{E}$ and $\mathbf{D}$, which will be done in the next subchapter.

As some preparation, let us assume now that a rapidly changing electric field with a completely arbitrary time dependence interacts with matter. One would naturally expect that the electric field tends to displace, in general, both negative and positive charges, thus creating a macroscopic dipole moment in the material system. As in the previous chapter, the polarization $\mathbf{P}$ is per definition the dipole moment per unit volume, and it will be, of course, time-dependent in a manner that is determined by the time dependence of $\mathbf{E}$. For the moment, we neglect the spatial dependence of $\mathbf{E}$ and $\mathbf{P}$, because it is not essential for the further derivation. Generally, the polarization is thus a possibly very involved functional $\mathbf{F}$ of the field $\mathbf{E}$:

$$\mathbf{P}(t) = \mathbf{F}\big[\mathbf{E}(t' \leq t)\big] \tag{13.3}$$

Of course, the polarization of the medium appears as the result of the action of the assumed electric field (here and in the following, we do *not* regard ferroelectrics!). Due to the causality principle, the polarization at a given time t can depend on the field at the same moment as well as at previous moments t', but not on the field behavior in the future. This is the meaning of the condition: $t' \leq t$. In order to comply with the requirement (13.3), we therefore postulate the following general relationship for the polarization as a functional of the electric field:

$$\mathbf{P}(t) = \varepsilon_0 \int_{-\infty}^{t} \kappa(t, t')\mathbf{E}(t')dt' \tag{13.4}$$

Equation (13.4) postulates that the polarization at any time t may principally depend on the first power of the field at the current and all previous moments, as it follows from the integration interval that is chosen in correspondence with the mentioned causality principle. The specific way, in which the system "remembers" the field strength at previous moments, is hidden in the response function $\kappa(t, t')$, which must be specific for any material. Equation (13.4) is in fact the first (linear) term of an expansion of (13.3) into a Taylor power series of $\mathbf{E}$. As we keep only the first (linear) term of the series, all optical effects that arise from (13.4) form the field of *linear optics*. Equation (13.4) thus represents a rather general writing of the material equation in linear optics. Note that κ is a real function, because the validity of (13.4) shall not depend on whether we use a real or complex (compare remark in Sect. 2.1) description of the oscillating fields and polarizations.

In general, when the materials are anisotropic, $\kappa(t, t')$ is a tensor. As we restrict our attention here to optically isotropic materials, $\mathbf{P}$ will always be parallel to $\mathbf{E}$, so that $\kappa(t,t')$ becomes a scalar function.

A further facilitation is possible. Due to the homogeneity of time, $\kappa(t,t')$ will in fact not depend on both individual times t and t' separately, but only on their difference $\xi \equiv t - t'$. Substituting t' by ξ, we obtain:

$$\mathbf{P}(t) = \varepsilon_0 \int_{0}^{\infty} \kappa(\xi)\mathbf{E}(t - \xi)d\xi \tag{13.5}$$

13.3 Theoretical Material

13.3.1 The Linear Dielectric Susceptibility

Let us now come to the utmost important model case of harmonic time dependence. Let us assume that the electric field performs rapid oscillations according to $\mathbf{E}(t) = \mathbf{E}_0 e^{-i\omega t}$. Correspondingly, we may write: $\mathbf{E}(t - \xi) = \mathbf{E}_0 e^{-i\omega t} e^{i\omega \xi}$. Note that we assume a completely monochromatic field. From (13.5), it is then obtained:

$$\mathbf{P}(t) = \varepsilon_0 \left[\int_0^\infty \kappa(\xi) e^{i\omega\xi} d\xi \right] \mathbf{E}_0 e^{-i\omega t} \tag{13.6}$$

We define the *linear dielectric susceptibility* χ according to:

$$\chi = \int_0^\infty \kappa(\xi) e^{i\omega\xi} d\xi = \chi(\omega) \tag{13.7}$$

The thus defined susceptibility must be complex (it has both real and imaginary parts), and it depends on the frequency of the field even after having performed the integration in (13.7). Both circumstances arise mathematically from (13.5) and physically from the finite inertness of any material system. Clearly, the charge carriers cannot instantaneously react on rapidly changing fields, so that their positions at a given time t depend on the history of the system, which is in fact the reason for the complicated behavior of the polarization with time. The information on the specific material properties is now carried by $\chi(\omega)$.

Remark Note that the response function $\kappa(\xi)$ as fixed in (13.5)–(13.7) has a clear and very general physical meaning. Indeed, let us write (13.5) in a more general symbolic manner:

$$\text{output}(t) = \int_0^\infty \kappa(\xi) \, \text{input}(t - \xi) d\xi$$

Here, the electric field has been associated with some time-dependent input to the system under consideration, while the resulting time-dependent polarization forms the output. In the special case that the input to the system is provided by a δ-kick: input$(t) \propto \delta(t)$ (Fig. 13.1 on left), we obtain the output for $t > 0$ according to:

$$\text{output}(t) = \int_0^\infty \kappa(\xi) \, \text{input}(t - \xi) d\xi \propto \int_0^\infty \kappa(\xi) \, \delta(t - \xi) d\xi = \kappa(t)$$

$$\Rightarrow \text{output}(t) \propto \kappa(t)$$

Hence, the response function reproduces the specific output of the system to a δ-kick. For a general type of input in (13.5), we therefore have to expect that the system output is defined by some convolution of the actual input and the systems reaction to a δ-kick (Fig. 13.1 on right). Hence, the situations shown in Fig. 13.1 will produce different, but not independent sounds: Knowledge of the sound obtained in the situation on left allows predicting what will happen on right.

Fig. 13.1 System response to a δ-kick (left) and to a periodic input (right). Cartoon by Dr. Alexander Stendal. Printed with permission

We are now able to formulate the relationship between **E** and **D** for *monochromatic electric fields*. Indeed, from (13.6) and (13.7) it follows that

$$\mathbf{P} = \varepsilon_0 \chi \mathbf{E} \tag{13.8}$$

In combination with the definition of **D** we have:

$$\mathbf{D} = \varepsilon_0 \mathbf{E} + \mathbf{P} = \varepsilon_0 [1 + \chi(\omega)]\mathbf{E} \equiv \varepsilon_0 \varepsilon(\omega)\mathbf{E} \tag{13.9}$$

where we defined the *dielectric function $\varepsilon(\omega)$*

$$\varepsilon(\omega) \equiv 1 + \chi(\omega)$$

Equation (13.9) is completely analogous to what is known from the electrostatics of dielectrics [compare (12.10)], with the only difference that ε is now complex and frequency dependent. That frequency dependence is called dispersion. So that we come to the conclusion that in optics we have a similar relationship between field and displacement vectors as in electrostatics, with the difference that in optics the dielectric constant has to be replaced by the dielectric function.

Remark In the case that the incident field is not monochromatic, in (13.8) and (13.9), all of the vectors must be replaced by the amplitudes of their corresponding Fourier components. We have:

$$\mathbf{D}_\omega = \varepsilon_0 \varepsilon(\omega)\mathbf{E}_\omega \tag{13.9a}$$

And so on. Here symbols like $\mathbf{D}_\omega$ denote Fourier components of the corresponding vectors.

13.3.2 Linear Optical Constants

We may now turn back to (13.2). Keeping in mind that our discussion is restricted to harmonic oscillations of the fields only, the second derivative with respect to time in (13.2) may be replaced by multiplying with $-\omega^2$. Replacing moreover **D** with (13.9), we obtain:

$$\text{curlcurl}\mathbf{E} - \frac{\omega^2 \varepsilon(\omega)}{c^2}\mathbf{E} = \mathbf{0} \tag{13.10}$$

Here we used the identity: $\varepsilon_0 \mu_0 = c^{-2}$.

where c is the velocity of light in vacuum. For polychromatic fields, the single Fourier components have to be treated separately in an analogous manner.

We now remember the vector identity: $\text{curlcurl}\mathbf{E} \equiv \text{graddiv}\,\mathbf{E} - \Delta\mathbf{E}$.

In the case that $\varepsilon \neq 0$, from $\text{div}\mathbf{D} = 0$ it follows that $\text{div}\mathbf{E} = 0$. Thus we finally have:

$$\Delta E + \frac{\omega^2 \varepsilon(\omega)}{c^2} E = 0 \tag{13.11}$$

where the field vector has been replaced by a scalar field due to the assumed isotropy. A completely identical equation may be obtained for the magnetic field.

Let us remark at this point that due to the assumed optical isotropy, we will often turn from the vectorial to the scalar mathematical description. Throughout this text, in these cases we will simply refrain from bold symbols without further notice.

Assuming that the dielectric function does not depend on the coordinates themselves (homogeneous media), we are looking for a solution in the form:

$$E(\mathbf{r}, t) = E_0 e^{-i(\omega t - \mathbf{kr})} \tag{13.12}$$

with **k** being the wavevector. Nontrivial solutions of (13.11) exist when

$$k = \pm \frac{\omega}{c}\sqrt{\varepsilon(\omega)} \tag{13.13}$$

is fulfilled. Assuming for simplicity, that **k** is parallel to the z-axis of a Cartesian coordinate system, (13.12) describes a plane wave traveling along the z-axis. It depends on the sign in (13.13) whether the wave is running into the positive or negative direction. We choose a wave running into the positive direction and obtain:

$$E = E_0 e^{-i\left(\omega t - \frac{\omega}{c}\sqrt{\varepsilon(\omega)}z\right)} \tag{13.14}$$

where E_0 is the field amplitude at $z = 0$. Let us look on (13.14) in some more detail.

As we obtained in Sect. 13.3.1, the dielectric function may be complex; hence, it may have an imaginary part. Of course, the square root will also be a complex function. We therefore have:

$$\sqrt{\varepsilon(\omega)} = \mathrm{Re}\sqrt{\varepsilon(\omega)} + i\,\mathrm{Im}\sqrt{\varepsilon(\omega)}$$

Equation (13.14) therefore describes a damped wave according to:

$$E = E_0 e^{-\frac{\omega}{c}\mathrm{Im}\sqrt{\varepsilon(\omega)}z} e^{-i\left(\omega t - \frac{\omega}{c}\mathrm{Re}\sqrt{\varepsilon(\omega)}z\right)} \tag{13.15}$$

with a z-dependent amplitude (see Fig. 13.2)

$$E_{ampl} = E_0 e^{-\frac{\omega}{c}\mathrm{Im}\sqrt{\varepsilon(\omega)}z} \tag{13.16}$$

and a real phase:phase $= \omega t - \frac{\omega}{c}\mathrm{Re}\sqrt{\varepsilon(\omega)}z$

Let us calculate the velocity dz/dt of any point at the surface of constant phase (which is a plane in our case). Regarding the phase as constant and differentiating the last equation with respect to time, we obtain the *phase velocity* of the wave according to:

$$\frac{\mathrm{d}z}{\mathrm{d}t} \equiv v_{\mathrm{ph}} = \frac{c}{\mathrm{Re}\sqrt{\varepsilon(\omega)}} \equiv \frac{c}{n(\omega)} \tag{13.17}$$

Here we introduced the *refractive index* $n(\omega)$ as the real part of the square root of the complex dielectric function. Naturally, the refractive index appears to be frequency dependent (so-called *dispersion* of the refractive index). In a medium with refractive index n, the phase velocity of a monochromatic electromagnetic wave changes with respect to vacuum according to (13.17).

Fig. 13.2 Electric field in a wave damping inside the medium. The z-dependent amplitude forms an envelope for the z dependence of the field

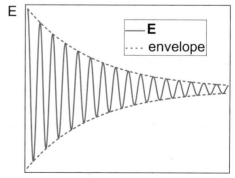

As a generalization to (13.17), one often defines the *complex index of refraction* as:

$$\hat{n}(\omega) = n(\omega) + i K(\omega) \equiv \sqrt{\varepsilon(\omega)} \tag{13.18}$$

Its real part is identical with the ordinary refractive index as defined in (13.17), while its imaginary part (the so-called *extinction coefficient*) K is responsible for the damping of a wave. Indeed, returning to (13.16), we obtain for the amplitude of the wave: $E_{\text{ampl}} = E_0 e^{-\frac{\omega}{c} K z}$.

Because the intensity I of the wave is proportional to the square of the field amplitude modulus [compare (2.6)], the intensity damps inside the medium as:

$$I = I(z = 0) e^{-2\frac{\omega}{c} K z} \equiv I(z = 0) e^{-\alpha z} \tag{13.19}$$

This exponential decay of light intensity for a wave traveling in a lossy medium is well known as Beer's law of absorption [compare (6.30)] with a frequency-dependent *absorption coefficient* α defined as:

$$\alpha(\omega) = 2\frac{\omega}{c} K(\omega) \tag{13.20}$$

In terms of the identities: $\nu \equiv \frac{1}{\lambda} = \frac{\omega}{2\pi c}$ we come to a more familiar expression:

$$\alpha(\nu) = 4\pi \nu K(\nu) \tag{13.20a}$$

Although the refractive index n and the extinction coefficient K are dimensionless, the absorption coefficient is given in reciprocal length units, usually in reciprocal centimeters. The reciprocal value of the absorption coefficient is sometimes called *penetration depth*. The pair of n and K forms the pair of *linear optical constants* of a material.

Remark Throughout this derivation, we supposed the time dependence of the fields according to $e^{-i\omega t}$. As a consequence, we defined the complex index of refraction as $n + iK$. The same kind of theory may be built postulating a time dependence of the fields as $e^{+i\omega t}$. However, in this case the index of refraction will be $n - iK$. Both approaches are equally correct and can be found in the literature, however, they shouldn't be confused with each other.

13.4 Consistency Considerations

In this chapter, so far we have derived or rederived several relations that have earlier been obtained in the frames of a somewhat different context. Let us recall those "convergencies" with our previous considerations:

- The material Eq. (13.9) for dielectric materials: It is a generalization of the electrostatic version (12.10) provided that we accept that the dielectric function is frequency-dependent and complex. In order to guarantee convergence of (13.9) to (12.10) when the frequency approaches zero, we must require that the imaginary part of the dielectric function of a dielectric material approaches zero when the frequency approaches zero.
- In deriving (13.19), in fact we rederived Beers law (6.30) as a side result.
- In Sect. 3.2.2, for a wave package propagating without damping, we established relation (3.9) for the phase velocity according to $v_{ph} = \frac{\omega_0}{k_0}$. For a monochromatic electromagnetic wave propagating without damping, wave, $\omega = \omega_0$, while $k = \frac{\omega}{c} n$. Hence, we find

$$v_{ph} = \frac{\omega_0}{k_0} = \frac{\omega}{k} = \frac{c}{n}$$

such that our new expression for the phase velocity according to (13.17) is consistent with the previously derived expression (3.9).

Before now turning to more practical problems, let us complete this section with another rather general remark. No matter whether the piece of matter under investigation is microscopic or macroscopic, whatever we assume to be an optical signal arriving at a detector must have traveled from its origin to the detecting unit. From the theoretical stuff described so far, we therefore find that the calculation of any optical signal that is to be compared with a measured signal will contain at least two different parts:

- First of all, one has to find a suitable model for the dielectric function (or other relevant material functions that describe the dynamics of the piece of matter to be investigated) that contain the specific information about the *material*.
- Secondly, having this model in hand, one has to solve a wave equation like (13.2) to account for the wave propagation in the particular *geometry* valid for the (given or assumed) experiment. Having solved the wave equation with realistic boundary conditions, we obtain electric and/or magnetic fields that may be converted for example into light intensities, which in turn may be compared with experimental data.

13.5 Application to Practical Problems

13.5.1 Energy Dissipation

A high extinction coefficient (strong damping) is not necessarily connected with a large imaginary part of the dielectric function. For example, a real but negative dielectric function will result in a purely imaginary refractive index, i.e. a possibly high extinction coefficient. This seemingly exotic situation is in fact a characteristic model assumption in metal optics. Here the penetrating wave is indeed damped, but the light is rather reflected than absorbed. Therefore, the generally accepted terminus "absorption coefficient" may be misleading in special cases. In fact, for light absorption (energy dissipation) it is essential that $\mathrm{Im}\varepsilon \neq 0$.

Let us illustrate this fact. Indeed, the volume density of the power dissipated from the electromagnetic field can be written as [1]:

$$\frac{\partial W_{\mathrm{diss}}}{V \partial t} = \mathbf{jE}$$

This notation corresponds to real fields $\mathbf{E}$ and current densities $\mathbf{j}$. It is a direct conclusion from the well-known relation that expresses the power as the product of electric current and applied voltage. In order to express the real functions by their complex counterparts, we simply make use of (2.5) and obtain:

$$\frac{\partial W_{\mathrm{diss}}}{V \partial t} = 4\mathrm{Re}\mathbf{j}\mathrm{Re}\mathbf{E}$$

Making further use of:

$$\frac{\partial \mathbf{P}}{\partial t} = \frac{1}{V} \sum_l q_l \dot{\mathbf{r}}_l = \mathbf{j} \tag{13.21}$$

and considering a harmonic time dependence of the oscillating electric field in the wave according to the convention defined in terms of (2.1), we can write:

$$\mathbf{j} = -i\varepsilon_0\omega(\varepsilon - 1)\mathbf{E}$$

Therefore we have:

$$\mathrm{Re}\mathbf{j} = \varepsilon_0\omega[(\mathrm{Re}\varepsilon - 1)\mathrm{Im}\mathbf{E} + \mathrm{Im}\varepsilon\mathrm{Re}\mathbf{E}]$$
$$\Rightarrow \mathrm{Re}\mathbf{j}\mathrm{Re}\mathbf{E} = \varepsilon_0\omega\left[(\mathrm{Re}\varepsilon - 1)\mathrm{Im}\mathbf{E}\mathrm{Re}\mathbf{E} + \mathrm{Im}\varepsilon(\mathrm{Re}\mathbf{E})^2\right]$$

Let us finally use expression (13.15) for the electric field. Once the real part of the electric field oscillates according to a cosine function, and the imaginary according to a sine function, the time average over a period results in:

$$\left\langle \frac{\partial W_{\mathrm{diss}}}{V \partial t} \right\rangle = 2\varepsilon_0\omega\mathrm{Im}\varepsilon|E_0|^2 e^{-\alpha z} \tag{13.22}$$

Thus, light absorption (or energy dissipation) is observed when $\mathrm{Im}\varepsilon \neq 0$ is fulfilled.

13.5.2 Interface Reflection

In practice, one often has to perform calculations of different spectra with the purpose to compare them with experimentally measured ones. One of the simplest tasks is the calculation of an absorption spectrum according to the definition (7.26). According to our considerations from Sect. 13.4, the calculation of any absorption spectrum will however contain at least two different parts: First of all, the optical constants of the material must be calculated. Secondly, one has to solve the wave Eq. (13.11) to account for the particular *geometry* valid for the (given or assumed) experiment. Having solved the wave equation with realistic boundary conditions, we obtain electric and/or magnetic fields that may be converted into light intensities, which in turn may be compared with experimental data. Changing the systems geometry will change the intensities obtained at the output, although the material might be the same. For example, in Sect. 13.3 we have solved the wave equation, assuming that the dielectric function is the same at any point. In other words, we assumed a completely homogeneous medium, particularly without any interfaces. That resulted in Beer's law (13.19), but the latter must not be applied in other geometries.

In summary, *both* material and geometry specifics must be considered in any spectra calculation. The situation is completely analogous to the approach used in electricity: In order to calculate the electric resistance of a resistor, you need to know a material parameter (the specific resistance or conductivity), as well as the geometry (in simplest cases, length and cross-section area of the resistor).

While the material information is contained in quantities like the dielectric function of the optical constants, the geometry of an object is connected to its surface and maybe internal interfaces. Any light incident to an object will first interact with its surface. Therefore, the specific phenomena occurring on surfaces are utmost important for a correct quantitative interpretation of any real optical spectrum. This short section is to make you familiar with the simplest relations that describe the reflection of electromagnetic radiation at plane interfaces.

Transmission and reflection of light at surfaces and interfaces of the objects surrounding us are utmost important for creating images of our environment by viewing—compare Fig. 13.3. It is therefore essential to extend our theoretical approach to quantifying interface reflections—at least in simplest geometries.

Let us restrict on the case of two *homogeneous, isotropic and non-absorbing media* with real refractive indices n_1 and n_2, separated by a plane and smooth interface. Assume that a plane light wave is incident from medium 1 (electric field strength E_e), while the interaction with the interface results in the formation of a reflected plane wave (electric field strength E_r) traveling back into medium 1, and a transmitted plane wave (electric field strength E_t) propagating into medium 2. Then, at normal incidence, the electric field strength will be parallel to the interface. According to Table 12.2, in this case, Maxwell's boundary conditions at the interface between media 1 and 2 result in: $\underbrace{E_e + E_r}_{\text{in medium 1}} = \underbrace{E_t}_{\text{in medium 2}}$.

Fig. 13.3 Thuringian
dung-beetle: note the dark
blue color obtained as a result
of daylight reflection at the
surface of the animal

On the other hand, energy conservation gives for the intensities I:

$$I_e = I_t + I_r$$

The intensity (energy per time and area) may be expressed as the product of the energy volume density and the propagation velocity. In a dense nonabsorbing medium, (2.6) is modified therefore. The energy volume density is proportional to $\varepsilon\varepsilon_0 E^2 \propto n^2 E^2$, while the propagation velocity is proportional to $\frac{c}{n}$. One of the n cancels out when forming the product, and instead of (2.6), we now have:

$$I = \frac{n}{2\mu_0 c}\left|E_{0,\text{real}}\right|^2 = \frac{2n}{\mu_0 c}|E_0|^2 \propto n|E_0|^2$$

Here E_0 is the amplitude of the electric field in the corresponding wave. Then, we have:

$$\left.\begin{array}{c} n_1 E_e^2 = n_1 E_r^2 + n_2 E_t^2 \\ E_e + E_r = E_t \end{array}\right\} \Rightarrow \frac{E_r^2}{E_e^2} = \left(\frac{n_1 - n_2}{n_1 + n_2}\right)^2 \tag{13.23}$$

The intensity reflection coefficient R (or *reflectance*) is therefore:

$$R = \frac{I_r}{I_e} = \left(\frac{n_1 - n_2}{n_1 + n_2}\right)^2 \tag{13.23a}$$

Thus, from (13.23a) it follows that the larger the refractive index contrast is, the larger the normal incidence reflectance will be. In the case of absorbing media, the refractive indices have to be replaced by their complex counterparts according to (13.18). Then, the real reflectance R is given by:

$$R = \left| \frac{n_1 - \hat{n}_2}{n_1 + \hat{n}_2} \right|^2 \tag{13.24}$$

It should be mentioned that as a result of the dispersion of the optical constants, the reflectance turns out to be frequency dependent as well.

Remark Formally, Eqs. (13.23)–(13.23a) describe a beam-splitting process, where an incident light beam (e) is split into a transmitted (t) and a reflected (r) light beam. Thereby, Eq. (13.23a) describes a classical splitting by light intensity. In a quantum mechanical picture that means, that a part of the incident photons is directed into the reflected beam, and the rest into the transmitted.

But what will happen if there is only one single photon incident to the interface? Obviously, our classical result is no more of use here, because it is impossible that a part of the photon is reflected, and the other part transmitted. Obviously, we have to turn to the quantum mechanical field description in such a case.

In fact there exists a quantum mechanical approach to beam splitting and optical mixing [2], which we may adapt to our situation. However, once we have only dealt with the description of the electric field in vacuum so far, we start with a simplified situation. Thus, (5.52) together with (5.61) gives the operator of the electric field in vacuum according to ($\mathbf{e}_p$ indicates a unit vector describing the polarization state):

$$\hat{\mathbf{E}} = \sqrt{\frac{\hbar\omega}{2\varepsilon_0 V}} \left[\mathbf{e}_p e^{-i(\omega t - \mathbf{kr})} \hat{a} + \mathbf{e}_p^* e^{+i(\omega t - \mathbf{kr})} \hat{a}^+ \right]$$

Then, if we would have vacuum on both sides of the beam-splitting interface, from.

$E_e + E_r = E_t \Rightarrow E_e = E_t - E_r$ we could write down the operator relation (assuming identical polarization in all beams) [2]:

$$\hat{a}_e^+ = t\hat{a}_t^+ - r\hat{a}_r^+$$

Here we have introduced the electric field transmission and reflection coefficients t and r according to:

$$t = \frac{E_t}{E_e}; \quad r = \frac{E_r}{E_e}$$

The photon creation operators $\hat{a}_e^+$, $\hat{a}_t^+$, $\hat{a}_r^+$ are responsible for photon creation in the incident, transmitted, or reflected beams, respectively.

Let us now assume that we have one single photon incident in path (e) and no photons in the paths (r) and (t). Such a state can be written as:

$$|1\rangle_e |0\rangle_r |0\rangle_t = \hat{a}_e^+ |0\rangle_e |0\rangle_r |0\rangle_t$$

Let us now write this state as a superposition of the two states $|0\rangle_e |1\rangle_r |0\rangle_t$ (one photon reflected, no photon in any of the other paths) and $|0\rangle_e |0\rangle_r |1\rangle_t$ (one photon transmitted, no photon in any of the other paths). We write:

$$|1\rangle_e |0\rangle_r |0\rangle_t = \hat{a}_e^+ |0\rangle_e |0\rangle_r |0\rangle_t = \left(-r\hat{a}_r^+ + t\hat{a}_t^+\right)|0\rangle_e |0\rangle_r |0\rangle_t$$
$$= \underbrace{-r|0\rangle_e |1\rangle_r |0\rangle_t}_{W(r)=|r|^2} + \underbrace{t|0\rangle_e |0\rangle_r |1\rangle_t}_{W(t)=|t|^2}$$

This relation makes it clear, that the state with the one photon incident represents a biased superposition of the states with one photon transmitted and one photon reflected. Therefore, in case of transmission or reflection measurements, we will never observe situations where *parts* of the photon are found in transmission or reflection, instead, we will observe the whole photon either reflected with the probability $W(r) = |r|^2 \equiv R$, or the photon is transmitted with the probability $W(t) = |t|^2 \equiv T$. Also, we have $T + R = 1$. This is a quite reasonable result.

Within a dielectric medium, this expression must be modified. First of all, from the corresponding expression for the energy density, we have to substitute:

$$\varepsilon_0 \rightarrow \varepsilon\varepsilon_0 = n^2\varepsilon_0$$

With n—refractive index. We must further consider that in a medium with refractive index n, the wavelength is modified $\left(\lambda \rightarrow \frac{\lambda}{n}\right)$. Therefore, in a medium with refractive index n, it is the reference volume V/n^3 that contains the same number of modes as the volume V in vacuum. We substitute:

$$V \rightarrow \frac{V}{n^3} \Rightarrow \sqrt{\frac{1}{\varepsilon_0 V}} \rightarrow \sqrt{n}\sqrt{\frac{1}{\varepsilon_0 V}}$$

Therefore, in a phenomenological fashion we further substitute

$$\hat{\mathbf{E}} \rightarrow \sqrt{n}\sqrt{\frac{\hbar\omega}{2\varepsilon_0 V}}\left[\mathbf{e}_p e^{-i(\omega t - \mathbf{kr})}\hat{a} + \mathbf{e}_p^* e^{+i(\omega t - \mathbf{kr})}\hat{a}^+\right]$$

Then, instead of the above-used operator relation $\hat{a}_e^+ = t\hat{a}_t^+ - r\hat{a}_r^+$ we get the modified relation:

$$\sqrt{n_1}\hat{a}_e^+ = \sqrt{n_2}t\hat{a}_t^+ - \sqrt{n_1}r\hat{a}_r^+$$

Here n_1 is the refractive index of medium 1, where the incident and reflected modes are propagating. Instead, n_2 is the refractive index of medium 2, where the transmitted mode may be observed. This results

$$|1\rangle_e|0\rangle_r|0\rangle_t = \hat{a}_e^+|0\rangle_e|0\rangle_r|0\rangle_t = \left(-r\hat{a}_r^+ + \sqrt{\frac{n_2}{n_1}}t\hat{a}_t^+\right)|0\rangle_e|0\rangle_r|0\rangle_t$$

in:
$$= \underbrace{-r|0\rangle_e|1\rangle_r|0\rangle_t}_{W(r)=|r|^2} + \underbrace{\sqrt{\frac{n_2}{n_1}}t|0\rangle_e|0\rangle_r|1\rangle_t}_{W(t)=\frac{n_2}{n_1}|t|^2} \qquad \text{Again, the}$$

state with the one photon incident represents a superposition of the states with one photon transmitted and one photon reflected. We will observe the photon either reflected with the probability $W(r) = |r|^2 \equiv R$, or the photon is transmitted with the probability $W(t) = \frac{n_2}{n_1}|t|^2 \equiv T$. Note that the expression $\frac{n_2}{n_1}|t|^2 \equiv T$ coincides with the usual expression for the classical intensity transmission coefficient at normal light incidence as it might be obtained from $T \equiv \frac{I_t}{I_e} = \frac{n_2|E_t|^2}{n_1|E_e|^2} \equiv \frac{n_2}{n_1}|t|^2$.

So far we have discussed normal incidence reflectance phenomena at plane smooth interfaces only. For a detailed discussion of oblique incidence phenomena, I would like to refer you, dear reader, to reference [3, pp. 103–109].

In many practical situations, diffuse reflection phenomena at rough surfaces or in turbid media are essential for understanding the optical appearance of an object. We will not go into details here, but finish this section with a rather qualitative discussion of the spectrally selective absorbers as introduced in Sect. 7.5.2. Spectrally selective absorbers usually show a frequency-dependent reflectance. This is especially simple to be understood if the absorber is thick enough that it does not transmit light. Then, (specular + diffuse) reflectance and absorptance must sum up to 1, and therefore, if the absorptance is frequency-dependent, the reflectance must also. Therefore, when being illuminated by white light, spectrally selective absorbers appear colored.

On the other hand, when the reflectance is nearly constant over the visible spectral range, the corresponding object will appear in gray (or white, if $R = 1$) when being illuminated with white light. Moreover, when being illuminated by colored light, the reflected light will have the same color. Therefore, gray bodies appear in the color of the light incident to them. Figure 13.4 shows corresponding examples of diffuse light reflection at a rock formation.

Qualitatively the same result will be obtained from (13.24). Indeed, when the extinction coefficient is frequency dependent, the reflectance according to (13.24) will be frequency dependent as well. As we will see later in Chap. 15, the situation is even more complex: a frequency-dependent absorption must be accompanied by a well-defined frequency-dependent refraction. Therefore, the reflectance according to (13.24) will show a very specific behavior in the vicinity of absorption features, such that reflection spectroscopy appears to be an important analytical tool. As an example, in Sect. 14.4 the reflectance in the vicinity of absorption lines will be illustrated.

Fig. 13.4 Gray bodies in different illumination conditions: The Schafalpenköpfe (view from the Kleinwalsertal, Austria, July 2020) in sunset conditions. The white numbers indicate the daytime

13.6   Advanced Material

13.6.1 Orientation Polarization

Let us assume a material that is built from permanent microscopic electric dipoles (compare Sect. 12.5). The dipoles are allowed to rotate with some damping. This is the typical situation in a liquid built from polar molecules (e.g., water). When no external electric field is applied, the stochastic thermally activated movement of the dipoles shall not be able to create a macroscopic polarization (compare

Sect. 12.5.2). However, in an external electric field, the dipoles will more or less align with the field, creating a resulting macroscopic polarization. We shall find the frequency dependence of the dielectric function (and consequently of the optical constants) of such a material.

We will solve this task by a direct application of (13.7). Because we still do not know the response function $\kappa(\xi)$, we start from the following thought experiment:

Let us assume that a static electric field has been applied to the system for a sufficiently long time, so that a static polarization of the liquid has been well established. Let us further assume that the field is switched off at the moment $t = 0$. We model this situation by means of the electric field:

$$E(t) = E_0[1 - \theta(t)]$$

where $\theta(t)$ is a step function that has the value one for $t \geq 0$ and zero elsewhere. It makes no sense to assume that the polarization will vanish instantaneously with a vanishing external field. On the contrary, in our model we shall assume that due to the thermal movement of the particles, the macroscopic polarization decreases smoothly and asymptotically approaches the value of zero. This situation may be approximated by an exponentially descending behavior with a time constant τ according to:

$$P(t) = P_0 e^{-\frac{t}{\tau}}; t > 0$$

Furthermore, from (13.5) we have:

$$P(t) = P_0 e^{-\frac{t}{\tau}} = \varepsilon_0 \int_0^\infty \kappa(\xi) E_0[1 - \theta(t - \xi)]d\xi$$

The only action of the step function is to reduce the integration interval:

$$P_0 e^{-\frac{t}{\tau}} = -\varepsilon_0 E_0 \int_\infty^t \kappa(\xi)d\xi$$

We differentiate with respect to time and make use of the identity:

$$f(x) = \frac{d}{dx}\left[\int_a^x f(\xi)d\xi\right]$$

That leads us to the following expression for the response function $\kappa(t)$:

$$\kappa(t) = \frac{P_0}{\varepsilon_0 E_0 \tau}e^{-\frac{t}{\tau}} \equiv \kappa_0 e^{-\frac{t}{\tau}} \tag{13.25}$$

Having found the response function, the further treatment is straightforward [4]. Equations (13.7) and (13.9) yield the dielectric function:

$$\varepsilon(\omega) = 1 + \chi(\omega) = 1 + \int_0^\infty \kappa(\xi)e^{i\omega\xi}\,d\xi = 1 + \kappa_0 \int_0^\infty e^{\left(i\omega - \frac{1}{\tau}\right)\xi}\,d\xi = 1 + \frac{\kappa_0\tau}{1 - i\omega\tau}$$

Or

$$\varepsilon(\omega) = 1 + \frac{\chi_{\text{stat}}}{1 - i\omega\tau} \tag{13.26}$$

where χ_{stat} is the static ($\omega = 0$) value of the susceptibility [the last term in (12.22)]. The real and imaginary parts of the dielectric function may be written as follows:

$$\text{Re}\varepsilon = 1 + \frac{\chi_{\text{stat}}}{1 + \omega^2\tau^2}; \text{Im}\varepsilon = \frac{\chi_{\text{stat}}\omega\tau}{1 + \omega^2\tau^2} \tag{13.27}$$

The thus obtained dielectric function represents a simplified version of Debye's equations valid for the dielectric function in polar viscous media. In Fig. 13.5, the spectral shapes of real and imaginary parts of this particular dielectric function are presented. Figure 13.6 shows the corresponding optical constants. In these Figures, a static susceptibility of $\chi_{\text{stat}} = 80$ has been assumed, similar to what is observed in ordinary water. Obviously, the presence of permanent dipoles in the medium results in a large static dielectric constant, while for higher frequencies, the real part of the dielectric function may be essentially lower. Thus, in the visible spectral range, water has a dielectric function with a real part of approximately 1.77 and a refractive index of 1.33. This behavior is in qualitative consistency with the predictions from Debye's equations, where the refractive index is expected to steadily decrease with increasing frequency.

There is another interesting fact that becomes obvious from Fig. 13.5. The imaginary part of the dielectric function has its maximum value exactly at the angular frequency $\omega = \tau^{-1}$. Consequently, the result of a *spectral* measurement

Fig. 13.5 Real and imaginary parts of the dielectric function according to (13.27)

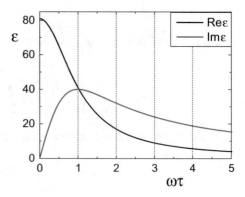

Fig. 13.6 Optical constants n and K for the dielectric function presented in Fig. 13.5, but in a broader spectral region

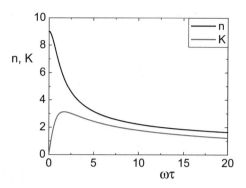

(determining the peak position of Imε) reveals information about the *dynamic* behavior of the system (the decay time of polarization). This is one example for the validity of a more general fundamental principle, that in optics the spectral ($\chi(\omega)$) and time domain ($\kappa(t)$) representations embody the same information and may be transferred into each other. Indeed, (13.7) is in fact a Fourier transformation of the response function, performed however only over a semi-infinite interval for reasons of causality. One may formally multiply the response function with a step function:

$$\tilde{\kappa}(\xi) = \kappa(\xi)\theta(\xi) \tag{13.28}$$

The thus obtained modified response function may be integrated over the full time interval, so that we have:

$$\chi(\omega) = \int_{-\infty}^{\infty} \tilde{\kappa}(\xi)e^{i\omega\xi}\,d\xi \tag{13.29}$$

Remark Note that (13.27) is nevertheless only a theoretical model of a dielectric function, because the exponential decay of the free polarization resulting in the time dependence of the response function (13.25) was simply postulated. We established a reasonable agreement with experimental findings in polar materials at lower frequencies, but at higher frequencies, peculiarities will occur. Thus, when combining the expression for energy dissipation (13.22) with the dielectric function (13.27), we observe that the energy dissipation approaches a constant positive value when the frequency approaches infinity [5]. This is really peculiar, because at highest frequencies, one should expect that all degrees of freedom within the material system are too inert to comply with the rapid oscillations of the electric field. Therefore, the wave should propagate through the medium without interaction, i.e., energy dissipation should not occur. Somewhere, in our model derivation, relevant physical interaction mechanisms must have been neglected. Let us mention at least two of them:

- Firstly, our assumption on rigid permanent dipoles is a rather crude model, applicable at low frequencies only. At higher frequencies, other degrees of freedom such as the oscillations of nuclei and electrons will dominate the interaction of the polar molecules with electromagnetic radiation.
- Secondly, our idea on forming a macroscopic polarization by the alignment of the permanent microscopic dipoles with the field does not work at smallest wavelength. In fact, reorientation will be caused by a torque acting on the dipole in a homogeneous electric field. This idea loses its sense when the field is inhomogeneous, what certainly happens when the wavelength of the propagating electromagnetic wave becomes comparable to the spatial extension of the dipole. Then, the dielectric function will explicitly depend on the wavevector, a phenomenon that is called spatial dispersion, but this is not taken into account in terms of a response function like (13.25). Some more information on spatial dispersion will be provided in Sect. 17.6.

13.6.2 Material Mixtures

Let us have a glance at a specific topic of applied material optics, namely the optical properties of mixtures. Let us assume that we have prepared an optical material as a physical mixture of a certain set of pure material, numbered by the subscript j. Let each of the constituents occupy a volume fraction V_j of the whole volume V. From $V = \sum_j V_j$ we find $1 = \sum_j \frac{V_j}{V} \equiv \sum_j p_j$. The p_j are called *volume filling factors* of the j-th component in the mixture.

It is a common practice to quantify the mixing ratio in optics in terms of volume filling factors, instead of working with mass contributions or molar fractions. The reason is simple: The macroscopic polarization, which is in the basis of all our dielectric theory, is defined as the dipole moment per volume, and not per mass or molar fraction [compare (12.7)].

Let us now assume that we have been able to prepare a mixture from a certain set of constituents, each of them being characterized by its dielectric function and a certain volume fraction. Are we able to predict the optical properties of the mixture?

In order to clarify this point, we shall return to the Clausius–Mossotti Eq. (12.16). In application to dielectric functions (instead of the static dielectric constant), we have:

$$\frac{\varepsilon - 1}{\varepsilon + 2} = \frac{N\beta}{3}$$

(12.16) thus relates the microscopic polarizability as well as the concentration of identical polarizing units to the dielectric function of the corresponding continuum in the frames of the so-called quasistatic approximation (compare Sect. 12.4). From a practical point of view, (12.16) offers a classical approach to modify the dielectric function (or correspondingly the optical constants) of a continuum by changing its density. This is of course practicable in only rather narrow limits, and therefore, any other mechanism would be welcome in order to tailor the optical constants of a material to a desired value.

Intermixing two or more materials (i.e., the preparation of material mixtures) offers the possibility to tailor optical constants in rather broad limits. Let us assume that we have several materials numbered by the subscript j, each of them having the polarizability β_j. Instead of (12.16), we then have:

$$\frac{\varepsilon - 1}{\varepsilon + 2} = \sum_j \frac{N_j \beta_j}{3} \tag{13.30}$$

ε has now to be regarded as the dielectric function of the mixture. Figure 13.7 on right illustrates such a situation for the special case of two mixing partners.

Usually, the polarizabilities β_j are unknown, instead we know the dielectric functions of the mixing partners when being prepared as a pure material (Fig. 13.7, left and center). We thus set:

$$\frac{\varepsilon_j - 1}{\varepsilon_j + 2} = \frac{N_{0j} \beta_j}{3} \Rightarrow \beta_j = \frac{3}{N_{0j}} \frac{\varepsilon_j - 1}{\varepsilon_j + 2} \tag{13.31}$$

Pure material 1: ε_1 Pure material 2: ε_2 Mixture: ε

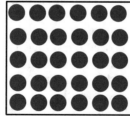
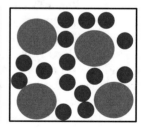

$$N_{01} = \frac{8}{V} \qquad N_{02} = \frac{30}{V} \qquad N_1 = \frac{4}{V}; \; N_2 = \frac{15}{V} \Rightarrow$$

$$p_1 = \frac{N_1}{N_{01}} = 0.5$$

$$p_2 = \frac{N_2}{N_{02}} = 0.5$$

Fig. 13.7 Illustration of a material mixture

Here, the N_{0j} represent the concentration of dipoles in the corresponding pure materials. When substituting (13.31) into (13.30), we finally have:

$$\frac{\varepsilon - 1}{\varepsilon + 2} = \sum_j \frac{N_j \beta_j}{3} = \sum_j \frac{N_j}{N_{0j}} \frac{\varepsilon_j - 1}{\varepsilon_j + 2} \equiv \sum_j p_j \frac{\varepsilon_j - 1}{\varepsilon_j + 2} \tag{13.32}$$

Here we have introduced the volume filling factors p_j according to:

$$p_j \equiv \frac{N_j}{N_{0j}} \tag{13.33}$$

This is again illustrated in Fig. 13.7, for the special case of a binary mixture with $p_1 = p_2 = 0.5$.

Note that all physical restrictions as discussed earlier in Sect. 12.4 remain valid. In particular, throughout our derivation we have assumed that the physical mixing process does <u>not</u> change the individual polarizabilities, in clear contrast to what would be observed in chemical bonding.

Equation (13.32) practically coincides with the Lorentz–Lorenz mixing formula (13.34):

$$\frac{\varepsilon - 1}{\varepsilon + 2} = \frac{\hat{n}^2 - 1}{\hat{n}^2 + 2} = \sum_j p_j \frac{\hat{n}_j^2 - 1}{\hat{n}_j^2 + 2} \tag{13.34}$$

Clearly, in such a mixture, the choice of a proper mixing ratio as expressed in terms of the filling factors allows tailoring the optical constants of the mixture within rather broad limits.

In practice, (13.34) provides a manageable expression for predicting the optical properties of loosely packed dielectric mixtures. Clearly, the sum of the filling factors in (13.34) must not exceed 1; if it is smaller than 1, then the rest of the volume is "automatically" occupied by a medium with a dielectric function equal to one, i.e., by air or vacuum. This is consistent with our derivation of the Clausius–Mossotti equation, where the dielectric sphere was assumed to polarize in a cavity that is empty, i.e., "filled with vacuum". Note that in this case, the local field was calculated according to (12.15):

$$E_{\text{loc}} = \frac{\varepsilon + 2}{3} E$$

Let us now have a glance at Fig. 13.8. It shows a transmission electron microscopy (TEM) image of a mixture of aluminum oxide Al_2O_3 (the bright fraction) and silver particles (the dark spots). The bar in the left corner on the bottom of the image indicates a length of 20 nm.

The silver particle size is well below the wavelength in the visible spectral region, so that we will treat the material as optically homogeneous. In particular that means that we may make use of the quasistatic approximation. In the

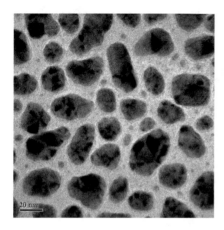 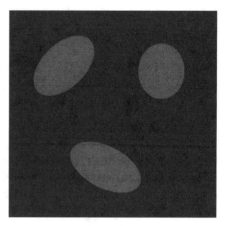

Fig. 13.8 On left: TEM-image of a composite thin film built from Al_2O_3 with embedded isolated silver particles; on right: schematic representation of a guest (red)—host (navy) system

present sample, the volume filling factor of the silver fraction is approximately 0.3. Accordingly, that of Al_2O_3 is 0.7.

In contrast to what has been illustrated in Fig. 13.7, Fig. 13.8 on left represents an example of a guest–host system: The dark inclusions obviously form a guest fraction that is embedded in the bright host. Note the difference between the schematic representations of mixtures shown on right of Figs. 13.7 and 13.8: We cannot expect that (13.34) is automatically a good approach for modeling the guest–host system.

The difference becomes clear when remembering the derivation of the Clausius–Mossotti equation. Let us return to Fig. 12.6 on bottom. It demonstrates the creation of a spherical cavity in a bulk material by removing a dielectric sphere from that bulk. Note that the surface charges at the border of the cavity and the sphere exactly compensate each other in the bulk. Therefore, when deriving the Clausius–Mossotti Equation, we could assume a homogeneous polarization, as well as homogeneous fields.

If, however, the sphere and the embedding material represent different materials, than those surface charges will no more exactly cancel out each other, and instead, an excess dipole moment is formed at the borders of the inclusion. That excess dipole moment may be calculated explicitly, formally replacing (12.15) by the corresponding expression that relates the electric field in a dielectric sphere embedded in another host material to the field applied in the host (task 13.7.5). The result is rather transparent: Instead of (12.15), we now obtain (13.35), where compared to (12.15) the dielectric function of the sphere is replaced by the ratio of the dielectric functions of sphere and host. Hence, we replace:

$$E_{loc} = \left.\frac{\varepsilon + 2}{3} E\right|_{\varepsilon \to \frac{\varepsilon}{\varepsilon_h}} \rightarrow \frac{\left(\frac{\varepsilon}{\varepsilon_h}\right) + 2}{3} E = \frac{\varepsilon + 2\varepsilon_h}{3\varepsilon_h} E \qquad (13.35)$$

This leads to a corresponding modification in (13.32). We now obtain the more general mixing formula (13.36):

$$\frac{\varepsilon - 1}{\varepsilon + 2} = \sum_j p_j \frac{\varepsilon_j - 1}{\varepsilon_j + 2} \Bigg|_{\varepsilon \to \frac{\varepsilon}{\varepsilon_h}} \quad \to \quad \frac{\varepsilon - \varepsilon_h}{\varepsilon + 2\varepsilon_h} = \sum_j p_j \frac{\varepsilon_j - \varepsilon_h}{\varepsilon_j + 2\varepsilon_h} \tag{13.36}$$

The previously derived Lorentz–Lorenz approach appears to be a particular case of (13.36), namely when setting $\varepsilon_h = 1$. In more detail, basing on the idea of the excess dipole moment, (13.36) is derived in [3, pp. 57–64].

Equation (13.36) may now easily be applied to the situation shown in Fig. 13.8. When associating the host material (aluminum oxide) with material 2, and the guest fraction (the silver nanoparticles) with material 1, from (13.36) we obtain:

$$\frac{\varepsilon - \varepsilon_2}{\varepsilon + 2\varepsilon_2} = p_1 \frac{\varepsilon_1 - \varepsilon_2}{\varepsilon_1 + 2\varepsilon_2} \tag{13.37}$$

(13.37) represents a special case of the Maxwell Garnett mixing formula.

Of course, as seen from Fig. 13.8, in a real material, the inclusions may be of nonspherical shapes. In this case, we may again make use of the depolarization factors earlier introduced in Sect. 12.4 (Table 12.3). This results in:

$$\frac{\varepsilon + 2\varepsilon_h}{3\varepsilon_h} E \to \frac{\varepsilon_h + (\varepsilon - \varepsilon_h)L}{\varepsilon_h} E$$

$$\Rightarrow \frac{\varepsilon - \varepsilon_h}{\varepsilon + 2\varepsilon_h} = \sum_j p_j \frac{\varepsilon_j - \varepsilon_h}{\varepsilon_j + 2\varepsilon_h} \to \frac{(\varepsilon - \varepsilon_h)}{\varepsilon_h + (\varepsilon - \varepsilon_h)L} = \sum_j p_j \frac{(\varepsilon_j - \varepsilon_h)}{\varepsilon_h + (\varepsilon_j - \varepsilon_h)L} \tag{13.38}$$

Equation (13.38) represents a rather general optical mixing formula. Of course, all dielectric functions here may be complex and frequency dependent. The (effective) dielectric function of the mixture ε appears to depend on the dielectric functions of the constituents, their filling factors, and the morphology (via L). According to the choice of ε_h, the following classification of mixing models is accepted:

Maxwell Garnett (MG) approach: It might be the most natural choice to regard one of the constituents (say, the lth one) as the host material, and the others as the inclusions. In the case shown in Fig. 13.8, it clearly makes sense to regard silver as inclusion and the dielectric as the host. That is the philosophy of the Maxwell Garnett approach. In this case, we have:

$$\frac{(\varepsilon - \varepsilon_l)}{\varepsilon_l + (\varepsilon - \varepsilon_l)L} = \sum_{j \neq l} p_j \frac{(\varepsilon_j - \varepsilon_l)}{\varepsilon_l + (\varepsilon_j - \varepsilon_l)L} \tag{13.38a}$$

Note that the sum of the filling factors on the right hand is less than 1. In application, one must keep in mind that (13.38a) depends on the choice of the host function: It makes a difference whether material 1 is embedded in material 2 or vice versa.

Lorentz–Lorenz (LL) approach: The Lorentz–Lorenz approach assumes that all inclusions polarize in vacuum ($\varepsilon_h = 1$). We therefore obtain:

$$\frac{(\varepsilon - 1)}{1 + (\varepsilon - 1)L} = \sum_j p_j \frac{(\varepsilon_j - 1)}{1 + (\varepsilon_j - 1)L} \tag{13.38b}$$

Effective Medium Approximation (EMA) or Bruggeman approach: Another possibility is to assume that the dielectric function of the mixture itself acts as the host medium for each of the inclusions. This leads to the following mixing formula:

$$0 = \sum_j p_j \frac{(\varepsilon_j - \varepsilon)}{\varepsilon + (\varepsilon_j - \varepsilon)L} \tag{13.38c}$$

There is no general recipe which of these approaches works best. As a rule, the MG theory works best when the constituents clearly may be subdivided into inclusions and one matrix material. On the contrary, in the presence of percolation or in molecular mixtures, the application of the EMA may lead to the best results. Finally, highly porous materials might be well fitted within the LL approach.

Note that L is confined between $L = 0$ and $L = 1$. When assuming $L = 0$ and $\sum_j p_j = 1$, from (13.38) we immediately obtain:

$$\varepsilon = \sum_j p_j \varepsilon_j \tag{13.38d}$$

On the other hand, when setting $L = 1$, we have:

$$\varepsilon^{-1} = \sum_j p_j \varepsilon_j^{-1} \tag{13.38e}$$

In binary mixtures with real positive dielectric functions, the pair of (13.38d) and (13.38e) forms the so-called Wiener bounds of the dielectric function of the mixture. Any physically reasonable mixing model must deliver dielectric functions that fall in-between these bounds.

13.7   Tasks for Self-check

13.7.1 From $\hat{n} = \sqrt{\varepsilon}$, find explicit expressions for the real and imaginary parts of the complex index of refraction as a function of Reε and Imε.

13.7.2 Imagine a continuous medium with a dielectric function $\varepsilon = 5 + 0.1i$. Calculate the phase velocity of an electromagnetic wave traveling through that medium. Assuming a vacuum wavelength of 400 nm, what would be the penetration depth of electromagnetic irradiation into that medium? (Note: The penetration depth is defined as the geometrical path necessary for intensity damping inside the medium to a level of 1/e).

13.7.3 Solve the left side of (13.23) explicitly in order to obtain the final expression for the interface reflectance.

13.7.4 Imagine two slabs (with parallel surfaces) of the same material, but with different thicknesses $d_1 = 1$ mm and $d_2 = 5$ mm. At 500nm wavelength and normal light incidence, the first slab transmits 91.8% of the incident light intensity, while the second one transmits only 90.5%. Basing on these data, estimate n and K @ 500 nm of the slab material when neglecting any multiple internal light reflections in the slab!

13.7.5 Assume a sphere with the dielectric constant ε_1, embedded into an extended medium with dielectric constant ε_2. Assume further that in the embedding medium, a static electric field E_2 is applied that is homogeneous at large distances from the sphere. Assuming a homogeneous field E_1 inside the sphere, calculate the relation between E_1 and E_2!

13.7.6 The graph shows the refractive index (black) and the extinction coefficient (red) of a fictive material as a function of the wavenumber.

(a) Sketch the principal behavior of the corresponding dielectric function (real and imaginary parts) in the given wavenumber region! (it is not necessary to perform a point-by-point calculation)
(b) Imagine a light wave incident from air ($n = 1.0$) on the surface of that fictive material (normal incidence). Without calculation, sketch the principal shape of the reflectance as a function of wavenumber!
(c) Assume that in part (b), the sample is now immersed in a liquid. Therefore, the incidence medium has now a refractive index of 1.5. Sketch the behavior of the interface reflectance in this modified case. Show both the reflectance curves corresponding to the tasks (b) and (c) *in one graph*!

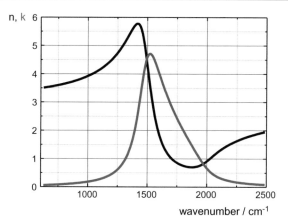

wavenumber / cm⁻¹

13.7.7 Imagine you have synthesized a new optical material, namely a binary mixture of metallic aluminum and dielectric stoichiometric aluminum oxide Al_2O_3. You wish to model the optical constants by a mixing model approach and need information on the volume filling factors of aluminum and aluminum oxide. From an elemental analysis, you know that the average ratio of atomic concentrations of aluminum and oxygen atoms is 50%/50%. What is the volume filling factor of the aluminum fraction?

References

Specific References

1. Gross/Marx, *Festkörperphysik*, 2nd edn (de Gruyter 2014), p. 582
2. H. Paul, *Photonen. Eine Einführung in die Quantenoptik* (Teubner Studienbücher Physik, 1995), pp. 261–265
3. O. Stenzel, *The Physics of Thin Film Optical Spectra. An Introduction* (Springer, Berlin, 2016)
4. R. Lenk (ed.), *Brockhaus Physik,* 2nd edn, vol.1 (VEB Brockhaus Verlag 1989z), p. 140
5. H.J. Goldsmid (ed.), *Problems in Solid State Physics* (Academic Press, 1968, task 7.5)

General Literature to this Chapter

6. O. Stenzel, *The Physics of Thin Film Optical Spectra. An Introduction* (Springer, Berlin, 2016).
7. M. Born, E. Wolf, *Principles of Optics* (Pergamon Press, Oxford London Edinburgh New York Paris Frankfurt, 1968)
8. L..D. Landau, E.M. Lifshitz, Electrodynamics of Continuous Media. *A Course of Theoretical Physics*, vol. 8 (Pergamon Press, 1960)
9. M. Schubert, B. Wilhelmi, *Einführung in die nichtlineare Optik I* (Teubner Verlagsgesellschaft Leipzig, BSB B. G, 1971)
10. C.F. Klingshirn, *Semiconductor Optics* (Springer, Berlin, Heidelberg, New York, 1997)

Linear Optical Constants II: Classical Dispersion Models

14

Abstract

Classical expressions for the dielectric function of continuous media in the presence of bound and/or free charge carriers are derived. The focus is on the single and multioscillator models. Application examples include modeling of metal and dielectric materials optical constants in various spectral regions.

14.1 Starting Point

We are now going to build a classical theory of the dispersion of optical constants in a continuous medium. Again we start from Eq. (2.13), i.e., Newton's law of motion. In agreement to our assumptions made in Sects. 2.2 and 12.3, we assume the existence of charge carriers in the medium that are accelerated by external forces. The full external force contains contributions of the Coulomb force arising from the electric field of the light wave, as well as some frictional force arising from collisions of the charge carriers with any kind of collision partner. As in Chap. 12, we formally discriminate between free and bound charge carriers: While bound charge carriers feel an additional restoring force when being elongated from their initial position, the free charge carriers do not. In agreement with the neglected magnetization of the medium, the action of any Lorentz forces will again be neglected.

14.2 Physical Idea

Of course, real materials may contain both free and bound electrons. Fortunately, as charges are additive, all the degrees of freedom present in real matter will contribute their dipole moments to the final polarization that is obtained as a sum over all dipole moments in the medium. Consequently, the susceptibilities that

© The Author(s), under exclusive license to Springer Nature Switzerland AG 2022   359
O. Stenzel, *Light–Matter Interaction*, UNITEXT for Physics,
https://doi.org/10.1007/978-3-030-87144-4_14

correspond to different degrees of freedom (numbered by j) sum up to the full susceptibility, so that the dielectric function will be:

$$\varepsilon(\omega) = 1 + \sum_j \chi_j(\omega) \qquad (14.1)$$

where the χ_j are the susceptibilities obtained for the corresponding groups of dipoles.

The idea is now to discuss the individual contributions to (14.1) separately. This is of course an approximation only, but in our classical treatment as provided here, it allows constructing a rather transparent physical picture on dispersion and absorption phenomena in continuous media.

We will start with the simpler case of free electrons that are accelerated by the electric field of the light waves. The positive atomic cores are much heavier than the electrons, and therefore, the cores will be considered as fixed. Consequently in our model, only the electrons are in motion when a harmonic electric field is applied. Nevertheless, in the following derivation, we will speak on charge carriers, keeping in mind that usually we deal with free electrons.

14.3 Theoretical Material

14.3.1 Free Charge Carriers and Drude Function

Assuming that the motion of electrons is confined to a region much smaller than the wavelength, we may write for the movement of a single charge carrier [compare (2.13) and (12.17)]:

$$qE = qE_0 e^{-i\omega t} = m\ddot{x} + 2\gamma m\dot{x} \qquad (14.2)$$

m and q are the mass and charge of the charge carrier, and γ is a damping constant necessary to consider the damping of the electrons movement. We assume that the electric field is polarized along the x-axis, hence we consider only movements of the charge carrier along the x-axis. As in Sect. 12.3.5, we consider only the drift motion of the charge carriers here, because the chaotic isotropic thermal movement does not create a macroscopic dipole moment. For nonrelativistic velocities, the Lorentz force may be neglected compared to the Coulomb force, so that only the latter is apparent in (14.2).

Assuming $x(t) = x_o e^{-i\omega t}$, we obtain from (14.2): $\frac{qE}{m} = -\omega^2 x - 2i\gamma\omega x$.

The oscillation of the charge carrier around its equilibrium position thus induces an oscillating dipole moment according to:

$$d = qx = -\frac{q^2 E}{m} \frac{1}{\omega^2 + 2i\gamma\omega}$$

If N is the number of free charge carriers per unit volume (the *concentration of free charge carriers*), then the polarization P is given by

$$P = Nd = -\frac{q^2 N E}{m} \frac{1}{\omega^2 + 2i\gamma\omega}$$

so that, according to (13.8), the susceptibility is:

$$\chi(\omega) = -\frac{Nq^2}{\varepsilon_0 m} \frac{1}{\omega^2 + 2i\gamma\omega} \tag{14.3}$$

where the term $\frac{Nq^2}{\varepsilon_0 m}$ represents the square of the plasma frequency defined as:

$$\omega_p = \sqrt{\frac{Nq^2}{\varepsilon_0 m}} \tag{14.4}$$

As predicted in Sect. 13.3, our derivation results in a complex and frequency-dependent susceptibility. The corresponding dielectric function is given by:

$$\varepsilon(\omega) = 1 - \frac{\omega_p^2}{\omega^2 + 2i\gamma\omega} \tag{14.5}$$

Figure 14.1 displays the principle shape of the real and imaginary parts of the dielectric function from (14.5), as well as the corresponding optical constants. The most striking feature appears in the refractive index, which is expected to be less than one in broad spectral regions. In fact, the imaginary part of the complex refractive index may be much larger than the real one. This is typical for metals, and it causes the well-known metallic brightness in reflection at a metal surface. Indeed, let us calculate the normal incidence reflection from a metallic surface according to (13.24). When assuming air as incidence medium ($n_1 = 1$) and the second medium as the metal ($\hat{n}_2 = n + iK$), from (13.24) we have:

$$R = \left|\frac{n_1 - \hat{n}_2}{n_1 + \hat{n}_2}\right|^2 = \frac{(1 - n)^2 + K^2}{(1 + n)^2 + K^2}$$

This reflectance approaches $R = 1$ when n becomes significantly smaller than 1.

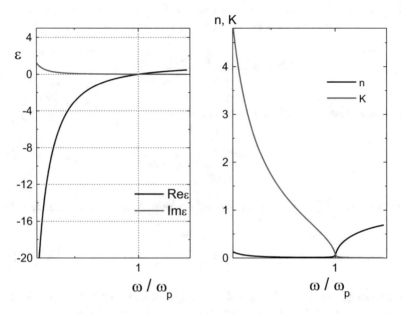

Fig. 14.1 Dielectric function and optical constants according to (14.5)

14.3.2 Specifics of Ultrathin Metal Films

So far we have assumed that the damping parameter γ is determined by the frequency of collisions the charge carrier suffers during its propagation in the medium. Thus, in a bulk metal, γ is somehow related to the mean free path of the charge carriers l_{free}. The smaller the mean free path is, the larger γ will be.

However, in ultrathin metal films, collisions between charge carriers and the film surface may dominate over bulk collision effects, and this may lead to a significant reduction in the average mean free path. According to the classical Drude theory, this will be accompanied by an increase in the Drude damping parameter. A further decrease in film thickness should be accompanied by a further increase in γ.

This is nevertheless a strongly simplified picture, because it makes a difference whether the charge carrier is specularly of diffusely scattered at the film surface. A specular reflection has no impact on the drift velocity parallel to the surface, while diffuse reflection has. Hence, only the latter has to be taken into account when calculating the effect of the film thickness on the Drude damping parameter.

As is shown in earlier studies [1, 2], the simplest model treatment of the reduction in the mean free path of the charge carriers results in a thickness-dependent average collision time τ given by:

$$\tau(d_{\text{film}}) = \frac{\tau_{\text{bulk}}}{\left[1 + 2\left(1 - p_{\text{spec}}\right)\frac{l_{\text{free}}}{d_{\text{film}}}\right]}$$

Fig. 14.2 $\Gamma \equiv \frac{\gamma}{2\pi c}$ versus film thickness for copper and gold films, prepared by evaporation on fused silica surfaces [3]. Symbols: experiment; lines: fit by (14.6)

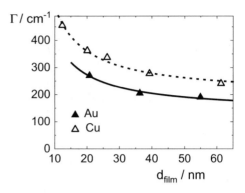

Here, p_{spec} is the relative amount of charge carriers that is specularly reflected at the film surface, and l_{free} the mean free path in the bulk material. d_{film} is the film thickness. Note that in this approach, only charge carriers reflected diffusely from the surface contribute to the mentioned thickness dependence. Then, we obtain a manageable expression for the thickness-dependent Drude damping parameters in thin films according to:

$$\gamma(d_{film}) = \gamma_{bulk}\left[1 + 2(1 - p_{spec})\frac{l_{free}}{d_{film}}\right] \qquad (14.6)$$

Obviously, $p_{spec} \to 1$ as well as $d_{film} \to \infty$ result in $\gamma(d_{film}) \to \gamma_{bulk}$. Otherwise, the Drude parameter in the film is expected to exceed the bulk value. Figure 14.2 demonstrates some corresponding experimental results.

14.3.3 Bound Charge Carriers and Lorentzian Oscillator Model

Even in metals, most of the electrons are bound, although the free electrons are utmost important for the specific optical behavior of metals. As everybody knows, metals like silver, gold, and copper have quite a different optical appearance, and this is a consequence of the response of the *bound* electron fraction. Of course, the optical properties of pure dielectrics are exclusively determined by the motion of bound charge carriers.

There is a more general question concerning the different role of negative electrons and positively charged cores. Generally, both electrons and cores may perform movements when being excited by external electric fields. But the cores are much heavier. In terms of classical physics, the vibrational eigenfrequencies of a system are determined by the restoring forces and the masses of the systems constituents. Assuming a typical core being 10^4 times heavier than an electron, one would expect the eigenfrequencies of the core motion approximately 100 times lower than that of electrons that are equally tight bound (in terms of quantum mechanics, these are the valence electrons). Therefore, at high frequencies, the movement of the cores may be neglected. At lower frequencies (and this is usually the infrared spectral region), the movements of the atomic nuclei determine the optical properties of the material.

On the other hand, not all electrons are equally tight bound. Although this is again rather a quantum mechanical matter, we may formally assume, that there are groups of electrons (*core* electrons) that suffer much higher restoring forces than the other (the *valence*) electrons. Consequently, there are different groups of electrons with different eigenfrequencies (compare Sect. 2.5).

The oscillator model derived in the following is very general. It may be applied to the intramolecular motion of nuclei (in infrared spectroscopy) as well as to bound electrons. So that we will simply speak in the following on induced dipole moments and not care about their physical origin.

So let us regard the motion of a charge carrier, which is bound to its equilibrium position ($x = 0$) by an elastic restoring force. An oscillating local field may lead to small ($x \ll \lambda$) movements of the charge carriers, thus inducing dipoles that interact with the field. The polarizability is now obtained according to (12.5):

$$\beta = \frac{q^2}{\varepsilon_0 m} \frac{1}{\omega_0^2 - \omega^2 - 2i\omega\gamma}$$

When assuming that the microscopic induced dipoles are much smaller than the wavelength of the incident light (quasistatic approximation), the relation between the polarizability and the dielectric function is identical to the result obtained earlier in the frames of electrostatics (12.16). Therefore, we find the dielectric function according to the Clausius–Mossotti equation:

$$\frac{\varepsilon - 1}{\varepsilon + 2} = \frac{N\beta}{3} \Leftrightarrow \varepsilon = 1 + \frac{N\beta}{1 - \frac{N\beta}{3}} \tag{14.7}$$

or the Lorentz–Lorenz equation.

$$\frac{\hat{n}^2 - 1}{\hat{n}^2 + 2} = \frac{N\beta}{3} \tag{14.7a}$$

The significance of these rather simple equations is in that they relate microscopic optical parameters (the polarizability β) to macroscopically measurable parameters (optical constants). In other words, measurements on the macroscopic scale, which yield the optical constants of a material, give further access to microscopic parameters such as molecular or atomic polarizabilities. In fact, this is the point from where analytical optical spectroscopy starts.

Let us have a look at the consequences. We have a microscopic polarizability according to (12.5) and a dielectric function from (14.7). In combination, that yields:

$$\varepsilon(\omega) = 1 + \frac{\omega_p^2}{\omega_0^2 - \omega^2 - 2i\gamma\omega - \frac{\omega_p^2}{3}} \equiv 1 + \frac{\omega_p^2}{\tilde{\omega}_0^2 - \omega^2 - 2i\gamma\omega} \tag{14.8}$$

where

$$\tilde{\omega}_0^2 \equiv \omega_0^2 - \frac{\omega_p^2}{3} \qquad (14.8a)$$

is the resonance frequency valid for the dielectric function. The angular frequency ω_p is again given by (14.4), but N has now the meaning of the concentration of bound electrons. The dielectric function has exactly the same spectral shape as the polarizability, but the resonance position in ε is red-shifted with respect to that of the polarizability. The larger the density, the larger is the red shift.

Figure 14.3 shows the real and imaginary parts of a dielectric function according to (14.8), and Fig. 14.4 the optical constants. We see that in the vicinity of the resonance frequency, the imaginary parts of both the dielectric function and the index of refraction show a local maximum. That means that at this frequency the light wave is effectively damped. The imaginary part of the dielectric function therefore describes an absorption line with a characteristic shape, which is called a Lorentzian line. In the region of strong damping, the refractive index n decreases with increasing frequency (anomalous dispersion). On the contrary, in the transparency regions, where damping is negligible, n increases with increasing frequency (normal dispersion). Note that at sufficiently large frequencies, the

Fig. 14.3 Dielectric function according to (14.8)

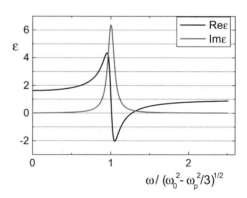

Fig. 14.4 Optical constants according to Fig. 14.3

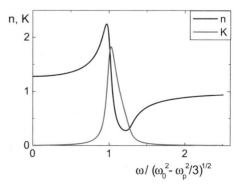

refractive index becomes smaller than 1. Therefore, refractive indices in the X-ray region are usually smaller than 1. For infinitely large frequencies, the refractive index approaches 1.

14.4   Consistency Considerations

All the classical dispersion models derived in the frames of this book have common features, namely:

- The dielectric function as well as the index of refraction appear to be complex, i.e., have real and imaginary parts
- The dielectric function as well as the index of refraction are frequency dependent, i.e., show dispersion

Both features are consistent with the conclusions from the general treatment provided in Sect. 13.3.

Concerning the oscillator model, so it might seem strange that the dielectric function is resonant at a frequency that is red-shifted with respect to that of the single oscillator. Indeed, while the microscopic polarizability according to (12.4) has a resonance at $\omega = \omega_0$, the resonance of the dielectric function given in (14.8) is red-shifted according to $\tilde{\omega}_0^2 \equiv \omega_0^2 - \frac{\omega_p^2}{3}$. The larger the density of the medium is, the larger the red shift is expected. The reason is in the self-consistent approach we have presented before: Clearly, provided that the electric field strength amplitude does not depend on the frequency, any microscopic dipole will absorb a maximum of light when the incident frequency coincides with its own resonance frequency. But when assuming that the externally applied field has a frequency-independent amplitude, the local fields according to Table 12.3 will not, because of the frequency dependence of the dielectric function. As an illustration one could imagine a microscopic dipole that would "like to absorb" at $\omega = \omega_0$ but cannot because the neighboring dipoles have already absorbed all the light. Therefore, the dipole must absorb at a somewhat shifted frequency—and in fact so must all of the dipoles. Then it is the compromise between the resonant behavior of the single dipole and the local electric field available in the medium that defines the frequency shift of the resonance position of the dipole response in a dense medium.

Equations (14.8) and (14.8a) are relevant when the depolarization factor is equal to 1/3. For other depolarization factors, (14.8a) generalizes to (14.8b):

$$\tilde{\omega}_0^2 = \omega_0^2 - L\omega_p^2 \qquad (14.8b)$$

Then, the Drude function appears to be a particular case of (14.8), when setting $L = 0$ and $\omega_0 = 0$ (compare the corresponding discussion in (12.4) in this regard).

Real materials, of course, have (much) more than one degree of freedom, and therefore the assumption of one single resonance frequency may be insufficient in many practical modeling situations. By making use of (14.1), we therefore generalize our classical dielectric functions to:

$$\varepsilon(\omega) = 1 - \frac{\omega_{p,\text{free}}^2}{\omega^2 + 2i\omega\gamma} + \omega_p^2 \sum_j \frac{f_j}{\tilde{\omega}_{0j}^2 - \omega^2 - 2i\omega\gamma_j} \tag{14.9}$$

We have now made the reasonable assumption that the medium has both free charge carriers (with concentration N_{free}) and bound charge carriers (with concentration N_{bound}). In order to not confuse these concentrations, in (14.9) we have introduced the "plasma frequencies" $\omega_{p,\text{free}}^2 = \frac{N_{\text{free}}q^2}{\varepsilon_0 m}$ and $\omega_p^2 = \frac{N_{\text{bound}}q^2}{\varepsilon_0 m}$. The index j counts the resonance frequencies characteristic for a given medium. If more than one resonance frequency is considered, it is common to speak on the multi oscillator model. The dimensionless parameters f_j are real and positive and may be regarded as classical analogous to the oscillator strength as relevant for light absorption.

When assuming the model case of a pure dielectric, from (14.9) we find:

$$\varepsilon(\omega) = 1 + \omega_p^2 \sum_j \frac{f_j}{\omega_{0j}^2 - \omega^2 - 2i\omega\gamma_j} \tag{14.9a}$$

Figure 14.5 shows the thus described dielectric function and optical constants in the vicinity of several resonance frequencies.

Note that again, apart from resonances, normal dispersion is observed, while in spectral regions with strong light absorption, the dispersion turns to be anomalous. Normal dispersion is easily demonstrated, for example, by the sequence of colors observed after white light has passed through a glass prism.

Note further that in terms of (14.9a), any dielectric has a static dielectric constant that is larger than one, while for infinitely large frequencies, the dielectric function approaches the value one from below. This is a physically rather transparent result, because at arbitrarily small frequencies, the charge carriers will always be able to comply with the oscillation of the field, such that the dielectric medium will always form a macroscopic polarization, and thus the dielectric constant of the medium will be larger than one. At highest frequencies, however, the inertness of the charge carriers will no more permit the charge carriers to oscillate with the field, such that the medium becomes transparent and, because of a lack of efficient light–matter interaction mechanisms, the light wave propagates through the medium almost like through vacuum. But that means, that whenever the refractive index is supposed to be a continuous function of the angular frequency, spectral regions with anomalous dispersion must necessarily exist. Moreover, when associating anomalous dispersion with spectral regions of significant light absorption, then the observation of a dielectric constant larger than one indicates the presence of spectral regions where the medium must absorb light. From here we can formulate the hypothesis that light refraction and absorption are physically interconnected phenomena, and in particular, that the dispersions of the real and

Fig. 14.5 Comparison
between the dielectric
functions (on top)—and the
optical constants (bottom) in
the multioscillator model

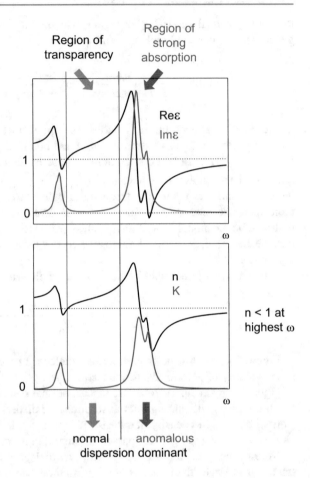

imaginary parts of the dielectric function must be correlated to each other. The formulation of that correlation will form the content of the next Chap. 15.

Let us now turn to simple examples on how the rather involved dispersion curves shown in Fig. 14.5 may manifest themselves in a real spectrum. Once we principally deal with absorbing materials, the transmission signal may be suppressed in many practical cases. Instead, let us have a look on the reflection spectrum recorded at normal incidence at the air-material surface. Then, Eq. (13.24) should apply, while the optical constants of the medium obey (14.9). Figure 14.6 shows the MIR reflection spectrum of a fused silica surface together with the associated optical constants. Obviously, the spectral features observed in the recorded reflection spectrum reproduce major spectral features apparent in both n and K.

Correspondingly, Fig. 14.7 shows the reflectance of an aluminum surface at air together with the optical constants of aluminum. The high reflection is clearly

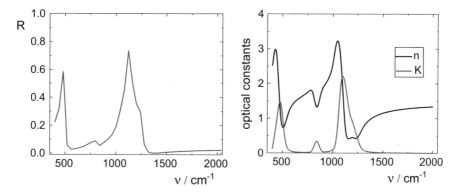

Fig. 14.6 Measured MIR reflection spectrum (on left) and associated MIR optical constants of SiO$_2$ after some smoothing

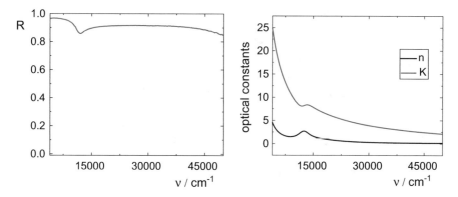

Fig. 14.7 Measured reflection spectrum (on left) and associated optical constants of aluminum

caused by the response of the free electrons in aluminum, while the spectral feature around 12,000 cm^{-1} is caused by the response of bound electrons. Once the reflectance of the aluminum surface is nearly constant over the whole VIS, an aluminum surface does not appear colored when illuminated by white light. This is in contrast to copper or gold, where the spectrally selective response of the bound electrons results in a colored appearance.

In summary, we are now familiar with three classical models that are useful to describe the optical properties of condensed matter. The orientation and reorientation of permanent molecular dipoles are very inert, and it will cause a remarkable optical response only in the microwave (MW) or far infrared (FIR) spectral regions. In liquids and also in some solids (for example ice), it may be tackled by means of Debye's equations. Drude's function describes the optical properties of free charge carriers, and depending on their concentration, it may be of use from the microwave up to the visible (VIS) or even ultraviolet (UV)

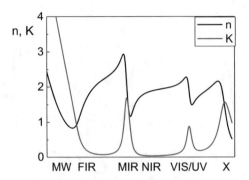

Fig. 14.8 Principle shape of the optical constants dispersion curves in different spectral regions

spectral regions. The Lorentzian oscillator model is suitable for the description of absorption and dispersion in the presence of distinct spectral lines. In the middle infrared (MIR), it may be used to describe the response of core vibrations in molecules and solids. The excitation of valence electrons in atoms or molecules causes absorption lines in the visible or ultraviolet (UV) spectral regions, while core electron excitation dominates the X-ray region. An overview on the possible optical spectrum of condensed matter is given in Fig. 14.8.

More quantitative information may be obtained from Table 14.1.

14.5 Application to Practical Problems

14.5.1 Sellmeier and Cauchy Formulas

Far from any resonances, (14.9a) may be written in a simplified fashion according to:

$$\varepsilon \approx \mathrm{Re}\varepsilon \approx n^2 \approx 1 + \frac{Nq^2}{\varepsilon_0 m} \sum_{j=1}^{M} \frac{f_j}{\tilde{\omega}_{0j}^2 - \omega^2} \tag{14.9b}$$

Replacing ω by λ via $\omega = \frac{2\pi c}{\lambda}$ and utilizing the identity

$$\frac{\lambda^2}{\lambda^2 - \tilde{\lambda}_{0j}^2} \equiv 1 + \frac{\tilde{\lambda}_{0j}^2}{\lambda^2 - \tilde{\lambda}_{0j}^2}$$

We can write:

$$\varepsilon - 1 = n^2 - 1 = a + \sum_j \frac{b_j}{\lambda^2 - \tilde{\lambda}_{0j}^2} \tag{14.9c}$$

Table 14.1 Overview on spectral regions

Spectral region	Vacuum wavelength λ nm	Wavenumber ν $= 1/\lambda$ cm^{-1}	Angular frequency ω $\omega = 2\pi\nu c$ s^{-1}	Origin of absorption (examples)
Microwave **MW**	10^9–10^6	0.01–10	1.9×10^9–1.9×10^{12}	Free carriers; orientation/ rotation
Terahertz **THz**	10^6–10^5	10–100	1.9×10^{12}–1.9×10^{13}	
Far Infrared **FIR**	10^6–5×10^4	10–200	1.9×10^{12}–3.8×10^{13}	
Middle Infrared **MIR**	5×10^4–2.5×10^3	200–4000	3.8×10^{13}–7.5×10^{14}	Free carriers; vibrations of nuclei
Near Infrared **NIR**	2.5×10^3–8×10^2	4000–12,500	7.5×10^{14}–2.4×10^{15}	Free carriers; vibrational overtones
Visible **VIS**	8×10^2–4×10^2	12,500–25,000	2.4×10^{15}–4.7×10^{15}	Excitation of valence electrons
Ultraviolet **UV**	4×10^2–10	25,000–10^6	4.7×10^{15}–1.9×10^{17}	
X-ray **X**	10–0.005	10^6–2×10^9 (unusual)	1.9×10^{17}–3.8×10^{20}	Excitation of core electrons

The wavelength (and related) data may slightly differ in different sources

where a and b_j are constant coefficients. They are interconnected with each other due to the requirement that the refractive index must approach one when the wavelength approaches zero. Equations (14.9b) and (14.9c) are possible writings of the Sellmeier dispersion formula. In Fig. 14.8, it may be successfully applied for describing the dispersion in spectral regions of relative transparency.

Another common dispersion formula is obtained when expanding (14.9b) into a series. This way one easily obtains the Cauchy formula according to:

$$n^2 = A + B\nu^2 + C\nu^4 + \cdots - B'\nu^{-2} - C'\nu^{-4} - \cdots \qquad (14.9d)$$

Here, the A-, B-, and C-values are new constants. Equation (14.9d) is again applicable in the regions of relative transparency.

14.5.2 Inhomogeneous Broadening

As we have already indicated in Sect. 2.6/Fig. 2.10, an inhomogeneously broadened absorption line may be understood as a superposition of a multiplicity of much narrower individual absorption lines. According to the underlying distribution in concentration and oscillator strength of these individual lines, the line shape of an inhomogeneously broadened spectral line may vary. We will exemplify two different possibilities here:

Brendel model: The Brendel model can be tackled as a particular case of Eq. (14.9a) [4]. It pursues the specifics of optical materials, which are characterized by fluctuations of the local density in the material, which gives rise to fluctuations in the resonance frequencies according to (14.8b) and thus provides an inhomogeneous line broadening mechanism (compare Fig. 2.10).

Let us for simplicity assume that the mentioned fluctuations result in a distribution of local resonance frequencies around a central frequency $\overline{\omega}_0$. When assuming a Gaussian distribution of these resonance frequencies, an approximate calculation of the "averaged" dielectric function is performed by the equation:

$$\varepsilon(\omega) = 1 + \frac{1}{\sqrt{2\pi}\sigma} \int_{-\infty}^{\infty} \exp\left[-\frac{(\xi - \overline{\omega}_0)^2}{2\sigma^2}\right] \frac{\omega_p^2}{\xi^2 - \omega^2 - 2i\gamma\omega} d\xi \qquad (14.10)$$

Here, σ is the standard deviation of the assumed Gaussian distribution, which again defines the inhomogeneous contribution to the width of the absorption line defined by the imaginary part of ε, while γ is the typical homogeneous linewidth of the Lorentzian oscillator. The shape of the absorption line is now defined by the relation between σ and γ: In the case of $\sigma \gg \gamma$, a Gaussian lineshape will be observed, while for $\sigma \ll \gamma$, we will find a rather Lorentzian behavior. When both linewidth contributions are comparable to each other, we have $\sigma \approx \gamma$, and then we obtain a so-called Voigt line. Voigt line or even Gaussian line fits may appear very useful in applied solid-state spectroscopy.

The beta-distributed oscillator (β_do) model: In the β_do model, it is the assumed that the envelope of the mentioned multiplicity of individual absorption is formed by a Beta-distribution being proportional to

$$w_{\beta\_do}(\omega) = \begin{cases} \dfrac{(\omega-\omega_a)^{A-1}(\omega_b-\omega)^{B-1}}{\int_{\omega_a}^{\omega_b}(\omega-\omega_a)^{A-1}(\omega_b-\omega)^{B-1}d\omega}; & A, B > 0; \ \omega \in (\omega_a, \omega_b) \\ 0; & \omega \notin (\omega_a, \omega_b) \end{cases} . \qquad (14.11)$$

The real parameters A, B, ω_a, and ω_b are free parameters within the β_do model [5, 6]. It is further assumed that all of the mentioned individual transitions may be modeled by Lorentzians with identical homogeneous FWHM linewidth 2γ. γ thus appears as the fifth model parameter. The dielectric function is then given by:

$$\varepsilon(\omega) = 1 + \frac{J}{\pi} \int_{\omega_a}^{\omega_b} w_{\beta\_do}(\xi) \left[\frac{1}{\xi - \omega - i\gamma} + \frac{1}{\xi + \omega + i\gamma}\right] d\xi. \qquad (14.12)$$

Here J is a sixth model parameter, it has the sense of an oscillator strength. Figure 14.9 shows an example on the optical constants dispersion as described in terms of the β_do model. Compared to the single oscillator model (Fig. 14.4), the β_do model clearly has more flexibility in modeling various (in particular asymmetric) shapes of the absorption feature, in particular in modeling what is called an *absorption edge*, defined by the parameter (or threshold frequency or absorption onset frequency) ω_a in (14.11) (Fig. 14.9).

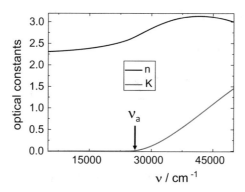

Fig. 14.9 Optical constants as modeled by the β_do model. The indicated value $v_a = \frac{\omega_a}{2\pi c}$ marks the absorption edge

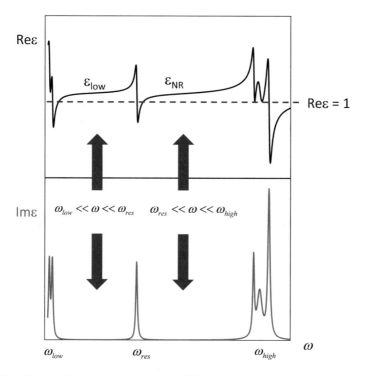

Fig. 14.10 Dielectric function according to (14.13)

14.6   Advanced Material

14.6.1 Oscillator Model and Dispersion Law

Let us now return to the electromagnetic wave propagating through a dielectric medium along the z-direction. According to (13.13) and (13.14) we have:

$$E = E_0 e^{-i\left(\omega t - \frac{\omega}{c}\sqrt{\varepsilon(\omega)}z\right)}$$

With $k = \pm\frac{\omega}{c}\sqrt{\varepsilon(\omega)}$.

Let us further assume that the angular frequency of the electromagnetic wave falls close to one of the resonance frequencies (say, ω_{res}) in (14.9) (see Fig. 14.10):

$$\varepsilon(\omega)|_{\omega\approx\omega_{res}} = 1 + \omega_p^2 \sum_j \frac{f_j}{\tilde{\omega}_{0j}^2 - \omega^2 - 2i\omega\gamma_j} \equiv 1 + \frac{\omega_p^2 f_{res}}{\omega_{res}^2 - \omega^2 - 2i\omega\gamma_{res}} + \chi_{NR} \equiv$$

$$\frac{\omega_p^2 f_{res}}{\omega_{res}^2 - \omega^2 - 2i\omega\gamma_{res}} + \varepsilon_{NR}$$

$$(14.13)$$

Here, for simplicity, all non-resonant contributions to the susceptibility of the dielectric function have been condensed into the merely frequency-dependent term ε_{NR}. This leads us to the expression for the wavevector (the dispersion law):

$$k(\omega) = \frac{\omega}{c}\sqrt{\frac{\omega_p^2 f_{res}}{\omega_{res}^2 - \omega^2 - 2i\omega\gamma_{res}} + \varepsilon_{NR}} \qquad (14.14)$$

This equation defines the dispersion relation for the propagating electromagnetic wave.

Note that for $\omega_{res} \ll \omega \ll \omega_{high}$, we have:

$$k\left(\omega_{res} \ll \omega \ll \omega_{high}\right) \approx \frac{\omega}{c}\sqrt{\varepsilon_{NR}} \equiv \frac{\omega}{c}\sqrt{\varepsilon_\infty}$$

Accordingly, for $\omega_{res} \gg \omega \gg \omega_{low}$, we have:

$$k(\omega_{res} \gg \omega \gg \omega_{low}) \approx \frac{\omega}{c}\sqrt{\frac{\omega_p^2 f_{res}}{\omega_{res}^2} + \varepsilon_\infty} \equiv \frac{\omega}{c}\sqrt{\varepsilon_{low}}$$

For weak damping, the resonance in k according to (14.14) is obviously observed at $\omega \approx \omega_{res}$. There is another characteristic frequency where k comes close to zero. From (14.14) we find:

$$k \approx 0 \Leftrightarrow \frac{\omega_p^2 f_{res}}{\omega_{res}^2 - \omega^2 - 2i\omega\gamma_{res}} + \varepsilon_{NR} \approx 0 \Rightarrow \omega^2(k \approx 0) \approx \omega_{res}^2 + \frac{\omega_p^2 f_{res}}{\varepsilon_\infty}$$

$$\Rightarrow \frac{\omega^2(k \approx 0)}{\omega_{\text{res}}^2} \approx \frac{\omega_{\text{res}}^2 \varepsilon_\infty + \omega_p^2 f_{\text{res}}}{\omega_{\text{res}}^2 \varepsilon_\infty} = \frac{\omega_{\text{res}}^2 \left(\varepsilon_\infty + \frac{\omega_p^2 f_{\text{res}}}{\omega_{\text{res}}^2} \right)}{\omega_{\text{res}}^2 \varepsilon_\infty}$$

$$= \frac{\varepsilon_{\text{low}}}{\varepsilon_\infty} \Rightarrow \frac{\omega(k \approx 0)}{\omega_{\text{res}}} \approx \sqrt{\frac{\varepsilon_{\text{low}}}{\varepsilon_\infty}} \qquad (14.15)$$

Note now that according to (13.8), the polarization is given by:

$$P = \varepsilon_0 [\varepsilon(\omega) - 1] E = \varepsilon_0 [\varepsilon(\omega) - 1] E_0 e^{-i(\omega t - kz)}$$

This way, in addition to the electromagnetic wave, there is a (mechanical) polarization wave propagating synchronously to the electromagnetic wave through the dielectric medium. This excitation, which represents a hybrid of the electromagnetic wave and the polarization wave, is called a propagating polariton. The dispersion relation of a polariton is visualized in Fig. 14.11.

More precisely, what we call a polariton is a quant of this hybrid phenomenon composed by an electromagnetic and a mechanical wave. Note that in off-resonance conditions, by means of (14.15), we may rewrite (14.14) in a more compact manner:

$$k(\omega) \approx \frac{\omega}{c} \sqrt{\frac{\omega_p^2 f_{\text{res}}}{\omega_{\text{res}}^2 - \omega^2} + \varepsilon_\infty} \Rightarrow \left(\frac{kc}{\omega} \right)^2 = \varepsilon(\omega) \approx \frac{\omega_p^2 f_{\text{res}}}{\omega_{\text{res}}^2 - \omega^2} + \varepsilon_\infty$$

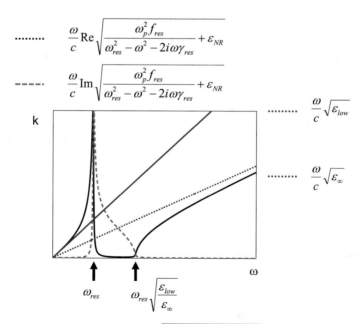

Fig. 14.11 Dispersion relation $k(\omega) = \frac{\omega}{c} \sqrt{\frac{\omega_p^2 f_{\text{res}}}{\omega_{\text{res}}^2 - \omega^2 - 2i\omega \gamma_{\text{res}}} + \varepsilon_{NR}}$ of a propagating polariton

$$= \varepsilon_\infty \frac{\omega^2(k \approx 0) - \omega^2}{\omega_{res}^2 - \omega^2} \tag{14.15a}$$

We will have to return to this equation and its interpretation in Sect. 17.3.2.

14.6.2 The Dielectric Function of a Gas of Diatomic Molecules in the FIR

In a gas, similar to (12.16a), we set

$$\varepsilon|_{N \to 0} \approx 1 + N\beta \Rightarrow \alpha = 4\pi \nu K = 4\pi \nu \text{Im}\sqrt{\varepsilon} \approx 4\pi \nu \text{Im}\sqrt{1 + N\beta}$$

$$\approx 4\pi \nu \text{Im}\left(1 + \frac{N}{2}\beta\right) \Rightarrow \alpha \approx 2\pi \nu N \text{Im}\beta \tag{14.16}$$

When focusing on the FIR, main contributions arise from the rotational degrees of freedom. A diatomic molecule has two rotational degrees of freedom perpendicular to its axis, because one of the diagonal elements of its tensor of inertia is zero. Assuming a chaotic distribution of rotation directions, only 2/3 of the molecules therefore contribute to the discussed interaction mechanism with an electromagnetic wave propagating along the z-axis. In average we expect:

$$\langle \beta \rangle = \frac{\beta_{xx} + \beta_{yy} + \beta_{zz}}{3}\Bigg|_{\substack{\beta_{xx} = \beta_{yy} \\ \beta_{zz} = 0}} = \frac{2}{3}\beta_{xx}$$

$$= \frac{4}{3}\frac{1}{\varepsilon_0 \hbar} \sum_l \sum_{n>l} \frac{|d_{x,nl}|^2 \omega_{nl}}{\omega_{nl}^2 - \omega^2 - 2i\omega\gamma_{nl}}[W(l) - W(n)] \equiv \beta$$

where β_{xx} is a microscopic polarizability as given by (12.26).

Assuming a purely rotational spectrum of an assembly of rigid rotors, from Chapters 8 and 11 we know (J-rotational quantum number, M-projection on a chosen axis):

$$|l\rangle = |J, M\rangle; \quad |n\rangle = |J+1, M'\rangle; \quad \omega = 2\pi c\nu; \quad \gamma = 2\pi c\Gamma$$

$$\Rightarrow \beta = \frac{2}{3\pi}\frac{1}{\varepsilon_0 \hbar c}\sum_J \sum_M \sum_{M'} \frac{|\langle J+1, M'|d_x|J, M\rangle|^2 \nu_{nl}}{\nu_{nl}^2 - \nu^2 - 2i\nu\Gamma_{nl}}[W(J, M) - W(J+1, M')] \tag{14.17}$$

According to (11.15), in (14.17) we set:

$$\nu_{n,l} = 2B(J+1)$$

In thermodynamic equilibrium, we have:

$$W(J, M) \equiv \frac{1}{Z}e^{-\frac{E_{rot}}{k_B T}} = \frac{1}{Z}e^{-\frac{BJ(J+1)}{\theta}}$$

independently of M. The statistical sum may be calculated by substituting summation by integration according to:

$$Z = \sum_{J=0}^{\infty} \sum_{M=-J}^{J} e^{-\frac{BJ(J+1)}{\theta}} \equiv \sum_{J=0}^{\infty} (2J+1)e^{-\frac{BJ(J+1)}{\theta}} = \int_{0}^{\infty} (2J+1)e^{-\frac{BJ(J+1)}{\theta}} dJ = \frac{\theta}{B}$$

Therefore,

$$W(J,M) - W(J+1,M') = \frac{B}{\theta}\left[e^{-\frac{BJ(J+1)}{\theta}} - e^{-\frac{B(J+1)(J+2)}{\theta}}\right]$$

Let us now make use of the oscillator strength sum rule (5.33). Then, for any state $|J,M\rangle$ we may write:

$$\frac{4\pi\mu c}{\hbar} \sum_{J'=J\pm 1} \sum_{M'} v_{J',J} |\langle J',M'|x|J,M\rangle|^2 = 1$$

$$= \frac{8\pi\mu cB}{\hbar}\left\{(J+1)\sum_{M'}|\langle J+1,M'|x|J,M\rangle|^2 - J\sum_{M'}|\langle J-1,M'|x|J,M\rangle|^2\right\}$$

$$\Rightarrow \frac{1}{J+1}\left\{\frac{\hbar}{8\pi\mu cB} + J\sum_{M'}|\langle J-1,M'|x|J,M\rangle|^2\right\} = \sum_{M'}|\langle J+1,M'|x|J,M\rangle|^2$$

$$\Rightarrow \sum_{M'}|\langle J+1,M'|d_x|J,M\rangle|^2 = \frac{1}{J+1}\left\{\frac{\hbar q^2}{8\pi\mu cB} + J\sum_{M'}|\langle J-1,M'|d_x|J,M\rangle|^2\right\}$$

$$\Rightarrow \sum_{M,M'}|\langle J+1,M'|d_x|J,M\rangle|^2$$

$$= \frac{1}{J+1}\left\{(2J+1)\frac{\hbar q^2}{8\pi\mu cB} + J\sum_{M,M'}|\langle J-1,M'|d_x|J,M\rangle|^2\right\}$$

Obviously,

$$\frac{\hbar q^2}{8\pi\mu cB} = \frac{\hbar q^2 r^2}{8\pi\mu r^2 cB} = \frac{\hbar d_{perm}^2}{8\pi IcB} = \frac{d_{perm}^2}{2\left(\frac{4\pi Ic}{\hbar}\right)B} = \frac{d_{perm}^2}{2}$$

We thus find the recursive recipe:

$$\sum_{M,M'}|\langle J+1,M'|d_x|J,M\rangle|^2 = \frac{1}{J+1}\left\{(2J+1)\frac{d_{perm}^2}{2}\right.$$

$$\left. + J\sum_{M,M'}|\langle J-1,M'|d_x|J,M\rangle|^2\right\}$$

That yields:

$$\sum_{M,M'} |\langle 1, M'|d_x|0, M\rangle|^2 = \frac{d_{\text{perm}}^2}{2}$$

$$\sum_{M,M'} |\langle 2, M'|d_x|1, M\rangle|^2 = \frac{1}{2}\left\{3\frac{d_{\text{perm}}^2}{2} + \frac{d_{\text{perm}}^2}{2}\right\} = d_{\text{perm}}^2$$

$$\sum_{M,M'} |\langle 3, M'|d_x|2, M\rangle|^2 = \frac{1}{3}\left\{5\frac{d_{\text{perm}}^2}{2} + 2d_{\text{perm}}^2\right\} = \frac{3}{2}d_{\text{perm}}^2 \quad \cdots$$

$$\Rightarrow \sum_{M,M'} |\langle J + 1, M'|d_x|J, M\rangle|^2 = \frac{J+1}{2}d_{\text{perm}}^2 \tag{14.18}$$

Remark Note that the proportionality of $\sum_{M,M'} |\langle J + 1, M'|d_x|J, M\rangle|^2$ to $(J + 1)$ in (14.18) can be understood as a particular case of the term $(2J + 1)A_{JK}$ in Eq. (11.35). Indeed, when regarding the diatomic molecule as a limiting case of a prolate symmetric top with ζ-symmetry axis and $I_C = 0$ (compare Sect. 11.6.1), the angular momentum will always be perpendicular to the symmetry axis, such that $K = \Delta K = 0$ (compare Fig. 11.14). The corresponding Hönl–London factor is (Table 11.2):

$$\Delta J = +1; \ \Delta K = 0 \Rightarrow A_{JK} = \left.\frac{(J+1)^2 - K^2}{(J+1)(2J+1)}\right|_{K=0} = \frac{(J+1)}{(2J+1)}$$

$$\Rightarrow (2J+1)A_{JK} = (J+1)$$

When finally postulating identical linewidth values ($\Gamma_{nl} = \Gamma$), we finally obtain:

$$\begin{aligned}
\beta &= \frac{2}{3\pi}\frac{1}{\varepsilon_0\hbar c}\sum_J\sum_M\sum_{M'}\frac{|\langle J+1, M'|d_x|J, M\rangle|^2 v_{nl}}{v_{nl}^2 - v^2 - 2iv\Gamma_{nl}}\left[W(J, M) - W(J+1, M')\right] \\
&= \frac{4B}{3\pi}\frac{1}{\varepsilon_0\hbar c}\frac{B}{\theta}\sum_J\frac{(J+1)}{(2B(J+1))^2 - v^2 - 2iv\Gamma}\left[e^{-\frac{BJ(J+1)}{\theta}}\right. \\
&\quad \left. -e^{-\frac{B(J+1)(J+2)}{\theta}}\right]\sum_M\sum_{M'}|\langle J+1, M'|d_x|J, M\rangle|^2 \\
&= \frac{2B^2}{3\pi}\frac{1}{\varepsilon_0\hbar c}\frac{1}{\theta}d_{\text{perm}}^2\sum_J\frac{(J+1)^2}{(2B(J+1))^2 - v^2 - 2iv\Gamma}\left[e^{-\frac{BJ(J+1)}{\theta}}\right. \\
&\quad \left. -e^{-\frac{B(J+1)(J+2)}{\theta}}\right]
\end{aligned} \tag{14.19}$$

In the static case, $v = 0$, and from (14.19) we find

$$\beta_{\text{stat}} = \frac{1}{6\pi}\frac{1}{\varepsilon_0\hbar c}\frac{1}{\theta}d_{\text{perm}}^2\sum_J\left[e^{-\frac{BJ(J+1)}{\theta}} - e^{-\frac{B(J+1)(J+2)}{\theta}}\right] =$$

$$= \frac{1}{3} \frac{1}{\varepsilon_0 hc} \frac{1}{\theta} d_{\text{perm}}^2 \underbrace{\left\{ \sum_{J=0}^{\infty} e^{-\frac{BJ(J+1)}{\theta}} - \sum_{J=1}^{\infty} e^{-\frac{BJ(J+1)}{\theta}} \right\}}_{=1} = \frac{d_{\text{perm}}^2}{3\varepsilon_0 k_B T}$$

which is in full consistency with the previously derived expression (12.21).

Let us finish this section by writing down the expression for the absorption coefficient (11.17). From (14.16) we have:

$$\alpha \approx 2\pi \nu N \text{Im}\beta =$$

$$= 2\pi \nu N \text{Im} \left\{ \frac{2B^2}{3\pi} \frac{1}{\varepsilon_0 \hbar c} \frac{1}{\theta} d_{\text{perm}}^2 \sum_J \frac{(J+1)^2}{(2B(J+1))^2 - \nu^2 - 2i\nu\Gamma} \right\} \left[e^{-\frac{BJ(J+1)}{\theta}} \right.$$

$$\left. - e^{-\frac{B(J+1)(J+2)}{\theta}} \right\} \right] =$$

$$= \frac{8B^2 d_{\text{perm}}^2 \nu^2}{3} \frac{N\Gamma}{\varepsilon_0 \hbar c\theta} \sum_J \frac{(J+1)^2}{\left[(2B(J+1))^2 - \nu^2\right]^2 + 4\nu^2\Gamma^2} \left[e^{-\frac{BJ(J+1)}{\theta}} \right.$$

$$\left. - e^{-\frac{B(J+1)(J+2)}{\theta}} \right]$$

This expression is in the basis of the model calculations demonstrated in Fig. 11.3.

14.7 Tasks for Self-check

14.7.1 Assume that at very high light frequencies ($\omega >> \omega_p$), the refractive index n of a medium may be regarded as real and is given by: $n^2(\omega) = 1 - \frac{\omega_p^2}{\omega^2}$. It is obvious that n is smaller than 1, so that the phase velocity of an electromagnetic wave propagating in such a medium will be *larger* than the velocity of light in vacuum c. However, in first-order dispersion theory, a short light pulse with a frequency spectrum centered at $\omega_0 >> \omega_p$ traveling through such a medium propagates with the group velocity. Assuming $n^2(\omega) = 1 - \frac{\omega_p^2}{\omega^2}$, derive an expression for the group velocity and show that it is *smaller* than c!

14.7.2. In the X-ray spectral region, the optical constants are usually almost real and close to one. Therefore, the complex refractive index in the X-ray region may be expressed through the small parameters δ and β (don't confuse with polarizability) via: $n + iK = 1 - \delta + i\beta$ with δ, $\beta \ll 1$. Consider now a plane surface between vacuum and a material with given δ and β and express the normal incidence reflectance R at this surface through these parameters!

14.7.3. Calculate the so-called dielectric loss function for a Drude metal and for the oscillator model. The loss function is defined as $-\text{Im}(1/\varepsilon)$.

14.7.4 Investigate the resonant behavior of the polarizability of a small metallic sphere embedded in vacuum. Make use of the quasistatic approximation, where the microscopic polarizability of the sphere is given by an expression like (12.20). In order to obtain a frequency dependence, instead of ε_{stat}, assume a Drude function for the dielectric response of the metal now!

References

Specific References

1. J.C. Anderson, Conduction in thin semiconductor films. Adv. Phys. **19**, 311–338 (1970)
2. C. Weißmantel, C. Hamann, *Grundlagen der Festkörperphysik* (VEB Deutscher Verlag der Wissenschaften Berlin, 1979), pp. 413–416
3. O. Stenzel, S. Wilbrandt, S. Stempfhuber, D. Gäbler, S.J. Wolleb, Spectrophotometric characterization of thin copper and gold films prepared by electron beam evaporation: thickness dependence of the drude damping parameter. Coatings **9**, 181 (2019). https://doi.org/10.3390/coatings9030181
4. R. Brendel, D. Bormann, An infrared dielectric function model for amorphous solids. J. Appl. Phys. **71**, 1–6 (1992)
5. S. Wilbrandt, O. Stenzel, Empirical extension to the multi oscillator model: the beta-distributed oscillator model. Appl. Opt. **56**, 9892–9899 (2017)
6. O. Stenzel, S. Wilbrandt, Beta-distributed oscillator model as an empirical extension to the Lorentzian oscillator model: physical interpretation of the β_do model parameters. Appl. Opt. **58**, 9318–9325 (2019)

General Literature to this Chapter

7. M. Fox: *Optical Properties of Solids* (Oxford University Press, 2010)
8. P.J.Yu, M. Cardona, *Fundamentals of Semiconductors. Physics and Material Properties*, 4th edn (Springer, Berlin, 2010)
9. C.F. Klingshirn, *Semiconductor Optics* (Springer, Berlin Heidelberg New York, 1997)
10. M. Born, E. Wolf, *Principles of Optics* (Pergamon Press, Oxford London Edinburgh New York Paris Frankfurt, 1968)
11. L.D. Landau, E.M. Lifshitz, Electrodynamics of continuous media. *A Course of Theoretical Physics*, vol. 8 (Pergamon Press, 1960)
12. O. Stenzel, *The Physics of thin Film Optical Spectra. An Introduction* (Springer, Berlin, 2016)

Linear Optical Constants III: The Kramers–Kronig Relations

15

ABSTRACT

The Kramers–Kronig relations establish a fundamental relationship between the dispersions of the real and imaginary parts of the susceptibility or the dielectric function. This way refraction of light and energy dissipation appear to be interconnected phenomena. Kramers–Kronig relations are derived for both insulators and electrical conductors. Simple sum rules are derived, too.

15.1 Starting Point

In the previous paragraphs, we have essentially introduced three different classical dispersion models, namely the Debye model for orientation polarization, the Drude model for the response of free charge carriers, and the Lorentzian oscillator model for the response of bound charge carriers. Let us summarize the observed types of dispersion in Fig. 15.1.

When comparing the dispersion curves shown in Fig. 15.1 and the underlying formulas it becomes evident, that all the models have certain common features. Thus, in the limiting case of infinitely large frequencies, in all models, the real part of the dielectric function approaches the value one, and the imaginary part the value zero. This is not astonishing, because all charge carriers have a certain mass and are thus inert. Hence, they are unable to follow the oscillations of an electric field when the frequency is too high. Therefore, the medium practically no more interact with the electromagnetic wave, such that the latter propagates through the medium like through vacuum. This is consistent with the experimental fact that γ-radiation is able to propagate through solid matter that does not transmit visible light (compare Sect. 14.4).

Another obvious feature is the correlation between the frequency dependence of the real and imaginary parts of the dielectric function. Indeed, prominent structures in the dispersion of Re ε occur, whenever Im ε is close to a local maximum. In

O. Stenzel, *Light–Matter Interaction*, UNITEXT for Physics,
https://doi.org/10.1007/978-3-030-87144-4_15

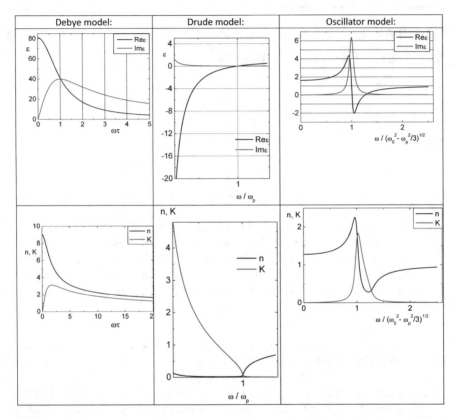

Fig. 15.1 Dispersion of the dielectric function and the complex index of refraction as predicted in terms of various classical models

regions of relative transparency, Im ε is negligible and Re ε shows a rather smooth frequency dependence. The same conclusion applies for the frequency dependence of the refractive index and the extinction coefficient. This correlation suggests that refraction and absorption are intercorrelated phenomena and cannot be modeled independently from each other. This empirical finding is the starting point for the material developed in this chapter.

15.2   Physical Idea

The physical idea is that (13.5) *completely* defines the polarization (i.e., both its amplitude and phase) of a homogeneous and isotropic medium in the time domain. This way the real response function $\kappa = \kappa(t)$ is expected to contain the full information on both refraction and absorption phenomena. The polarization is then given by:

$$\mathbf{P}(t) = \varepsilon_0 \int_0^\infty \kappa(\xi)\mathbf{E}(t - \xi)\mathrm{d}\xi$$

In agreement to the causality principle, the polarization thus depends on the field behavior in the past, but not in the future. For the special case of a monochromatic field, in particular we have from (13.6) and (13.7):

$$\mathbf{P}(t) = \mathbf{E}_0 e^{-i\omega t} \varepsilon_0 \int_0^\infty \kappa(\xi)e^{i\omega\xi}\mathrm{d}\xi \Rightarrow \mathbf{P} = \varepsilon_0\chi\mathbf{E}$$

Moreover, (13.7), (13.28), and (13.29) allow writing the expression for the susceptibility as:

$$\chi(\omega) = \int_0^\infty \kappa(\xi)e^{i\omega\xi}\mathrm{d}\xi = \int_{-\infty}^\infty \tilde{\kappa}(\xi)e^{i\omega\xi}\mathrm{d}\xi = \int_{-\infty}^\infty \tilde{\kappa}(\xi)\theta(\xi)e^{i\omega\xi}\mathrm{d}\xi \quad (15.1)$$

Because of causality, the response function is thus invariant with respect to multiple multiplication with the step function.

15.3 Theoretical Material: Kramers–Kronig Relations for Dielectrics

Let us start with some symmetry considerations. When expanding the exponential function in (13.29) or (15.1) into a series, we get:

$$\chi(\omega) = \int_{-\infty}^\infty \tilde{\kappa}(\xi)e^{i\omega\xi}\mathrm{d}\xi$$
$$= \int_{-\infty}^\infty \tilde{\kappa}(\xi)\mathrm{d}\xi + i\omega \int_{-\infty}^\infty \tilde{\kappa}(\xi)\xi\mathrm{d}\xi + \left(-\frac{\omega^2}{2}\right)\int_{-\infty}^\infty \tilde{\kappa}(\xi)\xi^2\mathrm{d}\xi + \cdots$$

Obviously, our ansatz for the susceptibility corresponds to an infinite series according to:

$$\chi(\omega) = A_0 + A_1\omega + A_2\omega^2 + A_3\omega^3 + \dots \quad (15.2)$$

where the A_j—values are constants. Thereby the even orders in ω compose the real part, while the odd orders compose the imaginary part of the susceptibility or the dielectric function. As the result, we obtain the important symmetry relations:

$$\mathrm{Re}\,\chi(-\omega) = \mathrm{Re}\,\chi(\omega); \quad \mathrm{Im}\,\chi(-\omega) = -\mathrm{Im}\,\chi(\omega)$$
$$\mathrm{Re}\,\varepsilon(-\omega) = \mathrm{Re}\,\varepsilon(\omega); \quad \mathrm{Im}\,\varepsilon(-\omega) = -\mathrm{Im}\,\varepsilon(\omega) \quad (15.3)$$

Hence, the real part of the dielectric function must be an even function of the frequency, while the imaginary part is an odd one. For $\omega = 0$, both the susceptibility and the dielectric function become real with $\chi = A_0$. This is clearly relevant for dielectric materials.

Let us now execute an inverse Fourier transformation according to:

$$\tilde{\kappa}(\xi) = \frac{1}{2\pi} \int_{-\infty}^{\infty} \chi(\omega) e^{-i\omega\xi}\, d\omega; \quad \theta(\xi) = \frac{1}{2\pi} \int_{-\infty}^{\infty} \Theta(\omega) e^{-i\omega\xi}\, d\omega.$$

With $\Theta(\omega)$—Fourier image of the step function. Substituting into (15.1) leads to [1]:

$$
\begin{aligned}
\chi(\omega) &= \frac{1}{(2\pi)^2} \int_{-\infty}^{\infty} e^{i\omega\xi}\, d\xi \int_{-\infty}^{\infty} \Theta(\omega_1) e^{-i\omega_1\xi}\, d\omega_1 \int_{-\infty}^{\infty} \chi(\omega_2) e^{-i\omega_2\xi}\, d\omega_2 \\
&= \frac{1}{(2\pi)^2} \int_{-\infty}^{\infty} \int_{-\infty}^{\infty} \Theta(\omega_1)\chi(\omega_2) d\omega_1 d\omega_2 \int_{-\infty}^{\infty} e^{i(\omega-\omega_1-\omega_2)\xi}\, d\xi \\
&= \frac{1}{2\pi} \int_{-\infty}^{\infty} \int_{-\infty}^{\infty} \Theta(\omega_1)\chi(\omega_2)\delta(\omega - \omega_1 - \omega_2) d\omega_1 d\omega_2 \\
&= \frac{1}{2\pi} \int_{-\infty}^{\infty} \Theta(\omega - \omega_2)\chi(\omega_2) d\omega_2 = \chi(\omega)
\end{aligned}
$$

where the identity:

$$\int_{-\infty}^{+\infty} e^{i(\omega-\omega_1-\omega_2)\xi}\, d\xi = 2\pi\delta(\omega - \omega_1 - \omega_2)$$

has been used with $\delta(x)$—Dirac's delta function. The Fourier spectrum of the step function should be calculated according to:

$$\Theta(\omega) = \int_{-\infty}^{\infty} \theta(\xi) e^{i\omega\xi}\, d\xi = \int_{0}^{\infty} e^{i\omega\xi}\, d\xi = ???$$

Because of the oscillatory behavior of the integrand, however, this improper integral does not exist in a strong mathematical sense. It exists only in the sense of so-called generalized functions, a branch of mathematics that is beyond the topics covered in this book. We will use the results obtained in the frames of this theory, but in order to make the background somewhat transparent, let us have a look at the following physical argumentation:

Return to (15.1). As a result of causality, we found it that the response function of the regarded system should be invariant with respect to multiplication with the step function. But this requirement may be relaxed. In any real physical system, the memory to an external perturbation will be restricted to certain characteristic relaxation times, and consequently, the

Fig. 15.2 Illustration of the mutual relation of the step function $\theta(\xi)$ (black dash), the function $e^{-\frac{\xi}{T}}$ (black solid), the term $\kappa(\xi)\theta(\xi) = \tilde{\kappa}(\xi)\theta(\xi)$ from (13.28) (red), and $\tilde{\kappa}(\xi)e^{-\frac{\xi}{T}}$ (navy). For large T, the navy and red curves coincide

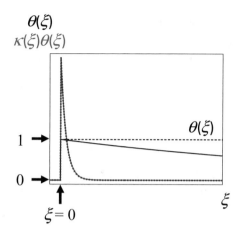

response function should approach zero for large positive arguments. Therefore, the concrete behavior of the step function for largest arguments is not so essential. In fact the true Fourier image of the step function (if it would exist) may be replaced by an alternative expression corresponding to another step-like function that leads to the same result when performing the integration over ξ in (15.1) (Fig. 15.2).

We therefore replace the Fourier image of the step function by:

$$\int_0^\infty e^{i\omega\xi}\,d\xi = \int_0^\infty \theta(\xi)e^{i\omega\xi}\,d\xi \to \lim_{T\to\infty}\int_0^\infty e^{-\frac{\xi}{T}}e^{i\omega\xi}\,d\xi$$

This limit may be calculated, and we obtain

$$\Theta(\omega) \to \lim_{T\to\infty}\int_0^\infty e^{-\frac{\xi}{T}}e^{i\omega\xi}\,d\xi$$

$$= \lim_{T\to\infty}\frac{1}{\left(-\frac{1}{T}+i\omega\right)} = \lim_{T\to\infty}\frac{1}{-i\omega+\frac{1}{T}} = \lim_{T\to\infty}\frac{T^{-1}}{T^{-2}+\omega^2} + \lim_{T\to\infty}\frac{i\omega}{T^{-2}+\omega^2}$$

$$= \pi\,\delta(\omega) + \frac{i}{\omega}$$

This coincides with the result obtained in the framework of generalized function theory in a mathematical strong sense. One then obtains:

$$\Theta(\omega-\omega_2) = \pi\,\delta(\omega-\omega_2) + \frac{i}{\omega-\omega_2} \to \chi(\omega)$$

$$= \frac{1}{2\pi}VP\int_{-\infty}^\infty\left[\pi\,\delta(\omega-\omega_2) + \frac{i}{\omega-\omega_2}\right]\chi(\omega_2)\,d\omega_2$$

We thus come to the relationship: $\chi(\omega) = \frac{i}{\pi} V P \int_{-\infty}^{\infty} \frac{\chi(\omega_2)}{\omega - \omega_2} d\omega_2$.

where "*VP*" denotes Cauchy's principal value of the integral. Separating the real (Re χ) and imaginary (Im χ) parts, we obtain the result:

$$\text{Re } \chi(\omega) = -\frac{1}{\pi} V P \int_{-\infty}^{\infty} \frac{\text{Im } \chi(\omega_2) d\omega_2}{\omega - \omega_2} = \frac{1}{\pi} V P \int_{-\infty}^{\infty} \frac{\text{Im } \chi(\omega_2) d\omega_2}{\omega_2 - \omega}$$

$$\text{Im } \chi(\omega) = \frac{1}{\pi} V P \int_{-\infty}^{\infty} \frac{\text{Re } \chi(\omega_2) d\omega_2}{\omega - \omega_2} = -\frac{1}{\pi} V P \int_{-\infty}^{\infty} \frac{\text{Re } \chi(\omega_2) d\omega_2}{\omega_2 - \omega} \quad (15.4)$$

In application to the real and imaginary parts of the dielectric function (Re ε and Im ε), we finally get the Kramers–Kronig Relations:

$$\text{Re } \varepsilon(\omega) = 1 + \frac{1}{\pi} V P \int_{-\infty}^{\infty} \frac{\text{Im } \varepsilon(\omega_2) d\omega_2}{\omega_2 - \omega}$$

$$\text{Im } \varepsilon(\omega) = -\frac{1}{\pi} V P \int_{-\infty}^{\infty} \frac{[\text{Re } \varepsilon(\omega_2) - 1] d\omega_2}{\omega_2 - \omega} \quad (15.4a)$$

The important conclusion is, that as a consequence of causality, the dispersion of the real and imaginary parts of the dielectric function are interconnected to each other via the integral transformations (15.4) and (15.4a).

In (15.4) and (15.4a), the susceptibility or the dielectric functions have to be defined for positive and negative frequencies. That does not cause any problems. According to (15.3), the imaginary part of the dielectric function has to be regarded as an odd function of the frequency, while the real part is an even one. Consequently, (15.4a) may be rewritten in the more familiar manner:

$$\text{Re } \varepsilon(\omega) = 1 + \frac{2}{\pi} V P \int_{0}^{\infty} \frac{\text{Im } \varepsilon(\omega_2) \omega_2 d\omega_2}{\omega_2^2 - \omega^2}$$

$$\text{Im } \varepsilon(\omega) = -\frac{2\omega}{\pi} V P \int_{0}^{\infty} \frac{[\text{Re } \varepsilon(\omega_2) - 1]}{\omega_2^2 - \omega^2} d\omega_2 \quad (15.4b)$$

15.4   Consistency Considerations

This paragraph may be kept short. Obviously, relation (15.4b) provides the correlations between refraction and absorption phenomena we have expected in Sect. 15.1. The extreme cases are well-described. Thus, for $\omega = 0$, we find a real dielectric constant that must be larger than one if the medium is dissipative, while for $\omega \to \infty$, the dielectric function approaches the value 1. In a denser medium, more dipols are present, and therefore Im ε in (15.4b) should be larger, which is expected to result in an increase in the static dielectric constant or the refractive index at low frequencies. All this sounds very reasonable and well consistent to what we have described so far.

However, when returning to Fig. 15.1, our formulas are unable to reproduce the optical behavior of conductors at lowest frequencies. Obviously, the treatment developed so far is only valid for ideal dielectrics. What is the reason for this discrepancy?

Let us return to the Drude function (14.3). For $\omega \to 0$, it behaves like:

$$\chi^{\mathrm{Drude}}(\omega)|_{\omega \to 0} \approx i \cdot \frac{Nq^2}{\varepsilon_0 m} \cdot \frac{1}{2\gamma\omega} = i\frac{\sigma_{\mathrm{stat}}}{\varepsilon_0\omega}$$

This is obviously inconsistent with (15.1) and (15.2), which was in the basis of our derivation of (15.4b). The physical reason is clear: For a conductor, (15.1) must simply diverge when the frequency is zero. The reason is, that in a static electric field, a free charge carrier could move for infinitely large distances from its starting position, which causes an infinitely large static dipole moment. Therefore, in order to correctly describe the optical behavior of conducting media, (15.2) must be generalized according to (15.5):

$$\chi^{\mathrm{conductor}}(\omega) = i\frac{\sigma_{\mathrm{stat}}}{\varepsilon_0\omega} + A_0 + A_1\omega + A_2\omega^2 + A_3\omega^3 + \dots \tag{15.5}$$

This is generally different from (15.2), but for sufficiently high (optical) frequencies, the term $\frac{\sigma_{\mathrm{stat}}}{\varepsilon_0\omega}$ has no significance, so that (15.2) remains valid. Indeed, already for $\omega > 2\gamma$, A_0 becomes larger by modulus than $\frac{\sigma_{\mathrm{stat}}}{\varepsilon_0\omega}$. Note further that (15.5) is still consistent with (15.3).

Our consistency considerations thus show that the Kramers–Kronig relations derived so far are restricted in their validity to aplications to dielectrics. The generalization to electric conductors will be topic of Sect. 15.6.1.

15.5 Application to Practical Problems

Let us use this short section to present some useful relationships that immediately follow from (15.5) and (15.6). We start with the derivation of a simple dispersion formula (Wemple's dispersion formula), which is obtained for the refractive index dispersion in a frequency region well below the region of absorption. We postulate that absorption (a nonzero imaginary part of the dielectric function) is restricted to a frequency range $[\omega_A, \omega_B]$. According to the mean value theorem, we have

$$
\begin{aligned}
\mathrm{Re}\,\varepsilon(\omega) = n^2(\omega) &= 1 + \frac{2}{\pi} V P \int_0^\infty \frac{\mathrm{Im}\,\varepsilon(\omega_2)\omega_2 d\omega_2}{\omega_2^2 - \omega^2} \\
&= 1 + \frac{2}{\pi} \int_{\omega_A}^{\omega_B} \frac{\mathrm{Im}\,\varepsilon(\omega_2)\omega_2 d\omega_2}{\omega_2^2 - \omega^2} \\
&= 1 + (\omega_B - \omega_A)\frac{\mathrm{Im}\,\varepsilon(\overline{\omega})\overline{\omega}}{\overline{\omega}^2 - \omega^2} \cong 1 + \frac{\mathrm{const.}\cdot\overline{\omega}}{\overline{\omega}^2 - \omega^2} = n^2(\omega)
\end{aligned}
$$

with $\overline{\omega} \in [\omega_A, \omega_B]$; $\omega << \omega_A < \omega_B$; Im $\varepsilon(\omega) = 0$.

By structure, this dispersion formula is similar to the Sellmeier formula, obtained for one single oscillator. In fact, in our derivation, the full absorption structure has been replaced by a single oscillator centered at $\overline{\omega} \in [\omega_A, \omega_B]$.

Another conclusion concerns the static dielectric constant for nonconductive materials. From (15.4b), for $\omega = 0$ we get:

$$\varepsilon_{\text{stat}} = 1 + \frac{2}{\pi} \int_0^\infty \frac{\text{Im } \varepsilon(\omega)}{\omega} d\omega \qquad (15.6)$$

So we see that the value of the static dielectric constant is directly connected to the high-frequency behavior of Im ε. In dielectrics, it will always be larger than one. On the other hand, for very high frequencies, we find from (15.4b) (provided, of course, that this integral exists!):

$$\text{Re } \varepsilon(\omega)|_{\omega \to \infty} \to 1 - \frac{2}{\pi \omega^2} \int_0^\infty \text{Im } \varepsilon(\omega_2) \omega_2 d\omega_2 \qquad (15.7)$$

This approach is valid when the current frequency is much higher than the frequencies where absorption occurs (it does not work for our writing of Debye's Eqs. (13.27), because of the slow descent of the imaginary part with increasing frequency). We see that for very high frequencies, we still have to expect normal dispersion, but the refractive index is smaller than one! This is a situation typical for the X-ray region—compare Sect. 14.4 in this regards.

From here, we may again conclude that the real part of the dielectric function of a dielectric *must* show anomalous dispersion in the vicinity of absorption structures, no matter whether or not the oscillator model is applicable. Indeed, in the static case it is larger than one (15.7). As far as we have no absorption, it further increases with frequency according to Wemple's formula. When the frequency is well above the absorption frequencies, we find normal dispersion again, but the refractive index is smaller than 1 (compare 14.9). Consequently, in the neighborhood of absorption structures, the refractive index must decrease with frequency (as far as it is regarded to be a continuous function of the frequency).

Let us finally come to an important sum rule. Let us return to the classical equation of motion of a charge carrier according to (14.11). Assuming that the electric field of the wave **E** leads to a displacement of the charge q along the x-axis, we write Newton's equation of motion according to:

$$qE = qE_0 e^{-i\omega t} = m\ddot{x} + 2\gamma m\dot{x} + m\omega_0^2 x$$

Let us further assume:

$$x \propto e^{-i\omega t}$$

In the limit of sufficiently high frequencies, we find:

$$\frac{qE}{m} = -\omega^2 x - 2i\gamma \omega x + \omega_0^2 x\big|_{\omega \to \infty} \to -\omega^2 x; \Rightarrow x \to -\frac{qE}{m\omega^2} \qquad (15.8)$$

This result holds for any assumed resonance frequency and, in particular, also for free charge carriers. From here, the expression for the dielectric function is easily found as:

$$\Rightarrow P \equiv \varepsilon_0(\varepsilon - 1)E = Nqx \rightarrow -\frac{Nq^2 E}{m\omega^2}$$

$$\varepsilon(\omega \rightarrow \infty) \rightarrow 1 - \frac{Nq^2}{\varepsilon_0 m\omega^2} \tag{15.9}$$

where N is the full concentration of charge carriers (electrons).

Comparing (15.7) and (15.9) leads us to the sum rule:

$$N = \frac{2\varepsilon_0 m}{\pi q^2} \int_0^\infty \operatorname{Im} \varepsilon(\omega)\omega d\omega \tag{15.10}$$

Hence, the integral absorption is connected to the concentration of dipoles that cause the absorption. Rewriting (15.10) in terms of the optical constants, one immediately obtains:

$$N = \frac{2\varepsilon_0 mc}{\pi q^2} \int_0^\infty n(\omega)\alpha(\omega)d\omega \tag{15.11}$$

Equation (15.11) is in the fundament of any quantitative spectroscopic analysis, where the integral absorption is measured in order to determine the concentration of any kind of absorption centers (molecules, impurities, and so on). Of course, in any practical application, one will always use a finite frequency interval where the integration in (15.11) is performed. Note the important result that the concentration is related to the integral absorption and *not* to the peak absorption.

15.6 Advanced Material

15.6.1 Kramers–Kronig Relations for Conductors

Equations (15.4–15.4b) are valid only for dielectrics. The reason is clear: The integration interval involves the argument $\omega_2 = 0$, but as we have mentioned in Sect. 15.4, (15.2) cannot be used to describe the low-frequency behavior of conductors. Instead, basing on (15.5) we postulate:

$$\chi^{\text{conductor}}(\omega) \equiv i\frac{\sigma_{\text{stat}}}{\varepsilon_0\omega} + \chi^{\text{opt}}(\omega)$$

Here, χ^{opt} behaves "regular" with respect to an expansion into a power series of the angular frequency. Therefore, for χ^{opt}, the Kramers–Kronig relations hold as derived so far:

$$\operatorname{Re} \chi^{\text{opt}}(\omega) = \frac{1}{\pi} VP \int_{-\infty}^\infty \frac{\operatorname{Im} \chi^{\text{opt}}(\omega_2)d\omega_2}{\omega_2 - \omega}$$

$$\operatorname{Im} \chi^{\mathrm{opt}}(\omega) = -\frac{1}{\pi} VP \int_{-\infty}^{\infty} \frac{\operatorname{Re} \chi^{\mathrm{opt}}(\omega_2) d\omega_2}{\omega_2 - \omega}$$

We have $\operatorname{Re} \chi^{\mathrm{conductor}}(\omega) = \operatorname{Re} \chi^{\mathrm{opt}}(\omega)$, and:

$$\operatorname{Im} \chi^{\mathrm{conductor}}(\omega) = \operatorname{Im} \chi^{\mathrm{opt}}(\omega) + \frac{\sigma_{\mathrm{stat}}}{\varepsilon_0 \omega}$$

$$\Rightarrow \operatorname{Im} \chi^{\mathrm{conductor}}(\omega) = -\frac{1}{\pi} VP \int_{-\infty}^{\infty} \frac{\operatorname{Re} \chi^{\mathrm{conductor}}(\omega_2) d\omega_2}{\omega_2 - \omega} + \frac{\sigma_{\mathrm{stat}}}{\varepsilon_0 \omega}$$

From here:

$$\operatorname{Re} \chi^{\mathrm{conductor}}(\omega) = \operatorname{Re} \chi^{\mathrm{opt}}(\omega) = \frac{1}{\pi} VP \int_{-\infty}^{\infty} \frac{\operatorname{Im} \chi^{\mathrm{opt}}(\omega_2) d\omega_2}{\omega_2 - \omega}$$

$$= \frac{1}{\pi} VP \int_{-\infty}^{\infty} \frac{\operatorname{Im} \chi^{\mathrm{conductor}}(\omega_2) d\omega_2}{\omega_2 - \omega} - \frac{\sigma_{\mathrm{stat}}}{\pi \varepsilon_0} VP \underbrace{\int_{-\infty}^{\infty} \frac{d\omega_2}{\omega_2(\omega_2 - \omega)}}_{=0 @ \omega \neq 0}$$

$$= \frac{1}{\pi} VP \int_{-\infty}^{\infty} \frac{\operatorname{Im} \chi^{\mathrm{conductor}}(\omega_2) d\omega_2}{\omega_2 - \omega}$$

When now making use of (15.3), we rewrite these equations according to:

$$\operatorname{Re} \varepsilon^{\mathrm{conductor}}(\omega) = 1 + \frac{2}{\pi} VP \int_{0}^{\infty} \frac{\operatorname{Im} \varepsilon^{\mathrm{conductor}}(\omega_2) \omega_2 d\omega_2}{\omega_2^2 - \omega^2}$$

$$\operatorname{Im} \varepsilon^{\mathrm{conductor}}(\omega) = -\frac{2\omega}{\pi} VP \int_{0}^{\infty} \frac{\left[\operatorname{Re} \varepsilon^{\mathrm{conductor}}(\omega_2) - 1\right]}{\omega_2^2 - \omega^2} d\omega_2 + \frac{\sigma_{\mathrm{stat}}}{\varepsilon_0 \omega}$$

(15.12)

Note that from here, (15.4b) is naturally obtained when setting $\sigma_{\mathrm{stat}} = 0$.

15.6.2   Once More: The f-sum Rule

According to (12.26), the quantum mechanical expression for the polarizability was:

$$\beta = \frac{2q^2}{\varepsilon_0 \hbar} \sum_{l} \sum_{n>l} |x_{nl}|^2 \omega_{nl} \frac{[W(l) - W(n)]}{\omega_{nl}^2 - \omega^2 - 2i\omega\gamma_{nl}}$$

Let us now assume the limiting case $\omega \to \infty$. Then the dielectric function approaches one, such that local field effects are of no significance anymore. We find:

$$\varepsilon(\omega \to \infty) \to 1 - \frac{2q^2 N}{\varepsilon_0 \hbar \omega^2} \sum_{l} \sum_{n>l} |x_{nl}|^2 \omega_{nl} [W(l) - W(n)]$$

On the other hand, from (15.9) we have

$$\varepsilon(\omega \to \infty) \to 1 - \frac{q^2 N}{\varepsilon_0 m \omega^2}$$

When comparing both these expressions, we find:

$$\frac{2m}{\hbar} \sum_l \sum_{n>l} |x_{nl}|^2 \omega_{nl} [W(l) - W(n)] = 1$$

Or, making use of the oscillator strength (5.32)

$$\sum_l \sum_{n>l} f_{nl} [W(l) - W(n)] = 1$$

This expression may be rewritten as:

$$1 = \sum_l \sum_{n>l} f_{nl} [W(l) - W(n)] = \sum_l \sum_{n>l} f_{nl} W(l) + \sum_l \sum_{n>l} f_{ln} W(n)$$
$$= \sum_l \sum_{n>l} f_{nl} W(l) + \sum_n \sum_{l<n} f_{ln} W(n)$$
$$= \sum_l \sum_{n>l} f_{nl} W(l) + \sum_l \sum_{n<l} f_{nl} W(l) = \sum_l \sum_{n \neq l} f_{nl} W(l) = 1$$

This should be valid for any time-independent assumed set of probabilities $\{W(l)\}$, particularly for any assumed set $\{W(l_0) = 1;\ W(l \neq l_0) = 0\}$, regardless on the choice of l_0. Therefore, the term $\sum_{n \neq l} f_{nl}$ should be equal to one for each l. Thus, $\sum_{n \neq l} f_{nl} = 1\ \forall l$. What we have rederived this way is the oscillator strength sum rule. Thereby, the basic idea was that the classical and quantum mechanical expressions coincide for infinitely large frequencies or photon energies. Practically, we have thus shown that the oscillator strength sum rule is consistent with the correspondence principle.

15.7   Taks for Self-check

15.7.1 Multiple-choice test: Mark all answers which seem you correct!

The linear absorption coefficient of a material is	Dimensionless
	Given in s^{-1}
	Given in cm^{-1}
In the X-ray spectral region, the refractive index of a nonmagnetic material is usually	Infinitely large
	Larger than zero

	Smaller than 1	
	Equal to one	
	Negative	
Permanent electric dipoles in a medium may result in	A high static dielectric constant	
	Orientation polarization	
	A temperature dependence of the static dielectric constant according to Curie's law	
Anomalous dispersion	Never occurs in liquids	
	Is observed in the vicinity of an absorption line	
	Violates causality	

15.7.2 True or wrong? Make your decision!

Assertion	True	Wrong
In linear optics, electromagnetic energy dissipation is related to a nonzero imaginary part of the dielectric function		
In condensed matter, the local (microscopic) electric field may significantly differ from the average field in the medium		
The real part of the complex index of refraction is always larger than the imaginary part		
The real part of the complex index of refraction is always larger than or equal to 1		
The selection rules relevant for Raman spectra are identical to those relevant for infrared absorption		
The concentration of absorbing species in a medium may be reliably estimated from the peak value of the absorption coefficient		
The concentration of absorbing species in a medium may be estimated from the imaginary part of the dielectric function integrated over the (angular) frequency		

15.7.3 In the special case of orientation polarization, the real part of the dielectric function of the material may be given by one of Debye's equations:

$$\operatorname{Re} \varepsilon(\omega) = 1 + \frac{\chi_{\text{stat}}}{1 + \omega^2 \tau^2}$$

Make use of the Kramers–Kronig Relation

$$\operatorname{Im} \varepsilon(\omega) = -\frac{1}{\pi} V P \int_{-\infty}^{\infty} \frac{[\operatorname{Re} \varepsilon(\omega_2) - 1] d\omega_2}{\omega_2 - \omega}$$

to derive the corresponding expression for the imaginary part of the dielectric function. You may make use of the integral [2]:

$$\int \frac{dx}{(x+b)(x^2+a^2)} = \frac{1}{a^2+b^2}\left[\ln|x+b| - \frac{1}{2}\ln|x^2+a^2| + \frac{b}{a}\arctan\frac{x}{a}\right]$$

15.7.4 As 15.7.3, but with the Drude function!

References

Specific References

1. М. Б. Виноградова, О. В. Руденко, А. П. Сухоруков: Теория Волн; Москва Наука, Главная Редакция, Физико-Математической Литературы, 1979, p. 62–65
2. А.П.Прудников, Ю.А.Брычков, О.И.Маричев: *Интегралы и ряды*, Moscow "Наука" 1981

General Literature to this Chapter

3. L.D. Landau, E.M. Lifshitz, Electrodynamics of continuous media, in *A Course of Theoretical Physics*, vol 8 (Pergamon Press, 1960)
4. O. Stenzel, An introduction, in *The physics of thin film optical spectra* (Springer, 2016)

Part VI
Optical Properties of Solids

"Schloss Tratzberg" (Tratzberg Castle, Tirol, Austria). Painting and Photograph by Astrid Leiterer, Jena, Germany (www.astrid-art.de). Photograph reproduced with permission

Solids form a subset of continuous media and are macroscopically character-ized by a well-defined volume and a well-defined shape. Often enough, they exhibit some short- or long-ranging atomic order, which gives rise to certain characteristic symmetry.

Introduction to Solid-State Physics

16

Abstract

Basic concepts of the physics of crystalline solids are introduced. In a single-electron picture, the characteristic energy band structure is obtained as a result of the spatial periodicity of the crystal lattice. Basing on the energy band occupation, a distinction between metals and dielectrics is introduced. The essence of partially filled bands for the formation of a DC electric conductivity is illustrated.

16.1 Starting Point

According to the classification provided in Sect. 12.1, solids belong to the category of condensed matter, while they are phenomenologically distinguished from liquids by a large shear modulus. The most popular class of solids is provided by the *crystalline solids* (c-solids), i.e., solids that exhibit a strongly periodic spatial arrangement of their atoms (Fig. 16.1). In this case, when the atoms are arranged in a strongly periodic manner, they are called to form a crystal lattice.

But solids do not need to be crystalline. Thus, glasses form a class of solids without a strongly periodic arrangement of their atoms ("frozen liquids"). Nevertheless, the atomic arrangement in glasses is not completely chaotic. Instead, it turns out that the positions of the nearest neighbors of a selected atom are more or less accurately predictable even in glasses, while with an increasing distance, the probability density to observe an atom in a certain space element tends to smear out more and more. This is considered in terms of the concept of short-range order, which is still observed in glasses. In this context, the class of solids that exhibit short-range order, while lacking long range order, forms the class of *amorphous solids* (a-solids). This way, amorphous materials have to be distinguished from completely disordered matter, which even lack any short-range order (Table 16.1).

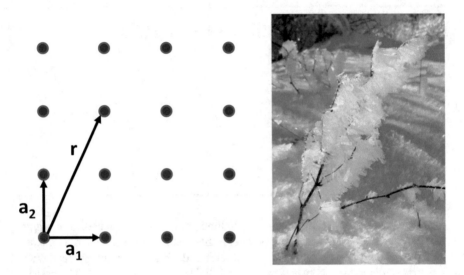

Fig. 16.1 Left: Schematic representation of a crystal lattice, navy circles represent the equilibrium positions of the atoms (=lattice points); right: ice crystals winter 2021 in Jena

Table 16.1 Classification of solids/materials

Type of solid/material	Short-range order	Long-range order
Crystalline	Present	Present
Amorphous	Present	Absent
Disordered	Absent	Absent

A quantitative measure of the degree of short-range order is possible in terms of the so-called radial distribution function RDF.

Zallen [1, p. 12] mentioned a very simple and helpful thought experiment to distinguish between an amorphous solid and a disordered system: Imagine a person with a bad memory, who removes exactly one atom from an amorphous structure and from a disordered system. Some days later he wants to reinsert the atoms into their correct positions. Clearly, the person has forgotten from where the atoms have been taken. But no doubt, a glance on the positions of the remaining atoms in the amorphous structure will enable him to identify the former neighbors of the removed atom, so that he will reinsert the atom approximately at the correct place. In a disordered system, however, the remaining atomic positions will give no clue about the missing one, and he will not be able to identify the former position.

In this context, Fig. 16.2 invites you, dear reader, to perform this experiment for yourself. The figure shows the positions of atoms in a perfect crystal (left), an amorphous solid (center), and a disordered system (right). But from each of the systems, exactly one atom has been removed. You are invited to look at the images carefully, before taking a pencil (see Fig. 16.3) and marking the position

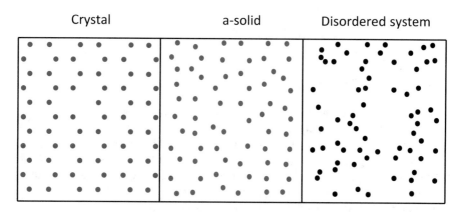

Fig. 16.2 Arrangement of atoms in a crystal (c-solid, left), amorphous solid (a-solid, center), and a disordered system. In each of the systems, one atom is "removed" (compare text). After an idea of Zallen [1, p. 12]

Fig. 16.3 Graphite lattice (on left) and diamond lattice (on right) as examples on different bonding types. Cartoon by Dr. Alexander Stendal. Printed with permission

of the missing atom according to your guess. Then, turn to Chap. 19 (the solutions to tasks to chapter 16) and compare with the "true" position.

Regarding bonding types, so basic mechanisms have already been discussed in Chap. 10. Bonding mechanisms in solids may be accomplished by:

- Covalent bonding
- Ionic bonding (compare also Fig. 10.6)
- Metallic bonding
- Van der Waals bonding
- Bonding through hydrogen bridges

Figure 16.3 illustrates simple examples of covalent and van der Waals bonding. In a diamond crystal (on right), all carbon atoms are covalently bound to each other. In graphite (on left), the two-dimensional honeycomb-like lattices form planes with covalent bonding within the planes, while in-between the planes, van der Waals bonding is dominant. In an ionic crystal like KCl, bonding occurs as a result of electrostatic interaction between differently charged ions (Fig. 10.6). The picture also demonstrates the ancient writing utensile called pencil.

16.2   Physical Idea

In condensed matter, atoms come close enough to each other that their electronic wavefunctions start to overlap. According to what we have learned in Chap. 10, that leads to a splitting of the previous atomic energy levels into a multiplicity of levels, thus forming energy intervals where the energy levels are arranged in a quasi-continuous manner. These energy intervals where "allowed" energy levels are concentrated will further be called energy bands (See schematic representation in Fig. 16.4). The closer the atoms come to each other, the broader these energy bands will be, and the narrower the energy spacing (forbidden zone, energy gap,

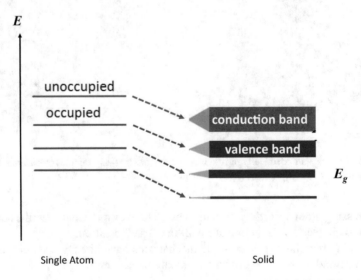

Fig. 16.4 Schematic comparison of atomic energy levels and energy bands in a dielectric solid

(energy) band gap) E_g will be. At the same time, new hybrid electronic orbitals are formed, such that an electron is no more necessarily localized in the vicinity of "its" atomic core, but has a chance to become delocalized.

Let us now remember that electrons are fermions, such that the Pauli's exclusion principle is valid. Then, at zero (or a rather low) temperature T, all electrons try to occupy the lowest by energy levels in the system. However, because of the Pauli principle, no more than one electron is allowed to occupy a given state (compare task 9.7.3 in this regard). The result is that that the electrons fill up the energy bands from bottom to top. We usually observe a certain set of completely filled bands of lowest energy (Fig. 16.4). One of them is that of the highest energy, and we will call that energy band, which has the highest energy among all completely filled bands, the valence band. Then we have two possibilities. The next highest energy band may be (at $T = 0$) either partially filled or completely empty. We will call that band the conduction band. If the conduction band is empty at $T = 0$, the material will be called a dielectric or a semiconductor (at the moment we will not distinguish between these two classes of materials). If the conduction band is partially filled, then the material is called a metal. This classification makes absolute sense, as we will see in Sect. 16.5.2.

This picture is of course very simplified, because in reality energy bands may overlap, which gives rise to another material class called semimetals. But the mentioned simple picture will be sufficient for our purposes here.

We have already mentioned that solids do not need to be crystalline. Nevertheless, c-solids represent superb model systems for demonstrating striking physical effects caused by specific symmetry properties. Thus, if the atomic arrangement is ideally periodic, the whole system must be exactly reproduced when it is shifted in space by a distance that coincides with an integer multiple of its period. We will then use the terminology that the system must obey a discrete translation invariance.

Let us illustrate this by a daily life example. Everybody knows tessellation solution that fill up a two-dimensional space in a periodic manner. An example is provided in Fig. 16.5.

An obvious feature of this tessellation is that it may be periodically extended to infinity in both directions; hence, it obeys discrete translation invariance. We will see in Sect. 16.3 that the translation invariance hast striking effects on the properties of the wavefunctions of particles that propagate in such a periodic potential. But this is obviously not the whole story. A further glance at the tessellation solution confirms us that it is obviously also invariant with respect to rotations for multiples of 90° (Fig. 16.6). In such a case, we speak on fourfold rotational symmetry, because $90° = 360°/4$. We come to the conclusion, that translational invariance may be compatible at least with certain types of rotational symmetry.

What about inversion symmetry, which is utmost important for the optical behavior? Obviously, inversion symmetry is present, as indicated by the black arrows in Fig. 16.7 on left. On the other hand, there is no mirror symmetry

Fig. 16.5 Tessellation
solution in the Doberan
Minster (Doberaner
Münster). The floor-stones
stem from the end of the
nineteenth century and have
been manufactured by
Villeroy and Boch. I am
grateful to Martin Heider for
the background information
and the permission to use the
photograph

(Fig. 16.7 in center and on right—the red arrows mark differently looking features at positions mirrored with respect to the axis indicated as the black dashed
line).

Let it now be the task to discuss the dynamics of the electrons in the potential
provided by the periodically arranged nuclei. Clearly, the potential $U(\mathbf{r})$ should
reflect the symmetry in the arrangement of the nuclei. Then we would have to
solve a many-body task, because there is a huge number of electrons in a real
solid, and in fact, similar to the situation in molecules, the nuclei may also perform oscillations (with frequencies corresponding to typical MIR frequencies). The
approximate solution of suchlike many-body tasks is beyond the frame of this
course, and therefore, we will restrict on a much easier model task. What we will
discuss here is the movement of a single electron in a periodic potential $U(\mathbf{r})$.
Then, the Schrödinger equation for that single electron is defined by the rather
simple single-particle Hamiltonian $\hat{H} = -\frac{\hbar^2 \nabla^2}{2m_e} + U(\mathbf{r})$, where $U(\mathbf{r})$ is now the
resulting potential provided by the nuclei *and all the other electrons*. And now, this
potential is *postulated* to be periodic in space and time-independent.

16.3 Theoretical Considerations: Basics of Crystalline Solids

16.3.1 Translation and Rotational Invariance

Let us imagine the atoms of a solid arranged in space in a completely periodic
manner. Such a periodic atomic arrangement is typical for crystals. Then, the position of any lattice point may be given as the superposition of three basis vectors,

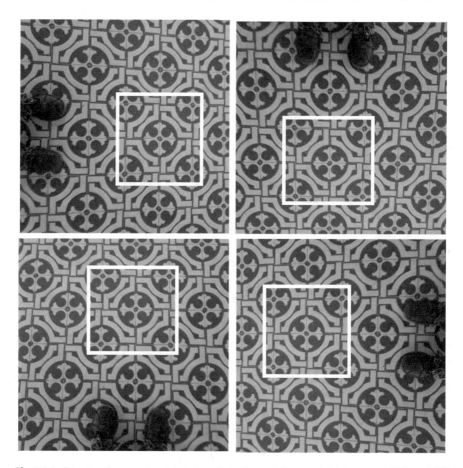

Fig. 16.6 Rotational symmetry. A rotation of the image left on top for integer multiples of 90° leads to an exact reproduction of the basic image in the white frame. My shoes are shown to indicate the reference direction

Fig. 16.7 Inversion symmetry of the tessellation (left), but lacking mirror symmetry with respect to the black dashed lines (center and right)

named $\mathbf{a}_1, \mathbf{a}_2, \mathbf{a}_3$. Any lattice point (compare Fig. 16.1) is thus given by

$$\mathbf{r}_n = n_1\mathbf{a}_1 + n_2\mathbf{a}_2 + n_3\mathbf{a}_3 \qquad (16.1)$$

n_1, n_2, n_3 integer.

The set of the shortest possible basis vectors spans a parallelepiped called the primitive cell of the lattice. The length of these basis vectors defines what we call the lattice constants.

Any vector as defined by (16.1) will be further called a lattice vector. The translation of all lattice points by any lattice vector results in a complete reproduction of the lattice (discrete translational invariance).

The rotation of the lattice by an angle φ may also result in the reproduction of the complete lattice. If so, the lattice is told to have a rotational symmetry axis. If $\varphi = 2\pi/n$ with n—a positive integer, we speak about an (at least) n-fold rotational symmetry axis. Obviously, the system shown in Fig. 16.5 has a fourfold rotational axis (Fig. 16.6).

It may be shown that in a lattice as defined by (16.1), n may accept values $n = 1, 2, 3, 4,$ and 6 [2, p. 375]. Indeed, let us assume, that rotation by an angle φ leads to a reproduction of the lattice defined by (16.1). Any new lattice vector obtained as a result of rotating (16.1) forms an angle φ with $\mathbf{r}_n$ and will be called $\mathbf{r}_{n,\varphi}$. Rotating into the reverse direction leads to the vector $\mathbf{r}_{n,-\varphi}$. Obviously, the vector $(\mathbf{r}_{n,\varphi} + \mathbf{r}_{n,-\varphi})$ must again be a lattice vector, and it must be parallel to $\mathbf{r}_n$. Thereby, the vector $(\mathbf{r}_{n,\varphi} + \mathbf{r}_{n,-\varphi})$ has the length of $2\mathbf{r}_n\cos\varphi$. On the other hand, it must be possible to represent it in terms of (16.1), while because of the parallelity to $\mathbf{r}_n$, it may be written as:

$$\mathbf{r}_{n,\varphi} + \mathbf{r}_{n,-\varphi} = m_1\mathbf{a}_1 + m_2\mathbf{a}_2 + m_3\mathbf{a}_3 = G\mathbf{r}_n$$
$$= G(n_1\mathbf{a}_1 + n_2\mathbf{a}_2 + n_3\mathbf{a}_3) = 2\cos\varphi(n_1\mathbf{a}_1 + n_2\mathbf{a}_2 + n_3\mathbf{a}_3)$$

Here, G is a constant. Once all m-values must be integer for any chosen integer n, we must require that G is also an integer number. In other words, we find $G = 2\cos\varphi$. On the other hand, $-1 \le \cos\varphi \le 1$. Therefore, G may accept the integer values $-2, -1, 0, +1, +2$ only. From $G = 2\cos\varphi = 2\cos(2\pi/n)$ we find the possible values $n = 1, 2, 3, 4, 6$. Therefore, 5-, sevenfold or higher rotational symmetry is impossible in crystal lattices with strong translational invariance.

A translational invariant tessellation solution combining elements of eight- and fourfold rotational symmetry is exemplified in Fig. 16.8. Obviously, the complete system nevertheless has only fourfold rotational symmetry. In Fig. 16.9, threefold rotational symmetry is exemplified in architectural applications. Even "total" spherical symmetry is observed in nature (Fig. 16.10).

Remark: Note that the so-called quasicrystals provide tessellation solutions with elements of fivefold rotational symmetry, but they are not strongly periodic (translational invariant). Much work in this field has been done by the 2020 Nobel Prize

Fig. 16.8 Detail from an Egyptian decorative wall plate. It has fourfold rotational symmetry, although it also contains elements of eightfold rotational symmetry

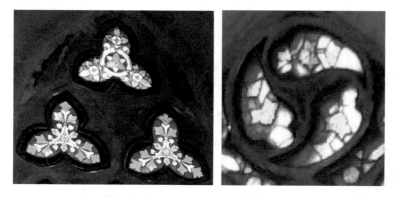

Fig. 16.9 Examples of threefold rotational symmetry. Details from church windows in the St. Aegidii Church Quedlinburg, Germany. Printed with permission

winner Roger Penrose, who is also very popular as a re-inventor of the "impossible triangle" (Penrose tribar—see picture below).

Fig. 16.10 Rotational symmetry in living nature. So-called fairy ring formed by a nearly circular arrangement of a multiplicity of mushrooms. Found in a meadow near Laasan, Thuringia

These were examples of two-dimensional geometries. Typical real crystals are, of course, three-dimensional. According to their crystalline structure, crystals may be optically isotropic or anisotropic. Table 16.2 visualizes the so-called Bravais lattices with the corresponding assignment to optical isotropy or anisotropy (uniaxial crystals and biaxial crystals).

16.3.2 Single-Electron Approximation

Let us now see what happens when a single electron is brought in such a strongly periodic potential. Let us assume that as the result of the periodic arrangement of the atoms, a single electron "feels" a spatially periodic potential (provided by the nuclei and all other electrons different from the discussed one) given by (Fig. 16.11):

$$U(\mathbf{r} + \mathbf{r}_n) = U(\mathbf{r}) \tag{16.2}$$

For the movement of the discussed electron, we thus have the Schrödinger equation:

$$\hat{H}\psi_j(\mathbf{r}) = E_j\psi_j(\mathbf{r}) \tag{16.3}$$

With the Hamiltonian:

$$\hat{H} = -\frac{\hbar^2\nabla^2}{2m_e} + U(\mathbf{r}) \tag{16.4}$$

Table 16.2 Bravais lattices (after [2, p. 377])

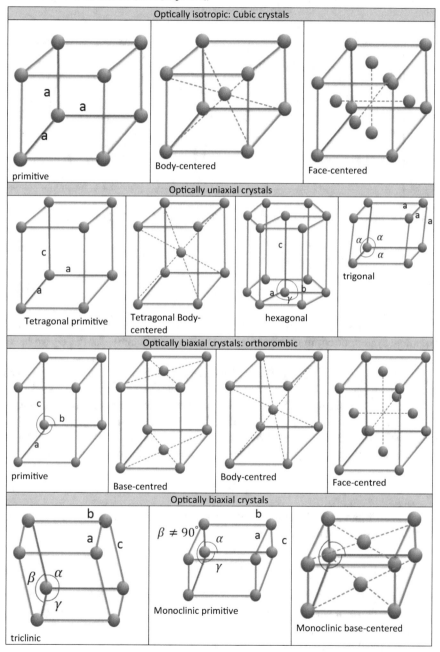

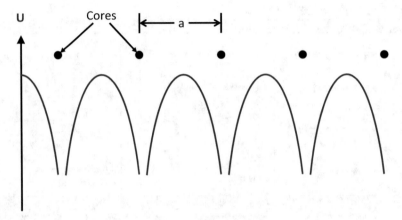

Fig. 16.11 Schematic representation of a 1D periodic potential. a is the lattice constant

Let us now define the translation operator $\hat{T}_n$ through

$$\hat{T}_n \psi(\mathbf{r}) \equiv \psi(\mathbf{r} + \mathbf{r}_n) \tag{16.5}$$

From (16.2) we may conclude that $\hat{T}_n$ and $\hat{H}$ commute and therefore have a complete system of common eigenfunctions [3]. Consequently, for the ψ_i we have:

$$\hat{T}_n \psi_j(\mathbf{r}) = t_n \psi_j(\mathbf{r}) = \psi_j(\mathbf{r} + \mathbf{r}_n) \tag{16.6}$$

From the definition (16.5) it should be clear that two translation operators $\hat{T}_{n_1}$ and $\hat{T}_{n_2}$ should commute, hence they have common eigenfunctions, too [4]. Once the normalization of ψ_i should not depend on the position of the origin of coordinates chosen, the eigenvalue t_n of the translation operator must fulfill:

$$|t_n| = 1$$

which is fulfilled for all complex numbers with an absolute value 1. For t_n we therefore chose the representation:

$$t_n = e^{i\mathbf{k}\mathbf{r}_n} \tag{16.7}$$

with $\mathbf{k}, \mathbf{r}_n$—real vectors.

Obviously, for a superposition of translations according to $\hat{T}_{n_1+n_2} = \hat{T}_{n_1} \hat{T}_{n_2} = \hat{T}_{n_2} \hat{T}_{n_1}$ we have the eigenvalues: $t_{n_1+n_2} = t_{n_1} t_{n_2}$.

From (16.6) and (16.7), for any given $\mathbf{r}_n$ we therefore obtain:

$$\psi_j(\mathbf{r} + \mathbf{r}_n) = e^{i\mathbf{k}\mathbf{r}_n} \psi_j(\mathbf{r}) \tag{16.8}$$

Equations (16.7) and (16.8) together indicate that for any eigenfunction of the periodic Hamiltonian according to (16.2) there exists at least one wavevector $\mathbf{k}$ such that a spatial displacement for an arbitrary lattice vector $\mathbf{r}_n$ is equivalent to multiplication of the wavefunction with the phase factor $e^{i\mathbf{k}\mathbf{r}_n}$. The wavevector $\mathbf{k}$ may be different for different wavefunctions, but in each case it should exist. All of the eigenfunctions of the Hamiltonian defined by (16.2) and (16.4) are thus expected to be characterized by a certain $\mathbf{k}$. Further, we therefore use the writing:

$$\hat{H}\psi_{\mathbf{k},j}(\mathbf{r}) = E_{\mathbf{k},j}\psi_{\mathbf{k},j}(\mathbf{r}) \tag{16.9}$$

with

$$E_{\mathbf{k},j} \equiv E_j(\mathbf{k})$$

Because $\mathbf{k}$ is continuous. The wavefunctions $\psi_{\mathbf{k},j}$ may be written in terms of a Bloch function according to:

$$\psi_{\mathbf{k},j}(\mathbf{r}) = e^{i\mathbf{k}\mathbf{r}}u_{\mathbf{k},j}(\mathbf{r}) \tag{16.10}$$

We then find:

$$\hat{T}_n\psi_{\mathbf{k},j}(\mathbf{r}) = \psi_{\mathbf{k},j}(\mathbf{r}+\mathbf{r}_n) = e^{i\mathbf{k}(\mathbf{r}+\mathbf{r}_n)}u_{\mathbf{k},j}(\mathbf{r}+\mathbf{r}_n)$$
$$\hat{T}_n\psi_{\mathbf{k},j}(\mathbf{r}) = t_n\psi_{\mathbf{k},j}(\mathbf{r}) = e^{i\mathbf{k}\mathbf{r}_n}\psi_{\mathbf{k},j}(\mathbf{r}) = e^{i\mathbf{k}\mathbf{r}_n}e^{i\mathbf{k}\mathbf{r}}u_{\mathbf{k},j}(\mathbf{r})$$
$$\rightarrow u_{\mathbf{k},j}(\mathbf{r}) = u_{\mathbf{k},j}(\mathbf{r}+\mathbf{r}_n), \tag{16.11}$$

It follows that the functions $u_{\mathbf{k}j}$ are periodic and have the same period as the lattice (Bloch theorem). The wavefunction of an electron in a spatially periodic potential is therefore the product of a propagating plane wave with a periodic function $u_{\mathbf{k},j}$. The parameter $\hbar\mathbf{k}$ is called the quasimomentum of the electron in the corresponding quantum state.

Note that $\mathbf{k}$ is ambiguous. Indeed, according to (16.7), a given eigenvalue t_n corresponds to an infinite number of $\mathbf{k}$.

16.3.3 Reciprocal Lattice

Let us shortly comment on the ambiguity of $\mathbf{k}$. Obviously any $\mathbf{k}'$—value, which differs from $\mathbf{k}$ by $\mathbf{g}$ according to

$$\mathbf{k}' = \mathbf{k} + \mathbf{g}; \quad \mathbf{g} = 2\pi\tau$$

with

$$\tau = m_1\mathbf{b}_1 + m_2\mathbf{b}_2 + m_3\mathbf{b}_3 \quad (m_1, m_2, m_3 \ldots \text{integer})$$

$$\mathbf{b}_1 = \frac{\mathbf{a}_2 \times \mathbf{a}_3}{\mathbf{a}_1 \cdot (\mathbf{a}_2 \times \mathbf{a}_3)} \cdots$$

corresponds to the same eigenvalue t_n. This is obvious from

$$e^{i\mathbf{k}'\mathbf{r}_n} = e^{i\mathbf{k}\mathbf{r}_n} e^{i\mathbf{g}\mathbf{r}_n}$$

$$\text{With } \mathbf{g}\mathbf{r}_n = 2\pi \mathbf{r}_n \boldsymbol{\tau} = 2\pi \sum_{i,j=1}^{3} n_i m_j \mathbf{a}_i \mathbf{b}_j =$$

$$= 2\pi \sum_{i,j=1}^{3} n_i m_j \delta_{ij} = 2\pi M \quad \left(\delta_{ij} = \begin{cases} 0 \ i \neq j \\ 1 \ i = j \end{cases} \right)$$

(M—integer), because of

$$\mathbf{a}_1 \mathbf{b}_1 = \frac{\mathbf{a}_1 \cdot (\mathbf{a}_2 \times \mathbf{a}_3)}{\mathbf{a}_1 \cdot (\mathbf{a}_2 \times \mathbf{a}_3)} = 1$$

$$\mathbf{a}_2 \mathbf{b}_1 = \frac{\mathbf{a}_2 \cdot (\mathbf{a}_2 \times \mathbf{a}_3)}{\mathbf{a}_1 \cdot (\mathbf{a}_2 \times \mathbf{a}_3)} = 0$$

And so on. So we find:

$$e^{i\mathbf{k}'\mathbf{r}_n} = e^{i\mathbf{k}\mathbf{r}_n}; \quad \mathbf{k}' = \mathbf{k} + \mathbf{g} \tag{16.12}$$

The vectors $\mathbf{g}$ mark points in the so-called reciprocal lattice. A special construction of the primitive cell in the reciprocal lattice (the so-called Wigner–Seitz cell) defines what we call the first Brillouin zone.

16.4 Consistency Considerations

The essence of (16.10) is that the assumed periodicity of the potential U defines a corresponding periodicity in the wavefunction of the electron propagating in that potential. This is nothing astonishing; we have earlier identified other examples where the symmetry of the potential (e.g., inversion symmetry) defined the symmetry of the wavefunction (odd or even parity). This is by the way not a specific quantum mechanical effect: Everybody knows from daily experience, that—for example—the picture of standing surface waves in a vessel filled with water depends on the specific symmetry of the vessel. It is therefore an effect rather connected to wave physics in general.

Let us now return to Fig. 16.4 and the accompanying argumentation in Sect. 16.2. Our idea was that the spatial overlap and the resulting interaction between the electronic wavefunctions in condensed matter results in a splitting of the atomic orbitals into many new orbitals, similarly as this has been demonstrated in the context of covalent bonding in molecules (Chap. 10). Because of

Table 16.3 Refractive index and band gap in selected crystals

Material	IR-refractive index	E_g/eV
c-Ge	4.0	≈ 0.7
c-Si	3.4	≈ 1.2
diamond	2.3	≈ 5.4

the large number of electrons, this is expected to finally result in the formation of quasi-continuous energy bands instead of discrete energy levels. But if so, then any increase in the electron density should result in a further broadening of the bands and, at the same time, in a narrowing of the forbidden zones. This is particularly relevant for the width of the forbidden zone between the valence and conduction band E_g, which is utmost important for the optical properties of a real material in the NIR/VIS/UV spectral region (compare Table 2.2 in this context). Thus, an increase in the electron concentration should result in a narrower band gap E_g.

But we also know that an increase in the electron concentration is expected to result in an increase in the refractive index at low frequencies (compare (14.7) or (14.7a)). Our argumentation therefore suggests that there should be a correlation observed between refractive index and band gap E_g. This is indeed observed as a general trend. As an example, Table 16.3 opposes the MIR refractive index to the band gap (here indirect band gap—compare later Sect. 17.3.3) in diamond, crystalline silicon, and crystalline germanium.

Thus, at least in this example, an increase in the band gap seems to be accompanied by a decrease in refractive index. This is by the way astonishing because in fact we provided a rather naïve argumentation, which reflects maybe 10% of the complexity of the interplay between refractive index and band gap. In fact, the density of silicon is *smaller* than that of diamond. Nevertheless, the qualitative result that we obtained from our naïve argumentation is reasonable. A more sophisticated quantitative formulation of this correlation is provided by the Moss rule established in semiconductor optics [5]:

$$n^4 E_g \approx \text{const.} \quad (\text{const.} \approx 95\,\text{eV})$$

16.5 Application to Practical Problems

16.5.1 Model Calculation: Energy Bands in a Tight Binding Approach

Let us now return to (16.4). In the proposed model calculation, the periodic potential (16.2) will be approximated by the sum of equidistantly spaced identical atomic potentials V_A as:

$$U(\mathbf{r}) = \sum_m V_A(\mathbf{r} - \mathbf{r}_m) \tag{16.13}$$

The use of potentials of the isolated atoms and the consequent use of a superposition of atomic eigenstates are characteristics for the tight binding approach: The effect of the atomic neighbors is tackled as a small perturbation. The Schrödinger equation (16.9) of a single electron in that potential results in:

$$\hat{H}\psi_{\mathbf{k},j}(\mathbf{r}) = \left\{ -\frac{\hbar^2}{2m_e}\nabla^2 + \sum_m V_A(\mathbf{r} - \mathbf{r}_m) \right\} \psi_{\mathbf{k},j}(\mathbf{r}) = E_j(\mathbf{k})\psi_{\mathbf{k},j}(\mathbf{r}) \quad (16.14)$$

Let us now multiply (16.14) from the left with $\psi_{\mathbf{k},j}^*$ and integrate over the whole space. We then find:

$$E_j(\mathbf{k}) = \int \psi_{\mathbf{k},j}^* \hat{H}\psi_{\mathbf{k},j} dV \quad (16.15)$$

Let us further assume that the solution of the Schrödinger equation for an electron in any of the single m-th atomic potentials

$$\hat{H}_A(\mathbf{r} - \mathbf{r}_m)\varphi_j(\mathbf{r} - \mathbf{r}_m) = \left\{ -\frac{\hbar^2}{2m_e}\nabla^2 + V_A(\mathbf{r} - \mathbf{r}_m) \right\} \varphi_j(\mathbf{r} - \mathbf{r}_m) = E_j\varphi_j(\mathbf{r} - \mathbf{r}_m)$$
$$(16.16)$$

is known. We will now search the solution of (16.14) as a linear superposition of single-atom wavefunctions φ [6] according to [compare (16.8)]

$$\psi_{\mathbf{k},j}(\mathbf{r}) = \sum_m a_m\varphi_j(\mathbf{r} - \mathbf{r}_m); \quad a_m = \frac{1}{\sqrt{N}}e^{i\mathbf{k}\mathbf{r}_m} \quad (16.17)$$

(N—number of atoms in the crystal). The ϕ_j are eigenfunctions of single atoms, and the E_j the corresponding atomic eigenenergies. Equation (16.17) corresponds to the ansatz for the wavefunction in the so-called LCAO (Linear Combination of Atomic Orbitals) approach (compare Chap. 10). Combining (16.15) and (16.17), we obtain:

$$E_j(\mathbf{k}) = \int \psi_{\mathbf{k},j}^* \hat{H}\psi_{\mathbf{k},j} dV$$
$$= \frac{1}{N}\sum_m\sum_n e^{i\mathbf{k}(\mathbf{r}_n-\mathbf{r}_m)} \int \varphi_j^*(\mathbf{r} - \mathbf{r}_m)\hat{H}\varphi_j(\mathbf{r} - \mathbf{r}_n)dV \quad (16.18)$$

where the Hamiltonian is given by (16.14). Let us further recognize that (16.13) may be written as:

$$U(\mathbf{r}) = \sum_{m'} V_A(\mathbf{r} - \mathbf{r}_{m'}) = V_A(\mathbf{r} - \mathbf{r}_n) + \sum_{m'\neq n} V_A(\mathbf{r} - \mathbf{r}_{m'}) \quad (16.19)$$

Here we choose summation over the independent summation index m', in order to not confuse the summation indices in (16.13) and (16.18). Moreover, we have

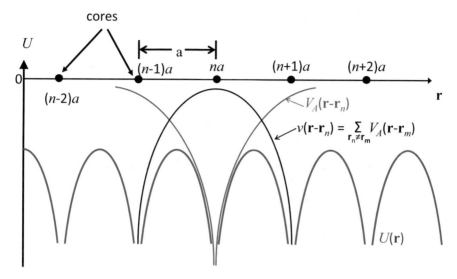

Fig. 16.12 Illustration to (16.20). At large distances from the n-th lattice point, the black and blue curves are practically coinciding

explicitly separated the potential of the n-th atom $V_A(\mathbf{r} - \mathbf{r}_n)$. The Hamiltonian in (16.18) is thus (compare Fig. 16.12):

$$\hat{H} = \left\{ \underbrace{-\frac{\hbar^2}{2m_e}\nabla^2 + V_A(\mathbf{r} - \mathbf{r}_n)}_{\equiv \hat{H}_A(\mathbf{r} - \mathbf{r}_n)} + \underbrace{\sum_{m' \neq n} V_A(\mathbf{r} - \mathbf{r}_{m'})}_{\equiv v(\mathbf{r} - \mathbf{r}_n)} \right\} \tag{16.20}$$

According to (16.16), we have

$$\hat{H}_A(\mathbf{r} - \mathbf{r}_n)\varphi_j(\mathbf{r} - \mathbf{r}_n) = \left\{ -\frac{\hbar^2}{2m_e}\nabla^2 + V_A(\mathbf{r} - \mathbf{r}_n) \right\}\varphi_j(\mathbf{r} - \mathbf{r}_n) = E_j\varphi_j(\mathbf{r} - \mathbf{r}_n)$$

Therefore, (16.18) and (16.20) together yield:

$$E_j(\mathbf{k}) = \int \psi^*_{\mathbf{k},j}\hat{H}\psi_{\mathbf{k},j}\,dV = \frac{1}{N}\sum_m\sum_n e^{i\mathbf{k}(\mathbf{r}_n - \mathbf{r}_m)}\int \varphi^*_j(\mathbf{r} - \mathbf{r}_m)\hat{H}_A\varphi_j(\mathbf{r} - \mathbf{r}_n)\,dV$$

$$+ \frac{1}{N}\sum_m\sum_n e^{i\mathbf{k}(\mathbf{r}_n - \mathbf{r}_m)}\int \varphi^*_j(\mathbf{r} - \mathbf{r}_m)v\varphi_j(\mathbf{r} - \mathbf{r}_n)\,dV$$

$$\equiv I_1 + I_2$$

The first term (I_1) approaches E_j when neglecting the spatial overlap of wavefunctions which belong to different atoms. Then:

$$I_1 = \frac{1}{N} \cdot E_j \sum_m \sum_n e^{i\mathbf{k}(\mathbf{r}_n - \mathbf{r}_m)} \underbrace{\int \varphi_j^*(\mathbf{r} - \mathbf{r}_m)\varphi_j(\mathbf{r} - \mathbf{r}_n)dV}_{\approx \delta_{mn}}$$

$$\approx \frac{E_j}{N} \sum_m \sum_n \delta_{mn} e^{i\mathbf{k}(\mathbf{r}_n - \mathbf{r}_m)} = E_j$$

In the second term, we separate the contributions arising from the products of wavefunctions of identical atoms from those arising from the products of wavefunctions of next neighbors, while neglecting all the others:

$$I_2 = \frac{1}{N} \sum_m \underbrace{\int \varphi_j^*(\mathbf{r} - \mathbf{r}_m)v\varphi_j(\mathbf{r} - \mathbf{r}_m)dV}_{-A_j}$$

$$+ \frac{1}{N} \sum_m \sum_{n_l = m_l \pm 1} e^{i\mathbf{k}(\mathbf{r}_n - \mathbf{r}_m)} \underbrace{\int \varphi_j^*(\mathbf{r} - \mathbf{r}_m)v\varphi_j(\mathbf{r} - \mathbf{r}_n)dV}_{-B_j} + \cdots$$

$$\approx -A_j - \frac{B_j}{N} \sum_m \sum_{n_l = m_l \pm 1} e^{i\mathbf{k}(\mathbf{r}_n - \mathbf{r}_m)}$$

$$= -A_j - B_j \sum_{n_l = m_{0l} \pm 1} e^{i\mathbf{k}(\mathbf{r}_n - \mathbf{r}_{m_0})} \tag{16.21}$$

Here, A_j and B_j are constants. The Index $l = 1,2,3$ indicates that summation over next neighbors goes along all three directions as defined by the corresponding 3D-lattice basis. The value of B_j again depends on the overlap of the wavefunctions. When assuming isotropic atomic wavefunctions (s-orbitals) and a cubic lattice, (16.21) results in:

$$I_2 \approx -A_j - 2B_j\left(\cos k_x a + \cos k_y a + \cos k_z a\right) \tag{16.22}$$

a—lattice constant

Figure 16.13 visualizes what may be obtained from (16.22). Instead of discrete energy levels, we find allowed and forbidden energy regions, in correspondence to what has guessed in Fig. 16.1. The width of the energy bands (allowed energy regions) is controlled by B_j, which is dependent on the degree of overlap of wavefunctions of neighbored atoms. As a trend, B_j will increase with higher j. Between the bands, we find forbidden energy zones. The $E(\mathbf{k})$—dependence appears to be periodic, while all information is contained in the first Brillouin zone defined here by $k \in \left[-\frac{\pi}{a}, +\frac{\pi}{a}\right]$.

Note that $E(k)$ in fact defines a particular dispersion law. Thereby, the derivative dE/dk is proportional to the group velocity for the electron in the corresponding

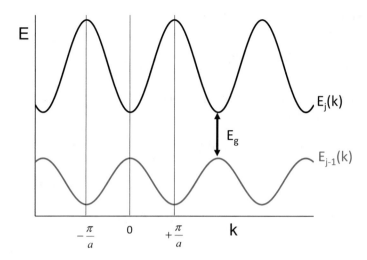

Fig. 16.13 Visualisation of $E(k)$ in a periodic zone

state. In a completely filled band, because of the symmetry of $E(k)$, each electron with a given group velocity will have a counterpart with exactly the opposite value of the group velocity. Therefore, macroscopic charge transport is impossible within a completely filled band. And therefore, in Sect. 16.1, materials with either completely filled or either completely empty bands have been classified as insulators (dielectric).

An illustration on the nature of forbidden zones may be obtained from the wave picture of the propagating electrons. When their wavevector approaches π/a, their wavelength corresponds to $2a$. Then, the reflections from two neighbored lattice points lead to constructive interference in reflection, hence the reflection is enhanced. The reflected waves themselves are back-reflected and so on, so that finally instead of a propagating wave, a standing wave is generated. The group velocity becomes zero, and propagation of electrons with corresponding k-vectors is impossible.

Note in this context that the qualitative picture illustrated in Fig. 16.13 is more general and not only applicable to the tight binding approach. In particular, it allows obtaining valuable qualitative information about the behavior of electrons in the conduction band as well.

16.5.2 Effective Masses

Let us now consider an extremum in a dispersion curve like shown in Fig. 16.13. Let the extremum be located at $k = k_0$. In the vicinity of that minimum, we have (for simplicity we write it in a 1D-version):

$$E(k) = E(k_0) + \frac{1}{2}(k - k_0)^2 \frac{d^2 E}{dk^2}\bigg|_{k_0} + \cdots$$

Let us further remember that the group velocity of the electron may be calculated according to:

$$v_{gr} = \frac{dE}{\hbar dk} = \frac{(k - k_0)}{\hbar} \frac{d^2 E}{dk^2}\bigg|_{k_0} \Rightarrow (k - k_0)^2 = \frac{\hbar^2 v_{gr}^2}{\left(\frac{d^2 E}{dk^2}\big|_{k_0}\right)^2}$$

$$\Rightarrow E(k) = E(k_0) + \frac{1}{2} \frac{\hbar^2}{\frac{d^2 E}{dk^2}\big|_{k_0}} v_{gr}^2 \equiv E(k_0) + \frac{1}{2} m^* v_{gr}^2$$

Here we have introduced the effective mass $m*$ of an electron in the periodic potential as:

$$m* = \frac{\hbar^2}{\frac{d^2 E}{dk^2}\big|_{k_0}} \tag{16.23}$$

Obviously, the effective mass is positive around a minimum, and negative around a maximum of the dispersion curve.

When applying an external force F to the charge carrier, its group velocity will change according to:

$$m* \frac{dv_{gr}}{dt} = F$$

On the other hand,

$$\frac{dv_{gr}}{dt} = \frac{1}{\hbar} \frac{d}{dt} \frac{dE}{dk} = \frac{1}{\hbar} \frac{d}{dk} \frac{dk}{dt} \frac{dE}{dk} = \frac{1}{\hbar^2} \frac{d^2 E}{dk^2} \frac{d(\hbar k)}{dt} = \frac{F}{m*} \tag{16.24}$$

Such that we obtain:

$$F = \frac{\hbar dk}{dt} \tag{16.25}$$

An external force thus causes a change in the quasimomentum of the electron, which means that the latter must occupy a new position in the dispersion curve like exemplified in Fig. 16.13. Correspondingly, it will change its group velocity. This gives us the key for understanding the appearance of macroscopic charge transport in a partially filled energy band, while macroscopic charge transport in a completely filled band remains forbidden.

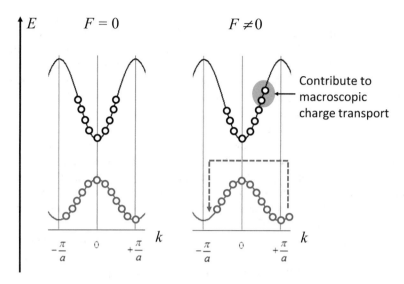

Fig. 16.14 Visualization of the response of completely (red) and partially (black) filled bands to an external field. Note that because of charge carrier scattering processes, for a time-independent applied electric field, the number of charge carriers contributing to electrical conductivity as highlighted right on top remains constant as well, such that a constant voltage results in a DC electrical current

Let us have a look at Fig. 16.14. In fact, it shows a detail from Fig. 16.13, while we restrict to the discussion of **k**-vectors within the first Brillouin zone, which contains all necessary physical information. The hollow circles indicate states occupied by electrons. The "red" dispersion curve now corresponds to a completely filled energy band, while the "black" dispersion curve illustrates the situation in a partially filled band. In equilibrium conditions, without applying an external force F to the charge carriers, each electron with a group velocity different from zero will have a counterpart with exactly the opposite value of the group velocity, so that there is no macroscopic charge transport, no matter whether the band is completely or only partially filled. This is the rather symmetric situation visualized on the left of Fig. 16.14. If, however, a constant electric field is applied, the resulting Coulomb force acting on each of the electrons will act as an external force different from zero. Then, according to (16.25), the "chain" of circles in the corresponding dispersion curve must move to the right if the force is positive. And now we see fundamental differences between the situations in the partially and completely filled bands. In the partially filled band (black curve), there appear charge carriers with uncompensated group velocities with the same sign. These charge carriers cause a macroscopic electric charge transport. In the completely filled band, however, because of the periodicity of the dispersion curve, each of the charge carriers still has a counterpart with the opposite group velocity, so that no charge transport is observed. Thus, completely filled bands cannot contribute to DC electrical conductivity. Moreover, because of the equivalence of k-vectors which differ from each other by a reciprocal lattice vector, the circle which has "left" the first Brillouin zone in Fig. 16.14 right on bottom, is in a physical situation that is equivalent to the "empty position" on the left side in the Brillouin zone indicated by the dashed red arrow. Therefore, when restricting the discussion to k-states in the first Brillouin zone only, any circle that leaves the first Brillouin zone on the right side is in fact re-entering that zone from left again. Therefore, the occupation situation in the filled band remains unchanged, no matter whether the field is different from zero or not.

Thus, if within a material all energy bands are either completely filled or completely empty, no DC electrical conductivity is observed. Those materials form the class of electric insulators. On the contrary, materials with partially filled energy bands are electrical conductors and usually assigned to metals. These are, for example, materials built from atoms that altogether contribute an odd number of valence electrons to the primitive cell of the crystal (such as in sodium or aluminum). In particular, the hydrogen atom has only one electron, although the $1s$ orbital could accept two electrons. Therefore, frozen hydrogen should form a solid with a half-filled band only, and therefore form a metal [7].

16.6 Advanced Material

16.6.1 The Kronig–Penney Model

In addition to the model calculation performed in Sect. 16.4, let us have a look at a periodic one-dimensional model potential, which allows to solve the Schrödinger equation for a single particle moving in that potential exactly [8]. In the so-called Kronig–Penney model, the particle (in our case the electron) is assumed to move in a periodic rectangular potential as sketched in Fig. 16.15. Let a be the period, and b be the width of the potential walls. The height of the potential walls should be U_0, while the potential is zero between the walls. Note the formal similarity with the situation of quantum tunneling through a rectangular potential barrier, discussed

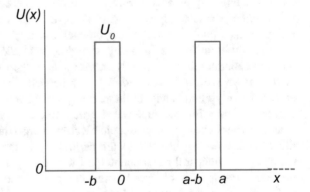

Fig. 16.15 Assumed model potential

in Sect. 4.6.2. According to Table 4.1, again we use the following ansatz for the wavefunction of the electron:

$$U = 0: \psi_1(x) = Ae^{i\alpha x} + Be^{-i\alpha x}; \alpha = \frac{\sqrt{2mE}}{\hbar}$$

$$U = U_0: \psi_2(x) = Ce^{\beta x} + De^{-\beta x}; \beta = \frac{\sqrt{2m(U_0 - E)}}{\hbar}$$

When requiring continuity of the wavefunction and its first derivative, we obtain at $x = 0$:

$$\psi_1(0) = \psi_2(0) \Rightarrow A + B = C + D$$

$$\frac{d}{dx}\psi_1(x)\Big|_{x=0} = \frac{d}{dx}\psi_2(x)\Big|_{x=0} \Rightarrow i\alpha(A - B) = \beta(C - D)$$

At $x = -b$, we find:

$$\psi_1(-b) = \psi_2(-b) \Rightarrow Ae^{-i\alpha b} + Be^{-i\alpha b} = Ce^{-\beta b} + De^{\beta b}$$

However, when making use of (16.8), we may write:

$$\psi_1(-b) = e^{-i\kappa a}\psi_1(a - b)$$

$$\Rightarrow Ce^{-\beta b} + De^{\beta b} = e^{-i\kappa a}\left[Ae^{i\alpha(a-b)} + Be^{-i\alpha(a-b)}\right] \tag{16.26}$$

With κ—real. In an identical manner, we have from the continuity of derivatives:

$$\beta\left(Ce^{-\beta b} - De^{\beta b}\right) = i\alpha e^{-i\kappa a}\left[Ae^{i\alpha(a-b)} - Be^{-i\alpha(a-b)}\right] \tag{16.27}$$

So that for the four coefficients, we have the four equations:

$$C + D = A + B$$

$$C - D = i\frac{\alpha}{\beta}(A - B)$$

$$Ce^{-\beta b} + De^{\beta b} = e^{-i\kappa a}\left[Ae^{i\alpha(a-b)} + Be^{-i\alpha(a-b)}\right]$$

$$Ce^{-\beta b} - De^{\beta b} = i\frac{\alpha}{\beta}e^{-i\kappa a}\left[Ae^{i\alpha(a-b)} - Be^{-i\alpha(a-b)}\right]$$

From the first two equations, C and D are easily expressed through $A + B$ and $A - B$. Substituting C and D into the last two equations, we obtain:

$$(A + B)\left[\cosh \beta b - e^{-i\kappa a} \cos \alpha(a - b)\right]$$

$$= i(A - B)\left[\frac{\alpha}{\beta} \sinh \beta b + e^{-i\kappa a} \sin \alpha (a - b)\right]$$

$$(A + B)\left[\sinh \beta b - \frac{\alpha}{\beta} e^{-i\kappa a} \sin \alpha (a - b)\right]$$

$$= i(A - B)\frac{\alpha}{\beta}\left[\cosh \beta b - e^{-i\kappa a} \cos \alpha (a - b)\right]$$

This is a homogeneous system of two equations in $A + B$ and $A - B$. In order to find nontrivial solutions, we must require:

$$\begin{vmatrix} \left[\cosh \beta b - e^{-i\kappa a} \cos \alpha (a - b)\right] & \left[\frac{\alpha}{\beta} \sinh \beta b + e^{-i\kappa a} \sin \alpha (a - b)\right] \\ \left[\sinh \beta b - \frac{\alpha}{\beta} e^{-i\kappa a} \sin \alpha (a - b)\right] & \frac{\alpha}{\beta}\left[\cosh \beta b - e^{-i\kappa a} \cos \alpha (a - b)\right] \end{vmatrix} = 0$$

$$\Rightarrow \cos \kappa a = \frac{\beta^2 - \alpha^2}{2\alpha\beta} \sinh \beta b \sin \alpha (a - b) + \cosh \beta b \cos \alpha (a - b)$$

Note that the right-hand part of this equation must be in-between the values -1 and 1 in order to have a real solution for κ. In fact, this restriction results in the formation of allowed and forbidden energy regions, i.e., the band structure. In order to visualize this more explicitly, let us introduce the parameter (do not confuse with a damping constant)

$$\gamma = \frac{mab}{\hbar^2} U_0 \tag{16.28}$$

It is proportional to the product $U_o b$, i.e., the "area" occupied by a single barrier in Fig. 16.15. Let us now consider the limiting case:

$$b \to 0 \text{ and } U_0 \to \infty, \text{ while } \gamma = \text{const.}$$

That results in:

$$\cos \kappa a = \gamma \frac{\sin \alpha a}{\alpha a} + \cos \alpha a \tag{16.29}$$

Figure 16.16 visualizes the dependence of $\gamma \frac{\sin \alpha a}{\alpha a} + \cos \alpha a$ as a function of αa. Note that α is directly related to the electron energy. Whenever $\gamma \frac{\sin \alpha a}{\alpha a} + \cos \alpha a$ becomes larger than $+1$ or smaller than -1, no solution for κ exists, and the corresponding energy region is classified as a forbidden energy zone. The other energy regions correspond to allowed energy bands.

Hence, this model calculation confirms us about the expected energy band structure obtained as a result of the periodicity of the assumed potential.

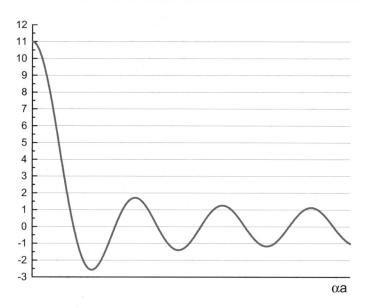

Fig. 16.16 Function $\gamma \frac{\sin \alpha a}{\alpha a} + \cos \alpha a$ for $\gamma = 10$

16.6.2 Once More About Consistency: Delocalized and Localized Electron States

Let us return in this section to some considerations on the consistency of the single-electron approximation introduced in Sect. 16.3.2. Let us first recall the qualitative picture we have developed in Sect. 10.3.2 when discussing the formation of covalent bonding in the hydrogen molecule ion. Our idea was there, that once the two hydrogen nuclei (protons) are close enough to each other, the electron might be considered to be common to both of the nuclei, which we associated with the formation of a molecule. On the other hand, when the nuclei are separated by a large distance, the natural situation is that the electron forms a hydrogen atom with one of the nuclei, while the other nucleus remains to be a positively charged hydrogen ion, and it makes no sense assuming that the electron "belongs" to both of the nuclei. Hence, the distance between the nuclei does matter.

What a about a hypothetical "crystal" with an interatomic spacing of 1 m? Will we find delocalized electrons described by Bloch functions, or will we simply obtain a periodic arrangement of well-isolated atoms that do not "feel" each other? Clearly, the second version appears more reasonable. But why?

Our idea expressed in Sect. 16.3.2 was that a single electron is moving in a spatially periodic potential. This potential is assumed to be provided by the (fixed in this picture) nuclei and all the other electrons. But that already means that we assume some cooperative behavior of the "other" electrons: If each of them is spatially delocalized with a wavefunction according to (16.10), then the considered electron feels the necessary periodic potential to also behave according to a Bloch function. This picture is internally consistent, but unfortunately this does not mean, that a considered physical system must behave accordingly.

The hypothetical picture of a crystal with an interatomic spacing of 1 m becomes especially impressive when a material with a partially filled conduction band (i.e., a metal) is discussed. Clearly, when the interatomic spacing becomes infinitely large, then the bandwidth B according to (16.21) approaches zero, such that the effective mass becomes infinitely large and the group velocity becomes zero. Hence, in this limit we arrive at an insulator which is reasonable. But as soon as the lattice constant becomes finite, the energy band picture we have derived so far predicts a metallic behavior of the material. This is not reasonable, and even worse, it is in contradiction with reality. In fact, there are materials that are electrically isolating, although the band theory described so far predicts a metallic behavior.

The point is that our single-electron picture neglects many-body effects [1, pp. 223–231]. In particular, correlations between the electrons resulting from their Coulomb interactions are not taken into account. This does not cause problems as long as the bandwidth B is sufficiently large. A partially filled broad energy band allows all the (delocalized in this picture) electrons occupying energy levels that correspond to a smaller kinetic energy as it would be the case if each electron would stay localized with "its" atom (compare with our discussion in Sect. 10.4). Thus, delocalization seems favored by energy, and this is absolutely true in a single-electron picture. However, in a many-electron picture, delocalization of electrons also means that electrons may come close to each other, which results in an increase in the potential energy because of the Coulomb interaction between the electrons. This Coulomb interaction between the electrons has however not been taken into account in Sect. 16.5.1, where we used a single-electron picture. As long as the gain in kinetic energy overcompensates the increase in potential energy (the case of a large B), everything is fine with our band theory, and the electrons stay delocalized. If however the bandwidth B is small such that the increase in Coulomb energy overcompensates the gain in kinetic energy, then the delocalization of electrons is no more favored, and localization takes place. In this case, the corresponding material behaves like an insulator although band theory would predict a partially filled band and consequently metallic behavior. Such materials are called Mott insulators, and the corresponding localization phenomenon is called Mott localization.

Note further, that in amorphous materials, long-range order is absent, such that our derivations that have based on strong periodicity of the potential are a priori invalid. Therefore, in amorphous materials, besides of delocalized electronic states one naturally observes also localized states. This occurs when the disorder-induced spread in the depth of the potential valleys in a potential like shown in Fig. 16.11 or Fig. 16.15 exceeds the nominal bandwidth B. The corresponding disorder-induced localization mechanism is called Anderson localization. As a rule, in the presence of some atomic disorder, spatially localized electronic states form band tails on top and bottom of the regarded energy band, while deeply inside the band, electronic states remain delocalized.

Figure 16.17 summarizes the cases of delocalized electrons, Mott localization and Anderson localization in the illustrative picture of a hungry wasp swarm in a beer garden.

Fig. 16.17 Localized and delocalized wavefunctions: On top: Illustration of delocalized wavefunctions: Any of the wasps cares on any of the guests. Center: Illustration of Mott localization: Each of the guests has its own wasp. On bottom: Illustration of Anderson localization: The table in the middle provides a stronger attractive potential than the other empty tables. The wasp swarm is strongly localized in the vicinity of that table. Cartoon by Dr. Alexander Stendal. Printed with permission.

## 16.7	Tasks for Self-check

16.7.1 The simplest model idea for a quantum mechanical description of conduction electrons in a solid is to treat them as freely moving within a potential box given by the extensions of the considered piece of metal. Assuming a temperature close to zero, determine the maximum electron energy occurring in the ground state, as well as the average kinetic electron energy [9].

16.7.2 In order to estimate the eigenenergies of the bound states of the so-called π-electrons in benzene, let us consider a set of 6 periodically arranged δ-potentials according to task 4.7.4. The periodical arrangement of the potentials brings the system close to the Kronig–Penney model that has been investigated in Sect. 16.6. In order to imitate the ring in terms of the one-dimensional treatment within the Kronig–Penney model, we introduce the extra requirement: $\psi(x) = \psi(x + 6a)\,\forall x$. a is now the distance between two adjacent carbon nuclei in the sixfold benzene ring. Investigate the position of the energy levels of a single-bound electron in that system! (after [10]).

References

Specific References

1. R. Zallen, *The Physics of Amorphous Solids* (Wiley, 1983)
2. W. Demtröder, *Experimentalphysik 3, Atome, Moleküle und Festkörper* (Springer, 2016)
3. А.С. Давыдов, Квантовая Механика, Москва Физматгиз (1963), pp. 579–582
4. M. Bartelmann, B. Feuerbacher, T. Krüger, D. Lüst, A. Rebhan, A. Wipf, *Theoretische Physik* (Springer Spektrum, 2015), pp. 894–895
5. H. Finkenrath, The Moss rule and the influence of doping on the optical dielectric constant of semiconductors—I. Infrared Phys. **28**, 327–332 (1988)
6. H. Ibach, H. Lüth, *Festkörperphysik. Einführung in die Grundlagen, 7. Auflage* (Springer, 2009), chapter 7
7. R. Gast, Wenn Wasserstoff zu Metall erstarrt. Spektrum der Wissenschaften **5**, 22–24 (2020)
8. А.А. Соколов, И.М. Тернов, В.Ч. Жуковский: Квантовая механика, Москва "Наука" (1979), pp. 464–466
9. S. Flügge, *Practical Quantum Mechanics* (Springer, 1999), problem 167
10. H. Haken, H.C. Wolf: *The Physics of Atoms and Quanta* (Springer 2005), task 24.3

General Literature

11. C. Kittel, *Introduction to Solid State Physics* (Wiley, any edition)
12. P.J. Yu, M. Cardona, *Fundamentals of Semiconductors. Physics and Material Properties*, 4th edn. (Springer, 2010)

13. M. Dresselhaus, G. Dresselhaus, S.B. Cronin, A.G.S. Filho, *Solid State Propertie. From Bulk to Nano* (Springer, Berlin, 2018)
14. J.M. Ziman, *Principles of the Theory of Solids* (Cambridge, 1972)

Introduction to Solid-State Optics

17

Abstract

Expressions for the dielectric function of optically isotropic solids are derived and discussed. Direct as well as indirect electronic absorption processes in crystals are considered, as well as basic features of the optical properties of amorphous solids. Typical power-law dependencies for describing the spectral shape of the absorption edge of solids are introduced.

17.1 Starting Point

A glance on the page number will confirm you, dear reader, that we are approaching the end of the course. Therefore, it is not astonishing that now we a drawing together more and more of the previously developed concepts in order to provide a quantitative description of more and more complex phenomena. So let us summarize what we will need from the previous paragraphs.

In this context, it is worth having a look at Fig. 17.1 on top. It is essentially a detail from Fig. 16.4, but symbolizing quantum transitions between discrete energy levels (on left) or energy bands (on right). The situation on left might correspond to an atom or molecule, while the situation on right illustrates the specifics of a solid.

The concepts we have derived so far enable us to describe the optical behavior of a quantum system with discrete energy levels in a semiclassical language. In particular, we have developed a perturbation approach for calculating the quantum transition rate and written down a corresponding semiclassical expression for the polarizability. This gives us access to the optical constants of the system shown on left of Fig. 17.1.

In the solid state, however, we have a continuous distribution of energy levels, and thus our treatment will need to be modified. In fact, we will develop two approaches to get access to the dielectric function of a solid. These two

© The Author(s), under exclusive license to Springer Nature Switzerland AG 2022

O. Stenzel, *Light–Matter Interaction*, UNITEXT for Physics,

https://doi.org/10.1007/978-3-030-87144-4_17

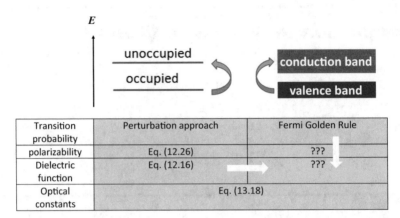

Fig. 17.1 To the organization of this chapter, discrete energy levels (on left) correspond to atoms or molecules (HOMO–LUMO transition), and energy bands (on right) to the situation in a solid

approaches are symbolized by the white arrows in the scheme presented on bottom of Fig. 17.1. The first one is to make use of the Fermi Golden rule introduced in Chap. 6. For that we will have to operate in termini of transition matrix elements as well as densities of states. The latter have already been introduced in Chap. 4, while transition matrix elements have accompanied us throughout the whole course. In particular, they define the selection rules for optical transitions. Therefore, Sect. 17.3 will start with the derivation of a specific selection rule relevant in crystalline solids. Then, a heuristic approach to the dielectric function will be provided that is based on the concept of the joint density of states.

The second approach, which starts from generalizing the dielectric function of a system with discrete energy levels to the continuous case, will be topic of Sect. 17.6. In both approaches, the behavior of the transition matrix elements will be of utmost importance.

17.2   Physical Idea

Starting from the concepts listed in Sect. 17.1, we will have to complement them by two further ideas that restrict the multiplicity of potentially possible quantum transitions between two energy bands in a solid as sketched in Fig. 17.1 right on top.

The first point is that in a crystal, according to Chap. 16, energy bands are defined by dispersion laws $E = E(\mathbf{k})$. The energy bands thus have an internal structure that is not obvious from simple illustrations such as shown in Fig. 17.1. The consequence is, that the relation beween the $\mathbf{k}$-values in the band from where the transition starts and the $\mathbf{k}$-values that are allowed in the band where the transition ends will be defined by specific selection rules.

The second point is that the occupation status of the energy bands involved in quantum transitions is essential. This immediately follows from expression (12.26). Moreover, once electrons are fermions, it is strongly forbidden that two electrons appear in the same quantum state (compare 9.12a). Therefore, we have to assume that any optical transition must start from a singly occupied state and must end in an empty state. This is possible in two entirely different situations:

- Both the starting and final energy levels are located in the same energy band. In this case, we speak on intraband transitions.
- Both the starting and final energy levels are located in different energy band. In this case, we speak on interband transitions.

Then, transitions between completely occupied energy bands do not give any contribution to optical absorption, while those between occupied and (partially) empty bands do.

Let us assume the situation like shown in Fig. 17.2 on left. Here we have a completely filled highest occupied energy band (the valence band), and a completely unoccupied next-higher energy band (the conduction band). Materials which such a band structure at $T = 0$ K will be called dielectrics or semiconductors (in contrast to conductors). Usually, in dielectrics we have $E_g \geq 3$ eV, and in semiconductors $E_g < 3$ eV. Light absorption between states of the valence and conduction bands may occur when the photon energy is sufficiently large to bridge the energy gap E_g.

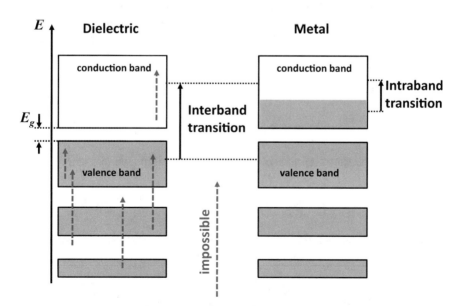

Fig 17.2 Interband and intraband transitions (black arrows) as induced by absorption of light. Gray regions correspond to filled quantum states, white to empty ones. Red dashed arrows indicate impossible combinations of initial and final energy levels in the solid

Thus, optical absorption in such a solid appears as a process with a well-defined threshold frequency of the photon.

Figure 17.2 on right demonstrates the situation in a metal. In the conduction band, intraband transitions are also expected to contribute to the optical behavior. Obviously, there is no "threshold photon energy" necessary to initiate an intraband transition. In fact, intraband transitions give rise to contributions to the dielectric function which have their classical analogon in the Drude function (= response of "free" electrons—see task 17.7.6). Interband transitions, instead, represent the quantum mechanical counterpart of what we called the response of bound electrons in the Lorentzian oscillator model.

17.3 Theoretical Material

17.3.1 Crystals: Direct Electronic Transitions

The k-Selection Rule
Let us start with deriving a general selection rule. From (16.10), we have the Bloch function describing a single electron in a periodic potential:

$$\psi_{\mathbf{k},j}(\mathbf{r}) = e^{i\mathbf{k}\mathbf{r}} u_{\mathbf{k},j}(\mathbf{r})$$

Let us now imagine a transition (by absorption of a photon) from the lth band to the nth band. We emphasize that the regarded transition shall only change the electronic state of the system without the excitation of vibrations of any of the nuclei. Such transitions are called direct transitions. The corresponding electronic wavefunctions are:

$$\text{State } l: \psi_{\mathbf{k}_l,l}(\mathbf{r}) = e^{i\mathbf{k}_l\mathbf{r}} u_{\mathbf{k}_l,l}(\mathbf{r}).$$

$$\text{State } n: \psi_{\mathbf{k}_n,n}(\mathbf{r}) = e^{i\mathbf{k}_n\mathbf{r}} u_{\mathbf{k}_n,n}(\mathbf{r}).$$

So we have the transition matrix element of the perturbation operator $\hat{V}$:

$$V_{nl} = \int \psi_{\mathbf{k}_n,n}^*(\mathbf{r}) \hat{V} \psi_{\mathbf{k}_l,l}(\mathbf{r}) dV = \int e^{-i\mathbf{k}_n\mathbf{r}} u_{\mathbf{k}_n,n}^*(\mathbf{r}) \hat{V} e^{i\mathbf{k}_l\mathbf{r}} u_{\mathbf{k}_l,l}(\mathbf{r}) dV$$

Assuming: $V = -qrE \propto re^{i\mathbf{k}_p\mathbf{r}}$ ($\mathbf{k}_p$—photon wavevector, compare [1, Chap. 3]), we have

$$V_{nl} \propto \int u_{\mathbf{k}_n,n}^*(\mathbf{r}) u_{\mathbf{k}_l,l}(\mathbf{r}) r e^{i(\mathbf{k}_l+\mathbf{k}_p-\mathbf{k}_n)\mathbf{r}} dV \propto \frac{\partial}{\partial k_p} \int u_{\mathbf{k}_n,n}^*(\mathbf{r}) u_{\mathbf{k}_l,l}(\mathbf{r}) e^{i(\mathbf{k}_l+\mathbf{k}_p-\mathbf{k}_n)\mathbf{r}} dV$$

The u-functions are periodic with the spatial lattice period, and hence, the term $u^*_{\mathbf{k}_n,n}(\mathbf{r})u_{\mathbf{k}_l,l}(\mathbf{r})$ is periodic and may be represented as a Fourier series of the spatial harmonics, i.e., the reciprocal lattice vectors $\mathbf{g}$. That yields:

$$u^*_{\mathbf{k}_n,n}(\mathbf{r})u_{\mathbf{k}_l,l}(\mathbf{r}) = \sum_{\vec{g}} \Phi_g e^{i\mathbf{g}\mathbf{r}}$$

Here we skipped all the other indices for simplicity. We then find:

$$\int u^*_{\mathbf{k}_n,n}(\mathbf{r})u_{\mathbf{k}_l,l}(\mathbf{r})e^{i(\mathbf{k}_l+\mathbf{k}_p-\mathbf{k}_n)\mathbf{r}}\mathrm{d}V = \sum_{\vec{g}} \Phi_g \int e^{i(\mathbf{k}_l+\mathbf{k}_p-\mathbf{k}_n+\mathbf{g})\mathbf{r}}\mathrm{d}V$$

$$\propto \sum_{\vec{g}} \Phi_g \delta(\mathbf{k}_l + \mathbf{k}_p - \mathbf{k}_n + \mathbf{g}) \qquad (17.1)$$

So the **k**-selection rule for direct transitions may be formulated as:

$$\mathbf{k}_l + \mathbf{k}_p = \mathbf{k}_n + \mathbf{g} \qquad (17.1a)$$

This approach is more or less reasonable as long as all **k**-vectors may be regarded as real (weak absorption), and the integration extends over a sufficiently large volume so that we may consider it as infinitely large. Equation (17.1) is an expression of the so-called **k**-selection rule and states, that in the case of a direct transition in a crystal, the electrons wavevector in the final state must be equal to the sum of the corresponding vector in the initial state plus the wavevector the absorbed photon had, and plus any arbitrary reciprocal lattice vector (quasimomentum conservation).

Once the wavelength of a photon in the VIS/UV is much larger than the lattice constant, $|\mathbf{k}_p| \ll \frac{\pi}{a}$ holds, and therefore, when using a reduced zone scheme (with $k \in \left[-\frac{\pi}{a}, +\frac{\pi}{a}\right]$, i.e., all **k**-vectors are restricted to the first Brillouin zone), the optical transitions in Fig. 17.3 are symbolized by vertical arrows. Therefore, direct transitions are also called vertical transitions. As seen from Fig. 17.3, in the reduced zone scheme we thus have:

$$\mathbf{k}_l \approx \mathbf{k}_n \qquad (17.1b)$$

Joint Density of States

Let us now come to a qualitative understanding of the shape of the dielectric function as caused by direct interband transitions according to the previous section. On a rather intuitive level, it is clear that the imaginary part of the dielectric function should be proportional to the rate of electronic transitions between the occupied energy band l and the empty band n caused by photon absorption. According to Fermi's golden rule, we should then assume that the imaginary part of the dielectric function is proportional to the square of the transition matrix element of the

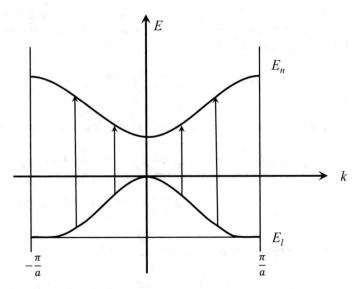

Fig. 17.3 Direct transitions between the lth and the nth energy band (*interband transitions*) in a reduced zone scheme, a is the lattice period

dipole operator, multiplied with the density of quantum states D which contribute to the transition at the given transition frequency. We write:

$$\text{Im } \varepsilon \propto D(\omega_{nl})|\mathbf{d}_{nl}|^2 \tag{17.2}$$

Note that once we have modeled the frequency dependence of the imaginary part of the dielectric function, the real part is straightforwardly obtained by making use of a Kramers–Kronig transformation according to (15.4b).

Remark Also, please note that the densities of states $D = dZ/dE$ introduced so far (see, e.g., 4.13) diverge for an infinite volume. Therefore, in this chapter, densities of states are always understood as as the number of states per energy interval per unit volume. This is also consistent with the definition of the dielectric function, because the latter provides information on the induced dipole moment per unit volume as well.

Let us now concentrate on the case of direct transitions, so that the electron wavevector practically does not change its value as a result of the quantum transition (compare 17.1b). Then, ω_{nl} is given by the energy spacing between two bands *at the same wavevector* $\mathbf{k}_l \approx \mathbf{k}_n \equiv \mathbf{k}$, and we obtain:

$$\text{Im } \varepsilon \propto D[E_n(\mathbf{k}) - E_l(\mathbf{k})]|\mathbf{d}_{nl}(\mathbf{k})|^2 \tag{17.3}$$

In order to simplify the task, let us at the beginning assume that the transition matrix element is constant and different from zero. In this case, the behavior of the

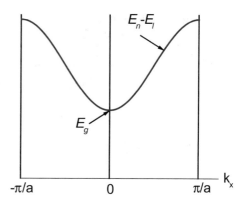

Fig. 17.4 Energy spacing between the bands from Fig. 17.3 as a function of the electron wavevector

dielectric function is determined by the *density of pairs of quantum states (i.e., the number of such pairs of states per energy interval), which have the same wavevector and are separated from each other by a given suitable energy spacing.* We will call this density of states a *joint density of states,* because it depends on features of both energy bands that are participating in the quantum transition.

In many practical situations, the occupied energy band l is generally associated with the valence band and the empty band n with the conduction band.

Let us have a closer look at the argument of the not yet quantitatively defined value D. For a band structure like shown in Fig. 17.3, the argument $E_n(\mathbf{k}) - E_l(\mathbf{k})$ looks like it is sketched in Fig. 17.4. In semiconductor physics, the minimal value of $E_n(\mathbf{k}) - E_l(\mathbf{k})$ is called the *direct gap* of the semiconductor, when the two mentioned bands are associated with the valence and conductive bands, respectively.

Let us first imagine, that a system characterized by a $E_n(\mathbf{k}) - E_l(\mathbf{k})$ behavior as given by Fig. 17.4 is illuminated with light of a sufficiently low frequency so that $\hbar\omega < E_g$ is fulfilled. There will clearly be no absorption of light. A rather sharp absorption onset is expected at $\hbar\omega = E_g$, which corresponds to transitions in the center of the Brillouin zone ($\mathbf{k} = \mathbf{0}$). The reason for this sharp absorption onset is, that at $\mathbf{k} = \mathbf{0}$, the derivative $\frac{d[E_n(\mathbf{k}) - E_l(\mathbf{k})]}{d\mathbf{k}} = \mathbf{0}$ as well. Consequently, a large amount of pairs of quantum states with suitable energy spacing becomes involved into the optical transition, which usually leads to prominent features in the optical absorption spectrum. The same is valid for the $\mathbf{k}$-values $\pm\pi/a$.

We come to the conclusion, that the characteristic features in the imaginary part of the dielectric function are determined by the behavior of the derivative $\frac{d[E_n(\mathbf{k}) - E_l(\mathbf{k})]}{d\mathbf{k}}$. Particularly, the points where this derivative is equal to zero, are called critical points or van Hove singularities. Practically this means, that in a van Hove singularity, the dispersion curves $E_n(\mathbf{k})$ and $E_l(\mathbf{k})$ are locally parallel to each other in the $\mathbf{k}$-space (Fig. 17.3). As a consequence, in a van Hove singularity, we observe a huge amount of pairs of quantum states which are separated by identical energy spacing. This results in characteristic features in the optical spectrum, when the photon energy comes into resonance with that particular energy spacing. Thus, the prominent NIR feature in the reflection spectrum of aluminum in Fig. 14.7 on left is caused by the joint effect of three different critical point interband transition contributions (compare [2]).

Let us now write down a quantitative expression for the joint density of states. The number of quantum states in a given **k**-interval is given by:

$$dZ \propto dk_x dk_y dk_z$$

In spherical coordinates (which makes sense in optically isotropic materials), we find:

$$3D : \ dk_x dk_y dk_z = 4\pi k^2 dk \Rightarrow dZ \propto k^2 dk$$

$$= \frac{k^2}{\frac{d[E_n(k) - E_l(k)]}{dk}} d[E_n(k) - E_l(k)] \tag{17.4}$$

The density of states $D(k)$ is then given by: $dZ \equiv D(k)dk \Rightarrow D(k) \propto k^2$.
In full analogy, the joint density of states $D[E_n(\mathbf{k}) - E_l(\mathbf{k})]$ will be defined as:

$$dZ = D[E_n(k) - E_l(k)]d[E_n(k) - E_l(k)] \tag{17.5}$$

Comparing finally (17.4) and (17.5), we find the expression:

$$D[E_n(k) - E_l(k)] \propto \frac{k^2}{\frac{d[E_n(k) - E_l(k)]}{dk}} \tag{17.6}$$

Expression (17.6) is obviously only valid in the three-dimensional case. It really may show a singular behavior at the *van Hove* singularities. In general, it is determined by the particular band structure valid for the material under consideration.

Let us now regard the special situation sketched in Fig. 17.4. At $k \to 0$, we obviously have:

$$E_n(k) - E_l(k) = E_g + \text{const.} * k^2 = \hbar\omega$$

Consequently,

$$k \propto \sqrt{\hbar\omega - E_g}$$

and

$$\frac{d[E_n(k) - E_l(k)]}{dk} \propto k \propto \sqrt{\hbar\omega - E_g}$$

We obtain from (17.6):

$$D[E_n(k) - E_l(k)] \propto \sqrt{\hbar\omega - E_g}; \quad \hbar\omega > E_g \tag{17.7a}$$

which is valid in the three-dimensional case for light frequencies slightly above the absorption edge. According to (17.3), we therefore have to expect that the shape of the imaginary part of the dielectric function resembles the square root of $\hbar\omega - E_g$.

The same type of discussion may be performed for the 2D and 1D cases. That may be easily done by the reader himself. We find ($\hbar\omega > E_g$):

$$3D : d^3\mathbf{k} = dk_x dk_y dk_z \rightarrow 4\pi k^2 dk \Rightarrow D[E_n(k) - E_l(k)] \propto \sqrt{\hbar\omega - E_g}$$

$$2D : d^2\mathbf{k} = dk_x dk_y \rightarrow 2\pi k dk \Rightarrow D[E_n(k) - E_l(k)] \propto \text{const}$$

$$1D : d\mathbf{k} = dk_x \rightarrow dk \Rightarrow D[E_n(k) - E_l(k)] \propto \frac{1}{\sqrt{\hbar\omega - E_g}} \tag{17.7b}$$

Note the similarity to the expression found earlier in Sect. 4.6.1.

Wannier–Mott excitons

So far, our discussion has only concerned the optical response of a single electron, propagating in a periodic potential. We will not deal with a many-electron theory which allows considering the effects caused by the Coulomb interaction between the electrons. But our knowledge obtained so far is sufficient to account for one additional effect which is most important in semiconductor optics: Imagine the situation sketched in Fig. 17.3. An electron (negatively charged) that is excited from the *l*th (the valence) band to the *n*th (the conduction) band is well-known to leave a (positively charged) hole in the valence band. In their respective bands, both the suddenly created conduction electron and the hole are expected to move with a group velocity determined by the first derivative of the band energy with respect to the wavevector. In the general case, these velocities are different, so that the electron and the hole are immediately separated from each other. However, for example at $k = 0$, the group velocities are identical (=0), so that the electron and the hole remain spatially close to each other and form a new quasiparticle, a *Wannier–Mott exciton*. Similar to a hydrogen atom, such an exciton has Rydberg-like energy levels, which contribute to the optical absorption behavior of the semiconductor. As a consequence, there may appear sharp absorption lines in the region of the absorption edge, corresponding to the excitation of different excitonic energy levels, as exemplified in Fig. 17.5.

Let us apply our knowledge on the hydrogen atom for estimating typical radii and binding energies of a Wannier–Mott exciton [1, Chap. 4]. In the hydrogen atom, we have (8.8), (8.10):

$$\text{Bohr's radius: } a_0 = \frac{4\pi\varepsilon_0}{e^2\mu}\hbar^2 = \frac{\varepsilon_0}{e^2\pi\mu}h^2 \approx 0.53 * 10^{-10}m$$

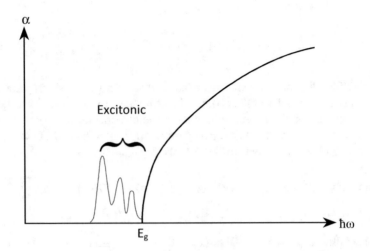

Fig. 17.5 Excitonic absorption in the region of the absorption edge of a direct semiconductor

while Ry denotes the Rydberg energy, given by:

$$Ry = \frac{1}{8\pi\varepsilon_0}\frac{e^2}{a_0} = \frac{e^4\mu}{8\varepsilon_0^2 h^2} \approx 13.6\,eV$$

In a Wannier–Mott exciton, the situation is somewhat different from the hydrogen atom. The reduced mass of a Wannier–Mott exciton is calculated from the effective masses of the electron and the hole. Assuming the conditions shown in Fig. 17.3 at $k = 0$, an electron (e) at the bottom of the conduction band n has a positive effective mass $m^*_{e,n} > 0$, while on top of the valence band l, the electron has a negative effective mass $m^*_{e,l} < 0$. The hole (h) in band l has an effective mass opposite to that of the electron, hence $m^*_{h,l} = -m^*_{e,l} > 0$ (see remark below). The reduced mass of the Wannier–Mott exciton is then given by:

$$\mu_{W-M} = \frac{m^*_{e,n}m^*_{h,l}}{m^*_{e,n} + m^*_{h,l}}$$

Replacing further $\varepsilon_0 \to \varepsilon_{stat}\varepsilon_0$, we find for the Wannier–Mott exciton:

$$a_{0,W-M} = \frac{\mu\varepsilon_{stat}}{\mu_{W-M}}a_0 \approx 10^1\,nm$$

$$Ry_{W-M} = \frac{\mu_{W-M}}{\mu\varepsilon_{stat}^2}Ry a_0 \approx 10^{-3}\ldots 10^{-2}eV$$

Hence, the binding energy of the Wannier–Mott exciton is much smaller than that in the hydrogen atom, while its ground-state radius is much larger than in a hydrogen atom.

As a consequence of the formation of the exciton, at $k = 0$, we have the modified resonance condition for light absorption:

$$\hbar\omega = E_g + E_{\text{exciton}} = E_g - \frac{R_{yW-M}}{n^2}; \quad n = 1, 2, 3, \ldots$$

which explains the additional absorption features at photon energies below the gap value E_g. Of course, an absorption spectrum like exemplified in Fig. 17.5 is observed only at very low temperatures, where the exciton is stable.

Remark Consider the constellation shown in Fig. 17.3. near the center of the Brillouin zone. When an electron–hole pair is generated by absorption of light, the effective mass of the electron in the conduction band is positive. An electric field would induce certain acceleration, i.e., a change in the electrons group velocity. The hole is positively charged and should therefore feel a force into the opposite direction. Therefore, its acceleration should have the opposite sign of that of the conduction electron.

On the other hand, from (16.24) we have: $\frac{dv_{gr}}{dt} = \frac{F}{m*}$. A change in the sign of F will lead to a change in the sign of the acceleration only if the sign of the effective mass is the same in both situations. Therefore, the effective mass of a hole in the valence band must have the same sign as the effective mass of the electron in the conduction band.

17.3.2 Crystals: Phonons

Let us now turn to the discussion of lattice vibrations, i.e., the oscillatory movement of atomic nuclei. According to our knowledge from molecule dynamics, we will expect that the corresponding eigenfrequencies correspond to the middle infrared (MIR) spectral range.

Let us discuss a simple classical model case of a crystal with *two* atoms in the primitive cell. We will use the following symbols (s is a counting index) (Table 17.1).

Figure 17.6 illustrates the model situation we a rediscussing now.

Table 17.1 Symbols used (compare Fig. 17.6)

Type of atom	Atom mass	Displacement of the sth atom of type 1 or 2
1	m_1	u_s
2	m_2	v_s

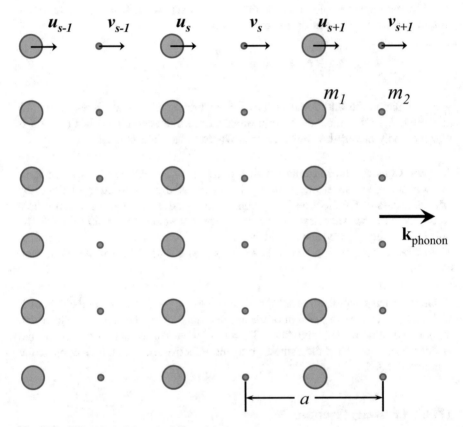

Fig. 17.6   Illustration of the regarded model system

When considering only next-neighbor interactions, we have the following classical equations of motion for the individual nuclei:

$$m_1 \ddot{u}_s = \kappa(v_s - u_s) + \kappa(v_{s-1} - u_s)$$
$$m_2 \ddot{v}_s = \kappa(u_{s+1} - v_s) + \kappa(u_s - v_s)$$

(17.8)

We will search the solution as:

$$u_s = u e^{i s k_{phonon} a} e^{-i\omega t}$$
$$v_s = v e^{i s k_{phonon} a} e^{-i\omega t}$$

(17.9)

The mentioned solutions are obviously delocalized in space. Such kind of lattice vibrations are called phonons. Here, k_{phonon} is the absolute value of the wavevector of the phonon, a—the lattice period along the propagation direction of the phonon, and u,v—corresponding oscillation amplitudes.

When substituting (17.9) into (17.8), we find for the amplitudes of the nuclei vibrations:

$$\left.\begin{aligned}-\omega^2 m_1 u = \kappa v\left(1 + e^{-ik_{phonon}a}\right) - 2\kappa u \\ -\omega^2 m_2 v = \kappa u\left(1 + e^{ik_{phonon}a}\right) - 2\kappa v\end{aligned}\right\}$$

$$\Rightarrow \quad \begin{aligned}\left(2\kappa - \omega^2 m_1\right)u - \kappa\left(1 + e^{-ik_{phonon}a}\right)v = 0 \\ -\kappa\left(1 + e^{ik_{phonon}a}\right)u + \left(2\kappa - \omega^2 m_2\right)v = 0\end{aligned}$$

(17.10)

The obvious trivial solution of (17.19) is $u = v = 0$; i.e., all nuclei are at rest. In order to have nontrivial solutions, we must require:

$$\begin{vmatrix}\left(2\kappa - \omega^2 m_1\right) & -\kappa\left(1 + e^{-ik_{phonon}a}\right) \\ -\kappa\left(1 + e^{ik_{phonon}a}\right) & \left(2\kappa - \omega^2 m_2\right)\end{vmatrix} = 0 =$$

$$= \left(2\kappa - \omega^2 m_1\right)\left(2\kappa - \omega^2 m_2\right) - \kappa^2\left(1 + e^{-ik_{phonon}a}\right)\left(1 + e^{ik_{phonon}a}\right)$$

From here we find the dispersion law $\left(\mu = \frac{m_1 m_2}{m_1 + m_2}\right)$:

$$\omega^2 = \frac{\kappa}{\mu} \pm \sqrt{\left(\frac{\kappa}{\mu}\right)^2 - 2\frac{\kappa^2}{m_1 m_2}\left(1 - \cos k_{phonon}a\right)} \qquad (17.11)$$

Note that $\omega^2\left(k_{phonon}\right) = \omega^2\left(k_{phonon} + \frac{2n\pi}{a}\right) = \omega^2\left(k_{phonon} + g\right)$.

With n—integer and $\mathbf{g}$—reciprocal lattice vector. So we may assign a quasi-momentum $\hbar k_{phonon}$ to the phonons as well, and again, the dispersion relation is invariant with respect to the addition of an arbitrary reciprocal lattice vector to the quasimomentum.

Let us now discuss the dispersion law (17.11).

We have a high-frequency solution, corresponding to what is called the optical phonon mode:

$$\omega^2_{opt} = \frac{\kappa}{\mu} + \sqrt{\left(\frac{\kappa}{\mu}\right)^2 - 2\frac{\kappa^2}{m_1 m_2}\left(1 - \cos k_{phonon}a\right)}$$

$$= \frac{\kappa}{\mu}\left(1 + \sqrt{1 - 2\frac{\mu^2}{m_1 m_2}\left(1 - \cos k_{phonon}a\right)}\right)$$

On the other hand, we have a low-frequency solution, corresponding to what is called the acoustic phonon mode:

$$\omega^2_{acoust} = \frac{\kappa}{\mu}\left(1 - \sqrt{1 - 2\frac{\mu^2}{m_1 m_2}\left(1 - \cos k_{phonon}a\right)}\right)$$

In order to discuss the physical sense of the two different phonon modes, let us look at the particular case $k_{phonon} \to 0$ (zone-center phonons). For the optical mode, we have:

$$\omega_{opt}^2 \to \frac{2\kappa}{\mu}$$

According to (17.11), we find:

$$-\omega^2 m_1 u = 2\kappa v - 2\kappa u \to -2\frac{\kappa}{\mu}m_1 u = 2\kappa v - 2\kappa u$$

$$\Rightarrow \left(1 - \frac{m_1}{\mu}\right)u = v \Rightarrow -\frac{m_1}{m_2}u = v$$

Hence, in the optical mode, the two different atoms oscillate in antiphase. The acoustic mode yields:

$$\omega_{acoust}^2 \to 0$$

And thus:

$$-\omega^2 m_1 u = 2\kappa v - 2\kappa u \to 0 = 2\kappa v - 2\kappa u \Rightarrow u = v$$

In the acoustic mode, the atoms oscillate in-phase.

If the atoms have different electric charges, in the optical mode, strong oscillating electric dipols are generated, which strongly interact with electromagnetic radiation and cause IR-active lattice vibrations. The reason is that the positive and negative charges move into opposite directions. In the acoustic mode, induced dipole moments are practically negligible, because the positive and negative charges move into the same direction.

So far, we have not yet specified our "spring constant" κ.

As mentioned in Sect. 12.1, solids—and thus crystals—have two independent elastic moduli—a bulk modulus and a shear modulus. Correspondingly, we find longitudinal and transversal phonon modes in a solid. The terminology is (Table 17.2):

Note that if you have a number of X atoms in the primitive cell, you find

- 3 acoustic modes (1 LA-mode, 2 TA-modes)
- $3X$ -3 optical modes (X -1 LO-modes, 2X-2 TO-modes)

Table 17.2 Accepted terminology

TA	Transversal acoustic mode
LA	Longitudinal acoustic mode
TO	Transversal optical mode
LO	Longitudinal optical mode

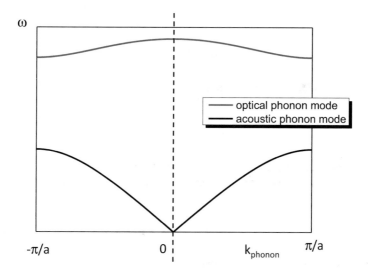

Fig. 17.7 Dispersion relation for optical and acoustic phonons according to (17.11) in the first Brillouin zone. The dashed line indicates the zone center. Note that the zone-center optical phonon has a vanishing group velocity, and an infinitely large phase velocity

Let us have a short look at the density of phonon states. Introducing the density of states via dZ/dE (compare 4.12a), we may write:

$$\frac{dZ}{dE} = \frac{\frac{dZ}{dk}}{\frac{dE}{dk}} = \frac{1}{\hbar \frac{d\omega}{dk}} \frac{dZ}{dk} = \frac{1}{\hbar v_{\text{group}}} \frac{dZ}{dk}$$

The zone-center optical phonon mode in Fig. 17.7 obviously corresponds to $\frac{dE}{dk} \to 0$ and is sometimes called a critical point phonon, in analogy to the terminology introduced in Sect. 17.3.1. Indeed, in this point, the denominator in the above-written expression becomes zero.

Let us now have a look at the interaction of such a phonon oscillation with an electromagnetic field. Imagine an electromagnetic wave propagating along the z-axis. When restricting on the Coulomb force, instead of (17.8), we now have:

$$m_1 \ddot{u}_s = \kappa(v_s - u_s) + \kappa(v_{s-1} - u_s) + q E_0 e^{-i(\omega t - k_p z)}$$

When setting (17.9):

$$u_s = u e^{i s k_{\text{phonon}} a} e^{-i\omega t}$$
$$v_s = v e^{i s k_{\text{phonon}} a} e^{-i\omega t}$$

And

$$z = sa$$

We have (compare 17.10):

$$\left\{-\omega^2 m_1 u - \kappa(v - u) - \kappa\left(v e^{-ik_{phonon}a} - u\right)\right\}e^{isk_{phonon}a} = q E_0 e^{isk_p a}$$

Which should be true for any given integer s. Hence,

$$e^{ik_p a} = e^{ik_{phonon}a} = e^{i(k_p a + 2n\pi)} = e^{i\left(k_p + \frac{2n\pi}{a}\right)a} = e^{i(k_p + g)a}$$

(n- any integer number). From here we conclude, that for the case of optical excitation of phonons by absorption of light, the **k**-selection rule (compare 17.1a) may be written as:

$$\mathbf{k}_{phonon} = \mathbf{k}_p + \mathbf{g} \qquad (17.12)$$

Thus, because $k_p \ll \frac{\pi}{a}$, only zone-center optical phonons may be excited by absorption of infrared electromagnetic radiation (one-phonon processes presumed; i.e., the absorption of one photon leads to the generation of one phonon only).

Note finally, that in a material with two atoms in the primitive unit cell, the atoms do not need to be of different type. Thus, in the diamond lattice, $X = 2$. Both these atoms are carbon atoms, and their displacement with respect to each other does not create any electric dipole moment. Therefore, the diamonds zone-center optical phonon is IR-inactive. On the other hand, the optical phonon oscillation leads to a modulation of the volume of the primitive unit cell, and thus to a modulation of the corresponding polarizability. The zone-center optical phonon is therefore Raman-active, and the position and width of the observed Raman line is used in practice to judge the quality of synthetic diamond materials (Fig. 17.8).

The fact that the zone-center optical phonon in diamond is IR-inactive but Raman-active again reflects the validity of the rule of mutual exclusion (the diamond lattice is centrosymmetric).

Let us finish this section by returning to the polariton dispersion law discussed earlier in Sect. 14.6.1. Apart from resonance, from (14.14) we find:

$$k(\omega) = \frac{\omega}{c}\sqrt{\varepsilon} \approx \frac{\omega}{c}\sqrt{\frac{\omega_p^2 f_{res}}{\omega_{res}^2 - \omega^2} + \varepsilon_{NR}}$$

Let us assume a tranversal optical phonon interacting with the incident electromagnetic wave at the single MIR angular frequency $\omega_{res} = \omega_T\left(\approx \sqrt{\frac{2\kappa}{\mu}}\right)$. Then, the dielectric function looks like shown in Fig. 17.9. When making use of (14.15a), we have:

$$\varepsilon(\omega) \approx \varepsilon_\infty \frac{\omega^2(k \approx 0) - \omega^2}{\omega_T^2 - \omega^2}$$

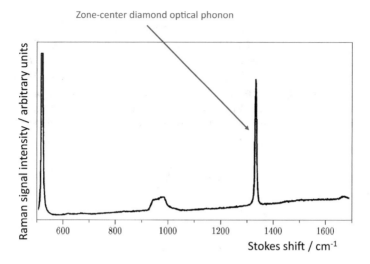

Fig. 17.8 Raman spectrum of a synthetic diamond film on a silicon substrate. Courtesy of Ralf Petrich, at that time Technical University of Chemnitz, Germany

Fig. 17.9 Dielectric function with a single IR resonance at ω_T

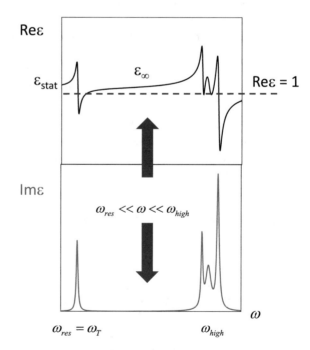

Note that $\varepsilon \to 0 \Leftrightarrow \omega \to \omega(k \approx 0)$.

In order to understand the physical sense of this particular frequency, we must remember the Maxwell equation: $\mathrm{div}\mathbf{D} = 0 = \varepsilon_0\varepsilon(\omega)\mathrm{div}\mathbf{E}$. Provided that $\varepsilon(\omega) \neq 0$, from here it follows that $\mathrm{div}\mathbf{E} = 0$, which is characteristic for transversal phonons, because the latter do not produce volume charge densities. However, if $\varepsilon(\omega) = 0$, $\mathrm{div}\mathbf{E}$ may be different from zero, which is characteristic for longitudinal optical phonons, because the latter may create a volume charge density that is different from zero. Therefore, $\omega(k \approx 0)$ has to be associated with the angular frequency of longitudinal phonons ω_L. That leads us to the more familiar writing:

$$\varepsilon(\omega) \approx \varepsilon_\infty \frac{\omega_L^2 - \omega^2}{\omega_T^2 - \omega^2} \tag{17.13}$$

Setting finally $\omega = 0$, we obtain:

$$\varepsilon_{\mathrm{stat}} \approx \varepsilon_\infty \frac{\omega_L^2}{\omega_T^2} \tag{17.13a}$$

This relation is known as the Lyddane–Sachs–Teller relation.

Let us finally look at two examples. In MgO, reported values for optical zone-center phonon frequencies are $f_{TO} \approx 12.05\,THz$; $f_{LO} \approx 21.52\,THz$ [3]. On the other hand, $\varepsilon_{\mathrm{stat}} \approx 9.83$; $\varepsilon_\infty \approx 2.944$. When substituting into (17.13a), a good agreement is observed.

On the other hand, in diamond we have $\nu_{TO} = \nu_{LO} \approx 1333\,\mathrm{cm}^{-1}$ (compare also Fig. 17.8). This is a consequence of the already mentioned infrared inactivity of the transversal optical phonon in the diamond crystal. Indeed, from (14.15) we have $\omega^2(k \approx 0) = \omega_L^2 \approx \omega_T^2 + \frac{\omega_p^2 f_{\mathrm{res}}}{\varepsilon_\infty}$. For an infrared inactive vibration, we have $\omega_p^2 f_{\mathrm{res}} = 0$, such that the zone-center longitudinal and transversal phonons have coinciding frequencies. Then of course, our distinction between $\varepsilon_{\mathrm{stat}}$ and ε_∞ makes no sense.

17.3.3 Crystals: Indirect Transitions

So far, in Sect. 17.3.1, we have considered direct electronic transitions. Additionally, in the previous Sect. 17.3.2, lattice vibrations have been discussed. Let us now regard a situation, where the absorption of a photon leads to the instantaneous excitation of both electronic and phonon states in a crystal. These so-called indirect transitions are particularly important in so-called indirect semiconductors.

In semiconductor practice, it appears that many of the semiconductors belong to the class of indirect semiconductors. In an indirect semiconductor, indirect interband transitions between the valence and the conduction bands may occur at photon energies which are smaller than the direct gap considered in Sect. 17.3.1. In other words, a semiconductor is indirect, when the condition:

indirect gap $E_{g,\mathrm{ind}} \equiv \min\left[E_n(\mathbf{k}_n) - E_l(\mathbf{k}_l)\right]|_{\mathbf{k}_n \neq \mathbf{k}_l} < \min\left[E_n(\mathbf{k}_n) - E_l(\mathbf{k}_l)\right]|_{\mathbf{k}_n = \mathbf{k}_l}$

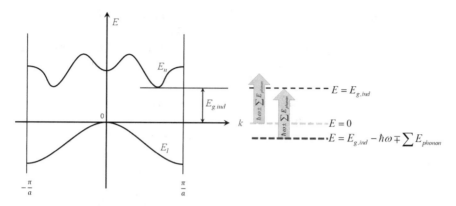

Fig. 17.10 Example of a band structure of an indirect semiconductor. The turquoise and violet dashed lines indicate the integration limits in (17.14a)

$$\equiv \text{direct gap } E_g$$

is fulfilled. Such a situation is shown in Fig. 17.10.

Let us now look how the shape of the absorption edge at an indirect gap looks like.

The main difference to the previously discussed case of direct transitions is in the violation of electron quasimomentum conservation. Indeed, when the absorption of light is accompanied by the generation (or annihilation) of one or several phonons, the electron wavevectors in the initial and final states may significantly differ from each other. When neglecting the light wavevector, the total (electron + phonon) quasimomentum conservation is obtained as a merger of (17.1a) and (17.12) according to:

$$\mathbf{k}_l - \mathbf{k}_n \approx \pm \sum \mathbf{k}_{\text{phonon}} + \mathbf{g}$$

Additionally, energy conservation leads to:

$$E_n - E_l = \hbar\omega \pm \sum E_{\text{phonon}}$$

Here, the sign "+" corresponds to phonon annihilation, while "−" denotes phonon creation. Due to the violation of *electron* quasimomentum conservation, the joint density of states is no more significant for the quantitative description of the absorption process. Instead, it is reasonable to consider the convolution of the densities of the initial and final quantum states, regardless on the quasimomentum. Hence, instead of (17.3), we make use of:

$$\text{Im } \varepsilon \propto |\mathbf{d}_{nl}|^2 \int_{-\infty}^{\infty} D_l(E) D_n \left(E + \hbar\omega \pm \sum E_{\text{phonon}} \right) dE \qquad (17.14)$$

where D is the usual density of states in the corresponding band as indicated by index in (17.14). Again, near the extremal values of the $E(k)$ dependence as shown in Fig. 17.10, the energy behaves proportional to the square of the wavevector. In analogy to the treatment in Sect. 17.3.1, we suppose for the 3D case:

$$dZ = D(k)dk = \frac{D(k)}{\frac{dE}{dk}}dE \equiv D(E)dE \Rightarrow D(E) \propto k(E)$$

$$\Rightarrow D_l(E) \propto \sqrt{-E}; \ E < 0; \ D_n(E) \propto \sqrt{E - E_{g,ind}}; \ E > E_{g,ind}$$

Then, from (17.14) we obtain:

$$\operatorname{Im} \varepsilon \propto |\mathbf{d}_{nl}|^2 \int_{E_{g,ind}-\hbar\omega\mp\sum E_{phonon}}^{0} \sqrt{-E}\sqrt{E - E_{g,ind} + \hbar\omega \pm \sum E_{phonon}}dE;$$

$$\hbar\omega > E_{g,ind} \mp \sum E_{phonon}$$

$$(17.14a)$$

Here we have assumed, that the top of the valence band in Fig. 17.10 corresponds to $E = 0$. The choice of the integration limits should become clear from the illustration in Fig. 17.10 on right.

We do not need to calculate this integral exactly. The only thing we want to know is the frequency dependence of the dielectric function. Performing the substitution:

$$-z = -E_{g,ind} + \hbar\omega \pm \sum E_{phonon}$$

we find

$$\operatorname{Im} \varepsilon \propto \int_0^z \sqrt{Ez - E^2}dE$$

The integrand itself represents half a circle with the diameter z, centered at $z/2$ on the abszissa. Hence it includes an area that is proportional to z^2. Consequently, the integral itself is proportional to z^2, and we find for the dielectric function:

$$\operatorname{Im} \varepsilon(\omega) \propto |\mathbf{d}_{nl}|^2 \left(\hbar\omega - E_{g,ind} \pm \sum E_{phonon} \right)^2; \ \hbar\omega > E_{g,ind} \mp \sum E_{phonon}$$

$$(17.15)$$

We see, that (17.15) is different from the expressions (17.7a) and (17.7b), the latter being valid for the direct transitions.

With this, we finish the the discussion of basic optical processes in crystalline matter. Before turning to another important class of solids in the next subchapter, namely amorphous solids, let us shortly resume what we have learned about crystalline solids and their optics so far.

Table 17.3 Overview on power laws for describing the shape of the absorption edge in crystals

Type of transition	Im $\varepsilon \propto$
3D, direct	$(\hbar\omega - E_g)^{\frac{1}{2}}$
2D, direct	$(\hbar\omega - E_g)^{0}$
1D, direct	$(\hbar\omega - E_g)^{-\frac{1}{2}}$
3D, indirect	$(\hbar\omega - E_{g,\text{ind}} \pm \sum E_{\text{phonon}})^{2}$

The main point is, that instead of atomic or molecular energy levels, in solid-state physics we deal with energy bands. In crystal physics, these energy bands are described by an $E(k)$-dependence. As in molecules, an electronic excitation may be accompanied by excitation of vibrational degrees of freedom, which gives rise to the division of optical transitions in a crystal into direct and indirect transitions. Both types of transitions differ from each other in their energy balance and the shape of the absorption structure near the absorption onset. In this context, Table 17.3 summarizes the types of power-law dependencies, derived in this chapter for describing the spectral behavior of the imaginary part of the crystal dielectric function in the region of the absorption edge (interband transitions).

17.3.4 Amorphous Solids

Let us now come to another kind of solids, namely amorphous solids. As it has already been discussed, amorphous solids lack long-range order in the atomic arrangement (which is characteristic for crystals), while short-range order is present. Optical glasses are prominent examples for the application of amorphous solids as optical materials. An amorphous solid therefore resembles some of the properties of its crystalline counterpart (namely those which are determined by the short-range order), while the properties basing on the long-range order will not be observed in amorphous solids.

In an amorphous solid, the interatomic distances are comparable to those in a crystal. Therefore, the spatial overlap of the atomic electronic wavefunctions gives rise to the formation of broad energy regions with allowed electron energy values, similar as in a crystal. On the other hand, the absence of translational invariance in the atomic arrangement does not allow us to use Bloch's theorem for the description of the electronic wavefunctions. This has several consequences:

- Although there are broad regions of allowed electron energy values, there is no $E(k)$-dependence as in crystals. Nevertheless it is common to speak on energy bands in amorphous semiconductor theory.
- In addition to the delocalised electronic states characteristic for a periodic potential, there may be localized electronic quantum states as well (Anderson localization). They may deeply extend into the forbidden zone and form what is called the energy band tails.
- There is no quasimomentum conservation in optical transitions.

- There is no joint density of states.

Even if we cannot define a joint density of states, we may introduce a conventional density of states defining (as usual):

$$dZ \equiv D(E)dE$$

where dZ again is again the full number of quantum states in the given E-interval (here per unit volume).

Figure 17.11 illustrates the differences between the density of electronic states as observed in a crystal and its amorphous counterpart.

When looking at Fig. 17.11 on top, it again becomes clear that our gap E_g has a clear physical meaning in the case of a crystalline solid. It is identical to the energy width of the forbidden zone, i.e., an energy interval where the density of

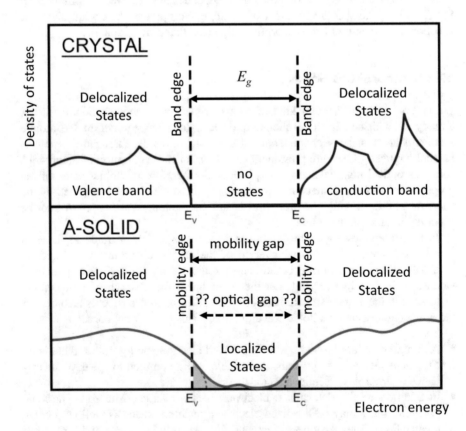

Fig. 17.11 Density of electronic states in a crystal and in an amorphous solid in the region of the valence (V) and conduction (C) bands. Picture adopted from [4], after an idea of Richard Zallen [5]. The dashed red lines indicate the assumption of "parabolic" band edges in an a-solid

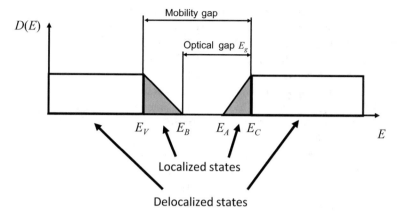

Fig. 17.12 Model shape of the density of states $D(E)$ in an amorphous semiconductor

states is zero. It appears as a parameter in dispersion dependencies like (17.7a) or (17.15) and is therefore also called the "optical gap." But how to define an "optical gap" in an amorphous semiconductor, as represented by the density of states dependence in Fig. 17.11 on bottom?

In amorphous semiconductor physics, there exist several models to describe the density of states in the region of the valence and conduction bands. Consequently, different definitions of optical gaps may be provided. We will focus here on a very simplified model dependence proposed by Davis and Mott [6]. It is illustrated in Fig. 17.12.

Figure 17.12 shows an example of the density of states in the vicinity of the energy gap between the valence and conduction bands. Note the qualitative similarity to the dependence visualized in Fig. 17.12 on bottom. It is characteristic, that in the vicinity of the band edges E_A and E_B, the electronic states are spatially localized (gray regions in Fig. 17.12). An electron in such a localized quantum state has only a small mobility, so that E_C and E_V are called *mobility edges*, while the value of E_C-E_V marks the so-called *mobility gap*. It is utmost important for the description of electrical properties of an amorphous semiconductor.

Concerning the optical properties (particularly the absorption of light), in a system like shown in Fig. 17.12, we have to distinguish two entirely different cases:

1. Both the initial and final states of the quantum transition are spatially localized.
2. At least one of the participating quantum states is delocalised.

In order to obtain nonvanishing transition matrix elements, it is necessary that the wavefunctions of the initial and final quantum states are spatially overlapping. This is automatically fulfilled for quantum transitions which involve delocalised states.

However, for localized quantum states, this requirement may make a quantum transition impossible even if the energy spacing between the states is suitable.

One should therefore expect that the transitions, which involve delocalised quantum states, give more intense contributions to the full absorption spectrum than the transitions between localized states.

Because of the absence of electronic quasimomentum conservation, the calculation of the imaginary part of the dielectric function in an amorphous material follows the philosophy from Sect. 17.3.3 (indirect transition). Practically, we will apply (17.14) again.

Let us now come to the system illustrated In Fig. 17.12. Let us assume, that in the band tails (the gray regions in Fig. 17.12, which symbolize the localized states), the density of states changes linearly with energy. For example, regarding the band tail of the valence band and setting $E_B = 0$ now, we have:

$$D_V(E) \propto -E$$

Which is valid in the valence band tail. On the other hand, we postulate that the conduction band is structureless:

$$D_C(E) \propto \theta(E - E_C) \Rightarrow D_C(E + \hbar\omega) \propto \theta(E + \hbar\omega - E_C) = \theta(E - (E_C - \hbar\omega))$$

Let us now consider transitions from the valence band tail into the conduction band. Neglecting transitions between localized states, we find:

$$\mathrm{Im}\,\varepsilon \propto |\mathbf{d}_{nl}|^2 \int_{E_C - \hbar\omega}^{0} D_V(E) \underbrace{D_C(E + \hbar\omega)}_{\propto \theta(E - (E_C - \hbar\omega))} \mathrm{d}E \propto |\mathbf{d}_{nl}|^2 \int_{E_C - \hbar\omega}^{0} D_V(E)\mathrm{d}E$$

$$\propto |\mathbf{d}_{nl}|^2 \int_{E_C - \hbar\omega}^{0} E\mathrm{d}E \propto |\mathbf{d}_{nl}|^2 (\hbar\omega - E_C)^2$$

$$(17.16)$$

Here, the integration limits correspond to those in (17.14), of course without any phonon contributions. Consequently, we obtain exactly the same type of frequency dependence as in (17.15) although the latter was obtained assuming "parabolic" band edges. The physical sense of the optical gap in (17.16) is different from that in (17.15), it is now identical to $(E_C - E_B)$. We may include transitions from the valence band into the conduction band tail as well, and in this case, the observed optical gap will correspond to the lower value of $(E_C - E_B)$ and $(E_A - E_V)$.

This way we found a definition of an "optical gap" in an amorphous semiconductor as:

$$E_{g,\mathrm{opt}} = \min\{(E_C - E_B); (E_A - E_V)\}$$

This definition is clearly related to the assumed density of states picture and must be modified when other shapes $D(E)$ are assumed. Note further that the optical gap turns out to be smaller than the mobility gap defined as:

$$E_{g,\text{mob}} = (E_C - E_V)$$

What about the matrix element in (17.16)? Again, different approaches are in use. The simplest possibility is to assume that the matrix element $|\mathbf{d}_{CV}|^2 = \text{const.}$ We then find:

$$\text{Im }\varepsilon = 2nK \propto \left(\hbar\omega - E_{g,\text{opt}}\right)^2 \Rightarrow K \propto \frac{\left(\hbar\omega - E_{g,\text{opt}}\right)^2}{n} \Rightarrow \sqrt{\frac{\alpha}{\omega}} \propto \frac{\left(\hbar\omega - E_{g,\text{opt}}\right)}{\sqrt{n}}$$

$$(17.16a)$$

This convenient expression relates the absorption coefficient to the optical gap; hence, the gap may be determined from experimental absorption coefficient data, fitting the data by means of (17.16a). Usually, the refractive index n is assumed to be constant. The thus determined optical gap is called the Cody gap. It is connected to the requirement of a constant transition matrix element of the dipole operator for transition between states in the valence and conduction bands.

A somewhat modified dependence is obtained, when the matrix element of the momentum operator is supposed to be constant. We must then require, that $|\mathbf{d}_{CV}|^2\omega^2 = \text{const.}$, and instead of (17.16a) we then find:

$$\text{Im }\varepsilon = 2nK \propto \frac{\left(\hbar\omega - E_{g,\text{opt}}\right)^2}{\omega^2} \Rightarrow K \propto \frac{\left(\hbar\omega - E_{g,\text{opt}}\right)^2}{n\omega^2} \Rightarrow \sqrt{\alpha\omega} \propto \frac{\left(\hbar\omega - E_{g,\text{opt}}\right)}{\sqrt{n}}$$

$$(17.16b)$$

The thus defined optical gap is called the Tauc gap. It is conveniently determined from the so-called Tauc plot, where $\sqrt{\alpha(\omega)\omega}$ is plotted against the photon energy. The Tauc gap is often applied in practice to characterize the optical properties of amorphous materials.

Bearing in mind the obvious ambiguity in the interpretation(s) of the optical gap(s) in amorphous materials, let us conclude that the optical gaps defined in this section are nothing else than fitting parameters in dependences like (17.16a) and (17.16b). This is a difference to the definition of forbidden zones as it is possible in the case of crystals, where the existence of energy gaps is a direct conclusion from the $E(\mathbf{k})$-dependence. On the contrary, in an amorphous solid it is quite possible that even in the "forbidden" zone in Fig. 17.11, there exist localized states with a density of states different from zero. For that reason, the terminus "optical gap" is not well-defined for amorphous semiconductors. On the other hand, the introduction of the optical gap by a dependence like (17.16b) gives at least a recipe for the unambiguous and convenient determination of a parameter that may be used to judge the quality of a prepared material with respect to certain optical applications. For that reason, these parameters are widely used in applied semiconductor optics. A bit more detail will be provided in Sect. 17.5.1.

17.3.5 Dielectric Function of a Crystal: Contributions of Direct Electronic Transitions

Let us finally return to the crystalline case and write down an expression for the dielectric function of a crystal involving direct transitions. This appears to be a special case of the general expressions found in Sect. 12.6 (12.26). From there we have:

$$\beta = \frac{2}{\varepsilon_0 \hbar} \sum_l \sum_{n>l} |\mathbf{d}_{nl}|^2 \omega_{nl} \frac{[W(l) - W(n)]}{\omega_{nl}^2 - \omega^2 - 2i\omega\gamma_{nl}}.$$

As we have pointed out in Chap. 16, a single-electron propagating in a periodic potential (single-electron approximation) has a continuous spectrum of energy eigenvalues instead of the discrete energy levels fixed in (12.26). Regions of "allowed" energy values (energy bands) appear to be separated from each other by "forbidden" energy zones. In each energy band, the electron energy is a continuous function of the wavevector of the electron $\mathbf{k}$. The general theoretical considerations given below are applicable to direct electronic transitions in different kinds of crystals, no matter whether they represent insulators, semiconductors, or metals. Nevertheless, we will often use a terminology which stems from the field of semiconductors. The reason is simple. The electronic transitions which are detected in optical spectroscopy do usually occur between the valence and the conduction bands of a crystal. Good insulators usually have a broad energy spacing between these bands, so that they may be regarded as transparent in the NIR/VIS regions. On the contrary, the absorption onset wavelength in semiconductors is considerably larger, so that the shape of the absorption bands has to be taken into account when performing optical spectroscopy with semiconductors. Therefore, in many cases we will apply a terminology which is usually relevant in semiconductor optics with regard to transitions between the valence and the conduction bands.

Caused by the mentioned band structure, we cannot further work with discrete energy levels E_n as assumed when having derived (12.26), but have to replace them by the functions $E_n(\mathbf{k}_n)$:

$$E_n \rightarrow E_n(\mathbf{k}_n)$$

The quantum number n is now to count the energy bands instead of the energy levels. Accordingly, the transition frequencies have to be replaced according to:

$$\omega_{nl} = \frac{E_n - E_l}{\hbar} \rightarrow \frac{E_n(\mathbf{k}_n) - E_l(\mathbf{k}_l)}{\hbar}$$

Thus, absorption of light may cause an electron from the lth energy band and with an initial wavevector $\mathbf{k}_l$ to perform a quantum transition into the nth energy band. In general, its wavevector may also change to $\mathbf{k}_n$ due to quasimomentum conservation. As long as $l \neq n$ holds, we deal with *interband transitions*.

In the case of the *direct* interband transitions, we have (compare Sect. 17.3.1)

$$\mathbf{k}_l \approx \mathbf{k}_n \equiv \mathbf{k} \Rightarrow \frac{E_n(\mathbf{k}_n) - E_l(\mathbf{k}_l)}{\hbar} = \frac{E_n(\mathbf{k}) - E_l(\mathbf{k})}{\hbar} \equiv \omega_{nl}(\mathbf{k}).$$

Similar to the transition frequencies, the other values encountering into (12.26) are now also dependent on the electrons wavevector, although the dependence may be weak. Let us further recall, that the electron wavefunctions in a periodic potential are delocalized. Then, the propagating electron may rather feel the average electric field than the local one. In such cases, a treatment in terms of the Lorentz–Lorenz formula does not make sense. We therefore assume:

$$\varepsilon = 1 + N\beta$$

Now, the wavevector $\mathbf{k}$ has to be regarded as one of the quantum numbers relevant in (12.26). Let us assume that N may be associated with the concentration of electrons. Moreover, due to Pauli's principle, any quantum state can only be occupied by a single electron. Summarizing over all occupied quantum states does therefore automatically sum up over the electrons. According to (12.7), their individual contributions to the full polarization sum up according to: $\mathbf{P} = \frac{\sum_j \mathbf{d}_j}{V}$. Instead of (14.7), we therefore obtain:

$$\varepsilon(\omega) = 1 + \frac{2}{\varepsilon_0 \hbar V} \sum_{\mathbf{k}} \sum_{l} \sum_{n>l} \frac{[W_l(\mathbf{k}) - W_n(\mathbf{k})]|\mathbf{d}_{nl}(\mathbf{k})|^2 \omega_{nl}(\mathbf{k})}{\omega_{nl}^2(\mathbf{k}) - \omega^2 - 2i\omega\gamma_{nl}(\mathbf{k})} \qquad (17.17)$$

Remark Electrons are fermions and obey the Fermi–Dirac statistics. Note that for the Fermi–Dirac statistics, the occupation probability of a given quantum state coincides with the average occupation of the state [7]. The W-values therefore may now be interpreted as the average occupation of the corresponding quantum states, as given by the Fermi–Dirac distribution. The indices l and n count the energy bands.

Exactly the same result is obtained in terms of an alternative approach. Imagine a polarizing piece of matter that is large enough such that long-range atomic order is observed. Then its polarizability according to (12.26) must include summation over the "quantum number" $\mathbf{k}$, such that instead of (12.26) we may write:

$$\beta(\omega) = \frac{2}{\varepsilon_0 \hbar} \sum_{\mathbf{k}} \sum_{l} \sum_{n>l} \frac{[W_l(\mathbf{k}) - W_n(\mathbf{k})]|\mathbf{d}_{nl}(\mathbf{k})|^2 \omega_{nl}(\mathbf{k})}{\omega_{nl}^2(\mathbf{k}) - \omega^2 - 2i\omega\gamma_{nl}(\mathbf{k})}$$

Then, from $\varepsilon = 1 + N\beta$ setting $N = 1/V$ we immediately obtain (17.17) again, when V is associated with the volume of exactly that peace of crystalline matter.

Remark In a strong sense, in addition to summation over $\mathbf{k}$, summation over $\mathbf{g}$ must also be included into expressions like (17.17). We did not explicitly write it down because $\omega_{nl}^2(\mathbf{k})$ and $W_n(\mathbf{k})$ are independent on $\mathbf{g}$. However, the square of the absolute value of the transition matrix element $|\mathbf{d}_{nl}(\mathbf{k})|^2$ in fact has to be understood as the result of summation over all $\mathbf{g}$ vectors $\sum_{\mathbf{g}} |\langle \mathbf{k} + \mathbf{g}, n|\mathbf{d}|\mathbf{k}, l\rangle|^2$, because according to the $\mathbf{k}$-selection rule (17.1a), all these individual transitions may be allowed.

Once $\mathbf{k}$ should be tackled as a continuous variable, it is reasonable in (17.17) replacing the sum over $\mathbf{k}$ by the corresponding integral. When remembering Sect. 4.6.1 and, in particular, (4.11) and (4.12), we have:

$$k = \frac{\pi}{L}n; \; n = \sqrt{n_x^2 + n_y^2 + n_z^2}; \; n_x; n_y; n_z = 1, 2, 3, \ldots$$

From $\frac{dZ}{dn} = \frac{\pi n^2}{2}$, we then have:

$$\frac{Z}{V} = \frac{1}{V}\sum_{\mathbf{k}} = \frac{1}{V}\int dZ = \frac{\pi}{2V}\int n^2 dn = \frac{\pi}{2V}\frac{L^3}{\pi^3}\int k^2 dk$$

$$= \frac{1}{8V}\frac{L^3}{\pi^3}\int 4\pi k^2 dk = \frac{1}{(2\pi)^3}\int d^3\mathbf{k}$$

We thus find the transformation recipe:

$$\frac{1}{V}\sum_{\mathbf{k}} \rightarrow \frac{1}{(2\pi)^3}\int d^3\mathbf{k}$$

When substituting this into (17.17) that results in:

$$\varepsilon(\omega) = 1 + \frac{1}{4\pi^3\varepsilon_0\hbar}\int d^3\mathbf{k}\sum_{l}\sum_{n>l}\frac{[W_l(\mathbf{k}) - W_n(\mathbf{k})]|\mathbf{d}_{nl}(\mathbf{k})|^2\omega_{nl}(\mathbf{k})}{\omega_{nl}^2(\mathbf{k}) - \omega^2 - 2i\omega\gamma_{nl}(\mathbf{k})} \quad (17.17a)$$

When assuming for simplicity that the bands l are fully occupied ($W_l(\mathbf{k}) = 1$), while n are completely empty ($W_l(\mathbf{k}) = 1$), we find for the corresponding material:

$$\varepsilon(\omega) = 1 + \frac{1}{4\pi^3\varepsilon_0\hbar}\int d^3\mathbf{k}\sum_{l}\sum_{n>l}\frac{|\mathbf{d}_{nl}(\mathbf{k})|^2\omega_{nl}(\mathbf{k})}{\omega_{nl}^2(\mathbf{k}) - \omega^2 - 2i\omega\gamma_{nl}(\mathbf{k})} \quad (17.17b)$$

Equation (17.17b) represents the expression for the electronic contribution to the dielectric function of a crystal at sufficiently low temperatures, as long as only direct transitions are involved.

In solid-state physics, it is common to express transition probabilities in terms of the transition matrix element of the electron momentum operator rather than of the dipole moment operator (as we do). In this case, (17.17–17.17b) hold as well by structure, but there is an additional pre-factor of $[e/(m_e\omega)]^2$.

Often, in practice, it is of particular interest to regard transitions between the valence (V) and conduction (C) bands only. Then the sum over l and n degenerates to a single term according to:

$$\varepsilon(\omega) = 1 + \frac{1}{4\pi^3\varepsilon_0\hbar} \int d^3\mathbf{k} \frac{|\mathbf{d}_{CV}(\mathbf{k})|^2\omega_{CV}(\mathbf{k})}{\omega_{CV}^2(\mathbf{k}) - \omega^2 - 2i\omega\gamma_{CV}(\mathbf{k})} \tag{17.17c}$$

17.4 Consistency Considerations

In Sects. 17.3.1 and 17.3.5, we have derived two alternative approaches for calculating the dispersion of the dielectric function of a crystal. It is of course necessary to clarify in how far these two approaches yield consistent results. Let us therefore look at an example in order to get an idea on the shape of the dielectric function as described by (17.17c).

Let us regard the contribution of an interband transition between the valence and conduction bands to the dielectric function. In a band scheme like sketched in Fig. 17.4 (and assuming isotropy), the resonance frequency might then be given by (compare also 16.22):

$$\omega_{CV}(\mathbf{k}) = \frac{1}{\hbar}\left[E_g + \frac{B}{2}(1 - \cos ka)\right] \tag{17.18}$$

where E_g marks the direct band gap, and B again is a constant that characterizes the energy band width. This way (17.17c) contains information about the bandwidth as well as about the homogeneous linewidth γ of any of the individual transitions entering into (17.7a). Hence, the approach presented in Sect. 17.3.5 seems to be the more general one, and the results of Sect. 17.3.1 (17.7b) should be obtainable as a particular case of (17.17c). This is what we want to exemplify in this section.

In order to preserve consistency with Sect. 17.3.1, let us again neglect the $\mathbf{k}$-dependence of the transition matric element, as well as of all the other parameters encountering into (17.17c) except the transition frequency which is set according to (17.18). Then, the contribution of the V $\rightarrow$ C transition to the dielectric function of the system may be directly calculated performing a numerical integration in (17.17c). It is interesting to perform this calculation with different assumed values of the bandwidth B and the homogeneous linewidth γ.

What could be expected from such a model calculation? Provided that we have $B \ll \hbar\gamma$, from (17.18) we might expect $\omega_{CV}(\mathbf{k}) \approx \frac{1}{\hbar}E_g = $ const., and in this case, (17.17c) should yield a single Lorentzian line. In the opposite case $B \gg \hbar\gamma$, from (17.17c) we should expect an output similar to (17.7a) at least for photon energies close to the gap value, because in deriving (17.7a) only a dependence like (17.18) has been taken into consideration, and any homogeneous linewidth was neglected.

Let us now look at the calculation results given in Fig. 17.13. In the case that $B \ll \hbar\gamma$, the band structure does not give any effect, and the imaginary part of the

Fig. 17.13 Shape of the imaginary part of the dielectric function as calculated from (17.17c) and (17.18), assuming $E_g = 3$ eV: solid: $B = 1$ eV, $\gamma/(2\pi c) = \Gamma = 1$ cm^{-1}; dash: $B = 0.01$ eV, $\gamma/(2\pi c) = \Gamma = 300$ cm^{-1}

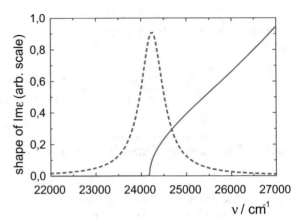

dielectric function appears as a typical Lorentzian line as known from the classical picture or the quantum mechanical treatment of systems with discrete energy levels. On the contrary, when the homogeneous linewidth is negligible compared to the bandwidth B, the imaginary part of the dielectric function shows a sharp onset at $\hbar\omega = E_g$, which is therefore also called the (fundamental) absorption edge energy. Consequently, optical measurements may be used to determine the direct band gap in crystalline solids by means of the absorption behavior (if the transition is allowed). For $\hbar\omega > E_g$, the imaginary part of the dielectric function indeed increases with frequency similar to:

$$\mathrm{Im}\ \varepsilon \propto \sqrt{\hbar\omega - E_g}$$

This behavior is typical for allowed electronic transitions in the vicinity of a direct band gap, and it coincides with the results obtained in Sect. 17.3.1.

On the other hand, it is interesting to check the behavior of the dielectric function depending on the dimensionality of the system. Figure 17.13 clearly corresponds to the three-dimensional case (3D) and has been calculated in spherical coordinates according to the upper relation in (17.7b). The same calculation may be carried out for assuming 2D and 1D cases in (17.17c). Figure 17.14 illustrates the shapes of the imaginary parts of the thus given dielectric functions in the vicinity of the absorption edge.

It is obvious, that the dependence $\mathrm{Im}\ \varepsilon \propto \sqrt{\hbar\omega - E_g}$ is only valid for the three-dimensional case. In the two-dimensional case (compare Sect. 4.6.1), we rather find $\mathrm{Im}\ \varepsilon \propto \mathrm{const}$, while in the the one-dimensional case a behavior similar to $\mathrm{Im}\ \varepsilon \propto \dfrac{1}{\sqrt{\hbar\omega - E_g}}$ is observed. This is in full consistency to the predictions (17.7b) derived in terms of the joint density of states concept.

The mentioned behavior of the dielectric function for a two-dimensional motion of an electron in a periodic potential is essential for the theory of so-called quantum well structures or superlattices. The one-dimensional case is practically relevant in so-called quantum wires. Particularly, the singularity at the direct

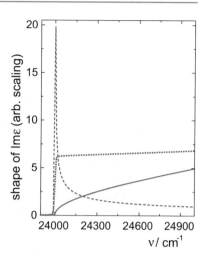

Fig. 17.14 Shape of the imaginary part of the dielectric function according to (17.17c) in the vicinity of the absorption edge, as calculated from (17.2) and (17.3). Solid: 3D; dot: 2D; dash: 1D. $E_g = 3$ eV; $B = 1$ eV; $\gamma/(2\pi c) = \Gamma = 1$ cm^{-1}

gap is of practical importance to achieve a high oscillator strength necessary for luminescent devices.

Again, the sharp absorption onset obvious from Fig. 17.14 is characteristic to crystalline materials and confirms us about the well-defined value of optical gaps in crystalline materials. In amorphous materials, in agreement with Fig. 17.11, those absorption features should smear out, leaving some freedom in ascribing a physical sense to what we call an optical gap in an amorphous solid. Strongly spoken, certain optical gap versions may simply be tackled as fitting parameters in dependencies like (17.16a) or (17.16b). Nevertheless, as suggested by Fig. 17.11, all those gaps might be close by numerical value to their crystalline counterparts. Section 17.5.1 will be used to oppose certain possible gap definitions in amorphous solids to each other while using amorphous silicon as a practical example.

What about highest laser light intensities? Everybody knows that intense laser radiation may destroy solid materials, which is in the basis of many industrial applications of high power lasers today. So far we have not yet spoken on such effects. Let us mention in this context that light–matter interactions may principally subdivided into destructive (Fig. 17.15) and nondestructive versions. This course is on nondestructive interactions only, and any effects concerned to laser damage are outside the scope of this book.

But even when restricting on nondestructive interaction mechanisms, the picture developed so far is far from being complete. Let us return to (12.2):

$$\mathbf{d} = \mathbf{d}(\mathbf{E}) = \sum_{j=0}^{\infty} a_j \mathbf{E}^j \equiv \underbrace{\mathbf{d}_{\text{perm}}}_{\substack{\text{permanent electric} \\ \text{dipole moment}}} + \underbrace{\varepsilon_0 \left(\beta \mathbf{E} + \beta^{(2)} \mathbf{EE} + \beta^{(3)} \mathbf{EEE} + \cdots \right)}_{\substack{\text{induced electric} \\ \text{dipole moment}}}$$

What we have discussed so far are effects related to induced electric dipole moments that are linearly dependent on the applied field strength. Correspondingly,

Fig. 17.15 Destructive light–matter interaction mechanism. Cartoon by Dr. Alexander Stendal. Printed with permission

the phenomena concerned to that linear approach from the field of linear optics. But obviously, this description is insufficient when the field strength becomes too large. Then, completely new optical phenomena have to be expected that arise from the nonlinear terms in (12.2). Those phenomena form the field on nonlinear optics. Chapter 18 will be devoted to an introduction into this field.

17.5   Application to Concrete Problems

17.5.1 Optical Properties of Amorphous Semiconductors: Parameters for a Practical Description

The rather vague assignment of what we call the optical gap of an amorphous solid to a concrete physical counterpart has resulted in a certain diversity of optical gap definitions that are convenient in a certain practical context. Let us have a look at Fig. 17.16 on left.

It shows the typical dispersion behavior of the optical constants of a nonmetallic solid material in a rather broad spectral range. Typically, we observe certain absorption in the MIR that is caused by core vibrations. Usually, in the NIR region we have some transparency window, where K is close to zero, while the refractive index n may be modeled in terms of Sellmeier or Cauchy approaches according to (14.9b–d). At even shorter wavelength, i.e., in the VIS or UV, we observe the onset of light absorption because of the optical excitation of valence electrons. It

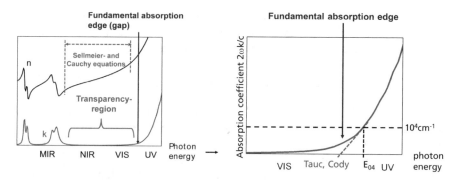

Fig. 17.16 Left: Typical dispersion picture of optical constants; right: visualization of Tauc, Cody, and E_{04} gaps

is the threshold photon energy of these electronic absorptions that is of exceptional importance for practical applications of the material in optics. This threshold energy is what is usually called the fundamental absorption edge, and it is of course related to the optical gap, no matter which particular gap definition is applied.

The most pragmatic definitions of optical gaps indeed make use of that threshold idea. Thus, the so-called E_{03} and E_{04} gaps are defined a the photon energies where the absorption coefficient in the fundamental absorption edge region reaches the values 10^3 cm^{-1} or 10^4 cm^{-1}, respectively (see Fig. 17.16 on right).

$$\alpha\left(\hbar\omega = E_{\text{phot}} = E_{03}\right) = 10^3 \text{cm}^{-1}; \; \alpha\left(\hbar\omega = E_{\text{phot}} = E_{04}\right) = 10^4 \text{cm}^{-1} \quad (17.19)$$

In other approaches, the optical gap rather appears as a fitting parameter when fitting the measured absorption coefficient in terms of approaches like (17.16–17.16b). This way we define gaps like the Cody or Tauc gaps. Note also that the earlier introduced β_do model (14.11) defines an optical gap as (compare also Fig. 14.9)

$$E_{g,\beta\_do} = \hbar\omega_a = hc\nu_a$$

Note that the Cody approach (17.16a):

$$\text{Im}\, \varepsilon(\omega) = 2nK \propto \left(\hbar\omega - E_{g,\text{opt}}\right)^2$$

appears to be a special case of the β_do envelope (14.11) $w_{\beta\_do}(\omega)$ when setting $A = 3$ and $B = 1$.

$$w_{\beta\_do}(\omega) \propto (\omega - \omega_a)^{A-1}(\omega_b - \omega)^{B-1}\Big|_{\substack{A = 3 \\ B = 1}} = (\omega - \omega_a)^2$$

Figure 17.16 on right vizualises the relation between those different gap definitions. The Cody and Tauc gaps are usually fitted in spectral regions where the absorption coefficient is larger than 10^4 cm^{-1}, and therefore, as a rule we have:

$$E_{g,\text{Tauc}}, E_{g,\text{Cody}} < E_{04} \qquad (17.20)$$

In finishing this section, in Table 17.4 we present some experimental data on optical gaps of (hydrogenated) amorphous silicon a-Si(:H), taken from reference [8]. The idea is to demonstrate the relation between different optical gaps, but also the dynamic range of optical properties of amorphous semiconductors. The latter thus appear as rather versatile materials, and in certain limits, their optical properties may be tailored by suitably adopting the preparation procedure.

Figure 17.17 demonstrates the a-Si(:H) Tauc gaps as depending on the mass density. Also, the reference value for c-Si is included. Note that the experimental data show a clear trend toward smaller gaps when the density is increased and thus well reproduce the qualitative behavior as expected from the discussion in Sect. 16.4. Concerning refractive indices, so they also increase with increasing mass density for the amorphous samples (as it was expected, too). They even fall higher than the crystalline value, which is attributed to additional internal surface states generated by subnanometer-sized voids in the amorphous atomic network [9, 10]. They should be absent in the crystal, which explains the seemingly inconsistent picture.

Table 17.4 Survey of reported optical parameters for amorphous (hydrogenated) silicon according to [8]

Type of sample	Density in gcm^{-3}	E_{03}/eV	E_{04}/eV	$E_{g,\text{Tauc}}$/eV	n@4.1 µm
a-Si:H by sputtering	1.96	1.98	2.12	1.98	2.97
	2.10	1.74	1.93	1.74	3.34
a-Si:H by glow charge decomposition	2.17	1.74	1.92	1.78	(3.45)
a-Si:H by chemical vapor deposition	2.26	1.1	1.7	1.45	(3.5)
a-Si by sputtering	2.29	0.9	1.44	1.26	3.84
Reference: c-Si (n, E_{03}, E_{04} estimated from the transmittance and reflectance of a commercial Si wafer)	2.336	≈1.5	≈2.3	$E_{g,ind}$≈ 1.2 eV	3.38

Note that (17.20) is always fulfilled. Refractive indices in brackets indicate that the wavelength was not specified in an obvious manner

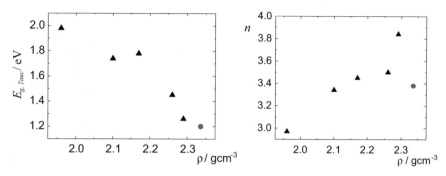

Fig. 17.17 Tauc gap (left) and refractive index (right) of amorphous silicon films as a function of the mass density

17.5.2 The Tauc–Lorentz Model

The Tauc–Lorentz model has been developed for convenient parametrization of the optical constants of amorphous materials in the interband absorption region. Let us start from the single Lorentzian oscillator model setting:

$$\chi \propto \frac{1}{\omega_0^2 - \omega^2 - 2i\omega\gamma} \Rightarrow \operatorname{Im}\varepsilon = \operatorname{Im}\chi \propto \frac{\omega\gamma}{\left(\omega_0^2 - \omega^2\right)^2 + 4\omega^2\gamma^2}$$

In the Tauc–Lorentz model, the imaginary part of the dielectric function of a single oscillator model is merged together with (17.16b) to generate the imaginary part of the dielectric function of the Tauc–Lorentz model according to [11]:

$$\operatorname{Im}\varepsilon(\omega)\begin{cases} \propto \dfrac{\left(\hbar\omega-E_{g,\text{opt}}\right)^2}{\omega\left[\left(\omega_0^2-\omega^2\right)^2+4\omega^2\gamma^2\right]}; & \hbar\omega \geq E_{g,\text{opt}} \\ = 0; & \hbar\omega < E_{g,\text{opt}} \end{cases} \tag{17.21}$$

The corresponding real part is calculated through (15.4b). Explicit expressions can be found, for example, in [12, 13]. A comparison of the spectral shape of a Lorentzian line according to (2.14a) and the spectral feature described by (17.21) is presented in Fig. 17.18. Note that (17.21) is an odd function with respect to the frequency as required by (15.3).

In contrast to the single Lorentzian oscillator model or the Brendel model (14.10), the Tauc–Lorentz model is able to describe strongly asymmetric absorption features. In particular, it allows modeling absorption thresholds. With a total of only four free parameters it is a convenient and often used tool for modeling optical constants in the vicinity of absorption edges in amorphous optical materials.

Fig. 17.18 Comparison of a
Lorentzian absorption
lineshape with the absorption
feature in the Tauc–Lorentz
model

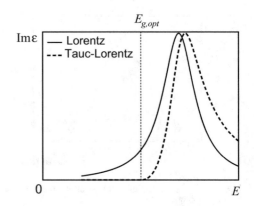

17.6   Advanced Material: Spatial Dispersion of the Dielectric Function

Let us for completeness extend our discussion to another phenomenon called *spatial dispersion*. In our discussion of dispersion as provided so far we have assumed that the polarizing system embodies some kind of memory, i.e., the polarization of the medium at time t depends on the field at that time as well as on the field at previous moments. As a natural conclusion, we found that the dielectric function, which describes the system response to a monochromatic field in the frequency domain, must necessarily be frequency dependent, a phenomenon which we have called dispersion. Moreover, from the examples discussed so far it appears obvious that strongest dispersion is observed when the periodic time of the electric field (from the electromagnetic wave) comes close to a characteristic intrinsic time of the medium, such as the Debye relaxation time or an eigenvibration period (resonance).

On the other hand, there are specific optical phenomena observed when the wavelength of the light becomes comparable to characteristic *spatial* dimensions of the medium. We do not mean inhomogeneities here, we still focus on optically homogeneous materials. But even in a homogeneous medium it is absolutely possible, that the polarization in a certain point $\mathbf{r}$ depends on the electric field strength in some spatial vicinity of that point. In the case that the dimensions of that "active" area become comparable with the wavelength of the light, we obtain phenomena concerned with what we will further call spatial dispersion.

Let us return to material equations like (13.4) and (13.5). They define the polarization at time t as a functional of the electric field at various previous times t' $< t$. But this is clearly not the most general writing. We obtain a more general version of this type of equation when allowing that the polarization in a certain point $\mathbf{r}$ depends on the electric field strength in neighboring points $\mathbf{r}'$. This is what is sometimes called nonlocality of the system response. Then, instead of (13.4), we may write:

$$\mathbf{P}(\mathbf{r}, t) = \varepsilon_0 \int_{-\infty}^{t} \int_{V} \kappa(t, t', \mathbf{r}, \mathbf{r}') \mathbf{E}(\mathbf{r}', t') \mathrm{d}t' \mathrm{d}\mathbf{r}' \tag{17.22}$$

Again, κ carries information about the material specifics, including the previously discussed "memory" of the system. At the same time, it now incorporates information on how far the electric field in neighboring points $\mathbf{r}'$ affects the polarization at point $\mathbf{r}$. The integration over $\mathbf{r}'$ is formally carried out in the full volume V where κ is different from zero.

We further proceed in the usual manner. Introducing homogeneity in time, we postulate that κ does not depend on t and t' explicitly, but only on their difference (retardation) ξ according to:

$$\xi = t - t'$$

In full analogy, we require spatial homogeneity by substituting the explicit dependence of κ on $\mathbf{r}$ and $\mathbf{r}'$ by a dependence on their difference (spatial distance) only:

$$\mathbf{R} = \mathbf{r} - \mathbf{r}'$$

In full analogy to our treatment in Sect. 13.3.1, we then obtain:

$$\mathbf{P}(\mathbf{r}, t) = \varepsilon_0 \mathbf{E}(\mathbf{r}, t) \int_{0}^{\infty} \int_{V} \kappa(\xi, \mathbf{R}) \mathrm{e}^{i(\omega\xi - \mathbf{k}_p \mathbf{R})} \mathrm{d}\xi \mathrm{d}\mathbf{R}$$

Note that in order to keep consistency with the symbols used in this chapter, the wavevector of the electromagnetic wave is here written as $\mathbf{k}_p$. We then have:

$$\chi = \int_{0}^{\infty} \int_{V} \kappa(\xi, \mathbf{R}) \mathrm{e}^{i(\omega\xi - \mathbf{k}_p \mathbf{R})} \mathrm{d}\xi \mathrm{d}\mathbf{R} \equiv \chi(\omega, \mathbf{k}_p) = \varepsilon(\omega, \mathbf{k}_p) - 1 \tag{17.23}$$

The outcome is that under the assumptions discussed here, i.e., the inclusion of nonlocality, the linear dielectric susceptibility does explicitly depend on the frequency *and* the wavevector of the assumed monochromatic electromagnetic wave propagating through the medium. It is that explicit dependence on the light wavevector that is called spatial dispersion.

When assuming that κ is different from zero at distances $R \ll \lambda$ only, in the effective integration volume we have:

$$\mathrm{e}^{-i\mathbf{k}_p \mathbf{R}} \approx 1 - i\mathbf{k}_p \mathbf{R} \tag{17.24}$$

In many situations, spatial dispersion phenomena contribute only small corrections to the otherwise dominating usual (time) dispersion. This is especially relevant in dielectrics, where the kernel of the integral in (17.23) is essentially different from zero only around **R**-values of the order of the interatomic spacing. In a classical picture, this is easy to understand, because the bound electrons are not "allowed" to leave "their" atomic cores; hence, their movement is restricted to distances of the order of the interatomic spacing. In metals, however, even classical electrons are allowed to propagate over much longer distances, and therefore, corrections to the Drude function that arise from spatial dispersion may become significant.

If, however, the frequency of the incident light comes close to frequencies corresponding to strong narrow absorption lines in dielectrics (i.e., in resonance conditions), the classical amplitude of the oscillations of bound electrons may increase. Also, close to resonance, the refractive index will increase remarkably (compare Fig. 14.5), which results in a corresponding decrease in the wavelength in the medium. Then, spatial dispersion may also become significant, and the explicit dependence of the susceptibility on the wavevector may even result in the appearance of birefringence in the relevant spectral regions, although the material appears to be optically isotropic elsewhere.

In a formal sense, the Doppler broadening in gases is also a manifestation of spatial dispersion. Indeed, (2.22) may be written as: $\Delta\omega_D \approx k_p \langle |v| \rangle$, where v is the thermal velocity of the gas particles. Thus, the width of the absorption line depends on k_p [14].

We will not derive a corresponding theory of the dielectric response of solids in the presence of spatial dispersion; this is outside the scope of this book. Let us only mention here that spatial dispersion is inherent to existing many-electron theories to the dielectric response of crystals, a convincent and at the same time accessible approach can be found in Chap. 5 of the excellent textbook [15]. In this source, the derivation of a more general theory of the dielectric function is sketched provided that the perturbation is time- and coordinate-dependent. Then, the Fourier image of the perturbation will contain different Fourier components with different angular frequencies and wavevectors, which we will for simplicity call $\mathbf{k}_p$ again. In the so-called random phase approximation, the response of the system to each of these Fourier-components is calculated independently. Naturally, in an expression like (6.21), the expansion coefficients a_n become $\mathbf{k}_p$-dependent then. Then, it naturally turns out that the dielectric function explicitly depends on both frequency and wavevector of the perturbation.

We will not derive the formulas here, but indicate the final expression for the dielectric function relevant for optical excitations (assuming direct electronic transitions between the valence and conduction bands and and a light frequency close to resonance):

$$\varepsilon\left(\omega; \mathbf{k}_p\right) = 1 + \frac{e^2}{\varepsilon_0 V k_p^2} \sum_{\mathbf{k}_V, \mathbf{g}} \frac{[W(\mathbf{k}_V) - W(\mathbf{k}_C)]\left|\langle \mathbf{k}_C | e^{i\mathbf{k}_p \mathbf{r}} | \mathbf{k}_V \rangle\right|^2}{[E(\mathbf{k}_C) - E(\mathbf{k}_V)] - \hbar\omega - i\hbar\gamma} \tag{17.25}$$

Here $\mathbf{k}_V$ denotes the electrons wavevector in the valence band, and $\mathbf{k}_C$ in the conduction band. Because of **k**-conservation, we have (17.1a) $\mathbf{k}_C = \mathbf{k}_V + \mathbf{k}_p + \mathbf{g}$.

$|\mathbf{k}_V\rangle$ and $|\mathbf{k}_C\rangle$ denote the corresponding quantum states in the valence and conduction bands. For small photon wavevectors (compare 17.24), we have $e^{i\mathbf{k}_p\mathbf{r}} \approx 1 + i\mathbf{k}_p\mathbf{r} \Rightarrow \langle \mathbf{k}_C|e^{i\mathbf{k}_p\mathbf{r}}|\mathbf{k}_V\rangle \approx \mathbf{k}_p\langle \mathbf{k}_C|\mathbf{r}|\mathbf{k}_V\rangle$, and from (17.25) we obtain

$$\varepsilon\left(\omega; \mathbf{k}_p \to 0\right) = 1 + \frac{e^2}{\varepsilon_0 V} \sum_{\mathbf{k}_V,\mathbf{g}} \frac{[W(\mathbf{k}_V) - W(\mathbf{k}_C)]|\langle \mathbf{k}_C|\mathbf{r}|\mathbf{k}_V\rangle|^2}{[E(\mathbf{k}_C) - E(\mathbf{k}_V)] - \hbar\omega - i\hbar\gamma}$$

This coincides with (17.17), when considering the approximation:

$$\frac{1}{\omega_0 - \omega - i\gamma} \underset{\omega \approx \omega_0 \gg \gamma}{\approx} \frac{1}{\omega_0 - \omega - i\gamma} + \frac{1}{\omega_0 + \omega + i\gamma} \approx \frac{2\omega_0}{\omega_0^2 - \omega^2 - 2i\omega\gamma}$$

$$(17.26)$$

Thus, in the case of small photon wavevectors $\mathbf{k}_p$, the dependence of the dielectric function on $\mathbf{k}_p$ cancels out, and we obtain our "standard" result (17.17) which has been derived neglecting any effects of spatial dispersion. Fortunately, everything is consistent here. In view of (17.26), apart from resonance, we therefore prefer the writing of (17.25) as:

$$\varepsilon\left(\omega; \mathbf{k}_p\right) = 1 + \frac{2e^2}{\varepsilon_0\hbar^2 V k_p^2} \sum_{\mathbf{k}_V,\mathbf{g}} \frac{[W(\mathbf{k}_V) - W(\mathbf{k}_C)][E(\mathbf{k}_C) - E(\mathbf{k}_V)]}{\left[\frac{E(\mathbf{k}_C)-E(\mathbf{k}_V)}{\hbar}\right]^2 - \omega^2 - 2i\omega\gamma}\left|\langle \mathbf{k}_C|e^{i\mathbf{k}_p\mathbf{r}}|\mathbf{k}_V\rangle\right|^2$$

$$(17.27)$$

Note that from the asymptotic behavior at $\omega \to \infty$, from (17.27) we now obtain the sum rule (compare Sect. 15.6.2):

$$\sum_{\mathbf{g}} [E(\mathbf{k}_C) - E(\mathbf{k}_V)]\left|\langle \mathbf{k}_C|e^{i\mathbf{k}_p\mathbf{r}}|\mathbf{k}_V\rangle\right|^2 = \frac{\hbar^2 k_p^2}{2m}$$

$$(17.28)$$

This is a generalization of the f-sum rule (5.33). Indeed, for small k_p, we have

$$e^{i\mathbf{k}_p\mathbf{r}} \approx 1 + i\mathbf{k}_p\mathbf{r} \Rightarrow \langle \mathbf{k}_C|e^{i\mathbf{k}_p\mathbf{r}}|\mathbf{k}_V\rangle \approx \mathbf{k}_p\langle \mathbf{k}_C|\mathbf{r}|\mathbf{k}_V\rangle$$

and (17.28) transforms into (5.33).

Note that the general stucture of (17.27) may be guessed from our previous result (17.17). If we chose the interaction operator like introduced in Sect. 17.3.1, instead of (17.17) we have:

$$\varepsilon(\omega; \mathbf{k}_p) = 1 + \frac{2e^2}{\varepsilon_0 \hbar^2 V} \sum_{\mathbf{k}_V, \mathbf{g}} \frac{[W(\mathbf{k}_V) - W(\mathbf{k}_C)][E(\mathbf{k}_C) - E(\mathbf{k}_V)]}{\left[\frac{E(\mathbf{k}_C) - E(\mathbf{k}_V)}{\hbar}\right]^2 - \omega^2 - 2i\omega\gamma} \left|\langle \mathbf{k}_C | \mathbf{r} e^{i\mathbf{k}_p\mathbf{r}} | \mathbf{k}_V \rangle\right|^2$$

For small $\mathbf{k}_p$ and close to resonance we approximate the above expression by (please do not show that to a mathematician!)

$$\left|\langle \mathbf{k}_C | \mathbf{r} e^{i\mathbf{k}_p\mathbf{r}} | \mathbf{k}_V \rangle\right|^2 \rightarrow \left|\frac{\partial}{\partial \mathbf{k}_p} \langle \mathbf{k}_C | e^{i\mathbf{k}_p\mathbf{r}} | \mathbf{k}_V \rangle\right|^2$$

$$\rightarrow \left|\lim_{\mathbf{k}_p \to 0} \frac{\langle \mathbf{k}_C | e^{i\mathbf{k}_p\mathbf{r}} | \mathbf{k}_V \rangle - \langle \mathbf{k}_C | 1 | \mathbf{k}_V \rangle}{\mathbf{k}_p - 0}\right|^2$$

$$\approx \left|\left|\frac{\langle \mathbf{k}_C | e^{i\mathbf{k}_p\mathbf{r}} | \mathbf{k}_V \rangle}{\mathbf{k}_p}\right|\right|^2_{\text{small } \mathbf{k}_p}$$

$$\Rightarrow \varepsilon(\omega; \mathbf{k}_p)$$

$$\approx 1 + \frac{2e^2}{\varepsilon_0 \hbar^2 V k_p^2} \sum_{\mathbf{k}_V, \mathbf{g}} \frac{[W(\mathbf{k}_V) - W(\mathbf{k}_C)][E(\mathbf{k}_C) - E(\mathbf{k}_V)]}{\left[\frac{E(\mathbf{k}_C) - E(\mathbf{k}_V)}{\hbar}\right]^2 - \omega^2 - 2i\omega\gamma} \left|\langle \mathbf{k}_C | e^{i\mathbf{k}_p\mathbf{r}} | \mathbf{k}_V \rangle\right|^2$$

$$\approx 1 + \frac{e^2}{\varepsilon_0 V k_p^2} \sum_{\mathbf{k}_V, \mathbf{g}} \frac{[W(\mathbf{k}_V) - W(\mathbf{k}_C)]\left|\langle \mathbf{k}_C | e^{i\mathbf{k}_p\mathbf{r}} | \mathbf{k}_V \rangle\right|^2}{[E(\mathbf{k}_C) - E(\mathbf{k}_V)] - \hbar\omega - i\hbar\gamma}$$

Which coincides with (17.25). This is an illustration on in how far the mathematical structure of (17.25) is related to (17.17), but it cannot be regarded a derivation. For the latter, again, please consult [15].

Let us finish this chapter by investigating two special cases of (17.25). Let us start with the response of a system of free electrons. In this case, both initial and final states belong to the same band, namely the conduction band. Accordingly, we rewrite (17.25) as:

$$\varepsilon(\omega; \mathbf{k}_p) = 1 + \frac{e^2}{\varepsilon_0 V k_p^2} \sum_{\mathbf{k}_C} \frac{[W(\mathbf{k}_C) - W(\mathbf{k}_C + \mathbf{k}_p)]\left|\langle \mathbf{k}_C | e^{i\mathbf{k}_p\mathbf{r}} | \mathbf{k}_C + \mathbf{k}_p \rangle\right|^2}{[E(\mathbf{k}_C + \mathbf{k}_p) - E(\mathbf{k}_C)] - \hbar\omega - i\hbar\gamma}$$

For a completely free electron, we identify the wavefunction with a plane wave according to (4.5). That means that the period in the Bloch function (16.10) is zero, so that $\mathbf{g}$ becomes infinitely large, and summation over $\mathbf{g}$ looses its sense.

Consequently, the matrix element in the above expression becomes equal to 1, and we obtain the Lindhard formula for the dielectric function [16]:

$$\varepsilon\left(\omega; \mathbf{k}_p\right) \approx 1 + \frac{e^2}{\varepsilon_0 V k_p^2} \sum_{\mathbf{k}_C} \frac{[W(\mathbf{k}_C) - W(\mathbf{k}_C + \mathbf{k}_p)]}{\left[E(\mathbf{k}_C + \mathbf{k}_p) - E(\mathbf{k}_C)\right] - \hbar\omega - i\hbar\gamma}.$$

The Lindhardt formula is important, for example, in the theory of energy electron loss spectra (EELS).

In application to a semiconductor, (17.25) (here in the writing of 17.27) yields a different result. Let us estimate what happens at $\omega \to 0$. For simplicity we again assume: $[W(\mathbf{k}_V) - W(\mathbf{k}_C)] = 1$.

Further we approximate: $E(\mathbf{k}_C) - E(\mathbf{k}_V) \approx \langle \Delta E \rangle$. From the sum rule (17.28), we then have:

$$\sum_{\mathbf{g}} [E(\mathbf{k}_C) - E(\mathbf{k}_V)] \left| \langle \mathbf{k}_C | e^{i\mathbf{k}_p\mathbf{r}} | \mathbf{k}_V \rangle \right|^2 \approx \sum_{\mathbf{g}} \langle \Delta E \rangle \left| \langle \mathbf{k}_C | e^{i\mathbf{k}_p\mathbf{r}} | \mathbf{k}_V \rangle \right|^2 = \frac{\hbar^2 k_p^2}{2m}$$

Therefore from (17.27), it follows:

$$\varepsilon\left(\omega; \mathbf{k}_p\right) = 1 + \frac{2e^2}{\varepsilon_0 \hbar^2 V k_p^2} \sum_{\mathbf{k}_V, \mathbf{g}} \frac{[W(\mathbf{k}_V) - W(\mathbf{k}_C)][E(\mathbf{k}_C) - E(\mathbf{k}_V)]}{\left[\frac{E(\mathbf{k}_C) - E(\mathbf{k}_V)}{\hbar}\right]^2 - \omega^2 - 2i\omega\gamma} \left| \langle \mathbf{k}_C | e^{i\mathbf{k}_p\mathbf{r}} | \mathbf{k}_V \rangle \right|^2$$

$$\to 1 + \frac{2e^2}{\varepsilon_0 \hbar^2 k_p^2} \frac{1}{\left[\frac{\langle \Delta E \rangle}{\hbar}\right]^2 - \omega^2 - 2i\omega\gamma} \frac{\hbar^2 k_p^2}{2m} N = 1 + \frac{Ne^2}{\varepsilon_0 m} \frac{1}{\left[\frac{\langle \Delta E \rangle}{\hbar}\right]^2 - \omega^2 - 2i\omega\gamma}$$

$$\omega = 0: \ \varepsilon_{\text{stat}} = 1 + \frac{\frac{Ne^2}{\varepsilon_0 m}}{\left[\frac{\langle \Delta E \rangle}{\hbar}\right]^2} = 1 + \frac{\omega_p^2}{\omega_0^2}; \ \omega \to \infty: \ \varepsilon \to 1 - \frac{\omega_p^2}{\omega^2}$$

Here we have included the concentration of electrons in the valence band by having set:

$$N \leftrightarrow \frac{1}{V} \sum_{\mathbf{k}_V}$$

In our approximation, we obtained a simple Lorentzian dielectric function for estimating the response of interband transitions in a semiconducting (or insulating) solid. This is an encouraging result, because it confirms us about the applicability of the classical single or multioscillator models for "quick and dirty" approximation of the dielectric function in a crystalline solid. Here the introduced value $\langle \Delta E \rangle$ corresponds to some average value of the difference $E(\mathbf{k}_C) - E(\mathbf{k}_V)$ in (17.25) or (17.27). When associating it with the central frequency of a single Lorentzian oscillator, we obtain the same asymptotes for $\omega \to 0$ and $\omega \to \infty$ as in the classical oscillator model.

17.7 Tasks for Self-check

17.7.1 Multiple-choice test: Mark all answers which seem you correct!

Solids usually have	A negligible mass density
	A negligible shear modulus
	A rather large bulk modulus
Reciprocal lattice vectors are given in	1/m
	m
	cm^{-1}
Bloch functions may be used to describe the wavefunctions of	diatomic molecules in diluted gases
	electrons in a spatially periodic potential
The binding energy of a Wannier–Mott exciton in a semiconductor like GaAs is typically of the order of	1...10 eV
	≈ 1 GeV
	1...10 meV
The radius of a Wannier–Mott exciton in GaAs (ground state) is of the order of	10 nm
	10^{-2} nm
	1 cm
Metals are characterized by	The presence of free electrons
	A good electrical conductivity
	a refractive index equal to 1 at any wavelength
The effective mass of an electron in a crystal lattice	is always positive
	is always negative
	May be negative or positive
All crystals are	good electric conductors
	free of absorption
	disordered
Amorphous solids exhibit	Short-range order
	Long-range order
	neither of them
Copper is	a dielectric
	electrically insulating
	optically transparent
Crystalline germanium	is transparent in the MIR
	has an MIR refractive index around 4.0
	is an indirect semiconductor

17.7.2 True or wrong? Make your decision!

Assertion:	true	wrong
All crystals are optically anisotropic		
The operator of discrete translations $\hat{T}_n$ defined by $\hat{T}_n \psi(\mathbf{r}) = \psi(\mathbf{r} + \mathbf{r}_n)$ ($\mathbf{r}_n$-lattice vector) with the eigenvalues $t_n = e^{i\mathbf{k}\mathbf{r}_n}$ is self-adjoint		
Indirect interband transitions in a crystal involve the creation or annihilation of phonons		
Amorphous solids lack any short-range order in their atomic configuration		
If a crystal lattice has inversion symmetry, the corresponding solid must be a metal		
The zone-centered optical phonon of diamond is Raman-active		

17.7.3 NaCl has a static dielectric constant $\varepsilon_{stat} = 5.9$. In the NIR, its dielectric function is $\varepsilon_{NIR} \equiv \varepsilon_\infty \approx 2.25$. We also know that the normal incidence reflectance of the air–NaCl interface is zero at a wavelength of 30.6 μm. Calculate the angular frequencies ω_T and ω_L! Sketch the normal incidence reflectance $R = R(\lambda)$ of the air–NaCl interface in the MIR! (after [17, task 15.5]).

17.7.4 Assume a doped InSb crystal with a concentration of conduction electrons $N = 10^{18}$ cm^{-3}. Their effective mass should be $0.015m_e$. Assuming a model dielectric function $\varepsilon(\omega) = \varepsilon_\infty - \frac{Ne^2}{\varepsilon_0 m^* \omega^2}$, obtain the wavelength where the normal incidence reflectance at the air-InSb surface is close to zero! $\varepsilon_\infty = 16$ (after [17, task 15.13]).

17.7.5 In the examples in Sect. 17.3.2, as well as in the result of task 17.7.3, we observe the relation $\omega_T \leq \omega_L$. Find a responsible physical mechanism behind this rule!

17.7.6 Starting from an expression like (17.27), rederive the Drude function by considering intraband transitions only and making use of $\mathbf{k}_p \rightarrow \mathbf{0}$!

References

Specific References

1. M. Fox, *Optical properties of solids* (Oxford University Press, 2010)
2. H. Ehrenreich, H.R. Philipp, B. Segall, Optical properties of aluminum. Phys. Rev. **132**, 1918–1928 (1963)
3. H. Warlimont, W. Martienssen (eds.), *Springer Handbook of Materials Data* (Springer 2018, p. 664–668)
4. O. Stenzel, Optical Catings. *Material aspects in theory and practice* (Springer, 2014, p. 36)

5. R. Zallen, *The Physics of Amorphous Solids* (Wiley, 1983, p. 235)
6. N.F. Mott, E.A. Davis, *Electronic Processes in Non-Crystalline Materials* (Clarendon Press, Oxford, 1979). (chapter 6)
7. C. Kittel, *Thermal Physics* (Wiley, 1969, chapter 9)
8. C. Eva, Freeman and William Paul: Optical constants of rf sputtered hydrogenated amorphous Si. Phys. Rev. B **20**, 716–728 (1979)
9. M.H. Brodsky (ed.), Amorphous Semiconductors; Springer-Verlag Berlin Heidelberg New York 1979, p. 84–87
10. O. Stenzel, *The Physics of Thin Film Optical Spectra* (Springer, An Introduction, 2016), pp. 72–83
11. G.E. Jellison, Spectroscopic ellipsometry data analysis: measured versus calculated quantities. Thin Solid Films **313**(314), 33–39 (1998)
12. V. Janicki, Design and optical characterization of hybrid thin film systems, PhD Thesis, Faculty of Science, University of Zagreb (2007)
13. D. Franta, D. Nečas, L. Zajíčková, I. Ohlídal, J. Stuchlík, D. Chvostová, Application of sum rule to the dispersion model of hydrogenated amorphous silicon. Thin Solid Films **539**, 233–244 (2013)
14. Л.Д.Ландау, Е.М.Лифшиц, Электродинамика Сплошных Сред Москва „Наука" 1982, p. 491–495
15. J.M. Ziman, *Principles of the theory of solids* (Cambridge, 1972, Chapter 5)
16. H. Kuzmany, *Festkörperspektroskopie* (Springer, Eine Einführung, 1989), p. 263
17. H.J. Goldsmid (ed.) *Problems in Solid State Physics* (Academic Press, 1968)

General Literature to this Chapter

18. C.F. Klingshirn, *Semiconductor Optics* (Springer-Verlag, Berlin Heidelberg New York, 1997)
19. P.J. Yu, M. Cardona, Fundamentals of semiconductors, in *Physics and Material Properties*, 4th edn (Springer, 2010)
20. M. Fox, Optical properties of solids (Oxford University Press, 2010)
21. C. Kittel, Introduction to Solid State Physics (Wiley, any edition)
22. M. Dresselhaus, G. Dresselhaus, S.B. Cronin, A.G.S. Filho, *Solid State Propertie* (Springer, From Bulk to Nano, 2018)
23. O. Stenzel, *The Physics of Thin Film Optical Spectra* (Springer, An Introduction, 2016)

Basic Effects of Nonlinear Optics

18

Abstract

A nonlinear material equation is introduced, and prominent examples of second- and third-order nonlinear optical effects are discussed in the framework of a classical approach. Emphasis is placed on nonlinear refraction, nonlinear absorption, and second harmonic generation. The treatment is mainly constricted to the classical picture.

18.1 Starting Point

Let us start again from the equation for the microscopic dipole moment, as given by (12.1) and (12.2):

$$\mathbf{d} = \mathbf{d}(\mathbf{E}) = \sum_{j=0}^{\infty} a_j \mathbf{E}^j \equiv \underbrace{\mathbf{d}_{\text{perm}}}_{\substack{\text{permanent electric} \\ \text{dipole moment}}} + \underbrace{\varepsilon_0 \left(\beta \mathbf{E} + \beta^{(2)} \mathbf{EE} + \beta^{(3)} \mathbf{EEE} + \cdots \right)}_{\text{induced electric dipole moment}}$$

In this representation, the dipole moment appears to be composed from three different contributions:

1. A permanent dipole moment
2. An induced dipole moment that depends linearly on the local electric field
3. An induced dipole moment that depends nonlinearly on the local electric field

In the previous chapters we have essentially considered effects arising from the first two contributions. This chapter will deal with effects that arise from the third contribution.

Also, we have to remember that (12.2) has been formulated for time-independent electric fields originally. The effects arising from a timedependence of the field have been introduced in Chap. 13 for the special case of the linear polarization. For isotropic media, we postulated there (13.5):

$$\mathbf{P}(t) = \varepsilon_0 \int_0^\infty \kappa(\xi)\mathbf{E}(t-\xi)d\xi$$

This is the linear material equation in the time-domain representation. Further we introduced the linear susceptibility via (13.7):

$$\chi = \int_0^\infty \kappa(\xi)e^{i\omega\xi}d\xi = \chi(\omega)$$

After having introduced the dielectric function, we then arrived at the material equation in the frequency domain representation as given by (13.9a):

$$\mathbf{D}_\omega = \varepsilon_0 \varepsilon(\omega)\mathbf{E}_\omega$$

It will be our task now to obtain a nonlinear material equation as a generalization to (13.9) or (13.9a). Although we have not dealt with optically anisotropic materials so far, let us write down the generalization of (13.9a) to the anisotropic case. We then have:

$$\mathbf{D} = \begin{pmatrix} D_x \\ D_y \\ D_z \end{pmatrix} = \varepsilon_0 \begin{pmatrix} 1+\chi_{xx}^{(1)} & \chi_{xy}^{(1)} & \chi_{xz}^{(1)} \\ \chi_{yx}^{(1)} & 1+\chi_{yy}^{(1)} & \chi_{yz}^{(1)} \\ \chi_{zx}^{(1)} & \chi_{zy}^{(1)} & 1+\chi_{zz}^{(1)} \end{pmatrix} \begin{pmatrix} E_x \\ E_y \\ E_z \end{pmatrix} = \varepsilon_0''(\omega)\mathbf{E} \quad (18.1)$$

Here, in order to distinguish from nonlinear susceptibilities, we have introduced the superscript (1) to each of the components of the linear susceptibility, which now form a 3×3 susceptibility tensor $\boldsymbol{\chi}$. Correspondingly, instead of the scalar dielectric function, we now have a 3×3 dielectric tensor $\boldsymbol{\varepsilon}$.

18.2 Physical Idea

In fact, the basic physical idea behind this chapter has already been disclosed at the end of Sect. 17.4. Instead of a linear dependence between electric field and induced dipole moment, we now assume a nonlinear dependence. This can be understood as resulting from a nonlinear dependence of the restoring force on the elongation of the carge carriers, an effect which is expected to become relevant at sufficiently large electric field strength.

In order to obtain the material equation relevant for nonlinear optics, essentially we have to reproduce the treatment from Chaps. 12 and 13 for the nonlinear terms in (12.2). In particular, we have to

- Generalize (13.5) with respect to the inclusion of nonlinear terms
- Formulate an analogon to the Clausius–Mossotti equation that is valid for non-linear susceptibilities. Then, from classical or semiclassical expressions for microscopic hyperpolarizabilities $\beta^{(j > 1)}$, we could conclude on corresponding nonlinear susceptibilities.
- Formulate and solve nonlinear wave equations in order to obtain recipes for the calculation of optical signals in the presence of optical nonlinearities.

The generalized material equation will have the formal structure of an expansion into a Tailor series according to:

$$\mathbf{D} = \varepsilon_0\mathbf{E}+\mathbf{P} = \varepsilon_0\mathbf{E} + \mathbf{P}^{(1)} + \mathbf{P}^{(2)} + \mathbf{P}^{(3)} + \cdots$$
$$= \varepsilon_0\left\{\mathbf{E} + \chi^{(1)}\mathbf{E} + \chi^{(2)}:\mathbf{EE} + \chi^{(3)}:\mathbf{EEE} + \cdots\right\} \qquad (18.2)$$

Equation (18.2) is written in a somewhat symbolic manner. In fact, the products at the right hand of (18.2) have to be understood as tensor products, the suscepti-bilities $\chi^{(1)}$, $\chi^{(2)}$, $\chi^{(3)}$, ... themselves represent tensors of different ranks. When P_i is the ith Cartesian component of the polarization with $i = x, y, z$, we may write:

$$\mathbf{P}^{(1)} = \varepsilon_0\chi^{(1)}\mathbf{E} \Leftrightarrow P_i^{(1)} = \varepsilon_0 \sum_{j=x,y,z} \chi_{ij}^{(1)} E_j;$$

$$\mathbf{P}^{(2)} = \varepsilon_0\chi^{(2)}\mathbf{EE} \Leftrightarrow P_i^{(2)} = \varepsilon_0 \sum_{j=x,y,z} \sum_{k=x,y,z} \chi_{ijk}^{(2)} E_j E_k;$$

$$\mathbf{P}^{(3)} = \varepsilon_0\chi^{(3)}\mathbf{EEE} \Leftrightarrow P_i^{(3)} = \varepsilon_0 \sum_{j=x,y,z} \sum_{k=x,y,z} \sum_{l=x,y,z} \chi_{ijkl}^{(3)} E_j E_k E_l \qquad (18.2a)$$

The first equation in (18.2a) is nothing else than the general (anisotropic) version of the linear material equation, as already introduced in Sect. 18.1. According to (18.1), the linear susceptibility $\chi^{(1)}$ may be regarded as a 3×3 quadratic matrix. The second-order nonlinear susceptibility, $\chi^{(2)}$, represents a $3 \times 3 \times 3$ tensor with 27 components, while the third-order nonlinear susceptibility $\chi^{(3)}$ has 81 components.

In any specialized treatment of nonlinear optics, one would now establish rela-tions between different components of the susceptibility tensors and show that there is only a limited number of independent components of the correspond-ing tensors. This is far beyond our treatment, but we will mention at least two important results:

- In centrosymmetric materials, all components of $\chi^{(2)}$ and other even-order susceptibilities are equal to zero in electric dipole approximation.
- There is no material symmetry, where all components of $\chi^{(3)}$ or other odd-order susceptibilities would be equal to zero (except, of course, vacuum).

In our further rather qualitative treatment, we will neglect the tensor character of the susceptibilities. Practically that means that we restrict on a single component of (18.2) and (18.2a) like according to:

$$P_x = \varepsilon_0 \left[\chi_{xx}^{(1)} E_x + \chi_{xxx}^{(2)} E_x^2 + \chi_{xxxx}^{(3)} E_x^3 + \cdots \right] \tag{18.2b}$$

And write for simplicity:

$$P = \varepsilon_0 \left[\chi^{(1)} E + \chi^{(2)} E^2 + \chi^{(3)} E^3 + \cdots \right] \tag{18.2c}$$

The thus defined susceptibilities have SI measurements unit according to:

$$\left[\chi^{(j)} \right] = \left(\frac{m}{V} \right)^{j-1} \tag{18.2d}$$

As a kind of slang, a medium which shows nonlinear optical properties is sometimes called a nonlinear medium.

18.3   Theoretical Material

18.3.1 Nonlinear Material Equation

In order to get some idea on the frequency arguments relevant for the nonlinear susceptibilities introduced previously, we will have to proceed the same way as we did in the case of the linear material equation. According to (18.2c) and (13.5), in the time domain we may now write:

$$P(t) = P^{(1)} + P^{(2)} + P^{(3)} + \cdots = \varepsilon + \varepsilon_0 \int_0^\infty \kappa^{(1)}(\xi) E(t-\xi) d\xi +$$

$$\varepsilon_0 \int_0^\infty \int_0^\infty \kappa^{(2)}(\xi_1, \xi_2) E(t-\xi_1) E(t-\xi_1-\xi_2) d\xi_1 d\xi_2 +$$

$$+ \varepsilon_0 \int_0^\infty \int_0^\infty \int_0^\infty \kappa^{(3)}(\xi_1, \xi_2, \xi_3) E(t-\xi_1) E(t-\xi_1-\xi_2)$$

$$E(t-\xi_1-\xi_2-\xi_3) d\xi_1 d\xi_2 d\xi_3 + \cdots \tag{18.3}$$

Equation (18.3) is nothing else than a nonlinear generalization of (13.5). We do not regard ferroelectrics, so that the constant, field-independent contribution $P^{(0)}$ is assumed to be zero. In the general case described by (18.3), the response functions $\kappa^{(i)}$ again represent tensors of a range according to (18.2a).

When comparing (13.5) and (18.3), we notice a further complication of the mathematical treatment of nonlinear optical processes. Because both the electric field strength and the polarization are real physical values, the response functions $\kappa^{(i)}$ must be real as well. Nevertheless, in linear optics, we were used to work with complex fields and polarizations. The reason is that in the linear equation (13.5), we may make use of the superposition principle. When in the final result the real polarization is required, one simply has to add the conjugated complex term, and everything will be fine (compare also (2.1)–(2.5) in this context).

This treatment is no more correct in nonlinear optics, because such a treatment would result in a loss of polarization terms. This may be demonstrated by a simple example. Let us regard the quadratic nonlinearity according to the simplified equation: $P^{(2)} = \varepsilon_0 \chi^{(2)} E^2$. The assumption $E = \frac{E_0}{2} e^{-i\omega t}$ leads us to

$$P^{(2)} = \varepsilon_0 \chi^{(2)} \frac{E_0^2}{4} e^{-2i\omega t} \qquad (18.4a)$$

For simplicity, the amplitude of the electric field strength should be real throughout this discussion. We find that the assumed time dependence of the electric field leads to a polarization in the medium which oscillates with twice the frequency of the incoming field. Of course, such an oscillating polarization gives rise to the generation of an electromagnetic wave at the angular frequency 2ω. This means that at least a part of the energy of the ingoing wave is transferred to a new wave with the doubled frequency, an effect which is called *second-harmonic generation* (SHG). SHG is the most prominent effect of nonlinear optics in media with a quadratic nonlinearity.

Let us now regard another case. We assume: $E = \frac{E_0}{2} e^{+i\omega t}$. The resulting second-order polarization is:

$$P^{(2)} = \varepsilon_0 \chi^{(2)} \frac{E_0^2}{4} e^{2i\omega t} \qquad (18.4b)$$

Both (18.4a) and (18.4b) describe a polarization that oscillates at a frequency 2ω. Summing up (18.4a) and (18.4b) will lead to a real value of the polarization, but it will not give any new physical effects.

Let us now regard a real field strength given by the algebraic sum of the versions discussed so far. We assume: $E = \frac{E_0}{2} \left(e^{+i\omega t} + e^{-i\omega t} \right) = E_0 \cos \omega t$. The corresponding second-order polarization becomes:

$$P^{(2)} = \varepsilon_0 \chi^{(2)} \frac{E_0^2}{4} \left(e^{2i\omega t} + e^{-2i\omega t} + 2 \right) \qquad (18.4c)$$

This is more than what has been predicted by the algebraic sum from (18.4a) and (18.4b). Although the SHG terms are present as expected, there is a further term, which corresponds to a time-independent (static) polarization, created by the quadratic nonlinearity as the result of a nonlinear effect called *optical rectification*. Equation (18.4c) makes it clear that as the result of the nonlinear interaction of a monochromatic wave with matter, second-order polarization terms occur that are constant or oscillating with twice the frequency of the incoming wave.

We see that the application of the simplified complex electric fields will lead to a significant loss of information, when we deal with nonlinear optics. Therefore, instead of (2.1) or the like, in this chapter we will always regard a real expression for the electric fields according to:

$$E(t) = \frac{1}{2} \sum_j E_{0j} e^{-i\omega_j t} + c.c. \tag{18.5}$$

with $c.c$—conjugate complex value.

18.3.2 Frequency Conversion Processes

The general conclusion from (18.3) and (18.5) is that the nonlinear second-order polarization oscillates with all sum and difference frequencies resulting from the primary frequencies of the incoming field (18.5). For a detailed treatment see [1]. Correspondingly, one speaks on *Sum Frequency Generation* (SFG) and *Difference Frequency Generation* (DFG) in nonlinear optics. Formally, from (18.2) when assuming only the second-order term, we obtain:

$$P^{(2)}(t) = \frac{1}{4}\varepsilon_0 \sum_j \sum_l E_{0j} E_{0l} e^{-i(\omega_j + \omega_l)t} \chi^{(2)}(\omega = \omega_j + \omega_l) + c.c$$
$$+ \frac{1}{4}\varepsilon_0 \sum_j \sum_l E_{0j}^* E_{0l} e^{-i(\omega_l - \omega_j)t} \chi^{(2)}(\omega = \omega_l - \omega_j) + c.c \tag{18.6}$$

The concrete expressions for $\chi^{(2)}$ follow from the comparison with (18.3).

The frequency arguments in (18.6) have to be understood in the following way: The first frequency indicates the resulting frequency of the second-order polarization. The following frequencies indicate the frequencies of the electric fields forming the polarization and the particular way of their combination. Thus, the effects described so far correspond to the following susceptibilities:

$$\chi^{(2)}(\omega = \omega_l + \omega_j) \Leftrightarrow \text{SFG}$$
$$\chi^{(2)}(\omega = \omega_l - \omega_j) \Leftrightarrow \text{DFG}$$
$$\chi^{(2)}(2\omega = \omega + \omega) \Leftrightarrow \text{SHG} \tag{18.7}$$
$$\chi^{(2)}(0 = \omega - \omega) \Leftrightarrow \text{optical rectification}$$

We will not perform a similarly detailed discussion of the properties of nonlinear susceptibilities as we did in the case of the linear susceptibility. We only repeat here that there exist several symmetry relations which may reduce the quantity of nonzero and independent components of the susceptibility.

But let us now have a closer look at an extremely important selection rule: In a medium with an inversion center (or inversion symmetry, or so-called centrosymmetric materials), all components of any even-order nonlinear susceptibility are zero in the dipole approximation. For the particular case of the second-order susceptibility, this is a direct conclusion from (18.6). Imagine an electric field like (18.5), which causes a certain second-order polarization. From the supposed inversion symmetry it is clear that an inversion of the electric field strength should be accompanied by an inversion of the polarization:

$$\mathbf{E} \to -\mathbf{E} \Rightarrow \mathbf{P} \to -\mathbf{P} \Rightarrow \mathbf{P}^{(2)}(-\mathbf{E}) = -\mathbf{P}^{(2)}(+\mathbf{E})$$

On the other hand, for all projections of $\mathbf{P}^{(2)}$ it follows from (18.6) that

$$P^{(2)}(-E) =$$

$$\frac{1}{4}\varepsilon_0 \sum_j \sum_l (-E_{0j})(-E_{0l})e^{-i(\omega_j+\omega_l)t} \chi^{(2)}(\omega = \omega_j + \omega_l) + c.c$$

$$+ \frac{1}{4}\varepsilon_0 \sum_j \sum_l \left(-E_{0j}^*\right)(-E_{0l})e^{-i(\omega_l-\omega_j)t} \chi^{(2)}(\omega = \omega_l - \omega_j) + c.c$$

$$= P^{(2)}(+E)$$

Both conditions together may only be fulfilled, when the second-order polarization is zero for any assumed electric field configuration. This means that the second-order susceptibility must be zero. This kind of discussion may be performed for any even-order susceptibility, but obviously not for the odd-order susceptibilities.

For us it is important to notice that second (and other even)-order nonlinear processes are only allowed in media, which lack inversion symmetry. On the other hand, odd-order processes are in principle allowed in any medium. For that reason, even-order processes are rarely used in analytic optical spectroscopy, because they cannot be applied to every bulk material. Nonlinear optical spectroscopy usually bases on odd-order (basically third order) optical effects. On the other hand, second-order processes are often applied for frequency conversion processes such as SHG, SFG or DFG, utilizing a couple of selected nonlinear materials which have the necessary nonzero components of the second-order susceptibility tensor. Prominent traditional examples are potassium dihydrogen phosphate KH_2PO_4 (KDP), ammonium dihydrogen phosphate $NH_4H_2PO_4$ (ADP), or lithium niobate $LiNbO_3$. Modern developments pursue the qualification of new second-order materials like $Li_2B_4O_7$ (LTB) or $YAl_3(BO_3)_4$ (YAB).

There is an important exclusion from this rule: At the interface between two materials, inversion symmetry is always violated, although both of the particular materials may be centrosymmetric. Therefore, second-order processes may be used for interface spectroscopy. Being applied to the surface or interfaces between centrosymmetric materials, their advantage is to supply a background-free second-order optical response of the interface region. Combined with resonant local electric field strength enhancement mechanisms, second-order processes supply highly interface-sensitive spectroscopic tools for interface and ultrathin adsorbate layer spectroscopy.

In complete analogy to the second-order polarization, one may discuss the third-order fundamental frequency terms in (18.3) in order to investigate optical effects that arise as a result of third-order nonlinearity in different media. It is particularly important for any spectroscopist to become familiar with third-order nonlinear effects, because the latter may be observed in any medium regardless on the concrete symmetry that is subject to the structure of the given sample. In this section we will restrict on a few remarks concerning the third-harmonic generation THG, but we will return to third-order processes is the Sects. 18.5.2 and 18.5.3.

Let us assume an oscillating electric field according to:

$$ E = \frac{E_0}{2} e^{-i\omega t} + c.c. $$

Proceeding the same way as for the quadratic nonlinearity, we obtain a third-order polarization term, which oscillates with the frequency 3ω:

$$ P^{(3)} = \cdots + \varepsilon_0 \chi^{(3)}(3\omega = \omega + \omega + \omega) \frac{E_0^3}{8} e^{-3i\omega t} + c.c. + \cdots \qquad (18.8) $$

Consequently, in a third-order nonlinear medium, an electromagnetic wave will be generated that has the frequency 3ω. The corresponding process is called *third-harmonic generation* THG. Nevertheless, in frequency conversion practice, THG is usually not accomplished by means of a third-order nonlinear frequency conversion. Instead, it turns out to be more efficient to use a cascade of second-order processes to generate higher order harmonics of a given fundamental frequency. Thus, the third harmonic may be generated by SHG of the fundamental frequency, followed by an SFG process between the second harmonic and the fundamental frequency.

18.4   Consistency Considerations

Regardless on the concrete values of the linear and nonlinear susceptibilities, it is clear that (18.2) converges to the linear equation (13.9) when the electric field strength becomes sufficiently small. Figure 18.1 demonstrates the principle dependence of the individual contributions to the polarization on the field strength. At weak fields, the linear contribution is dominating, and in this case the previously discussed effects of linear optics (LO) are sufficient to describe the optical properties of a material.

This raises the question on the value of a characteristic field strength when the nonlinear terms in (18.2) become significant. This question is difficult to be answered generally, because dispersion does matter, and off-resonance processes

Fig. 18.1 Linear and nonlinear contributions to the polarization of a nonlinear medium

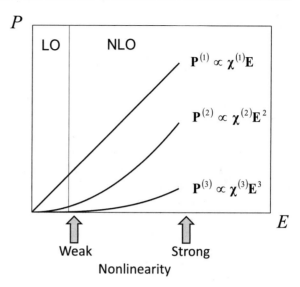

are as usually less efficient than resonant constellations. But when comparing typical off-resonant values of tabulated susceptibilities in solids [2, pp. 809–837], we arrive at the rule of thumb:

$$\chi^{(1)} \sim 10^0, \ \chi^{(2)} \sim 10^{-11} \ \text{m/V}, \ \chi^{(3)} \sim 10^{-22} \text{m}^2/\text{V}^2.$$

Then, according to (18.2c)

$$P = \varepsilon_0 \left[\chi^{(1)} E + \chi^{(2)} E^2 + \chi^{(3)} E^3 + \cdots \right]$$

We find that the different nonlinear contributions to the polarization become comparable to the linear one when $E \sim 10^{11}$ V/m is fulfilled. This is of the order of intraatomic electric fields (compare task 18.7.3). Hence, for applying linear optics, we should require that the local field induced by the incident light is significantly smaller than the intraatomic field strength.

At higher field strength, the nonlinear polarization becomes significant, so that we enter the field of nonlinear optics. Here, *both* linear and nonlinear contributions to the full polarization have to be taken into account. Practically, this is usually of significance when we deal with laser light. In our treatment, we use a representation of the polarization like (18.2c). When restricting on the quadratic and cubic terms, we implicitly assume that the series (18.2c) is quickly converging, such that higher order nonlinear terms do not matter. We will call this situation the regime of weak optical nonlinearity.

Let us finally note in this context that the relation between hyperpolarizabilities as introduced in (12.2) and the corresponding nonlinear susceptibilities is given by:

$$\chi^{(j>1)}\left(\omega_{j+1} = \sum_{l=1}^{j} \omega_l\right) = N \prod_{l=1}^{j+1}\left[\frac{\varepsilon(\omega_l)+2}{3}\right]\beta^{(j)}\left(\omega_{j+1} = \sum_{l=1}^{j}\omega_l\right) \quad (18.9)$$

Note the factor $\frac{\varepsilon(\omega_l)+2}{3}$, familiar from the local field treatment introduced in Sect. 12.3.4 [relation (12.15)]. Obviously, our local field treatment is again restricted to optically isotropic media.

Let us now have a closer look at the situation in centrosymmetric materials. As we have pointed out in Sect. 18.3.2, in a centrosymmetric material, the second-order (as well as any higher even order) polarization should be equal to zero for symmetry reasons. On the other hand, from Sect. 6.6.1 we know that centrosymmetric systems obey the parity selection rule, which means that in a centrosymmetric system the transition matrix element of the dipole moment becomes zero, whenever the participating quantum states have the same parity. This raises the question in how far these observations are interconnected.

In order to answer this question, we now would have to derive a semiclassical or quantum mechanical expression for the hyperpolarizabilities, similarly to our approach to the linear polarizability demonstrated in Sect. 12.6. This treatment is again beyond the scope of this book, but interested readers are referred to the density matrix approach demonstrated in [1]. We only mention here that in analogy to the already derived QM structure of the linear polarizability [compare (12.26)], the quantum mechanical expressions for higher order susceptibilities are dominated by terms like:

$$\chi^{(1)} \neq 0 \Leftrightarrow \sum_{n,m} d_{nm}d_{mn} \neq 0$$

$$\chi^{(2)} \neq 0 \Leftrightarrow \sum_{n,m,l} d_{nm}d_{ml}d_{ln} \neq 0$$

$$\chi^{(3)} \neq 0 \Leftrightarrow \sum_{n,m,l,k} d_{nm}d_{ml}d_{lk}d_{kn} \neq 0 \quad (18.10)$$

and so on. By the way, for the particular case of $\chi^{(2)}$, this may easily be guessed from elaborating (12.27a) in detail, while focusing on the terms proportional to E^2. Anyway, (18.10) leads us to the conclusion that in a quantum system with inversion symmetry, all even-order optical susceptibilities must be identical to zero in the dipole approximation.

Indeed, let us look at the second-order susceptibility in a system with inversion symmetry. In such systems, the parity selection rule is relevant. Supposing that state $|n\rangle$ is of even parity, we must require that $|m\rangle$ is odd, otherwise d_{nm} will be zero. For the same reason, $|l\rangle$ must be even. But if so, d_{ln} will be zero. So that there is no way to arrange the quantum states in a manner that the product $d_{nm}d_{ml}d_{ln}$ becomes different from zero whenever the system is centrosymmetric.

Fig. 18.2 Visualization of the effect of the parity selection rule on nonlinear optical susceptibilities of centrosymmetric materials: even-order processes are forbidden [right side, **(b)** and **(d)**], while odd-order processes are, in principle, allowed [left side, **(a)** and **(c)**]

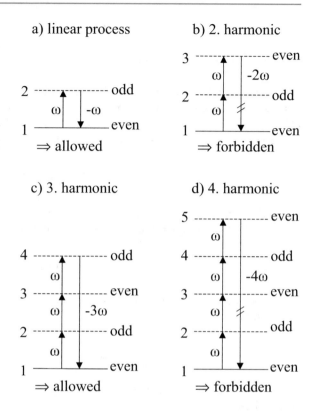

This argumentation applies to all even-order susceptibilities, but it is not applicable to the odd-order susceptibilities. This is exemplified in Fig. 18.2 for the particular cases of linear polarization (a), SHG (b), THG (c), and fourth-harmonic generation (d). The upwards directed arrows symbolize the frequency of the electric field, the downward directed arrow the resulting frequency of the (maybe nonlinear) polarization.

Remark: Just for the case: Please do not confuse inversion symmetry with optical isotropy!

Let us thus resume that compared to linear processes, nonlinear optical phenomena.

- Obey different selection rules
- Have different resonance conditions.

Therefore, in order to initiate a resonant nonlinear optical process, there is more flexibility in the choice of light sources than in a linear process.

18.5   Application to Concrete Problems

18.5.1 The Pockels Effect

Let us no turn to a further second-order optical effect, which is of practical importance for light modulation purposes. It is the so-called linear electrooptic effect or Pockels effect. Imagine a second-order material, externally excited by a monochromatic field and a static one (E_s). The field is thus given by:

$$E = \frac{E_0}{2} e^{-i\omega t} + c.c. + E_s$$

When calculating now the second-order polarization, it will contain a term that oscillates with the frequency ω. The latter is given by:

$$P^{(2)}\Big|_\omega = \varepsilon_0 E_0 E_s \chi^{(2)}(\omega = \omega + 0) e^{-i\omega t} + c.c.$$

Of course, the linear polarization will also contain terms that are oscillating with ω according to:

$$P^{(1)}\Big|_\omega = \frac{1}{2} \varepsilon_0 E_0 \chi^{(1)}(\omega) e^{-i\omega t} + c.c.$$

So that the full polarization at ω is given by (neglecting higher order polarization terms):

$$P\big|_\omega = P^{(1)}\Big|_\omega + P^{(2)}\Big|_\omega$$

$$= \varepsilon_0 \Big[\chi^{(1)}(\omega) + 2\chi^{(2)}(\omega = \omega + 0) E_s \Big] \frac{E_0}{2} e^{-i\omega t} + c.c. \qquad (18.11)$$

Equation (18.11) is completely analogous to a linear material equation, when regarding the term in parentheses as an effective susceptibility, which depends on the strength of the static field as a parameter. Hence, we may define:

$$\chi^{(\text{eff})}(\omega) \equiv \chi^{(1)}(\omega) + 2\chi^{(2)}(\omega = \omega + 0) E_s \qquad (18.12)$$

An electromagnetic wave with the frequency ω will propagate in such a non-linear medium in an identical manner as it would propagate in a linear one, when the linear susceptibility has the value as prescribed by (18.12). We may therefore define an effective refractive index analogously to the treatment in linear optics:

$$n^{(\text{eff})}(\omega) = \sqrt{\varepsilon^{(\text{eff})}(\omega)} \equiv \sqrt{1 + \chi^{(\text{eff})}(\omega)}$$

$$= \sqrt{1 + \chi^{(1)}(\omega) + 2\chi^{(2)}(\omega = \omega + 0) E_s} = n^{(\text{eff})}(\omega, E_s) \qquad (18.13)$$

According to (18.13), the value of the effective refractive index may be controlled by the strength of the applied static electric field. Hence, the propagation properties of the wave with frequency ω may be controlled by the static field. The name "linear electrooptical effect" arises from the linear dependence of the effective susceptibility on the field strength. In real world, the Pockels effect leads to the appearance of a field-induced birefringence or alters the already existent birefringence.

The mentioned effect is practically applied in so-called Pockels cells. They may be used to modulate the resonator properties of lasers in order to produce short laser pulses.

18.5.2 Nonlinear Refraction

Let us now regard a medium with cubic nonlinearity, illuminated by monochromatic light according to:

$$E = \frac{E_0}{2} e^{-i\omega t} + c.c.$$

In this writing, in a medium with refractive index n, the expression for the light intensity (2.6) generalizes to:

$$I = \frac{n}{2\mu_0 c} |E_0|^2$$

But let us return to (18.8), and in particular to terms which do not oscillate at the THG frequency. Indeed, as a result of the third-order frequency mixing, we obtain a third-order polarization term that oscillates at the fundamental frequency ω. It is given by:

$$P^{(3)}\Big|_{\omega} = \frac{3}{8}\varepsilon_0 \chi^{(3)}(\omega = \omega + \omega - \omega) E_0^2 E_0^* e^{-i\omega t} + c.c.$$

Of course, the linear polarization also yields a contribution to the fundamental frequency. So that the full polarization at the fundamental frequency will be given by:

$$P\big|_{\omega} = P^{(1)}\Big|_{\omega} + P^{(3)}\Big|_{\omega}$$
$$= \frac{1}{2}\varepsilon_0 \left(\chi^{(1)}(\omega) + \frac{3}{4}\chi^{(3)}(\omega = \omega + \omega - \omega)|E_0|^2 \right) E_0 e^{-i\omega t} + c.c. \quad (18.14)$$

We are now in a similar situation as in the previous section, when we discussed the linear electrooptical effect (18.11). Equation (18.14) is again equivalent to a linear material equation with an effective susceptibility, the latter depending on the intensity of the incoming light. The effective susceptibility is now given by:

$$\chi^{(\text{eff})}(\omega) = \chi^{(1)}(\omega) + \frac{3}{4}\chi^{(3)}(\omega = \omega + \omega - \omega)|E_0|^2 \qquad (18.15)$$

It depends on the square of the field strength and represents a special version of the optical Kerr effect. The square root of the effective dielectric function becomes:

$$\sqrt{\varepsilon^{(\text{eff})}(\omega)} = \sqrt{1 + \chi^{(1)}(\omega) + \frac{3}{4}\chi^{(3)}(\omega = \omega + \omega - \omega)|E_0|^2}$$

$$= \sqrt{\varepsilon}\sqrt{1 + \frac{3}{4}\frac{\chi^{(3)}(\omega = \omega + \omega - \omega)}{\varepsilon}|E_0|^2}$$

where, as usual, $\varepsilon = 1 + \chi^{(1)}$.

Let us now assume that the linear dielectric function is purely real at the given frequency. In linear optics, the medium would then be free of absorption, and the usual refractive index would be given by the square root of the dielectric function. The effective index of refraction is now given by:

$$\hat{n}^{(\text{eff})}(\omega) = n\sqrt{1 + \frac{3}{4}\frac{\chi^{(3)}(\omega = \omega + \omega - \omega)}{n^2}|E_0|^2}$$

If the nonlinear contribution is small compared to the linear one, this relationship may be rewritten as:

$$\hat{n}^{(\text{eff})}(\omega) \approx n + \frac{3}{8}\frac{\chi^{(3)}(\omega = \omega + \omega - \omega)}{n}|E_0|^2 \qquad (18.16)$$

Although the linear dielectric function has been assumed to be real, the effective index of refraction may be complex, depending on the properties of the third-order susceptibility at the given frequency. In any case, we obtain an intensity-dependent effective refractive index according to:

$$n^{(\text{eff})}(\omega) \approx n + \frac{3}{8}\frac{\text{Re}\chi^{(3)}(\omega = \omega + \omega - \omega)}{n}|E_0|^2 \equiv n + n_2 I$$

$$n_2 = \frac{3}{4}\frac{\mu_0 c}{n^2}\text{Re}\chi^{(3)}(\omega = \omega + \omega - \omega) \qquad (18.17)$$

It turns out to be dependent on the intensity of the electromagnetic wave. The value n_2 is called the *nonlinear refractive index* of the medium. The intensity dependence of the effective refractive index in a third-order (or even higher odd order) nonlinear optical medium is responsible for different self-interaction processes of highly intense light beams, such as self-focusing of laser beams, or self-phase-modulation processes in ultrashort light pulses, essential for white light continuum generation.

Let us now look at the imaginary part. From (18.16), it follows immediately that there is a nonlinear absorption coefficient given by:

$$\alpha_{nl}(\omega) \approx \frac{3}{4} \frac{\omega}{cn} \text{Im} \chi^{(3)}(\omega = \omega + \omega - \omega)|E_0|^2 = \beta I;$$

$$\beta = \frac{3}{2} \frac{\mu_0 \omega}{n^2} \text{Im} \chi^{(3)}(\omega = \omega + \omega - \omega) \tag{18.18}$$

With increasing intensity of the light, the medium may thus become absorbing. As it may be shown in terms of a semiclassical treatment of the nonlinear susceptibilities, the nonlinear absorption as described by (18.18) results from two-photon absorption (TPA) processes. Therefore, the nonlinear absorption coefficient β is also called the TPA coefficient.

Note that (18.17) and (18.18) need to be modified when the linear refractive index is complex. Then, in (18.16), n has to be replaced by $n + iK$. Instead of (18.17) and (18.18),we then have [3]:

$$n^{(\text{eff})}(\omega) \approx n + \frac{3}{8}\left[\frac{n\text{Re}\chi^{(3)} + K\text{Im}\chi^{(3)}}{n^2 + K^2}\right]|E_0|^2$$

$$\alpha^{(\text{eff})}(\omega) \approx 2\frac{\omega}{c}\left\{K + \frac{3}{8}\left[\frac{n\text{Im}\chi^{(3)} - K\text{Re}\chi^{(3)}}{n^2 + K^2}\right]|E_0|^2\right\} \tag{18.18a}$$

The mentioned effects shall give you, dear reader, a first idea on which kind of new optical effects may be expected in the field of nonlinear optics.

Note that in accordance the so-called Miller's rule, there appears to be a correlation between the (non-resonant) linear and nonlinear refractive indices in dielectrics and semiconductors. This is visualized in Fig. 18.3. The experimental points are collected from [4–6] as well as references cited therein. Although the data belong to different wavelength, there is a clear trend toward larger nonlinear indices when the linear refractive index increases. This correlation may be understood from the theoretical expressions for the susceptibilities as obtained in the semiclassical approach, the gray line in Fig. 18.3 shows a corresponding dependence predicted by the formula approximately valid for small frequencies [4]:

$$n_2 = \frac{3}{4}\frac{\mu_0 c}{n^2}\chi^{(3)} \approx \frac{1}{12cNE_g}\left[\frac{(n^2 + 2)(n^2 - 1)}{n}\right]^2$$

$$\approx \frac{1}{2.9}\frac{(n^2 - 1)^2(n^2 + 2)^2}{n^2\left[\frac{95}{n^4} + \frac{4.084-n}{0.62}\right]}10^{-15}\frac{\text{cm}^2}{\text{W}}$$

Figure 18.4 demonstrates the dispersion of the nonlinear absorption coefficient of fused silica in the UV (after [2, p. 819]).

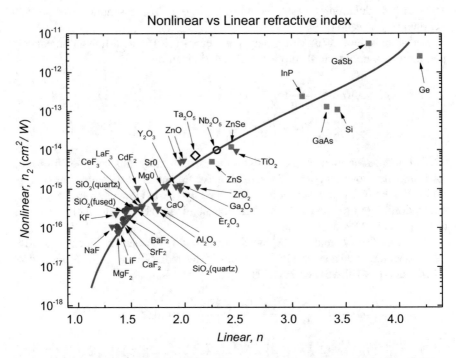

Fig. 18.3 Correlation between n_2 and n. Figure prepared by Abrar Fahim Liaf, a master student at Abbe School of Photonics, FSU Jena

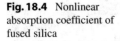

Fig. 18.4 Nonlinear absorption coefficient of fused silica

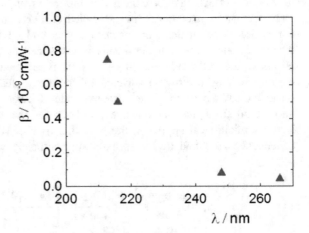

18.6 Advanced Material: Nonlinear Wave Equations

Prior to writing down nonlinear wave equations, let us return to (18.6) and (18.7). It turns out that in a second-order nonlinear material illuminated with a non-monochromatic input, electromagnetic waves at all possible sum and difference frequencies of the original input frequencies should be generated. And even more, the newly generated frequencies may again form new sum and difference frequencies and so on, such that one could expect that the energy of the incident light immediately spreads into an unmanageable multiplicity of waves with new frequencies. Of course, this is not observed in reality.

In order to understand the reason we have to recall an important point that we have already formulated in Sect. 13.4: Of course, knowledge of the nonlinear susceptibility is essential. But it is not only the nonlinear polarization that matters, but the propagation conditions for the waves generated by the nonlinear polarization term are important as well. In order to investigate those propagation conditions, the corresponding wave equations must be solved.

It turns out that only those conversion processes are efficient, which obey a so-called phase matching condition. Let us explain this condition turning to the example of SHG.

In a naive picture, the phase matching condition may be understood as a consequence of conservation of energy and momentum during a frequency conversion process. Let us illustrate this in the special case of an SHG process while neglecting any absorption losses. In the SHG process, energy conservation would require that the energy of two photons with angular frequency ω should be equal to the energy of a single photon with the doubled angular frequency:

$$2\omega = \omega + \omega$$

The same should be valid for the momentum. Then we have:

$$k(2\omega) = \frac{2\omega}{c}n(2\omega) = k(\omega) + k(\omega) = \frac{\omega}{c}n(\omega) + \frac{\omega}{c}n(\omega) = \frac{2\omega}{c}n(\omega)$$

From here we find the SHG phase matching condition:

$$n(2\omega) = n(\omega) \tag{18.19}$$

Essentially it requires that both the fundamental wave and its second harmonic propagate with the same phase velocity.

Because of normal dispersion, (18.19) is tricky to fulfill, and special precautions need to be met in order to achieve phase matching. The usual workaround is based on the application of optically anisotropic crystals with nonzero components of the quadratic susceptibility tensor, such that the ordinary refractive index at the input frequency equals the extraordinary one at the SHG frequency, or vice versa.

In this simple picture, violation of phase matching results in a violation of photon momentum conservation. In accordance to uncertainty relations like introduced in Chap. 3 that might be connected to a restricted geometrical path Δz where efficient SHG generation can occur. In fact we have a clear analogy to the earlier discussed properties of a wave packet [compare (3.15)]. According to (3.15), a vanishing momentum uncertainty requires $\Delta z \to \infty$, while a restricted Δz will surely result in a finite momentum uncertainty. Let us look at this effect in more detail and estimate the mentioned characteristic geometrical path.

We will now consider the process of interaction of an electromagnetic wave characterized by the angular frequency ω with the wave at the doubled frequency in a medium with second-order optical nonlinearity. For simplicity let us assume that both waves propagate along the z-axis.

Imagine now that (18.19) is violated. Assume further that at $z = 0$, an initial SHG signal has been generated that propagates with the wavevector $k\mathbf{k}(2\omega)$. At the same time, the fundamental wave propagates with the wavevector $\mathbf{k}(\omega)$. At each z, it generates an SHG signal proportional to:

$$P_{2\omega}^{(2)} \propto E^2 \propto e^{2ik(\omega)z}$$

If phase matching is violated, at a certain distance L, the newly generated SHG signal appears to be in exact antiphase to the initial SHG wave. Then, destructive interference in the SHG signal will occur, and the energy of the SHG signal will be transferred back to the wave with angular frequency ω. This antiphase condition will be accomplished when:

$$2k(\omega)L = k(2\omega)L \pm \pi$$

From here we obtain that characteristic length (called here the coherence length L_{coh}) by:

$$L_{\text{coh}} = \frac{\pi}{|2k(\omega) - k(2\omega)|} \equiv \frac{\pi}{|\Delta k|} = \frac{\lambda}{4|n(\omega) - n(2\omega)|} \tag{18.20}$$

Here, λ is the vacuum wavelength of the wave with angular frequency ω. Hence, the SHG process appears to be efficient only at a length scale defined by (18.20). After having traveled the geometrical path $2L_{\text{coh}}$, the SHG intensity is zero again, such that $2L_{\text{coh}}\Delta k = 2\pi$, which is in agreement to (3.15) [compare later (18.28)].

Let us now turn to a more quantitative treatment of the SHG conversion process. We start from the wave (13.2). From there we have:

$$\text{graddiv}\,\mathbf{E} - \Delta\mathbf{E} = -\mu_0 \frac{\partial^2 \mathbf{D}}{\partial t^2}$$

In contrast to our treatment in Chap. 13, however, we now have to consider non-linear terms in the dicplacement vector as according to (18.2). Therefore, instead of the linear wave (13.11) we now find its nonlinear counterpart according to (18.21):

$$\left.\begin{array}{l} \mathbf{D} = \varepsilon_0 \mathbf{E} + \mathbf{P}^{(1)} + \mathbf{P}^{(2)} = \varepsilon_0 \varepsilon \mathbf{E} + \mathbf{P}^{(2)} \\ \mathrm{div}\, \mathbf{E} = 0 \end{array}\right\} \Rightarrow \Delta \mathbf{E} - \frac{\varepsilon}{c^2}\frac{\partial^2 \mathbf{E}}{\partial t^2} = \mu_0 \frac{\partial^2 \mathbf{P}^{(2)}}{\partial t^2} \quad (18.21)$$

Again, we assume the nonlinear interaction of two waves only according to (18.22):

$$\left.\begin{array}{l} E(t) = \frac{1}{2}\sum_{j=1}^{2} E_{0j} e^{-i\omega_j t} + c.c. \\ P^{(2)}(t) = \frac{1}{2}\sum_{j=1}^{2} P_{0j}^{(2)} e^{-i\omega_j t} + c.c. \end{array}\right\} \Leftarrow \left\{\begin{array}{l} \omega_1 = \omega \\ \omega_2 = 2\omega \end{array}\right. \quad (18.22)$$

Hereby, the nonlinear interaction may turn the fundamental angular frequency ω to the doubled angular frequency by the SHG process, while the doubled frequency may be transferred back to the fundamental one by the DFG process. Assuming for simplicity again that both waves propagate along the z-axis, from (18.21), we find for the individual frequencies

$$\frac{\mathrm{d}^2}{\mathrm{d}z^2} E_{oj} + \underbrace{\varepsilon(\omega_j)\frac{\omega_j^2}{c^2}}_{k_j^2} E_{oj} = -\mu_0 \omega_j^2 P_{0j}^{(2)} \quad (18.23)$$

Let us introduce a somewhat simplified writing according to

$$E_{0j} = E_{0j}(z) \equiv A_j(z) e^{ik_j z}; \quad P_{0j}^{(2)} = P_{0j}^{(2)}(z) \equiv B_j(z) \quad (18.24)$$

When considering the two frequencies defined by (18.22) only, from (18.24) and (18.6) we have

$$B_1 = \varepsilon_0 \chi^{(2)}(\omega = 2\omega - \omega) A_2 A_1^* e^{i(k_2 - k_1)z}$$
$$B_2 = \frac{\varepsilon_0}{2}\chi^{(2)}(2\omega = \omega + \omega) A_1^2 e^{2ik_1 z} \quad (18.25)$$

Let us remember that only for the sake of simplicity, we have neglected the tensor character of the quadratic susceptibility, and used a scalar writing instead. In fact, for the corresponding tensor components relevant in (18.25), permutation symmetry requirements guarantee that both of the fixed susceptibility components are identical, such that we can skip the frequency arguments in our simplified writing (compare [7]).

Let us now notice that according to (18.24), the A_j terms represent z-dependent complex amplitudes of the corresponding waves. When now applying the approximation of slowly varying amplitudes, in (18.23) we will only consider the first derivatives of the A_j with respect to z, but will neglect the second ones. This leads to the equations for slowly varying amplitudes of both waves according to:

$$\frac{d^2}{dz^2} E_{oj} = \frac{d^2}{dz^2}\left[A_j(z)e^{ik_jz}\right] \approx \left(2ik_j\frac{d}{dz}A_j - k_j^2 A_j\right)e^{ik_jz}$$

$$\Rightarrow 2ik_j e^{ik_jz}\frac{d}{dz}A_j = -\mu_0\omega_j^2 B_j$$

$$\Rightarrow \left\{\begin{array}{l} \frac{d}{dz}A_1 = i\frac{\omega}{2cn_1}\chi^{(2)}A_2 A_1^* e^{i(k_2-2k_1)z} = i\frac{\omega}{2cn_1}\chi^{(2)}A_2 A_1^* e^{-i\Delta kz} \\ \frac{d}{dz}A_2 = i\frac{\omega}{2cn_2}\chi^{(2)}A_1^2 e^{-i(k_2-2k_1)z} = i\frac{\omega}{2cn_2}\chi^{(2)}A_1^2 e^{i\Delta kz} \end{array}\right\}$$

$$\Leftarrow \Delta k \equiv 2k_1 - k_2 \tag{18.26}$$

Equations (18.26) represent the final equations for slowly varying amplitudes of the two interacting waves. Let us discuss their solution for three special cases (Examples 1–3).

Example 1: Let us assume that the intensity of the SHG signal is much smaller than that of the fundamental frequency (low-conversion efficiency regime). Moreover, we assume that exact phase matching is guaranteed. We then set:

$$|A_2| << |A_1| \Rightarrow A_1 \approx \text{const.} = A_1(z=0) \equiv A_{10}; \ \Delta k = 0$$

From (18.26) we obtain:

$$\frac{d}{dz}A_2 = i\frac{\omega}{2cn_2}\chi^{(2)}A_{10}^2$$

Obviously, the term $\frac{2cn_2}{\omega\chi^{(2)}A_{10}}$ must have length dimension. Correspondingly, we define the so-called SHG nonlinear length as:

$$L_{nl} \equiv \left|\frac{2cn_2}{\omega\chi^{(2)}A_{10}}\right| \tag{18.27}$$

When setting $A_2(z=0) \equiv A_{20} = 0$, we immediately find:

$$\Rightarrow |A_2|^2 = |A_{10}|^2\left(\frac{z}{L_{nl}}\right)^2$$

This equation is approximately valid at distances $z \ll L_{nl}$, but clearly violates energy conservation at larger distances.

Example 2: Same as example 1, but with phase matching violated. We set:

$$|A_2| \ll |A_1| \Rightarrow A_1 \approx A_{10}; \ \Delta k \neq 0; \ A_{20} = 0$$

$$\frac{d}{dz} A_2 \propto e^{i\Delta kz} \Rightarrow A_2(z) = \int_0^z e^{i\Delta k\xi} d\xi \propto \frac{e^{i\Delta kz} - 1}{\Delta k}$$

$$\Rightarrow |A_2(z)|^2 \propto (1 - \cos \Delta kz) \tag{18.28}$$

We thus obtain oscillating behavior of the SHG amplitude. The first maximum occurs at:

$$|A_2(z)|^2 = \max \Rightarrow \Delta kz = \pm\pi \Rightarrow z = \frac{\pi}{|\Delta k|} = \frac{\lambda}{4|n_1 - n_2|} \equiv L_{coh}$$

We find that at a distance $z = L_{coh}$, which coincides with the coherence length earlier introduced in (18.20), the SHG signal intensity reaches its maximum and then drops down again. At $z = 2L_{coh}$, according to (8.28), the SHG intensity is zero again, and the process starts from a new. The reason has already been identified: Destructive interference leads to a suppression of the SHG signal, and accordingly, the energy is pumped back into the fundamental wave.

Example 3: Let us now suppose phase matching as well as energy conservation. We set:

$$\Delta k = 0; \quad A_{20} = 0; \quad |A_1|^2 + |A_2|^2 = |A_{10}|^2 \neq 0$$

We then obtain:

$$\left.\begin{array}{l} \dfrac{d}{dz} A_1 = i\dfrac{\omega}{2cn} \chi^{(2)} A_2 A_1^* \\[2mm] \dfrac{d}{dz} A_2 = i\dfrac{\omega}{2cn} \chi^{(2)} A_1^2 \end{array}\right\} \Rightarrow |A_1| = \frac{|A_{10}|}{\cosh \frac{z}{L_{nl}}}; \quad |A_2| = |A_{10}| \tanh \frac{z}{L_{nl}}$$

$$L_{nl} = \frac{2nc}{|\omega\chi^{(2)} A_{10}|}$$

Let us estimate the nonlinear length by setting:

$$n = 2, \lambda = 1000\,\text{nm}; \quad A_{10} = 10^7 \frac{V}{m}; \quad \chi^{(2)} = 10^{-11} \frac{m}{V}$$

$$\Rightarrow L_{nl} = \frac{2nc}{|\omega\chi^{(2)} A_{10}|} = \frac{n\lambda}{|\pi\chi^{(2)} A_{10}|} \approx 0.63\,\text{cm}$$

Therefore, characteristic dimensions of frequency conversion nonlinear crystals are in the cm range.

Figure 18.5 visualizes numerical solutions of (18.26) with and without phase matching (solid lines) as well as with inclusion of weak linear absorption (dot). The Python calculations have been performed by Jian Ying He, a master student of

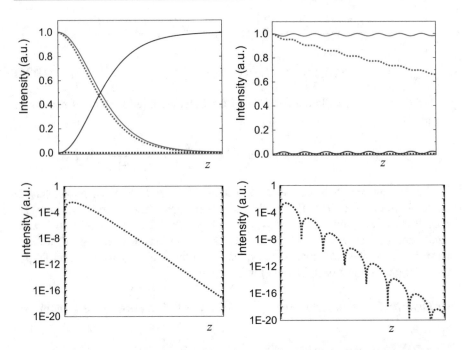

Fig. 18.5 Numerical solution of (18.26). Red lines correspond to the wave at the fundamental frequency, navy lines to the SHG signal. Solid lines = no linear absorption; dot = linear absorption included. Left side: in phase matching conditions; right side: phase matching violated. Note that the absorption coefficient for the SHG wave has been assumed to be 100 times larger than that of the fundamental wave

photonics at Abbe School of Photonics, FSU Jena. For efficient SHG generation, obviously, the nonlinear length should be as small as possible, and the coherence length as large as possible. The oscillatory behavior of the intensities in the absence of phase matching is easily recognized and confirms our qualitative discussion at the beginning of this section.

18.7   Tasks for Self-check

18.7.1 Multiple-choice test: Mark all answers which seem you correct!

A diamond crystal	Is rather transparent in the visible spectral range
	Shows a pronounced Raman line
	Has inversion symmetry
	Has two atoms in its primitive cell

	Has a cubic optical nonlinearity different from zero
The third harmonic of the Nd: YAG laser line ($\lambda \approx 1064$ nm) corresponds to a wavelength of	≈ 3192 nm
	≈ 355 nm
	≈ 0.355 μm
Third-order nonlinear optical susceptibilities	Are dimensionless in SI units
	Are observed in gases only
	May describe two-photon absorption processes
Nonlinear optical effects may be observed	At high light intensities
	Only in the middle infrared
	At interfaces
In electric dipole approximation, centrosymmetric materials	Obey the parity selection rule
	Do not show even-order nonlinearities
	Are always optically anisotropic

18.7.2 True or wrong? Make your decision!

Assertion	True	Wrong
Several crystal classes may show second-order nonlinear optical effects		
SHG is particularly efficient in centrosymmetric materials		
SHG is never observed in optically isotropic materials		
The nonlinear refractive index is dimensionless		

18.7.3 Calculate the electric field strength at a distance of a_0 from a single proton. Compare this field strength with the electric field strength in an electromagnetic wave with the Intensity $I = 0.15$ W cm^{-2}. Repeat the comparison for the case of intense laser light with $I = 10^{12}$ W cm^{-2}.

18.7.4 Return to our perturbative treatment of quantum transitions from Sect. 6.3.2. In the expression for the expansion coefficient (6.21) we identify the term $\frac{d_{ml}E_0}{\hbar}$, which obviously has angular frequency measurement units. The so-called Rabi frequency $\Omega = \left| \frac{d_{ml}E_{0,real}}{\hbar} \right|$ as well as its magnetic counterpart $\left| \frac{d_{magn,ml}B_{0,real}}{\hbar} \right|$ play an important role in coherent laser spectroscopy. In the conditions of the previous task, estimate the corresponding Rabi frequencies and compare them to typical electronic transition frequencies! For estimation, set $d_{ml} = 1e*a_0$.

18.7.5 Derive (18.18a)!

References

Specific References

1. O. Stenzel, *The Physics of Thin Film Optical Spectra. An Interoduction* (Springer, 2016) (chapter 13)
2. H. Warlimont, W. Martienssen (eds.), *Springer Handbook of Materials Data* (Springer, 2018)
3. R. del Coso, J. Solis, Relation between nonlinear refractive index and third-order susceptibility in absorbing media. J. Opt. Soc. Am. B **21**, 640–644 (2004)
4. O. Stenzel, Simplified expression for estimating the nonlinear refractive index of typical optical coating materials. Appl. Opt. **56**, C21–C23 (2017)
5. T.R. Ensley, N.K. Bambha, Ultrafast nonlinear refraction measurements of infrared transmitting materials in the mid-wave infrared. Opt. Express **27**, 37940 (2019)
6. K. Werner et al., Ultrafast mid-infrared high harmonic and supercontinuum generation with n_2 characterization in zinc selenide. Opt. Express **27**, 2867 (2019)
7. Y.R. Shen, *The Principles of Nonlinear Optics* (Wiley, 1984), chapter 2

General Literature

8. M. Schubert, B. Wilhelmi, Einführung in die nichtlineare Optik I und II; BSB B. G. Teubner Verlagsgesellschaft Leipzig (1971) (engl.: Introduction in Nonlinear Optics I and II)
9. N. Bloembergen, *Nonlinear Optics* (Addison-Wesley Publishing Company, Inc., 1992)

Solutions to Tasks

19

"Es irrt der Mensch, solang er strebt." (engl.: "Man fails as long as he strives")

Johann Wolfgang von Goethe: Faust: The first part of the tragedy, 1808 (Prologue in Heaven)

"Pukará de Quitor" (**Atacama desert, Chile**)

Painting and Photo by Astrid Leiterer, Jena, Germany (www.astrid-art.de). Photo reproduced with permission.

If you, dear reader, are a student, then it is highly recommended that you try to solve the tasks independently before consulting this chapter. Note that usually several

© The Author(s), under exclusive license to Springer Nature Switzerland AG 2022
O. Stenzel, *Light–Matter Interaction*, UNITEXT for Physics,
https://doi.org/10.1007/978-3-030-87144-4_19

approaches are possible and lead to the same result. The approaches presented in this chapter are solution proposals.

Chapter 1

1.5.1 measurement units

Expression	Solution
$1\frac{VAs}{m^2 Pa} =?$	1 m
$1\frac{kgm^2}{s^2 V^2} =?$	1 F
$1\frac{JF}{As} =?$	1 C
$1\frac{A^2 s^5}{m^2 kg F^2} =?$	1 Ω
$1\frac{VAs^2}{JF} =?$	1 Ω
$1 s^2\, mPa =?$	1 kg
$1\frac{VAs^3}{m^2} =?$	1 kg
$1\frac{Ws^2}{\Omega^2 FA^2} =?$	1 s

Information	Task	Your answer
$\int_V |\psi|^2 dV = 1$ with V—volume	$[dV] =?$	m^3
	$[\psi] =?$	$m^{-3/2}$
$\psi = \sqrt{\frac{2}{L}} \sin kx$ with $[L] = m$ and $[x] = m$	$[\psi] =?$	$m^{-1/2}$
	$[k] =?$	m^{-1}
$I = I_0 e^{-\alpha x}$ with $[I] = Wm^{-2}$ and $[x] = m$	$[I_0] =?$	Wm^{-2}
	$[\alpha] =?$	m^{-1}
$\psi = \frac{1}{\sqrt{\pi}}\left(\frac{Z}{a_0}\right)^{\frac{3}{2}} e^{-Zr/a_0}$ with $[Z] = 1$ and $[r] = m$	$[a_0] =?$	m
	$[\psi] =?$	$m^{-3/2}$
$Re\varepsilon(\omega) = 1 + \frac{1}{\pi} V P \int_{-\infty}^{\infty} \frac{Im\varepsilon(\xi)d\xi}{\xi-\omega}$ with $[Im\varepsilon] = 1$ and $[\xi] = s^{-1}$	$[\omega] =?$	s^{-1}
	$[Re\varepsilon(\omega)] =?$	Unitless

1.5.2

- $\sqrt{\varepsilon_0 \mu_0} = \frac{1}{c}$
- $\frac{e^2}{4\pi \varepsilon_0 \hbar c} = 0.053$ nm (Bohrs radius)
- $\sqrt{\frac{\hbar G}{c^3}} = 1.6 \times 10^{-35}$ m (the Planck length)
- $\frac{2e}{h} \approx 4.836 \times 10^{14} \frac{Hz}{V}$ (the Josephson constant)
- $\frac{h}{e^2} \approx 25812.8\ \Omega$ (the von Klitzing constant).

1.5.3 straightforward from $f(x) = \sum_{n=0}^{\infty} \frac{1}{n!}\left[\frac{d^n}{dx^n} f(x)\right]_{x=0} x^n$, consult textbooks on mathematics if necessary

1.5.4 straightforward, consult textbooks on mathematics if necessary.

1.5.5 Assuming a density of 1 kg/l, one liter of water contains approximately 3.32×10^{25} water molecules. Then, per square centimeter, $\approx 6.5 \times 10^6$ molecules will be observed.

1.5.6

$$\left.\begin{array}{l} \rho = Nm_{\text{atom}} \approx 4\,Nm_p \\[6pt] N = \dfrac{\text{number of atoms}}{\text{Volume}} = \dfrac{1}{\text{volume available per atom}} \end{array}\right\}$$

$$\Rightarrow \text{volume available per atom} \approx \frac{4m_p}{\rho}$$

$$\text{space filling} = 0.74 \Rightarrow V_{\text{atom}} \approx 0.74\frac{4m_p}{\rho} \approx \frac{4\pi}{3}r^3$$

$$\Rightarrow r \approx \sqrt[3]{0.74\frac{3m_p}{\pi\rho}} \approx 0.21\,\text{nm}$$

1.5.7 see the following table. Note that although the momentum of the tectonic plate is huge because of its mass, its kinetic energy is marginal. The reason is in the quadratic dependence of the kinetic energy on the velocity.

Object	m/kg	v/(m/s)	p/(kgm/s)	T_{kin}/(Ws)
Earth	6×10^{24}	3×10^4	1.8×10^{29}	2.7×10^{33}
Tectonic plate	3×10^{20}	3.2×10^{-10}	9.6×10^{10}	15.4
Train	10^6	27.8	2.8×10^7	3.9×10^8
Runner	80	5.6	448	1254
Electron	9.1×10^{-31}	3×10^7	2.73×10^{-23}	4.1×10^{-16}

Chapter 2
2.7.1 Multiple-choice test

The light wavelength 1064 nm belongs to the	Ultraviolet spectral range	
	Visible spectral range	
	Infrared spectral range	x
The wavelength of 500 nm belongs to the	Microwave spectral range	
	Middle infrared spectral range	
	X-ray spectral range	
Vibrations of nuclei in molecules and solids cause light absorption primarily in the	Ultraviolet spectral range	
	Visible spectral range	
	Infrared spectral range	x
The emission line of an assembly of resting light emitters appears to be naturally broadened. Its spectral shape may therefore be described by a	Lorentzian lineshape	x
	Gaussian lineshape	
	Triangular lineshape	

A Doppler-broadened emission line as emitted from a hot gas has a	Gaussian lineshape	x
	Lorentzian Lineshape	
	Rectangular lineshape	
The mass moment of inertia of a diatomic molecule is typically of the order	10^{-27} kgm^2	
	10^{-47} kgm^2	x
	10^{-67} kgm^2	

2.7.2 True or wrong?

Assertion	True	Wrong
The reduced mass of a system of two bodies is always in-between the masses of the individual two bodies		x
Molecular fingerprint spectra are typically recorded in the middle infrared spectral range	x	

2.7.3 $r_0 = \sqrt[6]{2}\sigma$: For a graph see Fig. 10.7

2.7.4 $M = m_{Cl-atom} + m_{H-atom} \approx m_{Cl-atom}$; $\mu = \frac{m_{Cl-atom} m_{H-atom}}{m_{Cl-atom} + m_{H-atom}} \approx m_{H-atom}$; $I \approx 2.7 \times 10^{-47}$ kg m^2

2.7.5 $E_{phot, min} = 4$ eV

2.7.6 $E_{0,real} = \sqrt{2\mu_0 c I} = 2743\sqrt{\frac{I cm^2}{W} \frac{V}{m}}$; or $E_{0,real}$ (in V/m) $= 2743\sqrt{I \, (\text{in } W/cm^2)}$

Chapter 3

3.7.1 $r_e \approx 2.82 \times 10^{-15}$ m

3.7.2

(a) $\lambda = 0.62$ pm

(b) For an e^-/e^+: $mc^2 = 0.511$ MeV. Then

$$E = \frac{mc^2}{\sqrt{1 - \frac{v^2}{c^2}}} = 1 \, \text{MeV} = \frac{0.511 \, \text{MeV}}{\sqrt{1 - \frac{v^2}{c^2}}} \Rightarrow \frac{v}{c} = \sqrt{1 - 0.511^2} = 0.86$$

Note that momentum conservation in such a conversion process will not be satisfied if the process is to occur in empty space. The presence of a heavy nucleus is indeed necessary for momentum conservation. However, its impact (kinetic recoil energy) on the energy balance is in most cases negligible. Therefore, the task may be solved in terms of the energy balance only.

3.7.3 $\lambda_2 - \lambda_1 = \frac{h}{m_e c}(1 - \cos\theta) \equiv \Lambda(1 - \cos\theta) \rightarrow$ max? $\Rightarrow \theta = 180°$; $(\lambda_2 - \lambda_1)_{max} = 2\Lambda = 4.8$ pm

Chapter 4

4.7.1 Direct calculation yields: $v_{ph} = \frac{\omega}{k} = \frac{E}{p} = \frac{v}{2} + \frac{U_0}{p}$; $v_{gr} = \frac{\mathrm{d}\omega}{\mathrm{d}k} = \frac{\mathrm{d}E}{\mathrm{d}p} = \frac{\hbar k}{m} = \frac{p}{m} = v$

4.7.2 Use (4.12) and assume that each of those states may be occupied by two electrons (with opposite spin directions). Together with (4.10):

$\frac{h^2}{8mL^2}n^2 = E_n$ we find: $Z = g\frac{\pi}{6}n^3 = \frac{g}{6}\frac{L^3}{\pi^2}\left(\frac{2mE}{\hbar^2}\right)^{\frac{3}{2}} \approx 36$

4.7.3 KBr. For a deeper understanding, compare later with task 6.7.2, too.

4.7.4 For $E < 0$, from Table 4.1 we write the wavefunction of the particle as:

$$\psi(x) \propto \begin{cases} e^{\beta x}; & x < 0 \\ e^{-\beta x}; & x \geq 0 \end{cases}$$

With

$$\beta = \frac{\sqrt{-2mE}}{\hbar}$$

The Schrödinger equation is:

$$-\frac{\hbar^2}{2m}\frac{\mathrm{d}^2}{\mathrm{d}x^2}\psi(x) - [A\delta(x) + E]\psi(x) = 0$$

Integrating in the ε-environment of $x = 0$ yields:

$$-\frac{\hbar^2}{2m}\left[\frac{\mathrm{d}}{\mathrm{d}x}\psi(x)\Big|_{x=\varepsilon} - \frac{\mathrm{d}}{\mathrm{d}x}\psi(x)\Big|_{x=-\varepsilon}\right] - A\psi(0)$$

$$= 0 \Rightarrow \left[\frac{\mathrm{d}}{\mathrm{d}x}\psi(x)\Big|_{x=\varepsilon} - \frac{\mathrm{d}}{\mathrm{d}x}\psi(x)\Big|_{x=-\varepsilon}\right] = -\frac{2mA}{\hbar^2}\psi(0)$$

Obviously, the first derivative of the wavefunction must have a discontinuity at $x = 0$ in order to compensate the effect of the delta function in the Schrödinger equation. Substituting $\psi(x)$, we have: $\left[2\beta e^{-\beta\varepsilon}\right] = \frac{2mA}{\hbar^2}$. If $\varepsilon \to 0$, that yields $\beta = \frac{mA}{\hbar^2} = \frac{\sqrt{-2mE}}{\hbar} \Rightarrow E = -\frac{mA^2}{2\hbar^2}$.

The obvious conclusion is that the delta potential has only one stationary bound state with $E = -\frac{mA^2}{2\hbar^2}$.

What about the constant A? From $\int_{-\infty}^{\infty}\delta(x)\mathrm{d}x = 1 \Rightarrow \int_{-\infty}^{\infty}U(x)\mathrm{d}x = -A$. When interpreting $U(x)$ as the limit of a rectangular potential box with height $-U_0$ and width L, while requiring $U_0 \to +\infty$ and $L \to 0$ with $LU_0 = \text{const.}$, then it becomes clear that $A = LU_0$.

Chapter 5

5.7.1

$$\left[\hat{A}\hat{B}, \hat{C}\right] = \hat{A}\hat{B}\hat{C} - \hat{C}\hat{A}\hat{B} = \hat{A}\hat{B}\hat{C} - \hat{A}\hat{C}\hat{B} + \hat{A}\hat{C}\hat{B} - \hat{C}\hat{A}\hat{B}$$

$$= \hat{A}\left(\hat{B}\hat{C} - \hat{C}\hat{B}\right) + \left(\hat{A}\hat{C} - \hat{C}\hat{A}\right)\hat{B} = \hat{A}\left[\hat{B},\hat{C}\right] + \left[\hat{A},\hat{C}\right]\hat{B}$$

5.7.2

$$[\hat{x}, \hat{p}_x^2] = 2i\hbar\hat{p}_x = 2\hbar^2 \frac{\partial}{\partial x}$$

(a) direct calculation using $\hat{p}_x = -i\hbar\frac{\partial}{\partial x}$
$$[\hat{x}, \hat{p}_x^2] = \hat{x}\hat{p}_x^2 - \hat{p}_x^2\hat{x} = \hat{x}\hat{p}_x^2 - \hat{p}_x^2\hat{x} + \hat{p}_x\hat{x}\hat{p}_x - \hat{p}_x\hat{x}\hat{p}_x$$
(b)
$$= \hat{p}_x\hat{x}\hat{p}_x - \hat{p}_x^2\hat{x} + \hat{x}\hat{p}_x^2 - \hat{p}_x\hat{x}\hat{p}_x$$
$$= \hat{p}_x\left(\hat{x}\hat{p}_x - \hat{p}_x\hat{x}\right) + \left(\hat{x}\hat{p}_x - \hat{p}_x\hat{x}\right)\hat{p}_x$$
$$= \hat{p}_x(i\hbar) + (i\hbar)\hat{p}_x = 2i\hbar\hat{p}_x$$
(c) $[\hat{x}, \hat{p}_x^2] = \hat{x}\hat{p}_x^2 - \hat{p}_x^2\hat{x} = \left(i\hbar + \hat{p}_x\hat{x}\right)\hat{p}_x - \hat{p}_x\left(\hat{x}\hat{p}_x - i\hbar\right) = 2i\hbar\hat{p}_x$

5.7.3 all the operators are self-adjoint
 5.7.4

$$I = \frac{2}{L}\int_0^L x\sin^2\frac{n\pi}{L}x\,\mathrm{d}x$$

$$\xi = x - \frac{L}{2} \Rightarrow I = \frac{2}{L}\int_{-\frac{L}{2}}^{\frac{L}{2}} \overbrace{\xi}^{\text{odd}}\underbrace{\sin^2\left[\frac{n\pi}{L}\xi + \frac{n\pi}{2}\right]}_{\text{even}}\mathrm{d}\xi$$

$$\underbrace{\phantom{\frac{2}{L}\int_{-\frac{L}{2}}^{\frac{L}{2}}}}_{=0}$$

$$+ \int_{-\frac{L}{2}}^{\frac{L}{2}}\sin^2\left[\frac{n\pi}{L}\xi + \frac{n\pi}{2}\right]\mathrm{d}\xi = \int_{-\frac{L}{2}}^{\frac{L}{2}}\sin^2\left[\frac{n\pi}{L}\xi + \frac{n\pi}{2}\right]\mathrm{d}\xi$$

$$\zeta = \frac{n\pi}{L}\xi + \frac{n\pi}{2} \Rightarrow \mathrm{d}\xi = \frac{L}{n\pi}\mathrm{d}\zeta; \ \xi = \begin{cases} -\frac{L}{2} \Rightarrow \zeta = 0 \\ +\frac{L}{2} \Rightarrow \zeta = n\pi \end{cases}$$

$$I = \frac{L}{n\pi}\underbrace{\int_0^{n\pi}\sin^2\zeta\,\mathrm{d}\zeta}_{=\frac{n\pi}{2}} = \frac{L}{2}$$

Or alternatively:

$$I = \frac{2}{L}\int_0^L x\sin^2\frac{n\pi}{L}x\,\mathrm{d}x$$

$$\xi = \frac{n\pi}{L}x \Rightarrow I = \frac{2L}{(n\pi)^2}\int_0^{n\pi}\xi\sin^2\xi\,\mathrm{d}\xi$$

$$\zeta = \xi - \frac{n\pi}{2} \Rightarrow I = \frac{2L}{(n\pi)^2}\int_{-\frac{n\pi}{2}}^{\frac{n\pi}{2}}\overbrace{\zeta}^{\text{odd}}\underbrace{\sin^2\left(\zeta + \frac{n\pi}{2}\right)}_{\text{even}}\mathrm{d}\zeta$$

$$\underbrace{\phantom{\frac{2L}{(n\pi)^2}\int_{-\frac{n\pi}{2}}^{\frac{n\pi}{2}}}}_{=0}$$

$$+ \frac{L}{n\pi} \underbrace{\int_{-\frac{n\pi}{2}}^{\frac{n\pi}{2}} \sin^2\left(\zeta + \frac{n\pi}{2}\right) d\zeta}_{=\frac{n\pi}{2}} = \frac{L}{2}$$

5.7.5 $\mathbf{j} = \frac{i\hbar}{2m}(\Psi\nabla\Psi^* - \Psi^*\nabla\Psi)$

Once the ψ are real, $\mathbf{j}$ must be zero (compare task 5.7.6)

5.7.6 $\hat{p}-$ self-adjoint $\rightarrow$ so $\langle p \rangle$ must be real. Moreover, $\Psi(\mathbf{r}, t) = e^{-i\frac{E_n}{\hbar}t}\psi(\mathbf{r})$ with $\Psi(\mathbf{r})$—real, Therefore, $\Psi^*\nabla\Psi = \psi^*\nabla\psi = \psi\nabla\psi$, and the straightforward calculation yields an imaginary result: $\langle p \rangle = -i\hbar \int \Psi^*\nabla\Psi dV = -i\hbar \int \psi\nabla\psi dV$. Therefore, $\langle p \rangle$ should be zero.

Alternative approaches:

$$\langle p \rangle = -i\hbar \int \Psi^*\nabla\Psi dV = -i\hbar \int \psi\nabla\psi dV$$

$$= -\frac{i\hbar}{2} \int \nabla\psi^2 dV = -\frac{i\hbar}{2} \psi^2\Big|_{-\infty}^{\infty} = 0$$

$$\langle p \rangle = -i\hbar \int \Psi^*\nabla\Psi dV = -i\hbar \int \psi\nabla\psi dV$$

$$= -i\hbar \underbrace{\psi^2\Big|_{-\infty}^{\infty}}_{=0} + i\hbar \int \psi\nabla\psi dV = 0$$

$$\mathbf{j} = \frac{i\hbar}{2m}(\Psi\nabla\Psi^* - \Psi^*\nabla\Psi) = \frac{i\hbar}{2m}(\psi\nabla\psi^* - \psi^*\nabla\psi)$$

$$= \frac{i\hbar}{2m}(\psi\nabla\psi - \psi\nabla\psi) = 0$$

5.7.7

$$\langle p_x \rangle_n = 0 \text{ (see task 5.7.6)}$$

$$\langle p_x^2 \rangle_n = \langle 2mT_{\text{kin}} \rangle_n = 2 \, \text{mE}_n = \left(\frac{\hbar\pi n}{L}\right)^2$$

$$\langle x \rangle_n = \frac{L}{2} \text{ (see task 5.7.4}$$

$$\langle x^2 \rangle_n :$$

The basic integral is of type: $\int \xi^2 \sin^2 \xi d\xi$. Twofold application of partial integrating results in:

$\int \xi^2 \sin^2 \xi d\xi = \frac{\xi^3}{6} - \frac{1}{4}\left[(\xi^2 - 1)\sin 2\xi + \xi \cos 2\xi\right]$. From here:

$$\langle x^2 \rangle_n = L^2\left[\frac{1}{3} - \frac{1}{2(n\pi)^2}\right]$$

Therefore, $\langle (\Delta x)^2 \rangle_n = \langle x^2 \rangle_n - \langle x \rangle_n^2 = L^2\left[\frac{1}{3} - \frac{1}{2(n\pi)^2}\right] - \frac{L^2}{4} = \frac{L^2}{12}\left[1 - \frac{6}{(n\pi)^2}\right]$.

and $\langle (\Delta x)^2 \rangle_n \langle (\Delta p_x)^2 \rangle_n = \frac{\hbar^2}{12}\left[(n\pi)^2 - 6\right] = \frac{\hbar^2}{4}\left[\frac{(n\pi)^2}{3} - 2\right] > \frac{\hbar^2}{4} \forall n > 0$.

5.7.8

$$\widehat{H} = \left[\frac{\hat{p}^2}{2m} + \frac{m\omega_0^2}{2}x^2\right] \Rightarrow \langle E \rangle = \left[\frac{\langle p^2 \rangle}{2m} + \frac{m\omega_0^2}{2}\langle x^2 \rangle\right]$$

From

$$\langle x^2 \rangle = \langle x \rangle^2 + (\Delta x)^2; \langle p^2 \rangle = \langle p \rangle^2 + (\Delta p)^2$$

It follows:

$$\langle E \rangle = \left[\frac{\langle p^2 \rangle}{2m} + \frac{m\omega_0^2}{2}\langle x^2 \rangle\right] \geq \left[\frac{(\Delta p)^2}{2m} + \frac{m\omega_0^2}{2}(\Delta x)^2\right]$$

$$\geq \frac{(\Delta p)^2}{2m} + \frac{m\omega_0^2\hbar^2}{8(\Delta p)^2}$$

The term on right achieves its minimum at $(\Delta p)^2 = \frac{m\omega_0\hbar}{2}$. Substituting this value, we get

$$\langle E \rangle = \left[\frac{\langle p^2 \rangle}{2m} + \frac{m\omega_0^2}{2}\langle x^2 \rangle\right] \geq \left[\frac{(\Delta p)^2}{2m} + \frac{m\omega_0^2}{2}(\Delta x)^2\right]$$

$$\geq \frac{(\Delta p)^2}{2m} + \frac{m\omega_0^2\hbar^2}{8(\Delta p)^2} \geq \frac{\omega_0\hbar}{2}$$

5.7.9. Basically, $\langle T_{kin} \rangle = \frac{\langle p^2 \rangle}{2m}$
setting $\langle p^2 \rangle - \langle p \rangle^2 = \langle (\Delta p)^2 \rangle$.

we have the estimation: $\langle T_{kin} \rangle \geq \frac{\langle (\Delta p)^2 \rangle}{2m}$.

But T_{kin} should not be much larger than $\langle (\Delta p)^2 \rangle/2 m$; because in a physically meaningful state, the energy is minimized. Thus, for estimation we set: $\langle T_{kin} \rangle \approx \frac{\langle (\Delta p)^2 \rangle}{2m}$.

In order to decide whether relativistic or nonrelativistic calculus is relevant, the kinetic energy should be compared with the electron energy at rest!

$$E = mc^2 \approx 0.5 MeV$$

Nonrelativistic approach possible if $T_{kin} \ll mc^2 \approx 0.5\,\text{MeV}$ is fulfilled (compare Sect. 3.3.2).

Solution: Uncertainty: $(\Delta p)^2 (\Delta x)^2 \geq \frac{\hbar^2}{4}$.

We set for estimation:

$$(\Delta p)^2 (\Delta x)^2 \approx \hbar^2$$

Nonrelativistic approach: $T_{kin} = \frac{p^2}{2m} \rightarrow \frac{(\Delta p)^2}{2m} \approx \frac{\hbar^2}{2mx^2} \approx T_{kin}$.

1. $x = 10^{-10}\,\text{m} \rightarrow T_{kin} \approx \frac{\hbar^2}{2mx^2} \approx \frac{1}{2} 10^{-18}\,Ws \approx 3\,\text{eV} \ll mc^2$ (nonrelativistic approach consistent)

2. $x = 10^{-15}\,m \rightarrow T_{kin} \approx \frac{\hbar^2}{2mx^2} \approx \frac{1}{2} 10^{-8}\,Ws \approx 3 \times 10^{10}\,\text{eV} \gg mc^2$ (nonrelativistic approach inconsistent, relativistic calculus needs to be applied!).

Relativistic approach: $E = \sqrt{p^2 c^2 + m^2 c^4} \rightarrow$

$$E = mc^2 \sqrt{1 + \frac{p^2}{m^2 c^2}} \approx mc^2 \sqrt{1 + \frac{\hbar^2}{m^2 c^2 x^2}}$$

$$x = 10^{-15}\,m \rightarrow \frac{\hbar}{mcx} \approx 300 \gg 1 \Rightarrow E \approx 300\,mc^2 \approx 150\,\text{MeV}$$

5.7.10.

$$\left[\left(\widehat{A}\widehat{B} \right)^+ \right]_{nm} = \left(\widehat{A}\widehat{B} \right)^*_{mn}$$

$$\left(\widehat{B}^+ \widehat{A}^+ \right)_{nm} = \sum_l B^+_{nl} A^+_{lm} = \sum_l B^*_{ln} A^*_{ml} = \sum_l A^*_{ml} B^*_{ln} = \left(\widehat{A}\widehat{B} \right)^*_{mn}$$

$$\Rightarrow \left[\left(\widehat{A}\widehat{B} \right)^+ \right]_{nm} = \left(\widehat{B}^+ \widehat{A}^+ \right)_{nm} \ \forall \, n, m \Rightarrow \left(\widehat{A}\widehat{B} \right)^+ = \widehat{B}^+ \widehat{A}^+$$

5.7.11 from (5.35), we have

$$H_n(\xi) = (-1)^n e^{\xi^2} \frac{d^n e^{-\xi^2}}{d\xi^n} \Rightarrow \frac{d}{d\xi} H_n(\xi)$$

$$= (-1)^n \left[(2\xi) e^{\xi^2} \frac{d^n e^{-\xi^2}}{d\xi^n} + e^{\xi^2} \frac{d^{n+1} e^{-\xi^2}}{d\xi^{n+1}} \right]$$

$$= 2\xi H_n(\xi) - H_{n+1}(\xi) \Rightarrow H_{n+1}(\xi)$$

$$= \left(2\xi - \frac{d}{d\xi} \right) H_n(\xi)$$

Odd n correspond to odd functions, even n to even functions (compare Table 5.1).

5.7.12

$$\langle x^2 \rangle_n = \left(x^2 \right)_{nn} = \frac{\hbar}{2m\omega_0}(2n+1) = \frac{\hbar}{m\omega_0}\left(n + \frac{1}{2} \right)$$

$$U(x) = \frac{m\omega_0^2}{2}x^2 \Rightarrow \langle U \rangle_n = \frac{m\omega_0^2}{2}\langle x^2 \rangle_n = \frac{\hbar\omega_0}{2}\left(n + \frac{1}{2} \right) = \frac{1}{2}E_n$$

Alternatively: from (5.20e) and $U \propto x^2$

$$\langle U \rangle = \langle T_{kin} \rangle \Rightarrow \langle E \rangle = \langle U \rangle + \langle T_{kin} \rangle = 2\langle U \rangle$$

$$\Rightarrow \langle U \rangle_n = \frac{1}{2}\langle E \rangle_n = \frac{1}{2}E_n = \frac{\hbar\omega_0}{2}\left(n + \frac{1}{2} \right)$$

5.7.13 $f_{n+1,n} = n+1, f_{n-1,n} = -n$, all other oscillator strength are 0
5.7.14 Directly from (5.41), we have:

$$\left(\hat{a} \right)_{nm} = \frac{1}{\sqrt{2}}\left[\frac{x_{nm}}{x_0} + i\frac{x_0}{\hbar}p_{nm} \right]; \quad \left(\hat{a}^+ \right)_{nm} = \frac{1}{\sqrt{2}}\left[\frac{x_{nm}}{x_0} - i\frac{x_0}{\hbar}p_{nm} \right]$$

When using explicit expressions of the harmonic oscillator matrix elements, we obtain the matrix writing

$$\hat{a} = \begin{pmatrix} 0 & \sqrt{1} & 0 & 0 & \dots \\ 0 & 0 & \sqrt{2} & 0 & \dots \\ 0 & 0 & 0 & \sqrt{3} & \dots \\ 0 & 0 & 0 & 0 & \dots \\ \dots & \dots & \dots & \dots \end{pmatrix}; \quad \hat{a}^+ = \begin{pmatrix} 0 & 0 & 0 & 0 & \dots \\ \sqrt{1} & 0 & 0 & 0 & \dots \\ 0 & \sqrt{2} & 0 & 0 & \dots \\ 0 & 0 & \sqrt{3} & 0 & \dots \\ \dots & \dots & \dots & \dots \end{pmatrix}$$

$$\hat{a}\hat{a}^+ = \begin{pmatrix} 1 & 0 & 0 & 0 & \dots \\ 0 & 2 & 0 & 0 & \dots \\ 0 & 0 & 3 & 0 & \dots \\ 0 & 0 & 0 & 4 & \dots \\ \dots & \dots & \dots & \dots \end{pmatrix}; \quad \hat{a}^+\hat{a} = \begin{pmatrix} 0 & 0 & 0 & 0 & \dots \\ 0 & 1 & 0 & 0 & \dots \\ 0 & 0 & 2 & 0 & \dots \\ 0 & 0 & 0 & 3 & \dots \\ \dots & \dots & \dots & \dots \end{pmatrix};$$

$$[\hat{a}, \hat{a}^+] = \begin{pmatrix} 1 & 0 & 0 & 0 & \dots \\ 0 & 1 & 0 & 0 & \dots \\ 0 & 0 & 1 & 0 & \dots \\ 0 & 0 & 0 & 1 & \dots \\ \dots & \dots & \dots & \dots \end{pmatrix}$$

with

$$|n\rangle = \begin{pmatrix} c_0 \\ c_1 \\ c_2 \\ c_3 \\ \dots \end{pmatrix} \quad \text{with } c_j = \delta_{jn}$$

By multiplying the matrices according to the usual rules, we find:

$$\hat{a}^+|n\rangle = \sqrt{n+1}|n+1\rangle$$

$$\Rightarrow \begin{pmatrix} 0 & 0 & 0 & 0 & \dots \\ \sqrt{1} & 0 & 0 & 0 & \dots \\ 0 & \sqrt{2} & 0 & 0 & \dots \\ 0 & 0 & \sqrt{3} & 0 & \dots \\ \dots & \dots & \dots & & \end{pmatrix} \begin{pmatrix} \delta_{0n} \\ \delta_{1n} \\ \delta_{2n} \\ \delta_{3n} \\ \dots \end{pmatrix} = \sqrt{n+1} \begin{pmatrix} 0 \\ \delta_{1,n+1} \\ \delta_{2,n+1} \\ \delta_{3,n+1} \\ \dots \end{pmatrix}$$

$$\hat{a}|n\rangle = \sqrt{n}|n-1\rangle$$

$$\Rightarrow \begin{pmatrix} 0 & \sqrt{1} & 0 & 0 & \dots \\ 0 & 0 & \sqrt{2} & 0 & \dots \\ 0 & 0 & 0 & \sqrt{3} & \dots \\ 0 & 0 & 0 & 0 & \dots \\ \dots & \dots & \dots & & \end{pmatrix} \begin{pmatrix} \delta_{0n} \\ \delta_{1n} \\ \delta_{2n} \\ \delta_{3n} \\ \dots \end{pmatrix} = \sqrt{n} \begin{pmatrix} \delta_{0,n-1} \\ \delta_{1,n-1} \\ \delta_{2,n-1} \\ \delta_{3,n-1} \\ \dots \end{pmatrix}$$

5.7.15 (5.58) $\rightarrow |\alpha\rangle = e^{-\frac{|\alpha|^2}{2}} \sum_{n=0}^{\infty} \frac{\alpha^n}{\sqrt{n!}}|n\rangle$

Substituting into (5.56) and keeping in mind that $\hat{a}|0\rangle = 0$, we have:

$$\hat{a}|\alpha\rangle = e^{-\frac{|\alpha|^2}{2}} \sum_{n=0}^{\infty} \frac{\alpha^n}{\sqrt{n!}}\hat{a}|n\rangle = e^{-\frac{|\alpha|^2}{2}} \sum_{n=1}^{\infty} \frac{\alpha^n}{\sqrt{n!}}\sqrt{n}|n-1\rangle$$

$$= \alpha e^{-\frac{|\alpha|^2}{2}} \sum_{n=1}^{\infty} \frac{\alpha^{n-1}}{\sqrt{(n-1)!}}|n-1\rangle$$

$$= \alpha e^{-\frac{|\alpha|^2}{2}} \sum_{(n-1)=0}^{\infty} \frac{\alpha^{n-1}}{\sqrt{(n-1)!}}|n-1\rangle = \alpha|\alpha\rangle$$

Normalization: Fock states are orthonormalized, such that $\langle n \mid n'\rangle = \delta_{n,n'} \Rightarrow$

$$\langle\alpha \mid \alpha\rangle = \sum_{n=0}^{\infty} \left| \frac{\alpha^n}{\sqrt{n!}} e^{-\frac{|\alpha|^2}{2}} \right|^2 = e^{-|\alpha|^2} \underbrace{\sum_{n=0}^{\infty} \frac{\left(|\alpha|^2\right)^n}{n!}}_{=e^{|\alpha|^2}} = 1$$

5.7.16

$$\langle N\rangle = \langle\alpha \left| \widehat{N} \right| \alpha\rangle = \langle\alpha \left| \hat{a}^+\hat{a} \right| \alpha\rangle = |\alpha|^2 \Rightarrow \langle N\rangle^2 = |\alpha|^4$$

$$\langle N^2\rangle = \langle\alpha \left| \widehat{N}\widehat{N} \right| \alpha\rangle = \langle\alpha \left| \hat{a}^+ \underbrace{\hat{a}\hat{a}^+}_{=\hat{a}^+\hat{a}+1} \hat{a} \right| \alpha\rangle$$

$$= \langle\alpha \left| \hat{a}^+\hat{a}^+\hat{a}\hat{a} \right| \alpha\rangle + \langle\alpha \left| \hat{a}^+\hat{a} \right| \alpha\rangle = |\alpha|^4 + |\alpha|^2$$

$$\Rightarrow \left\langle (\Delta N)^2 \right\rangle = \left\langle N^2 \right\rangle - \langle N \rangle^2 = |\alpha|^2 = \langle N \rangle$$

5.7.17 from

$$\widehat{H}\psi_n = E_n\psi_n \Rightarrow E_n = \int_V \psi_n^* \widehat{H}\psi_n dV$$

$$= \left\langle \widehat{H} \right\rangle_n \Rightarrow \frac{\partial E_n}{\partial \Lambda} = \frac{\partial}{\partial \Lambda} \left\langle \widehat{H} \right\rangle_n = \frac{\partial}{\partial \Lambda} \langle n|\widehat{H}|n \rangle =$$

$$= \left\langle \frac{\partial}{\partial \Lambda} n \middle| \underbrace{\widehat{H}|n\rangle}_{=E_n|n\rangle} + \underbrace{\langle n|\frac{\partial}{\partial \Lambda}\widehat{H}|n\rangle}_{=\left\langle \frac{\partial}{\partial \Lambda}\widehat{H} \right\rangle_n} + \underbrace{\langle n|\widehat{H}}_{=E_n\langle n|} \middle| \frac{\partial}{\partial \Lambda} n \right\rangle$$

$$= \left\langle \frac{\partial}{\partial \Lambda}\widehat{H} \right\rangle_n + E_n \left(\left\langle \frac{\partial}{\partial \Lambda} n \middle| n \right\rangle + \left\langle n \middle| \frac{\partial}{\partial \Lambda} n \right\rangle \right)$$

$$= \left\langle \frac{\partial}{\partial \Lambda}\widehat{H} \right\rangle_n + E_n \left(\underbrace{\frac{\partial}{\partial \Lambda} \underbrace{\langle n \mid n \rangle}_{=1}}_{=0} \right) = \left\langle \frac{\partial}{\partial \Lambda}\widehat{H} \right\rangle_n$$

Which coincides with (5.20f)

Chapter 6

6.7.1 From (5.32) and the selection rule for the harmonic oscillator, we immediately find that the oscillator strength is equal to zero. Compare also task 5.7.13.

6.7.2 From (1.2), we have:

$$n - m \text{ odd} \Rightarrow \frac{2}{L} \int_0^L x \sin\left(\frac{n\pi x}{L}\right) \sin\left(\frac{m\pi x}{L}\right) dx = \frac{2L}{\pi^2}\left[\frac{1}{(n+m)^2} - \frac{1}{(n-m)^2}\right]$$

$$n - m \text{ even} \Rightarrow \frac{2}{L} \int_0^L x \sin\left(\frac{n\pi x}{L}\right) \sin\left(\frac{m\pi x}{L}\right) dx = 0$$

$j = 2$:

$$j - k \text{ odd} \Rightarrow x_{jk} = \frac{2}{L} \int_0^L x \sin\left(\frac{j\pi x}{L}\right) \sin\left(\frac{k\pi x}{L}\right) dx$$

$$= \frac{2L}{\pi^2}\left[\frac{1}{(j+k)^2} - \frac{1}{(j-k)^2}\right]$$

$$\Rightarrow x_{21} \approx -0.18 \, \text{nm} \quad \text{or} |x_{21}| \approx 0.18 \, \text{nm}$$

$$f_{jk} = \frac{2m}{\hbar}\omega_{jk}|x_{jk}|^2 = \frac{2m}{\hbar}\frac{(E_j - E_k)}{\hbar}\left\{\frac{2L}{\pi^2}\left[\frac{1}{(j+k)^2} - \frac{1}{(j-k)^2}\right]\right\}^2$$

$$
= \frac{2m}{\hbar} \frac{\pi^2 \hbar^2}{2mL^2 \hbar} (j^2 - k^2) \left\{ \frac{2L}{\pi^2} \left[\frac{1}{(j+k)^2} - \frac{1}{(j-k)^2} \right] \right\}^2
$$

$$
= \frac{64}{\pi^2} \frac{j^2 k^2}{(j+k)^3 (j-k)^3}
$$

$$
\Rightarrow f_{21} \approx 0.96
$$

$j = 3$: $j - k$ even : $x_{jk} = 0 \Rightarrow f_{jk} = 0 \Rightarrow x_{31} = 0$; $f_{31} = 0$.
$j = 4$: $x_{41} \approx -0.0144$ nm $\Rightarrow |x_{41}| \approx 0.0144$ nm; $f_{41} \approx 0.031$. Note that $\sum_{j=2}^{4} f_{j1} \approx 0.99 < 1$. That restricts the possible absolute values of transition matrix elements with larger j.

6.7.3 Each of the eigenfunctions is of either even or odd parity (compare Fig. 4.1). $k = 1$ and $j = 3$ correspond to even eigenfunctions, while $j = 2$ and $j = 4$ are odd. Therefore, x_{31} must be equal to zero.

Chapter 7
7.7.1

A photon with a photon energy of 2000 eV corresponds to the	Visible spectral range	
	Infrared spectral range	
	X-ray spectral range	x
A photon with a photon energy of 0.5 eV corresponds to the	Visible spectral range	
	Infrared spectral range	x
	X-ray spectral range	
A resting particle has a de Broglie wavelength that is	Zero	
	Infinitely large	x
	Equal to the particle dimension	
$[\hat{x}, \hat{p}_y] = ?$	0	x
	$i\hbar$	
	$-i\hbar$	
The spectral energy density (u) of a radiation field may be given in	Js/m^3	x
	Ws^2/m^3	x
	kg/(sm)	x
The Einstein coefficient A_{21} responsible for spontaneous light emission is given in	s^{-1}	x
	W/m^2	
	m/s	
The ground state ($n = 1$) wavefunction of a particle in a rectangular potential box is	Even with respect to symmetry center	x
	Odd with respect to symmetry center	
	Neither even or odd	
The oscillator strength is	Dimensionless	x
	Given in cm$^{-3/2}$	
	Given in s^{-1}	
Which of the following quantum transitions in a 1D harmonic oscillator are dipole-allowed ($d_{nm} \neq 0$)?	$m = 2 \rightarrow n = 1$	x

	$m = 2 \rightarrow n = 3$	x
	$m = 2 \rightarrow n = 5$	
	$m = 2 \rightarrow n = 7$	
	$m = 2 \rightarrow n = 0$	
Consider the second excited state ($E_2 = \frac{5}{2}\hbar\omega_0$) of a one-dimensional harmonic oscillator. In this state, the expectation value of the momentum will be	$+\sqrt{5\hbar\omega_0 m}$	
	$-\sqrt{5\hbar\omega_0 m}$	
	$+\sqrt{3\hbar\omega_0 m}$	
	$-\sqrt{3\hbar\omega_0 m}$	
	0	x
The parity selection rule allows dipole transitions	Between quantum states of even parity	
	Between quantum states of odd parity	
	Between states that have different parity	x

7.7.2

Assertion	True	Wrong
The phase velocity may never exceed c		x
A rocket traveling with a velocity 1000 m/s has the same de Broglie wavelength like a proton traveling with the same velocity		x
All linear operators are self-adjoint		x
The operator $-i\frac{\mathrm{d}}{\mathrm{d}x}$ is self-adjoint	x	
The operator $-\frac{\mathrm{d}^2}{\mathrm{d}x^2}$ is self-adjoint	x	
All eigenvalues of a self-adjoint operator are real	x	
From $\left[\widehat{A}, \widehat{B}\right] = 0$ and $\left[\widehat{A}, \widehat{C}\right] = 0$ it follows that $\left[\widehat{B}, \widehat{C}\right] = 0$		x[a]
The Einstein coefficients are all dimensionless		x
The Einstein coefficient A_{21} is given in s^{-1}	x	
The harmonic oscillator has equidistant energy levels	x	
All eigenfunctions of the harmonic oscillator are even functions of x		x
In the ground state of a harmonic oscillator with $U \propto x^2$, $\langle x^2 \rangle = 0$		x
In the ground state of a harmonic oscillator with $U \propto x^2$, $\langle x^3 \rangle = 0$	x	
In a harmonic oscillator, electric dipole-allowed quantum transitions occur only between adjacent energy levels	x	

[a]For example, set $\widehat{A} = \widehat{y}$; $\widehat{B} = \widehat{x}$; $\widehat{C} = \widehat{p}_x$

7.7.3 $u(\omega, T) = \frac{\hbar\omega^3}{c^3\pi^2}\frac{1}{e^{\frac{\hbar\omega}{k_B T}} - 1} \Rightarrow \begin{cases} u(\omega \rightarrow 0, T) = \frac{\omega^2}{c^3\pi^2}k_B T \\ u(\omega \rightarrow \infty, T) = \frac{\hbar\omega^3}{c^3\pi^2}e^{-\frac{\hbar\omega}{k_B T}} \end{cases}$

7.7.4 From (7.28), we have: $I_{\omega,em,BB} \equiv u_{Planck}\frac{c}{4} = \frac{\hbar\omega^3}{4c^2\pi^2}\frac{1}{e^{\frac{\hbar\omega}{k_B T}} - 1}$.

And therefore:

$$\int_0^\infty I_{\omega,em,BB}\,d\omega$$

$$= \int_0^\infty \frac{\hbar\omega^3}{4c^2\pi^2}\frac{1}{e^{\frac{\hbar\omega}{k_B T}}-1}\,d\omega \propto T^4 \int_0^\infty \zeta^3\frac{1}{e^\xi-1}\,d\xi\Bigg|_{\xi\equiv\frac{\hbar\omega}{k_B T}} \propto T^4$$

7.7.5 From

$$\int_0^\infty u(\lambda,T)\,d\lambda = \int_0^\infty u(\omega,T)\,d\omega$$

And $\omega = \dfrac{2\pi c}{\lambda} \Rightarrow \begin{cases} d\omega = d\left(\frac{2\pi c}{\lambda}\right) = -\frac{2\pi c}{\lambda^2}\,d\lambda \\ \omega = 0 \Rightarrow \lambda \to \infty;\ \omega \to \infty \Rightarrow \lambda \to 0 \end{cases}$

$$\int_0^\infty u(\lambda,T)\,d\lambda = \int_0^\infty u(\omega,T)\,d\omega = -\int_\infty^0 u\left(\omega = \frac{2\pi c}{\lambda},T\right)\frac{2\pi c}{\lambda^2}\,d\lambda$$

And therefore $= \displaystyle\int_0^\infty u\left(\omega = \frac{2\pi c}{\lambda},T\right)\frac{2\pi c}{\lambda^2}\,d\lambda = \int_0^\infty u(\lambda,T)\,d\lambda;$

i.e. $u(\lambda,T) = \dfrac{8hc\pi}{\lambda^5}\dfrac{1}{e^{\frac{hc}{\lambda k_B T}}-1}$ if $d\lambda > 0$

If $u(\lambda,T)$ has a maximum, then it corresponds to a minimum of $u(\lambda,T)^{-1}$. Defining:

$$x \equiv \frac{hc}{\lambda k_B T} \Rightarrow u(\lambda,T) = \frac{8hc\pi}{\lambda^5}\frac{1}{e^{\frac{hc}{\lambda k_B T}}-1} \propto \frac{x^5}{e^x-1}$$

$$\Rightarrow [u(\lambda,T)]^{-1} \propto \frac{e^x-1}{x^5}$$

$$\frac{d}{dx}[u(\lambda,T)]^{-1} = 0 \Rightarrow 0 = \frac{d}{dx}\frac{e^x-1}{x^5}$$

$$\Rightarrow e^{-x} + \frac{x}{5} - 1 = 0 \Rightarrow x \approx 4.97$$

$$\Rightarrow 4.97 \approx x = \frac{hc}{\lambda k_B T} \Rightarrow \lambda T \approx \frac{hc}{4.97 k_B}$$

$$\approx 2.898 \times 10^{-3}\ \text{Km}$$

i.e., Wiens displacement law: $\lambda T = \text{const.} \approx 2.898 \times 10^{-3}$ Km

7.7.6

From $B_{21}u = A_{21}$, we have:

$$u = \frac{A_{21}}{B_{21}} = \frac{\hbar\omega^3}{c^3\pi^2}$$

In equilibrium, we also have:

$$u = \frac{\hbar\omega^3}{c^3\pi^2}\frac{1}{e^{\frac{\hbar\omega}{k_BT}} - 1}$$

That leads to $e^{\frac{\hbar\omega}{k_BT}} = 2$ or $\frac{\hbar\omega}{k_BT} = \ln 2$. Therefore, because of $\omega = \frac{2\pi c}{\lambda}$ we have

$$\lambda = \frac{1}{\ln 2}\frac{hc}{k_BT} \approx 70\,\mu m$$

7.7.7

$$\tau = A_{21}^{-1} \approx 1.7 \times 10^{-8}\,s$$
$$\frac{\Delta\omega}{\omega} = \frac{\Delta\lambda}{\lambda}\text{ with } \omega = \frac{2\pi c}{\lambda}\text{ and }\Delta\omega \approx \tau^{-1}$$
$$\Rightarrow \Delta\lambda \approx 10^{-5}\,nm$$

Chapter 8
8.7.1

Assertion	1D harmonic oscillator	Particle in 1D box potential with impermeable walls	H atom
Example: $U(r) = -\frac{1}{4\pi\varepsilon_0}\frac{e^2}{r}$			X
$E_n = \hbar\omega_0(n + \frac{1}{2})$	X		
$\psi_1 = \frac{1}{\sqrt{\pi}}\left(\frac{1}{a_0}\right)^{\frac{3}{2}}e^{-r/a_0}$			X
$E_n \propto n^2$		X	
$x_{n,n+2} = 0\,\forall n$	X	X	
$E_n \propto -n^{-2}$			X
$\psi_n(x) = \sqrt{\frac{2}{L}}\sin\frac{n\pi}{L}x$		X	
$\left(x_{n,n-1}\right)^2 = \frac{n\hbar}{2m\omega_0}$	X		
$\nu_{nm} = R_\infty\left(\frac{1}{m^2} - \frac{1}{n^2}\right)$			X
$\widehat{H} = \left[-\frac{\hbar^2}{2m}\frac{d^2}{dx^2} + \frac{m\omega_0^2}{2}x^2\right]$	X		
$\psi_{311} =$			X
$\frac{\sqrt{2}}{81\sqrt{\pi}}\left(\frac{1}{a_0}\right)^{\frac{3}{2}}\left(6 - \frac{r}{a_0}\right)\frac{r}{a_0}e^{-r/3a_0}\sin\vartheta\, e^{i\phi}$			

8.7.2 Answers:

- $F = G\frac{m_n m_e}{r^2}\left(=\frac{\mu v^2}{r}\right)$
- $T_{\text{kin}} = G\frac{m_n m_e}{2r}$; $U = -G\frac{m_n m_e}{r}$; $E = T_{\text{kin}} + U = -G\frac{m_n m_e}{2r}$
- $\hbar n = \mu v r$
- $r_n = \frac{\hbar^2 n^2}{G\mu m_n m_e}$; $r_{n=1} \approx 1.2 \times 10^{29}$ m

When comparing with the specific spatial dimensions presented in Table 1.1, it becomes clear that you will certainly never detect such a kind of "atom."

8.7.3 From (8.7) and (8.9), we have:

$$\mu = \frac{m_e}{2} \Rightarrow \begin{cases} E_n \approx -\dfrac{Ry}{2n^2} \\ r_n \approx 2n^2 a_0 \end{cases}$$

8.7.4 Of course you may directly calculate both wavelength and obtain their difference. The more elegant way is:

$$\nu = R_\infty\left(1 - \frac{1}{4}\right) = \frac{3}{4}R_\infty; \quad R_\infty \approx 10^5 \text{ cm}^{-1} \Rightarrow \Delta\nu$$

$$= \frac{3}{4}\Delta R_\infty \approx \frac{3}{4}\Delta\mu\frac{\partial}{\partial\mu}R_\infty = \frac{3}{4}R_\infty\frac{\Delta\mu}{\mu}$$

$$\Rightarrow |\Delta\lambda| = \frac{|\Delta\nu|}{\nu^2} = \frac{4}{3}R_\infty^{-1}\frac{|\Delta\mu|}{\mu}; \quad \mu - \text{reduced mass}$$

$$\mu_H = \frac{m_e m_p}{m_e + m_p}; \mu_D = \frac{2m_e m_p}{m_e + 2m_p} \Rightarrow \frac{|\Delta\mu|}{\mu} \approx \frac{m_e}{2m_p} \approx \frac{1}{4000}$$

$$\Rightarrow |\Delta\lambda| = \frac{4}{3}R_\infty^{-1}\frac{|\Delta\mu|}{\mu} \approx 0.033 \text{ nm}$$

8.7.5 The commutation relations are easily obtained making use of the combined application of (5.3) and (5.5).

8.7.6
$\left[\hat{L}_z, \hat{\mathbf{L}}^2\right] = 0$ is trivially obtained from (8.21) and (8.30)

8.7.7 The probability dw to observe the electron in the volume element dV is:

$$dw = |\Psi|^2 dV = |\Psi|^2 r^2 \sin\theta dr d\phi d\theta$$

From here by integrating:

$$w(R) = \frac{1}{\pi a_0^3}\int_0^{2\pi} d\phi \int_0^\pi \sin\theta d\theta \int_0^R e^{-\frac{2r}{a_0}} r^2 dr = \frac{4}{a_0^3}\int_0^R e^{-\frac{2r}{a_0}} r^2 dr$$

$$= \frac{1}{2}\int_0^{\xi_0} e^{-\xi}\xi^2 d\xi = w(\xi_0); \xi = \frac{2r}{a_0}; \xi_0 = \frac{2R}{a_0}$$

Partial integrating results in:

$$w(\xi_0) = 1 - \left[1 + \xi_0 + \frac{\xi_0^2}{2}\right] e^{-\xi_0}$$

$$w(R) = 1 - \left[1 + \frac{2R}{a_0} + \frac{2R^2}{a_0^2}\right] e^{-\frac{2R}{a_0}}$$

Obviously, $w(0) = 0$, while $w(R \rightarrow \infty) \rightarrow 1$. $w(\xi_0)$ may be numerically calculated (figure below!), hence $w = 1/2$ corresponds to $\xi_0 \approx 2.68$ or $R \approx 1.34a_0$.

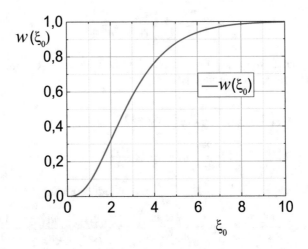

8.7.8 If you are very lazy, you may use Table 8.4. Otherwise you may calculate:

$$U(r) = -\frac{1}{4\pi\varepsilon_0}\frac{e^2}{r}$$

$$|2\,1\,0\rangle = \frac{1}{4\sqrt{2\pi}a_0^{\frac{3}{2}}}\frac{r}{a_0}e^{-\frac{r}{2a_0}}\cos\theta \quad \text{compare Table 8.3}$$

$$\Rightarrow 4\pi\varepsilon_0\langle U(r)\rangle$$

$$= -\frac{e^2}{32\pi a_0^5}\underbrace{\int_0^\infty r^3 e^{-\frac{r}{a_0}}\,dr}_{6a_0^4}\underbrace{\int_0^{2\pi}d\varphi}_{2\pi}\underbrace{\int_0^\pi \cos^2\theta\sin\theta\,d\theta}_{\frac{2}{3}}$$

$$\underbrace{\hspace{6cm}}_{8\pi a_0^4}$$

$$= -\frac{e^2}{4a_0} \Rightarrow \langle U(r)\rangle = -\frac{e^2}{16\pi\varepsilon_0 a_0}$$

8.7.9 From (8.42)

$$\Rightarrow \frac{d^2u}{dr^2} + 2\frac{du}{dr}\left(\frac{1}{r} - \kappa\right) + u\left[\left(\frac{2a - 2\kappa}{r}\right) - \frac{l(l+1)}{r^2}\right] = 0$$

The 3s-state ($n = 3$ and $l = 0$) is obtained from:

$$R(r) = u(r)e^{-\kappa r} \text{ with } u = \left(b_0 + b_1 r + b_2 r^2\right); \quad \frac{du}{dr}$$

$$= b_1 + 2b_2 r; \quad \frac{d^2u}{dr^2} = 2b_2$$

Substitution into (8.42) yields:

$$\Rightarrow b_2 + (b_1 + 2b_2 r)\left(\frac{1}{r} - \kappa\right) + \left(b_0 + b_1 r + b_2 r^2\right)\left(\frac{a - \kappa}{r}\right) = 0$$

Equating coefficients results in:

$$r^1 : -2b_2\kappa + b_2(a - \kappa) = 0 \Rightarrow \kappa = \frac{a}{3} = \frac{1}{3a_0} \Rightarrow (a - \kappa) = \frac{2}{3a_0}$$

$$r^{-1} : b_1 + b_0(a - \kappa) = b_1 + b_0\frac{2}{3a_0} = 0 \Rightarrow b_1 = -\frac{2}{3a_0}b_0$$

$$r^0 : 3b_2 - \kappa b_1 + b_1(a - \kappa) = 3b_2 + \frac{1}{3a_0}b_1 = 0$$

$$\Rightarrow b_2 = -\frac{1}{9a_0}b_1 = +\frac{2}{27}\frac{1}{a_0^2}b_0$$

Correspondingly, $R = b_0\left(1 - \frac{2r}{3a_0} + \frac{2r^2}{27a_0^2}\right)e^{-\frac{r}{3a_0}}$, where b_0 follows from the normalization requirement.

8.7.10 Check the commutation relations:

$$\left[\hat{L}_x, x\right] = 0$$

$$\left[\hat{L}_z, V(r)\right] = 0$$

$$\left[\hat{J}^2, \hat{L}^2\right] = 0$$

The first one is immediately obtained when writing L_x in Cartesian coordinates. For the second one, make use of spherical coordinates. Then, L_z is expressed in terms of a polar angle only, while V depends only on the distance. For the third one, we have (making use of task 5.7.1, (8.57) and (8.60)):

$$\left[\hat{J}^2, \hat{L}^2\right] = \underbrace{\left[\hat{L}^2, \hat{L}^2\right]}_{=0} + \underbrace{\left[\hat{S}^2, \hat{L}^2\right]}_{=0} + 2\left[\hat{S}\hat{L}, \hat{L}^2\right]$$

$$= 2\left[\hat{\mathbf{S}}\hat{\mathbf{L}}, \hat{\mathbf{L}}^2\right] = 2\hat{S}\left[\hat{\mathbf{L}}, \hat{\mathbf{L}}^2\right]$$

$$= 2\hat{S}_x\left[\hat{L}_x, \hat{\mathbf{L}}^2\right] + 2\hat{S}_y\left[\hat{L}_y, \hat{\mathbf{L}}^2\right] + 2\hat{S}_z\left[\hat{L}_z, \hat{\mathbf{L}}^2\right] = 0$$

8.7.11 (8.52) and (8.55) together yield:
$U = -\mathbf{d}_{\mathrm{magn}}\mathbf{B} = -\left(\mathbf{d}_{\mathrm{magn},l} + \mathbf{d}_{\mathrm{magn},s}\right)\mathbf{B} = \mu_B \underbrace{(m + 2m_s)}_{\equiv m_{PB}} B.$ Here, $m = -l$,

..., l and $m_{\mathrm{s}} = -1/2, +1/2$.

If $l = 0$, we have $m_{\mathrm{PB}} = -1, 1$ such that the external magnetic field leads to a splitting into two levels.

If $l \neq 0$, we have $m_{\mathrm{PB}} = -(m + 1), -m, -(m - 1), ..., (m - 1), m, (m + 1)$, such that the external magnetic field leads to a splitting into $2l + 3$ levels.

Chapter 9
9.7.1 Multiple-choice test: Mark all answers which seem you correct!

The emission lines of the Lyman spectral series are observed in the	Infrared	
	Visible	
	Ultraviolet	x
	γ-range	
The emission lines of the Paschen spectral series are observed in the	Infrared	x
	Visible	
	Ultraviolet	
	γ-range	
In a hydrogen atom, Bohrs radius is approximately equal to	0.05 nm	x
	5×10^{-11} m	x
	10 nm	
In spherical coordinates, the volume element dV is given by the expression (θ is the angle between $\mathbf{r}$ and the z-axis):	$dV = dr d\phi d\theta$	
	$dV = r^2 \sin\theta dr d\phi d\theta$	x
	$dV = 2\hbar^3 \pi r dr$	
	$dV = \sin\theta d\phi d\theta$	
In a hydrogen atom, a state $|n = 2; l = 4; m = -7\rangle$ is	Possible	
	Impossible	x
The s-orbital	Has spherical symmetry	x
	Corresponds to $l = 0$	x
	Corresponds to $l = 2$	
The Pauli exclusion principle is valid for	Bosons	
	Fermions	x

Electrons belong to the class of	Fermions	x
	Bosons	
	Neither of them	
Slater determinants describe the wavefunctions of systems of noninteracting	Fermions	x
	Bosons	
	Neither of them	
The electronic configuration $^1S_7$ is	Possible	
	Impossible	x

9.7.2 True or wrong? Make your decision!

Assertion	True	Wrong
$\left[\hat{x}, \hat{p}_z^2\right] = 0$?	x	
$\left[\hat{z}, \widehat{L}_z\right] = 0$?	x	
$\left[\widehat{L}_x, \widehat{L}_y\right] = 0$?		x
$\left[\widehat{L}_z, \widehat{\mathbf{L}}^2\right] = i\hbar\left(\widehat{L}_x + \widehat{L}_y\right)$?		x
$\left[\widehat{L}_z, U(r)\right] = 0$?	x	
In any circular Bohr orbit, the electrons kinetic energy is equal to its potential one		x
From the hydrogen emission spectrum, only certain lines of the Balmer series fall into the visible spectral range	x	
In its ground state, the electron in a hydrogen atom can be observed in both s- and p-orbitals		x
Dipole-allowed quantum transitions in a hydrogen atom are only observed when the principal quantum number changes for a value of ± 1		x
The wavefunction of a system of two fermions is always symmetric with respect to interchanging the particles		x
The wavefunction of a system of noninteracting bosons may be constructed in terms of a Slater determinant		x

9.7.3 We have:

Operator of number of photons	Operator of the occupation number of a fermionic state
$\widehat{N} \equiv \hat{a}^+\hat{a} \Rightarrow \widehat{N}\lvert n\rangle = n\lvert n\rangle$	$\widehat{M} \equiv \hat{b}^+\hat{b} \Rightarrow \widehat{M}\lvert n\rangle = n\lvert n\rangle$
$[\hat{a}, \hat{a}^+] = \hat{a}\hat{a}^+ - \hat{a}^+\hat{a} = 1$	$\left[\hat{b}, \hat{b}^+\right]_+ = \hat{b}\hat{b}^+ + \hat{b}^+\hat{b} = 1$
$\hat{a}^+\lvert n\rangle = \sqrt{n+1}\lvert n+1\rangle$	$\hat{b}^+\lvert n\rangle = \sqrt{n+1}\lvert n+1\rangle$
$\hat{a}\lvert n\rangle = \sqrt{n}\lvert n-1\rangle$	$\hat{b}\lvert n\rangle = \sqrt{n}\lvert n-1\rangle$

Let us start from repeating the treatment of bosons. We have the obvious relation:

$$\langle n|\widehat{N}|n\rangle = \langle n|\hat{a}^{+}\hat{a}|n\rangle = \sqrt{n}\langle n|\hat{a}^{+}|n-1\rangle = n\langle n \mid n\rangle = n$$

This is obvious from (5.48) for any $n > 0$, and from the ground-state definition $\hat{a}|0\rangle = 0$ for $n = 0$. For bosons, we further have:

$$\hat{a}\hat{a}^{+} - \hat{a}^{+}\hat{a} = 1$$

And therefore:

$$\langle n|\widehat{N}|n\rangle = \langle n|\hat{a}^{+}\hat{a}|n\rangle = \langle n|\hat{a}\hat{a}^{+} - 1|n\rangle$$
$$= \langle n|\hat{a}\hat{a}^{+}|n\rangle - \langle n \mid n\rangle = \sqrt{n+1}\langle n|\hat{a}|n+1\rangle - 1$$
$$= (n+1)\langle n \mid n\rangle - 1 = n + 1 - 1 = n$$

which coincides with the previously obtained result, i.e., does not constrain the allowed values of n. Therefore, for example, a given photon mode may be occupied by an arbitrary number of photons.

Another result will be obtained for fermions. We start from:

$$\langle n|\widehat{M}|n\rangle = \langle n|\hat{b}^{+}\hat{b}|n\rangle = \sqrt{n}\langle n|\hat{b}^{+}|n-1\rangle = n\langle n \mid n\rangle = n$$

Let us now make use of the anticommutation relation as valid for fermions. We write:
$\hat{b}\hat{b}^{+} + \hat{b}^{+}\hat{b} = 1$ and therefore obtain:

$$\langle n|\widehat{M}|n\rangle = \langle n|\hat{b}^{+}\hat{b}|n\rangle = \langle n|1 - \hat{b}\hat{b}^{+}|n\rangle = 1 - \langle n|\hat{b}\hat{b}^{+}|n\rangle$$

Let us analyze this relation. For $n = 0$, we obtain:
$\langle 0|\widehat{M}|0\rangle = 1 - \langle 0|\hat{b}\hat{b}^{+}|0\rangle = 1 - \langle 0|\hat{b}|1\rangle = 1 - \langle 0 \mid 0\rangle = 0$, which is a reasonable result and coincides which what we have expected. But for $n = 1$, we find:

$$\langle 1|\widehat{M}|1\rangle = 1 - \langle 1|\hat{b}\hat{b}^{+}|1\rangle = 1 - \sqrt{2}\langle 1|\hat{b}|2\rangle = 1 - 2\langle 1 \mid 1\rangle = -1$$

This result is obviously senseless. In order to obtain the expected value $\langle 1|\widehat{N}|1\rangle = 1$, the only chance is to postulate:

$$\hat{b}^{+}|1\rangle = 0 \Rightarrow \langle 1|\widehat{M}|1\rangle = 1 - \langle 1|\hat{b}\underbrace{\hat{b}^{+}|1\rangle}_{=0} = 1$$

But this has further consequences. From $\hat{b}^{+}|1\rangle = 0$, we immediately obtain:

$0 = \hat{b}^+|1\rangle = \sqrt{2}|2\rangle \Rightarrow |2\rangle = 0 \Rightarrow |3\rangle = 0$ and so on. Thus, the assumed anticommutation relation forbids occupying the corresponding states with more than one identical particle. This is exactly what we have learned from the Pauli principle; i.e., the fermionic quantum state may be either empty ($n = 0$), or singly occupied ($n = 1$).

Another approach is based on the repeated operation of $\widehat{M}$ on its eigenstate. Then,

$$\widehat{M}^2|n\rangle = n\widehat{M}|n\rangle = n^2|n\rangle \text{ (a)}$$

On the other hand, we have:

$$\widehat{M}^2|n\rangle = \hat{b}^+\hat{b}\hat{b}^+\hat{b}|n\rangle = \hat{b}^+\left(1 - \hat{b}^+\hat{b}\right)\hat{b}|n\rangle$$
$$= \hat{b}^+\hat{b}|n\rangle - \hat{b}^+\hat{b}^+\hat{b}\hat{b}|n\rangle = n|n\rangle - \hat{b}^+\hat{b}^+\hat{b}\hat{b}|n\rangle \quad \text{(b)}$$

For the interesting values $n = 0$ and $n = 1$, the last term in (b) is definitely zero. But formally, it may be written as:

$$\hat{b}^+\hat{b}^+\hat{b}\hat{b}|n\rangle = \hat{b}^+\hat{b}^+\hat{b}\sqrt{n}|n - 1\rangle = \hat{b}^+\hat{b}^+\sqrt{n - 1}\sqrt{n}|n - 2\rangle$$
$$= \hat{b}^+(n - 1)\sqrt{n}|n - 1\rangle = \left(n^2 - n\right)|n\rangle$$

Anyway, from (a) and (b) we therefore find: $\widehat{M}^2|n\rangle = \widehat{M}|n\rangle \Leftrightarrow n^2 = n \Rightarrow n = 0, 1$

The thus defined operator $\widehat{M}$ indeed has only two eigenvalues $n = 0$ and $n = 1$.

Consequently, $\hat{b}^+|1\rangle = 0$, $\hat{b}^+|0\rangle = |1\rangle$; $\hat{b}|1\rangle = |0\rangle$; $\hat{b}|0\rangle = 0$.

Also, $\hat{b}^+\hat{b}^+\hat{b}\hat{b}|n\rangle = 0 \,\forall n$.

9.7.4 In a d-state, we have $l = 2$. Therfore, $\cos\vartheta = \frac{2}{\sqrt{2(2+1))}} = 0.8165 \Rightarrow \vartheta \approx 35.3°$.

9.7.5 In this state, **S**, **L** and **J** form an isosceles triangle, while $J(J + 1) = S(S + 1) = 3/2*5/2$, and $L(L + 1) = 6$. From there $\cos\varphi = \sqrt{\frac{6}{15}} = \sqrt{0.4} \Rightarrow \varphi = 50.77°$.

9.7.6 Let us regard the special case $L > S$ (other cases are calculated in an analogous way). Then $\sum_{\{J\}} (2J + 1) = \sum_{J=L-S}^{L+S} (2J + 1)$.

When introducing: $k = J - (L - S)$.

We have:

$$\sum_{\{J\}} (2J + 1) = \sum_{J=L-S}^{L+S} (2J + 1) = \sum_{k=0}^{2S} (2k + 2L - 2S + 1)$$

$$= \sum_{k=0}^{2S} (2L - 2S + 1) + 2\sum_{k=0}^{2S} k = (2L - 2S + 1)(2S + 1) + 2\underbrace{\sum_{k=1}^{2S} k}_{=S(2S+1)}$$

$$= (2S + 1)(2L + 1) - 2S(2S + 1) + 2S(2S + 1) = (2S + 1)(2L + 1)$$

Chapter 10

10.7.1 Possible approaches are:

- In densely packed structures: estimation from the mass density (compare task 1.5.6)
- In densily packed structures: diffraction images like in Fig. 3.1
- In real gases: parameters of van der Waals isotopes
- Diluted molecular gases: Investigation of rotational energy levels (compare Sect. 8.5.1).

10.7.2 When assuming that the individual single-electron wavefuctions in (10.34) are normalized and real, we directly find when using (10.35):

$$\iint_{V_1,V_2} dV_1 dV_2 [\psi_a(1)\psi_b(2) \pm \psi_a(2)\psi_b(1)]^2 = 2 \pm 2S$$

From there the necessary normalization factor is obtained.

10.7.3 Solution: we make use of the necessary parity of the wavefunction and regard the situation left from the inversion center:

Then: $\psi = A \sin kx; \quad 0 \le x \le L$

$$\psi = B\left(e^{-\beta(x-L)} \pm e^{\beta(x-L-b)}\right); \quad L < x < L+b$$

Here, the choice in the sign distinguishes between symmetric and antisymmetric solutions with respect to the inversion center.

Continuity of the wavefunction at $x = L$ leads to:

$$A \sin kL = B(1 \pm e^{-\beta b})$$

$$Ak \cos kL = \beta B(-1 \pm e^{-\beta b}) \Rightarrow \tan kL = \frac{k}{\beta}\left(\frac{1 \pm e^{-\beta b}}{-1 \pm e^{-\beta b}}\right)$$

$$\Rightarrow kL = n\pi + \arctan \frac{k}{\beta}\left(\frac{1 \pm e^{-\beta b}}{-1 \pm e^{-\beta b}}\right) \equiv f(k)$$

($n = 1, 2, 3, \ldots$). From here allowed k-values are obtained, when considering.
$\beta = \frac{\sqrt{2m(U_0-E)}}{\hbar}$ (compare Table 4.1 for $E < U_0$)
The bound state energy levels correspond to (compare Table 4.1 for $U_0 = 0$):

$$E = \frac{\hbar^2 k^2}{2m}$$

A numerical solution is presented in the figure below (intersection points mark solutions for k). The black line denotes kL, red (dash) and navy (dot) show $f(k)$ according to the asymmetric and symmetric wavefunctions, accordingly.

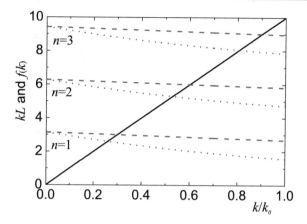

The value k_0 is the k-vector corresponding to $E = U_0$.

Note that for $U_0 \to \infty$; $kL \to n\pi$, which corresponds to the simple rectangular box potential with box legth L. For a finite barrier hight and width, both box potentials correspond with each other via quantum tunneling. The presence of the barrier with finite hight U_0 leads to a decrease in the position of the energy levels with respect to each of the single boxes, as well as to their splitting into two sublevels, corresponding to the symmetric and antisymmetric wavefunctions. Hereby, the antisymmetric solution corresponds to energy values closer to those of the simple particle-in-the box task, because in the center of the barrier, the wavefunction must be zero. Thus, in the antisymmetric case, the probability to observe the particle within the barrier is smaller than in the case of the symmetric solution, so the influence of the barrier is smallest in the antisymmetric case. Note that the symmetric state corresponds to a smaller energy than the anrisymmetric one, as earlier discussed in connection with the physics behind covalent bonding.

Practical applications of the investigated model potential concern, for example, the ammonia molecule. The ammonia molecule has a trigonal pyramidal shape, with the three hydrogen atoms defining a plane, while the nitrogen atom is placed outside this plane. The ammonia molecule readily undergoes nitrogen inversion at room temperature, as an analogy, consider an umbrella turning itself inside out in a strong wind. The effect is clearly dominated by quantum tunneling, because in the ammonia case, the "barrier height" U_0 (corresponding to an in-plane position of the nitrogen atom) is 0.4 eV, much higher than the thermal energy at room temperature.

10.7.4 We start by noting that the potential is inversion symmetric, Hence, $\psi(-x) = \pm\psi(x)$ (compare Sect. 6.6.1). Because of this required symmetry, we make the following ansatz for the wavefunction:

$$\psi(x) = C_1 e^{\beta x}; \quad x \leq -x_0$$
$$\psi(x) = C_2\left(e^{-\beta x} \pm e^{\beta x}\right); \quad -x_0 \leq x \leq x_0$$
$$\psi(x) = \pm C_1 e^{-\beta x}; \quad x_0 \leq x$$

Here, $\beta = \frac{\sqrt{-2mE}}{\hbar}$ Note that in the "$\pm$" and "$\mp$" symbols, the upper one corresponds to the symmetric solution, and the lower one to the antisymmetric. Because of the obvious symmetry of the task, it is sufficient to analyze the behavior at one of the singularities, say at $x = -x_0$. From task 4.7.4 we already know, that at the singularity, the wavefunction is continuous, while its first derivative is discontinuous. From there:

$$x = -x_0 \Rightarrow \begin{cases} C_1 e^{-\beta x_0} = C_2 \left(e^{\beta x_0} \pm e^{-\beta x_0} \right) \\ -\frac{2mA}{\hbar^2} C_1 e^{-\beta x_0} = C_2 \beta \left(-e^{\beta x_0} \pm e^{-\beta x_0} \right) - \beta C_1 e^{-\beta x_0} \end{cases}$$

Nontrivial solutions for the coefficients C_1 and C_2 exist when the following condition is fulfilled:

$$\begin{vmatrix} \left(e^{\beta x_0} \pm e^{-\beta x_0} \right) & -1 \\ \beta \left(e^{\beta x_0} \mp e^{-\beta x_0} \right) & \beta - \frac{2mA}{\hbar^2} \end{vmatrix} = 0$$

From here:

$$\beta = \frac{\frac{2mA}{\hbar^2}}{1 + \frac{e^{\beta x_0} \mp e^{-\beta x_0}}{e^{\beta x_0} \pm e^{-\beta x_0}}}$$

Let us analyze this expression. If the distance between the singularities approaches infinity, we obviously find: $\beta|_{x_0 \to \infty} \to \frac{mA}{\hbar^2}$ no matter whether the solution is symmetric or antisymmetric. From

$\beta|_{x_0 \to \infty} \to \frac{mA}{\hbar^2} = \frac{\sqrt{-2m\,E|_{x_0 \to \infty}}}{\hbar}$ we find $E|_{x_0 \to \infty} \to -\frac{mA^2}{2\hbar^2}$, which coincides with the result from task 4.7.4. Hence,

$$\beta = \frac{2\beta|_{x_0 \to \infty}}{1 + \frac{e^{\beta x_0} \mp e^{-\beta x_0}}{e^{\beta x_0} \pm e^{-\beta x_0}}}$$

In the case of the symmetric solution, that yields:

$$\beta_{sym} = \frac{2\beta|_{x_0 \to \infty}}{1 + \frac{e^{\beta x_0} - e^{-\beta x_0}}{e^{\beta x_0} + e^{-\beta x_0}}} = \frac{2\beta|_{x_0 \to \infty}}{1 + \tanh \beta x_0} > \beta|_{x_0 \to \infty}$$

Then:

$$E_{sym} = -\frac{\hbar^2 \beta_{sym}^2}{2m} < -\frac{\hbar^2 \left(\beta|_{x_0 \to \infty} \right)^2}{2m} = -\frac{mA^2}{2\hbar^2}$$

Such that the symmetric state has a smaller energy than the single "δ-atom," similar to what has been obtained in Sect. 10.3.2. Hence, it is expected to correspond to the bounding state.

In the case of the antisymmetric solution, we find:

$$\beta_{asym} = \frac{2\beta|_{x_0 \to \infty}}{1 + \frac{e^{\beta x_0} + e^{-\beta x_0}}{e^{\beta x_0} - e^{-\beta x_0}}} = \frac{2\beta|_{x_0 \to \infty}}{1 + \coth \beta x_0} < \beta|_{x_0 \to \infty}$$

Then:

$$E_{asym} = -\frac{\hbar^2 \beta_{asym}^2}{2m} > -\frac{\hbar^2 \left(\beta|_{x_0 \to \infty}\right)^2}{2m} = -\frac{mA^2}{2\hbar^2}$$

Such that the antisymmetric state has a larger energy than the single "δ-atom," similar to what has been obtained in Sect. 10.3.2. Hence, it is expected to correspond to the anti-bounding state.

Finally, the figure below symbolizes the dependence

$$E = -\frac{\hbar^2 \beta^2}{2m} = -\frac{\hbar^2}{2m} \left(\frac{\frac{2mA}{\hbar^2}}{1 + \frac{e^{\beta x_0} \mp e^{-\beta x_0}}{e^{\beta x_0} \pm e^{-\beta x_0}}} \right)^2$$

$$= -\frac{2mA^2}{\hbar^2} \left(\frac{1}{1 + \frac{e^{\beta x_0} \mp e^{-\beta x_0}}{e^{\beta x_0} \pm e^{-\beta x_0}}} \right)^2 = E(x_0)$$

Here, the navy line corresponds to the antisymmetric solution and the red one to the symmetric solution.

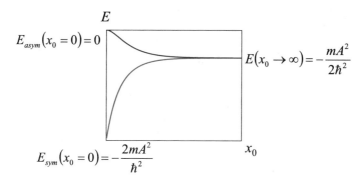

10.7.5 According to our general postulates, the substitution $\psi_b \to -\psi_b$ should not change the physical situations. In fact, the signs in (10.23) would change synchroneously with the curves in Fig. 10.3. Thus, the general physical conclusions remain valid.

Chapter 11
11.7.1

In the bonding configuration of H_2, the electron spins are	Parallel	
	Antiparallel	x
The mass moment of inertia of a diatomic molecule is typically of the order	10^{-27} kgm^2	
	10^{-47} kgm^2	x
	10^{-67} kgm^2	
In an anharmonic oscillator, vibrational overtone absorption	Is strongly forbidden	
	May be observed	x
Purely rotational spectra of diatomic molecules are typically observed in the	Far Infrared	x
	Ultraviolet	
In a C_{60} molecule, the number of vibrational degrees of freedom is equal to	180	
	186	
	174	x
The rotational–vibrational absorption spectrum of a gas of diatomic molecules may	Show a P-branch	x
	Be temperature-dependent	x
	Be observed in the infrared spectral region	x

11.7.2

Assertion	True	Wrong
Molecular fingerprint spectra are typically recorded in the middle infrared spectral range	x	
The oxygen molecule O_2 shows an intense purely rotational absorption spectrum in the FIR		x
While fundamental molecular vibration modes lead to absorptions in the MIR, overtones may also contribute to absorption phenomena in the NIR	x	

11.7.3 Note first that the centrifugal force enlarges the mass moment of inertia and thus reduces the rotational constant. The rotational absorption lines are therefore no more equidistant, but their wavenumber spacing must decrease when the rotational quantum nuber increases. The rest is calculations. Let us start from the centrifugal force: $F = \frac{\mu v^2}{r} = \mu r \Omega^2$.

With Ω rotational angular frequency.

$$\Omega = \frac{L}{I} \Rightarrow F = \mu r \Omega^2 = \mu r \left(\frac{L}{I}\right)^2 = \frac{\mu r \hbar^2 J(J+1)}{I^2}$$

This centrifugal force causes a change in the distance between the atoms of a diatomic molecule according to: $r = r_0 \rightarrow r = r_0 + x$ with

$$x = \frac{F}{\mu \omega_0^2} = \frac{r_0 \hbar^2 J(J+1)}{I^2 \omega_0^2}$$

With ω_0—vibrational angular frequency. The energy correction term is:

$$\Delta E = \frac{\mu\omega_0^2}{2}x^2 = \frac{\mu\omega_0^2}{2}\left[\frac{r_0\hbar^2 J(J+1)}{I^2\omega_0^2}\right]^2$$

$$= \frac{\hbar^4}{2I^3\omega_0^2}J^2(J+1)^2 = \frac{\hbar^4}{2I^34\pi^2c^2v_0^2}J^2(J+1)^2$$

$$\Rightarrow \Delta G = \frac{\Delta E}{hc} = \frac{\hbar^3}{2I^38\pi^3c^3v_0^2}J^2(J+1)^2$$

$$= \frac{4B^3}{v_0^2}J^2(J+1)^2 \Rightarrow G(J)|_{non-rigid}$$

$$= G(J)|_{rigid} - \frac{4B^3}{v_0^2}J^2(J+1)^2$$

$$= BJ(J+1)\left[1 - \frac{4B^2}{v_0^2}J(J+1)\right]$$

With $\frac{4B^2}{v_0^2} \ll 1$, corresponding to $\Omega_{rot} \ll \omega_0$.

Hence, the correction with respect to the action of centrifugal forces is only relevant at highly excited (high J) rotational states.

11.7.4 When referring to Fig. 5.1, one immediately obtains: $\frac{\mu\omega_0^2}{2}x^2 = \hbar\omega_0\left(v + \frac{1}{2}\right) \Rightarrow x = \sqrt{\frac{\hbar}{\mu\omega_0}}\sqrt{(2v+1)} = x_0\sqrt{(2v+1)}$

Therefore, for $v = 1$ we have $x \approx 0.18 \times 10^{-10}$ m $\ll r_0$.

11.7.5 Answer: 3.34 μm.

11.7.6 Answer:

$$\langle r \rangle = r_0 + \langle x \rangle = \frac{\int_{-\infty}^{\infty} xe^{-\frac{U(x)}{k_BT}}\,dx}{\int_{-\infty}^{\infty} e^{-\frac{U(x)}{k_BT}}\,dx} = \cdots \approx r_0 + \frac{3\gamma}{\kappa^2}k_BT$$

In this approximation, the linear dimensions of a solid change with temperature in a linear manner, this gives rise to a temperature-independent thermal expansion coefficient.

11.7.7 The Central Mass moment of inertia of a sphere is:

$$I = \frac{2}{5}mr^2; \quad m = \rho V = \frac{4\pi}{3}\rho r^3 \Rightarrow I = \frac{8\pi}{15}\rho r^5$$

The hollow sphere mass moment of inertia is then calculated as:

$$I = \frac{8\pi}{15}\rho\left[(r + \Delta r)^5 - r^5\right] \approx \frac{8\pi}{15}\rho * 5r^4\Delta r = \frac{8\pi}{3}\rho r^4\Delta r$$

The unknown term $\rho\Delta r$ is eliminated by:

$$V \approx 4\pi r^2 \Delta r = \frac{m}{\rho} \Rightarrow \Delta r = \frac{m}{4\pi r^2 \rho} \Rightarrow I = \frac{8\pi}{3}\rho r^4 \frac{m}{4\pi r^2 \rho} = \frac{2}{3}mr^2$$

Setting $r = (0.71/2)$nm and $m = 60 \times 12 * m_p$, we find

$$I = \frac{2}{3}mr^2 \approx 1 \times 10^{-43} \text{ kgm}^2$$

11.7.8 Cuvette 1 contains HBr

11.7.9 Cuvette 2 contains DCl, because DCl has a larger reduced mass, while the bond strength should be nearly the same in both molecule. Therefore, the vibration frequency in DCl must be smaller compared to HCl. Further, $\frac{v_1}{v_2} \approx \sqrt{\frac{\mu_2}{\mu_1}}$.

11.7.10 From (11.31), we have:

$$\left.\begin{array}{l} v_{n,0} = v_e\left[(1-x_e)n - x_e n^2\right] \\ v_{1,0} = v_e[(1-x_e) - x_e] \\ v_{2,0} = v_e[2(1-x_e) - 4x_e] \end{array}\right\} \Rightarrow \left\{\begin{array}{l} v_e(1-x_e) = \frac{4v_{1,0}-v_{2,0}}{2} \\ v_e x_e = \frac{2v_{1,0}-v_{2,0}}{2} \end{array}\right.$$

$$\Rightarrow v_{n,0} = v_e\left[(1-x_e)n - x_e n^2\right] = nv_{1,0} - \frac{n(n-1)}{2}(2v_{1,0} - v_{2,0})$$

$$\Rightarrow v_{3,0} = [3*2992 - 3(2*2992 - 5882)]\text{cm}^{-1} = 8670\,\text{cm}^{-1}$$

Experiment: 8685 cm^{-1}.

Chapter 12

12.7.1

(a) The field in a homogeneously charged sphere is (Gaussian theorem): $\mathbf{E}_1 = \frac{1}{3\varepsilon_0}\rho\mathbf{r}_1$.

with ρ charge volume density.

Consider the superposition of two oppositely charged spheres, shifted by a distance r:

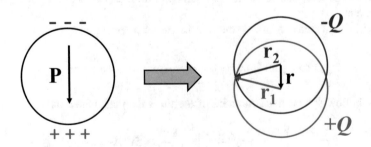

$$E = E_1 + E_2 = \frac{1}{3\varepsilon_0}\rho(\mathbf{r}_1 - \mathbf{r}_2) = \frac{1}{3\varepsilon_0}\rho(\mathbf{r}_1 - (\mathbf{d} + \mathbf{r}_1))$$

$$= -\frac{1}{3\varepsilon_0}\rho\mathbf{d} = -\frac{\mathbf{P}}{3\varepsilon_0}$$

(b) Field in a homogeneously charged cylinder (Gaussian theorem): $E_1 = \frac{1}{2\varepsilon_0}\rho\mathbf{r}_1$

$$E = E_1 + E_2 = \frac{1}{2\varepsilon_0}\rho(\mathbf{r}_1 - \mathbf{r}_2) = \frac{1}{2\varepsilon_0}\rho(\mathbf{r}_1 - (\mathbf{d} + \mathbf{r}_1)) = -\frac{1}{2\varepsilon_0}\rho\mathbf{d} = -\frac{\mathbf{P}}{2\varepsilon_0}$$

Task 12.7.2

$|j| = \frac{I}{A}$ with A—cross-sectional area of the wire. We further obtain:

$$|j| = \frac{I}{A} = N e v_D \Rightarrow v_D = \frac{I}{A N e}$$

$$A = \frac{\pi}{4}d^2; \quad N = \frac{\rho}{63.5 m_p} \Rightarrow v_D$$

$$= \frac{I}{A N e} \approx 3.5 \times 10^{-2} \frac{mm}{s}$$

12.7.3 526 nm

12.7.4 Answer: From (12.28), we find assuming a harmonic oscillator in the quantum state $|k\rangle$:

$$\beta = \frac{2}{\varepsilon_0 \hbar}\sum_{l \neq k}\frac{|d_{lk}|^2 \omega_{lk}}{\omega_{lk}^2 - \omega^2} \Rightarrow \beta = \frac{2}{\varepsilon_0 \hbar}\sum_{l=k\pm1}\frac{|d_{lk}|^2 \omega_{lk}}{\omega_0^2 - \omega^2} \Rightarrow$$

$$\beta_{stat} = \frac{2}{\varepsilon_0 \hbar \omega_0^2}\sum_{l=k\pm1}|d_{lk}|^2 \omega_{lk} = \frac{e^2}{\varepsilon_0 \underbrace{\omega_0^2 m}_{=\kappa}}\underbrace{\sum_{l=k\pm1}\frac{2m}{\hbar}|x_{lk}|^2 \omega_{lk}}_{=\sum_l f_{lk}=1} = \frac{e^2}{\varepsilon_0 \kappa}$$

Hence, the static polarizability is independent on the quantum number k, but is inversely proportional to the "spring constant" κ!

Note that this result coincides with the classical expression directly following from (12.5):

$$\beta = \frac{q^2}{\varepsilon_0 m}\frac{1}{\omega_0^2 - \omega^2 - 2i\omega\gamma} \Rightarrow \beta_{stat} = \frac{q^2}{\varepsilon_0 m \omega_0^2} = \frac{q^2}{\varepsilon_0 \kappa}$$

Chapter 13

13.7.1 From $\varepsilon = (n + iK)^2 \Rightarrow \begin{cases} \mathrm{Re}\varepsilon = n^2 - K^2 \\ \mathrm{Im}\varepsilon = 2nK \end{cases}$

Directly solving this system of equation results in:

$$n = \frac{1}{\sqrt{2}}\sqrt{\sqrt{(\mathrm{Re}\varepsilon)^2 + (\mathrm{Im}\varepsilon)^2} + \mathrm{Re}\varepsilon}; \quad K = \frac{1}{\sqrt{2}}\sqrt{\sqrt{(\mathrm{Re}\varepsilon)^2 + (\mathrm{Im}\varepsilon)^2} - \mathrm{Re}\varepsilon}$$

13.7.2 Making use of the result of the previous task, you find $n \approx 2.24$. According to (13.17), we find $v_{\mathrm{ph}} = c/n \approx 134160$ km/s. The penetration depth is $d_{penetr} = \frac{1}{\alpha} = \frac{\lambda}{4\pi K} \cong 1.424$ μm.

13.7.3 You sould arrive at a quadratic equation. The physically meaningful solution corresponds to the required expression.

13.7.4 Let R be the reflectance per interface, and d the slab thickness. Neglecting multiple internal reflections, we may write:

$$\Rightarrow T(d) \approx (1 - R)^2 e^{-\alpha d} \Rightarrow \alpha \approx \frac{\ln T(d_1) - \ln T(d_2)}{d_2 - d_1} = 0.0357 cm^{-1}$$

$$\Rightarrow K = \frac{\alpha \lambda}{4\pi} \approx 1.4 \times 10^{-7}$$

$$\Rightarrow R \approx 0.04 \ (\stackrel{\wedge}{=} 4\%)$$

$$R = \frac{(1 - n)^2 + K^2}{(1 + n)^2 + K^2} \approx \frac{(1 - n)^2}{(1 + n)^2} \approx 0.04 \Rightarrow n \approx \frac{1 + \sqrt{R}}{1 - \sqrt{R}} \approx 1.5$$

More precise expressions taking into account all multiple internal reflections may be found in [O. Stenzel, The physics of thin film optical spectra. An introduction, Springer 2016, Chap. 7].

13.7.5 (compare L.D. Landau, E.M. Lifshitz: Electrodynamics of continuous media, Pergamon Press 1960). Let the sphere be centered in the origin of a Cartesian coordinate system. Let the field $\mathbf{E}_2$ be directed along the x-axis. In order to observe a homogenous field $\mathbf{E}_1$ in the sphere, the potential φ_1 in the sphere should be:

$$\phi_1(\mathbf{r}) = -\mathbf{E}_1 \mathbf{r}$$

The total field outside the sphere is given by the sum of the applied field $\mathbf{E}_2$ and the field of the dipole represented by the sphere. We set:

$$\phi_2(\mathbf{r}) = -\mathbf{E}_2 \mathbf{r} + A \frac{\mathbf{E}_2 \mathbf{r}}{r^3} = -\mathbf{E}_2 \mathbf{r} \left(1 - \frac{A}{r^3} \right)$$

With A—some constant.

Remark Generally, the potential of a point dipole $\mathbf{d}$ may be written as: $\phi \propto \frac{\mathbf{dr}}{r^3}$ ([L. D. Landau, E. M. Lifshitz, Klassische Feldtheorie, Akademie-Verlag Berlin, 1976, p. 113–114]). Once we deal with induced dipoles, it is reasonable setting $\mathbf{d} \propto \mathbf{E}_2$, which results in the above-mentioned ansatz.

Once the potential is continuous at the border of the sphere, we have:

$$\phi_2(\mathbf{r} = \mathbf{R}) = \phi_1(\mathbf{r} = \mathbf{R}) \Rightarrow \mathbf{E}_1 = \mathbf{E}_2 \left(1 - \frac{A}{R^3} \right) \Rightarrow \frac{A}{R^3} = 1 - \frac{E_1}{E_2}$$

Moreover, the vertical components of the displacement vector $\mathbf{D}$ should be continuous. From

$\mathbf{D} = -\varepsilon_0 \varepsilon \nabla \phi$, we have at the points of the sphere which intersect the x-axis:

$$\left.\begin{array}{l} \mathbf{D}_{1\perp} = -\varepsilon_0 \varepsilon_1 \nabla \phi_1 = \varepsilon_0 \varepsilon_1 \mathbf{E}_1 \\[2mm] \mathbf{D}_{2\perp} = -\varepsilon_0 \varepsilon_2 \nabla \phi_2 = \varepsilon_0 \varepsilon_2 \left(\mathbf{E}_2 + 2\frac{A}{r^3}\mathbf{E}_2\right)\Big|_{r=R} \end{array}\right\} \Rightarrow \varepsilon_1 E_1 = \varepsilon_2 \left(1 + 2\frac{A}{R^3}\right) E_2$$

Remembering $\frac{A}{R^3} = 1 - \frac{E_1}{E_2}$, from here we quickly find:

$$E_2 = \frac{\varepsilon_1 + 2\varepsilon_2}{3\varepsilon_2} E_1$$

This equation naturally coincides with (13.35).

13.7.6 see next figure. Qualitatively, the following rules are useful to consider:

- According to task 13.7.1, we have:

$$n > K : \operatorname{Re}\varepsilon > 0$$
$$n = K : \operatorname{Re}\varepsilon = 0$$
$$n < K : \operatorname{Re}\varepsilon < 0$$

Violet arrows mark the position where the real part of the dielectric function changes its sign.

- The reflectance according to (13.24) has a minimum close to the wavenumber where n is equal to the refractive index of the ambient medium. This is indicated by the black and magenta arrows.
- A larger refractive index contrast results in a larger reflectance.

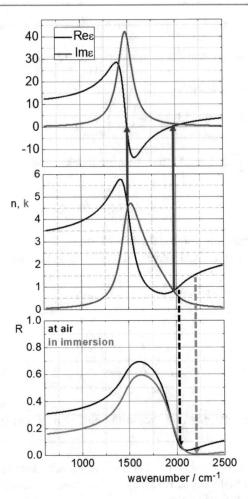

13.7.7 Let V_{Al} be the the volume occupied by metallic aluminum, and V_{ox} that of the aluminum oxide. The required filling factor is given by:

$$p_{Al} = \frac{V_{Al}}{V_{Al} + V_{ox}} = \frac{\frac{M_{Al}}{\rho_{Al}}}{\frac{M_{Al}}{\rho_{Al}} + \frac{M_{ox}}{\rho_{ox}}}$$

Here M indicates the total mass of the corresponding fraction in the mixture, and ρ the corresponding mass density. The mass M_{Al} may now be expressed through the number of aluminum atoms bound in the metal fraction N_{Al} and the corresponding atomic mass m_{Al}. In the same manner, M_{ox} is given by the number of Al_2O_3 units N_{ox}, multiplied with the mass of such a molecular unit m_{ox}. Thereby, the mass of an Al_2O_3 unit is obviously given by (m_O is the mass of an oxygen atom)

$$m_{ox} = 2m_{Al} + 3m_O$$

We then have:

$$M_{Al} = N_{Al} m_{Al}$$
$$M_{ox} = N_{ox}(2m_{Al} + 3m_O)$$

And therefore:

$$p_{Al} = \frac{V_{Al}}{V_{Al} + V_{ox}} = \frac{\frac{M_{Al}}{\rho_{Al}}}{\frac{M_{Al}}{\rho_{Al}} + \frac{M_{ox}}{\rho_{ox}}} = \frac{1}{1 + \frac{\rho_{Al}}{\rho_{ox}} \frac{2m_{Al} + 3m_O}{m_{Al}} \frac{N_{ox}}{N_{Al}}}$$

The atomic masses as well as the bulk densities for aluminum and alumina (we take data relevant for corundum for simplicity here) may be found in standard references. We set:

$$m_O \approx 16 m_p$$
$$m_{Al} \approx 27 m_p$$
$$\rho_{Al} \approx 2.7 g cm^{-3}$$
$$\rho_{ox} \approx 4.0 g cm^{-3}$$

We have not yet used the information about the atomic concentrations. It will be necessary to find the ratio $\frac{N_{ox}}{N_{Al}}$. Obviously, each Al_2O_3 unit contributes 2 aluminum atoms and 3 oxygen atoms to the atomic concentration. Therefore, the corresponding number of atoms n in the regarded volume is:

$$n_{Al} = N_{Al} + 2N_{ox}$$
$$n_O = 3N_{ox}$$

From here:

$$\frac{N_{ox}}{N_{Al}} = \frac{1}{3\frac{n_{Al}}{n_O} - 2}$$

And finally:

$$p_{Al} = \frac{V_{Al}}{V_{Al} + V_{ox}} = \frac{\frac{M_{Al}}{\rho_{Al}}}{\frac{M_{Al}}{\rho_{Al}} + \frac{M_{ox}}{\rho_{ox}}} = \frac{1}{1 + \frac{\rho_{Al}}{\rho_{ox}} \frac{2m_{Al} + 3m_O}{m_{Al}} \frac{N_{ox}}{N_{Al}}} = \frac{1}{1 + \frac{\rho_{Al}}{\rho_{ox}} \frac{2m_{Al} + 3m_O}{m_{Al} \left[3\frac{n_{Al}}{n_O} - 2\right]}}$$

Let us check this expression. Setting $n_O = 0$ (pure aluminum), we immediately obtain the reasonable result $p_{Al} = 1$. Setting $n_O = 3/2\, n_{Al}$, we have stoichiometric alumina, and immediately obtain $p_{Al} = 0$. In a mixture with an average atomic ratio 50%/50%, we obviously have $n_O = n_{Al}$, and obtain an intermediate volume filling factor $p_{Al} \approx 0.282$.

Chapter 14

14.7.1 Answer: $v_{gr} = cn < c$

Derivation: $k^2 = \frac{\omega^2}{c^2} n^2 = \frac{\omega^2}{c^2} - \frac{\omega_p^2}{c^2} \Rightarrow k\,dk = \frac{\omega\,d\omega}{c^2} \Rightarrow \frac{d\omega}{dk} = k\frac{c^2}{\omega} = \frac{\omega}{c} n\frac{c^2}{\omega} = cn.$

14.7.2 $R = \frac{(1-n)^2 + K^2}{(1+n)^2 + K^2} \approx \frac{\delta^2 + \beta^2}{4} \ll 1$

The intensity reflection coefficient R (or *reflectance*) is:

$$R = \frac{I_r}{I_e} = \left(\frac{n_1 - n_2}{n_1 + n_2}\right)^2$$

In the case of absorbing media, the refractive indices have to be replaced by their complex counterparts according to (13.18). Then, r becomes complex, while the real reflectance R is now given by: $R = \left|\frac{n_1 - \hat{n}_2}{n_1 + \hat{n}_2}\right|^2$.

From here, the result is immediately obtained.

14.7.3 Answer: $-\mathrm{Im}\frac{1}{\varepsilon} = \frac{2\omega\gamma\omega_p^2}{\left(\tilde{\omega}_0^2 + \omega_p^2 - \omega^2\right)^2 + 4\omega^2\gamma^2}$

A resonance appears at $\omega \approx \sqrt{\tilde{\omega}_0^2 + \omega_p^2}$. The resonance of the loss function is thus always blue-shifted with respect to that of the dielectric function. When assuming a Drude function, we have

$$\varepsilon = 1 - \frac{\omega_p^2}{\omega^2 + 2i\omega\gamma} \Rightarrow -\mathrm{Im}\frac{1}{\varepsilon} = \frac{2\omega\gamma\omega_p^2}{\left(\omega_p^2 - \omega^2\right)^2 + 4\omega^2\gamma^2}$$

Correspondingly, in a Drude metal, the resonance in the loss function occurs at $\omega \approx \omega_p$.

14.7.4 Instead of (12.20) we now have $\beta(\omega) = 4\pi r^3 \frac{\varepsilon(\omega) - 1}{\varepsilon(\omega) + 2}$.

When substituting (14.5) into this equation, we quickly find:

$\beta(\omega) = \frac{4\pi}{3} r^3 \frac{\omega_p^2}{\frac{\omega_p^2}{3} - \omega^2 - 2i\omega\gamma}$; Therefore, a resonance occurs at $\omega_{res} = \frac{\omega_p}{\sqrt{3}}$. This is the classical localized surface plasmon resonance condition in a small metal sphere embedded in vacuum.

Note that if the metal sphere is embedded in a dielectric host different from vacuum, the resonance condition will change. Instead of (12.20), according to Sect. 13.6.2, we should now write:

$$\beta(\omega) = 4\pi r^3 \frac{\varepsilon(\omega) - \varepsilon_h}{\varepsilon(\omega) + 2\varepsilon_h}$$

Then, the resonance condition changes to: $\omega_{res} = \frac{\omega_p}{\sqrt{1 + 2\varepsilon_h}}$.

The localized surface plasmon resonance in a small metal sphere may thus be tuned by a proper choice of the host medium. The optically denser the host medium is, the smaller the resonance frequency becomes (immersion effect). When being surrounded by air ($\varepsilon_h \approx 1$), we again observe the "classical" plasmon resonance in a metal sphere according to $\omega_{res} \approx \frac{\omega_p}{\sqrt{3}}$. Once in noble metals, ω_p is usually located in the UV, the resonance may be tuned into the visible spectral range by a proper choice of the host medium, which explains the colored appearance of silver nanocolloids or silver island films. The physics behind is in the collective oscillation of the free electrons within the metal sphere, which creates oscillating surface charges that are able to interact with external oscillating electric fields and, of course, also with the electrons of the embedding medium.

Chapter 15
15.7.1

The linear absorption coefficient of a material is	Dimensionless	
	Given in s^{-1}	
	Given in cm^{-1}	x
In the X-ray spectral region, the refractive index of a nonmagnetic material is usually	Infinitely large	
	Larger than zero	x
	Smaller than 1	x
	Equal to one	
	Negative	
Permanent electric dipoles in a medium may result in	A high static dielectric constant	x
	Orientation polarization	x
	A temperature dependence of the static dielectric constant according to Curie's law	x
Anomalous dispersion	Never occurs in liquids	
	Is observed in the vicinity of an absorption line	x
	Violates causality	

15.7.2 True or wrong? Make your decision!

Assertion	True	Wrong
In linear optics, electromagnetic energy dissipation is related to an imaginary part of the dielectric function different from zero	x	
In condensed matter, the local (microscopic) electric field may significantly differ from the average field in the medium	x	
The real part of the complex index of refraction is always larger than the imaginary part		x
The real part of the complex index of refraction is always larger than or equal to 1		x
The selection rules relevant for Raman spectra are identical to those relevant for infrared absorption		x
The concentration of absorbing species in a medium may be reliably estimated from the peak value of the absorption coefficient		x

Assertion	True	Wrong
The concentration of absorbing species in a medium may be estimated from the imaginary part of the dielectric function integrated over the (angular) frequency	x	

15.7.3 $\mathrm{Re}\varepsilon(\omega) - 1 = \frac{\chi_{stat}}{1+\omega^2\tau^2}$.

Let us for convenience take ξ as the integration variable (instead of ω_2 as in (15.4a)). Then

$$\mathrm{Im}\varepsilon(\omega) = -\frac{\chi_{stat}}{\pi} VP \int_{-\infty}^{\infty} \frac{d\xi}{(\xi - \omega)(1 + \xi^2\tau^2)}$$

When using the substitution: $x = \tau\xi$, we have

$$\mathrm{Im}\varepsilon(\omega) = -\frac{\chi_{stat}}{\pi} VP \int_{-\infty}^{\infty} \frac{dx}{(x - \omega\tau)(1 + x^2)} =$$
$$-\frac{\chi_{stat}}{\pi} \lim_{A\to\infty} \lim_{B\to 0} \left[\int_{-A}^{\omega\tau - B} \frac{dx}{(x - \omega\tau)(1 + x^2)} + \int_{\omega\tau+B}^{A} \frac{dx}{(x - \omega\tau)(1 + x^2)} \right]$$
$$(19.1)$$

The integral itsself may be solved by means of partial fraction decomposition, or simply by using the integral:

$$\int \frac{dx}{(x + b)(x^2 + a^2)} = \frac{1}{a^2 + b^2} \left[\ln|x + b| - \frac{1}{2} \ln|x^2 + a^2| + \frac{b}{a} \arctan\frac{x}{a} \right]$$
$$(19.2)$$

while setting $b = -\omega\tau$ and $a = 1$. Obviously, $a^2 + b^2 = (1 + \omega^2\tau^2)$.

When substituting (2) into (1), many terms cancel out immediately. The surviving terms yield:

$$\mathrm{Im}\varepsilon(\omega) = -\frac{\chi_{stat}\omega\tau}{\pi(1 + \omega^2\tau^2)} \lim_{A\to\infty} [\arctan(-A) - \arctan(A)]$$

$$-\frac{\chi_{stat}}{\pi(1 + \omega^2\tau^2)} \lim_{B\to 0} \left[\frac{1}{2} \ln\left|\frac{(\omega\tau + B)^2 + 1}{(\omega\tau - B)^2 + 1}\right| + \omega\tau(\arctan(\omega\tau + B) - \arctan(\omega\tau - B)) \right]$$

We obviously have:

$$\lim_{A\to\infty} [\arctan(-A) - \arctan(A)] = -\pi$$

$$\lim_{B\to 0} \left[\frac{1}{2} \ln\left|\frac{(\omega\tau + B)^2 + 1}{(\omega\tau - B)^2 + 1}\right| + \omega\tau(\arctan(\omega\tau + B) - \arctan(\omega\tau - B)) \right] = 0$$

And finally:

$$\mathrm{Im}\varepsilon(\omega) = -\frac{\chi_{stat}\omega\tau}{\pi\left(1+\omega^2\tau^2\right)}(-\pi) = \frac{\chi_{stat}\omega\tau}{1+\omega^2\tau^2}$$

An analogous calculation may be performed in the reverse direction:

$$\mathrm{Re}\varepsilon(\omega) = 1 + \frac{1}{\pi}VP\int_{-\infty}^{\infty}\frac{\mathrm{Im}\varepsilon(\xi)d\xi}{(\xi-\omega)} = 1$$

$$+\frac{\chi_{stat}}{\pi}VP\int_{-\infty}^{\infty}\frac{\xi\tau d\xi}{(\xi-\omega)\left(1+\xi^2\tau^2\right)} = 1 + \frac{\chi_{stat}}{\pi}VP\int_{-\infty}^{\infty}\frac{xdx}{(x-\omega\tau)\left(1+x^2\right)}$$

$$= 1 + \frac{\chi_{stat}}{\pi}\left[VP\int_{-\infty}^{\infty}\frac{dx}{\left(1+x^2\right)} + \omega\tau VP\int_{-\infty}^{\infty}\frac{dx}{(x-\omega\tau)\left(1+x^2\right)}\right]$$

$$= 1 + \frac{\chi_{stat}}{\pi}\left[\pi - \frac{\pi\omega^2\tau^2}{1+\omega^2\tau^2}\right] = 1 + \frac{\chi_{stat}}{1+\omega^2\tau^2}$$

15.7.4 From (14.5), we know: $\varepsilon(\omega) = 1 - \frac{\omega_p^2}{\omega^2+2i\gamma\omega}$

$\Rightarrow \mathrm{Re}\varepsilon = 1 - \frac{\omega_p^2}{\omega^2+4\gamma^2}$; $\mathrm{Im}\varepsilon = \frac{2\gamma\omega_p^2}{\omega(\omega^2+4\gamma^2)}$. The task is thus to derive the expression for the imaginary part through a Kramers–Kronig transformation of the real part. However, because of the presence of free electrons, (15.4a) is not applicable. Instead, we have to apply (15.12) or

$$\mathrm{Im}\varepsilon(\omega) = -\frac{1}{\pi}VP\int_{-\infty}^{\infty}\frac{[\mathrm{Re}\varepsilon(\omega_2)-1]d\omega_2}{\omega_2-\omega} + \frac{\sigma_{stat}}{\varepsilon_0\omega}$$

Combining now (12.19) and (14.4), and comparing (12.17) and (14.2), we find:
$\frac{\sigma_{stat}}{\varepsilon_0\omega} = \frac{\omega_p^2\tau}{\omega} = \frac{\omega_p^2}{2\gamma\omega}$.

Hence, $\mathrm{Im}\varepsilon(\omega) = \frac{\omega_p^2}{2\gamma}\left[\frac{1}{2\pi\gamma}VP\int_{-\infty}^{\infty}\frac{d\xi}{\left(1+\frac{\xi^2}{4\gamma^2}\right)(\xi-\omega)} + \frac{1}{\omega}\right]$.

The integral in this expression coincides by structure with the integral calculated in the previous task 15.7.3. Proceeding in the same manner, we obtain the correct result.

$$\mathrm{Im}\varepsilon = \frac{\omega_p^2}{2\gamma}\left[\frac{1}{2\pi\gamma}VP\int_{-\infty}^{\infty}\frac{d\xi}{\left(1+\frac{\xi^2}{4\gamma^2}\right)(\xi-\omega)} + \frac{1}{\omega}\right] = \frac{\omega_p^2}{2\gamma}\left[\frac{1}{\omega} - \frac{\omega}{(\omega^2+4\gamma^2)}\right] =$$

$\frac{2\gamma\omega_p^2}{\omega(\omega^2+4\gamma^2)}$.

For checking the KK-consistency in the reverse direction, it is recommended to start from (15.12)

$$\mathrm{Re}\varepsilon(\omega) = 1 + \frac{2}{\pi}VP\int_0^{\infty}\frac{\mathrm{Im}\varepsilon(\omega_2)\omega_2 d\omega_2}{\omega_2^2-\omega^2}$$

And to make use of:

$$\frac{1}{\left(x^2 - a^2\right)\left(x^2 + b^2\right)} = \frac{1}{\left(a^2 + b^2\right)}\left[\frac{1}{\left(x^2 - a^2\right)} - \frac{1}{\left(x^2 + b^2\right)}\right]$$

as well as of the integrals:

$$\int \frac{dx}{\left(x^2 - a^2\right)} = \frac{1}{2a} \ln\left|\frac{x - a}{x + a}\right|$$

$$\int \frac{dx}{\left(x^2 - a^2\right)\left(x^2 + b^2\right)} = \frac{1}{a} \arctan \frac{x}{a}$$

Chapter 16

"The missing atoms" in Sect. 16.1/Fig. 16.2: the missing atoms are indicated by a hollow circle.

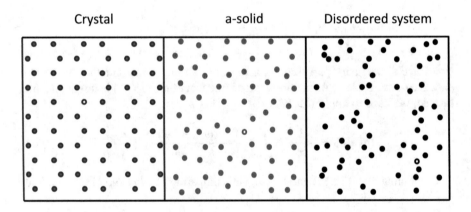

16.7.1 From (4.13), we know the density of states:

$D(E) = 2\pi V \left(\frac{2m}{h^2}\right)^{\frac{3}{2}} \sqrt{E}$. When deriving this equation, however, spin degeneration of electrons has not been taken into account. When now considering that each of the states encountering into (4.14) is twofold degenerated with respect to the electron spin directions, instead of (4.13) we have:

$D(E) = 4\pi V \left(\frac{2m}{h^2}\right)^{\frac{3}{2}} \sqrt{E}$. When assuming that in the ground state of the metal, all those individual electronic states are filled from bottom up until a maximum energy by one electron only, the total number of electrons is given by:

$$NV = \int_0^{E_{max}} D(E)dE = 4\pi V \left(\frac{2m}{h^2}\right)^{\frac{3}{2}} \int_0^{E_{max}} \sqrt{E}dE = 4\pi V \left(\frac{2m}{h^2}\right)^{\frac{3}{2}} \frac{2}{3} E_{max}^{\frac{3}{2}} \Rightarrow$$

$$E_{max} = \frac{h^2}{2m}\left(\frac{3N}{8\pi}\right)^{\frac{2}{3}} = \frac{h^2}{2m}\left(3\pi^2 N\right)^{\frac{2}{3}}$$

In quantum statistics, this maximum energy is called the Fermi energy E_F. At zero temperature, none of the electrons has a kinetic energy that exceeds E_F.

The average kinetic energy is simply calculated by:

$$\langle E \rangle = \frac{\int_0^{E_F} E D(E) dE}{\int_0^{E_F} D(E) dE} = \frac{\int_0^{E_F} E \sqrt{E} dE}{\int_0^{E_F} \sqrt{E} dE} = \frac{\frac{2}{5} E_F^{\frac{5}{2}}}{\frac{2}{3} E_F^{\frac{3}{2}}} = \frac{3}{5} E_F$$

In typical metals, Fermi energies are of the order of several eV. Thus, when assuming example an electron concentration $N = 10^{23}$ cm^{-3}, the Fermi energy turns out to be around 7.1 eV.

16.7.2 Let us return to the stuff from Sect. 16.6, which needs to be only slightly modified in order to solve the task. Thus, once after 6 periods the wavefunction must have reproduced itself, the application of (16.8) in (16.26 and 16.27) must be complemented by:

$$\psi_1(a - b) = e^{i\kappa a} \psi_1(-b) \Rightarrow \psi_1(6a - b) = \left(e^{i\kappa a}\right)^6 \psi_1(-b)$$

$$= \psi_1(-b) \Rightarrow 6\kappa a = 2j\pi \Rightarrow$$

$$\kappa a = \frac{\pi}{3};\ 2\frac{\pi}{3};\ \pi;\ 4\frac{\pi}{3};\ 5\frac{\pi}{3};\ 2\pi$$

Thus, in contrast to the situation in an infinitely large crystal, we now have only few discrete possible κ-values. Moreover, U_0 is negative. The product $-bU_0$ corresponds to the parameter A, as introduced in tasks 4.7.4 and 10.7.4. Instead of (16.28), we find:

$$\gamma = \frac{mab}{\hbar^2} U_0 \rightarrow -\frac{ma}{\hbar^2} A$$

Then, instead of (16.29), we have:

$$\cos \kappa a = \gamma \frac{\sin \alpha a}{\alpha a} + \cos \alpha a \rightarrow$$

$$\cos \kappa a = -\frac{mA}{\hbar^2} \frac{\sin \alpha a}{\alpha} + \cos \alpha a$$

Because of $E < 0$, we now recognize that α as defined in Sect. 16.6 is imaginary. It is therefore convenient to replace the trigonometric functions of imaginary arguments by hyperbolic functions of real arguments by using $\sin ix = i \sinh x$ and $\cos ix = \cosh x$. This results in:

$\cos \kappa a = -\frac{mA}{\hbar^2} \frac{\sinh \alpha a}{\alpha} + \cosh \alpha a$, where α is now given by $\alpha = \frac{\sqrt{-2mE}}{\hbar}$.

The set of equations:

$$\cos \kappa a = -\frac{mA}{\hbar^2} \frac{\sinh \alpha a}{\alpha} + \cosh \alpha a$$

$$\alpha = \frac{\sqrt{-2mE}}{\hbar}; \ E < 0$$

$$\kappa a = \frac{\pi}{3}; \ 2\frac{\pi}{3}; \ \pi; \ 4\frac{\pi}{3}; \ 5\frac{\pi}{3}; \ 2\pi \Rightarrow \cos \kappa a = -1; \ -0.5; \ 0.5; \ 1$$

together defines the allowed energy levels as predicted by the model. The figure below shows an example of a graphical solution of these equations for a set of suitably chosen fictive model parameters.

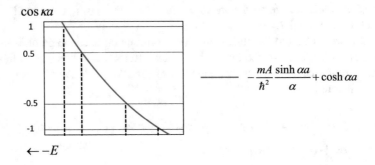

$$\underline{\hspace{2cm}} \qquad -\frac{mA}{\hbar^2}\frac{\sinh \alpha a}{\alpha} + \cosh \alpha a$$

Chapter 17
17.7.1

Solids usually have	A negligible mass density	
	A negligible shear modulus	
	A rather large bulk modulus	x
Reciprocal lattice vectors are given in	1/m	x
	m	
	cm^{-1}	x
Bloch functions may be used to describe the wavefunctions of	Diatomic molecules in diluted gases	
	Electrons in a spatially periodic potential	x
The binding energy of a Wannier–Mott exciton in a semiconductor like GaAs is typically of the order of	1...10 eV	
	≈ 1 GeV	
	1...10 meV	x
The radius of a Wannier–Mott exciton in GaAs (ground state) is of the order of	10 nm	x
	10^{-2} nm	
	1 cm	
Metals are characterized by	The presence of free electrons	x
	A good electrical conductivity	x
	A refractive index equal to 1 at any wavelength	

The effective mass of an electron in a crystal lattice	Is always positive	
	Is always negative	
	May be negative or positive	x
All crystals are	Good electric conductors	
	Free of absorption	
	Disordered	
Amorphous solids exhibit	Short-range order	x
	Long-range order	
	Neither of them	
Copper is	A dielectric	
	Electrically insulating	
	Optically transparent	
Crystalline germanium	Is transparent in the MIR	x
	Has an MIR refractive index around 4.0	x
	Is an indirect semiconductor	x

17.7.2 True or wrong? Make your decision!

Assertion:	True	Wrong
All crystals are optically anisotropic		x
The operator of discrete translations $\widehat{T}_n$ defined by $\hat{T}_n \psi(\mathbf{r}) = \psi(\mathbf{r} + \mathbf{r}_n)$ ($\mathbf{r}_n$-lattice vector) with the eigenvalues $t_n = e^{i\mathbf{k}\mathbf{r}_n}$ is self-adjoint		x
Indirect interband transitions in a crystal involve the creation or annihilation of phonons	x	
Amorphous solids lack any short-range order in their atomic configuration		x
If a crystal lattice has inversion symmetry, the corresponding solid must be a metal		x
The zone-center optical phonon of diamond is Raman active	x	

17.7.3 We will make use of (17.13) and (17.13a). The reflectance becomes zero when the dielectric function is equal to 1. Equations (17.13) and (17.13a) together with the latter condition may be written as:

$$\varepsilon(\omega) \approx \varepsilon_\infty \frac{\omega_L^2 - \omega^2}{\omega_T^2 - \omega^2} = \varepsilon_\infty \frac{\frac{\varepsilon_{stat}}{\varepsilon_\infty}\omega_T^2 - \omega^2}{\omega_T^2 - \omega^2} = 1$$

Here ω is the angular frequency where the normal incidence reflectance is zero. From there:

$$\omega_T = \omega\sqrt{\frac{\varepsilon_\infty - 1}{\varepsilon_{stat} - 1}}$$

From $\omega = 6.15 \times 10^{13}$ s^{-1}, we obtain $\omega_T = 3.1 \times 10^{13}$ s^{-1}. Then, from the Lyddane–Sachs–Teller relation (17.13a) we find $\omega_L = 5.03 \times 10^{13}$ s^{-1}.

The figure below shows the IR reflectance as a function of the wavenumber. Note that whenever the dielectric function is real but negative, the corresponding refractive index is purely imaginary. According to (17.13), this is obviously the case for angular frequencies between ω_T and ω_L. Then, according to (13.24), the reflectance at the interface to air must be equal to one. In the figure, therefore, the transversal and longitudinal phonon wavenumbers are well-recognized as minimum and maximum wavenumber of the prominent high reflection band. The corresponding IR reflection feature is called a reststrahlen band. In experimental practice, this reflection feature appears to be smeared out because of the imaginary part of the dielectric function that has been neglected in our model calculation. For more realistic reflection features see task 13.6.7 again.

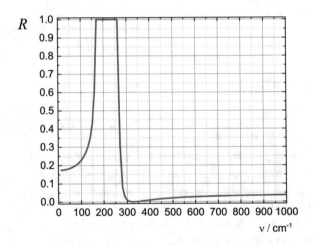

17.7.4 Again, the reflectance comes close to zero when the refractive index of the InSb comes close to 1, i.e. the refractive index of the ambient medium (air). From $\varepsilon(\omega) = \varepsilon_\infty - \frac{Ne^2}{\varepsilon_0 m^* \omega^2}$, we the find the condition $\varepsilon(\omega) = 1 \Rightarrow \lambda = 2\pi c \sqrt{\frac{\varepsilon_0 (\varepsilon_\infty - 1) m^*}{Ne^2}} \approx 15.8 \mu m$. The figure below shows the IR reflectance as a function of the wavenumber.

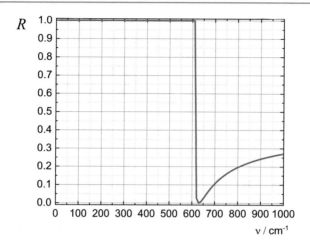

17.7.5 This rule may be illustrated in classical terms. Indeed, the transversal phonons do not create a volume charge density. Therefore, the restoring force acting on the elongated ions is defined by the bond strength only. In longitudinal phonons, on the contrary, volume charge densities are generated, such that the restoring force is obtained as a superposition of bond strength and electrostatic attraction between differently charged ions. This leads to an enhancement of the effective elasticity constant, and consequently to a larger resonance frequency (compare P. J. Yu, M. Cardona, Fundamentals of semiconductors. Physics and material properties, 4th edition, Springer 2010, p. 294–295).

17.7.6 Practically, we start with the same modifications in (17.27) as when deriving the Lindhard formula in Sect. 17.6. Instead of considering interband transitions by summarizing over valence band states, we now summarize over occupied conduction band states and assume that the matrix element for the intraband transitions is equal to 1 by absolute value. Then, instead of (17.27) we find:

$$\varepsilon(\omega; \mathbf{k}_p) = 1 + \frac{2e^2}{\varepsilon_0\hbar^2 V k_p^2} \sum_{\mathbf{k}_V, \mathbf{g}} \frac{[W(\mathbf{k}_V) - W(\mathbf{k}_C)][E(\mathbf{k}_C) - E(\mathbf{k}_V)]}{\left[\frac{E(\mathbf{k}_C) - E(\mathbf{k}_V)}{\hbar}\right]^2 - \omega^2 - 2i\omega\gamma} \left|\langle \mathbf{k}_C | e^{i\mathbf{k}_p\mathbf{r}} | \mathbf{k}_V \rangle\right|^2$$

$$\to 1 + \frac{2e^2}{\varepsilon_0\hbar^2 V k_p^2} \sum_{\mathbf{k}_C} \frac{[W(\mathbf{k}_C) - W(\mathbf{k}_C + \mathbf{k}_p)][E(\mathbf{k}_C + \mathbf{k}_p) - E(\mathbf{k}_C)]}{\left[\frac{E(\mathbf{k}_C + \mathbf{k}_p) - E(\mathbf{k}_C)}{\hbar}\right]^2 - \omega^2 - 2i\omega\gamma} \approx$$

$$\approx 1 + \frac{2e^2}{\varepsilon_0\hbar^2 V k_p^2} \sum_{\mathbf{k}_C} \frac{[E(\mathbf{k}_C + \mathbf{k}_p) - E(\mathbf{k}_C)]}{\left[\frac{E(\mathbf{k}_C + \mathbf{k}_p) - E(\mathbf{k}_C)}{\hbar}\right]^2 - \omega^2 - 2i\omega\gamma}$$

For $\mathbf{k}_p \to 0$, we make use of the expansion

$$E(\mathbf{k}_C) = const + \frac{1}{2} \frac{\partial^2 E}{\partial k_C^2}\bigg|_0 k_C^2 \Rightarrow E(\mathbf{k}_C + \mathbf{k}_p)$$

$$= const + \frac{1}{2} \frac{\partial^2 E}{\partial k_C^2}\bigg|_0 \left(k_C^2 + k_p^2 + 2\mathbf{k}_C\mathbf{k}_p\right)$$

$$\Rightarrow E(\mathbf{k}_C + \mathbf{k}_p) - E(\mathbf{k}_C) = \frac{1}{2}\frac{\partial^2 E}{\partial k_C^2}\bigg|_0 \left(k_p^2 + 2\mathbf{k}_C\mathbf{k}_p\right)$$

Hence, for any $\omega \neq 0$, we have:

$$\varepsilon\left(\omega; \mathbf{k}_p \to 0\right) \to 1 - \frac{e^2}{\varepsilon_0 \hbar^2 V k_p^2} \sum_{\mathbf{k}_C} \frac{\frac{\partial^2 E}{\partial k_C^2}\Big|_0 \left(k_p^2 + 2\mathbf{k}_C\mathbf{k}_p\right)}{\omega^2 + 2i\omega\gamma}$$

$$= 1 - \frac{\partial^2 E}{\partial k_C^2}\bigg|_0 \frac{e^2}{\varepsilon_0 \hbar^2}\left(\underbrace{\frac{1}{V}\sum_{\mathbf{k}_C}\frac{1}{\omega^2 + 2i\omega\gamma}}_{=\frac{N}{\omega^2 + 2i\omega\Gamma}} - \underbrace{\frac{2\mathbf{k}_p}{V k_p^2}\sum_{\mathbf{k}_C}\frac{\mathbf{k}_C}{\omega^2 + 2i\omega\gamma}}_{=0}\right)$$

$$= 1 - \frac{\frac{Ne^2}{\varepsilon_0 m^*}}{\omega^2 + 2i\omega\gamma} = 1 - \frac{\omega_p^2}{\omega^2 + 2i\omega\gamma}$$

Here we have used the definition of the effective mass of the conduction electrons according to (16.23). Also, because of the symmetry of the dispersion relation, $\sum_{\mathbf{k}_C} \mathbf{k}_C = 0$.

Chapter 18
18.7.1

A diamond crystal	Is rather transparent in the visible spectral range	x
	Shows a pronounced Raman line	x
	Has inversion symmetry	x
	Has two atoms in its primitive cell	x
	Has a cubic optical nonlinearity different from zero	x
The third harmonic of the Nd:YAG laser line ($\lambda \approx 1064$ nm) corresponds to a wavelength of	≈ 3192 nm	
	≈ 355 nm	x
	≈ 0.355 μm	x
Third-order nonlinear optical susceptibilities	Are dimensionless in SI units	
	Are observed in gases only	
	May describe two-photon absorption processes	x
Nonlinear optical effects may be observed	At high light intensities	x
	Only in the middle infrared	

	At interfaces	x
In electric dipole approximation, centrosymmetric materials	Obey the parity selection rule	x
	Do not show even-order nonlinearities	x
	Are always optically anisotropic	

18.7.2 True or wrong? Make your decision!

Assertion	True	Wrong
Several crystal classes may show second-order nonlinear optical effects	x	
SHG is particularly efficient in centrosymmetric materials		x
SHG is never observed in optically isotropic materials		x
The nonlinear refractive index is dimensionless		x

18.7.3 nearby the proton, we find: $E = \frac{e}{4\pi\varepsilon_0 a_0^2} \approx 5.7 \times 10^{11}\frac{V}{m}$. Of course, at larger orbital radii, the field strength will be smaller.

For the field strength in the electromagnetic wave, we use (compare task 2.7.6). $E_{0,\mathrm{real}}(\mathrm{in}\ V/m) = 2743\sqrt{I(\mathrm{in}\ W/cm^2)} \Rightarrow E_{0,\mathrm{real}} \approx 1060\frac{V}{m}$, which is much smaller than any intraatomic field strength. For a laser intensity around 10^{12} Wcm^{-2}, however, we obtain:

$E_{0,real}(in\ V/m) = 2743\sqrt{I(in\ W/cm^2)} \Rightarrow E_{0,real} \approx 2.743 \times 10^9\frac{V}{m}$, which comes much closer to intraatomic orders of magnitude.

18.7.4 $I = 0.15$ Wcm^{-2}: $\Omega \approx 8.5 \times 10^7$ s$^{-1} < < \omega_{\mathrm{electr}}$
$I = 10^{12}$ Wcm^{-2}: $\Omega \approx 2.5 \times 10^{14}$ s$^{-1} \sim \omega_{\mathrm{electr}}$

18.7.5 Straightforwardly by substituting n by $n + iK$ in (18.16) and honestly calculating the real and imaginary parts of $\hat{n}^{(\mathrm{eff})}$.

Index

Printed in the United States
by Baker & Taylor Publisher Services